Coastal
and Estuarine Studies

58

Nancy N. Rabalais and R. Eugene Turner (Eds.)

Coastal Hypoxia
Consequences for Living Resources and Ecosystems

American Geophysical Union
Washington, D.C.

Library of Congress Cataloging-in-Publication Data
Coastal Hypoxia : consequences for living resources and ecosystems / Nancy N. Rabalais and R. Eugene Turner, editors.
     p. cm -- (Coastal and estuarine studies ; 58) (Coastal and estuarine studies ; 58)
    Based on papers from a workshop held in Baton Rouge, La., Mar. 1998.
    Includes bibliographical references.
    ISBN 0-87590-272-3
    1. Coastal ecology--Congresses. 2. Water--Dissolved oxygen--Environmental aspects. I. Rabalais, Nancy N., 1950- II. Turner, R. E. (Robert Eugene), 1945- III. American Geophysical Union. IV. Series. V. Series: Coastal and estuarine studies ; 58
QH541.5.C65 C5914 2001
577.5'1--dc21
                00-065060

ISSN 0733-9569
ISBN 0-87590-272-3

CONTENTS

PREFACE

You must not just live on the earth.
You must live with the Earth.
Will you leave it a little better than when you found it?
W. Niering, Commencement Convocation, Connecticut College, 1993

Hypoxia is a condition that occurs when dissolved oxygen falls below the level necessary to sustain most animal life. In U.S. coastal waters, and in the entire western Atlantic, we find the largest hypoxic zone in the northern Gulf of Mexico on the Louisiana/Texas continental shelf. The area affected, which is about the size of the state of New Jersey at its maximal extent, has increased since regular measurements began in 1985. Sediment cores from the hypoxic zone also show that algal production and deposition, as well as oxygen stress, were much lower earlier in the 1900s and that significant increases occurred in the latter half of the twentieth century. We publish this book against the background of such measurements, and to review how the developing and expanding hypoxic zone has affected living resources on this continental shelf.

Human alterations of the landscape within the Mississippi River watershed and increases in nutrient loading particularly from agricultural activities are directly linked to the changing coastal ecosystem, including the worsening of hypoxia. As these changes parallel global patterns of increasing nutrient loads to estuarine and coastal waters, we consider the northern Gulf of Mexico hypoxic area as symptomatic of similar situations worldwide. The continental scale of the watershed, the immense size of the Gulf hypoxic zone (up to 20,000 km²), and the relatively open coastal system into which the Mississippi River discharges, however, create different physical and biological dimensions that bound the ecological effects.

Fishery resources of the Gulf are among the most valuable in the United States. Gulf commercial landings of fish and shellfish have an annual dockside value of about $700 million (approximately $1.4 billion when processed), while recreational and commercial fisheries together generate around $2.8 billion per year. But defining the ecological and economic consequences of hypoxia on the living resources of the northern Gulf has proven difficult. Lacking here is an historic data base against which to detect environmental changes, or no changes, in the fisheries themselves. Also lacking is current data specific to the distribution and abundance of living resources. Although ecological systems have changed and living resources are affected in the Gulf as a result of nutrient over-enrichment and hypoxia, available economic indicators do not necessarily translate excess nutrients to loss of fishing revenue. If experiences in other coastal and marine systems are applicable to the Gulf of Mexico, however, the potential impact of worsening hypoxic conditions may prompt the decline (perhaps precipitous) of ecologically and commercially important species.

v

Initially, we designed this book to identify the state of knowledge for the Gulf of Mexico with regard to the effects of nutrient over-enrichment and hypoxia on living resources, from phytoplankton to marine mammals. Ambitious as this goal may seem, there were many individuals within the Gulf and elsewhere working on just these aspects. It soon became clear that we would be able to define the limits, and more clearly define the numerous uncertainties, of our knowledge. Although relevant data existed, they were insufficient for a complete analysis of the Gulf. In this respect, the information and experience of those working outside the Gulf became quite valuable in our attempt to craft a synthetic idea of what might be happening in the Gulf. Our interests in the subject, and in completing this book, did not take place in an administrative or managerial vacuum, nor were they confined to academics.

An effort to embark on nutrient management within the Mississippi River watershed began in earnest in 1994, following a dramatic increase in the size of the Gulf hypoxic zone and legal actions taken by environmental organizations. Mobilization of the necessary political and social forces to support a nutrient management plan that might affect 41% of the lower 48 United States unfortunately necessitated the identification of a "smoking gun" of environmental degradation in the Gulf. (Elsewhere in the world -- e.g., Chesapeake Bay, Tampa Bay, the Baltic Sea, North Sea, and Long Island Sound -- states and nations have moved forward with nutrient reduction plans for the benefit of improving coastal water quality.) As part of the process towards implementing a nutrient management plan, the Mississippi River/Gulf of Mexico Task Force commissioned a series of studies through the White House Committee on Environment and Natural Resources (CENR), including a study on the "Ecological and Economic Consequences of Hypoxia." The CENR effort was a broad-brush synthesis of available published information without new data analyses. In a parallel effort, we obtained support through the competitive Sea Grant Program for a synthesis of the effects of hypoxia on living resources in the Gulf of Mexico. The two efforts came together in a workshop on "Effects of Hypoxia on Living Resources in the Northern Gulf of Mexico" in March 1998 in Baton Rouge, Louisiana. Nearly 70 Gulf of Mexico, U.S. and international scientists met to determine our current level of understanding on the Gulf, draw comparisons from other areas of the world, discuss necessary research to better define the effects of hypoxia on Gulf resources, and plan this book.

Contributing authors provide reviews and analyses from different perspectives, ranging from plankton to benthos and fish, from an organism's perspective to an ecosystem view, and from a scientist's analytical and empirical experience to that describing the complex interactions of management, politics and administration. This is a complex subject to cover thoroughly, and we thank the authors for their scientific leadership and useful insights in this regard. Chapter by chapter, our authors identify much to accomplish, so as to thoroughly address all aspects of concern and to identify options for future work toward our common goal: to sustain the natural resource base and societal needs within a well-ventilated, informed, and timely decision-making process. Certainly, we do not believe that the problems we face here can be "solved" in a brief, few years. We also understand that they will never be addressed satisfactorily without the efforts of authors, such as those we present, who have written for an audience of informed citizens, college students, practicing scientists, and resource managers.

The March 1998 workshop was co-sponsored by the Louisiana Sea Grant College Program (Grant No. NA86RG0073, R166753A&B to N. N. Rabalais and R. E. Turner); the Hypoxia Working Group, Committee on Environmental and Natural Resources, White House Office of Science and Technology Policy (grant to R. Diaz and A. Solow); Louisiana Universities Marine Consortium; and Louisiana State University, Coastal Ecology Institute. The editors gratefully acknowledge funds for completion of the book from the Louisiana Sea Grant College Program and the Department of Energy (DE-FG02-97ER12220 to N. N. Rabalais), and from the San Diego Foundation for support of color plates and photographs. We particularly appreciate the work of the three-to-four reviewers for each chapter who provided critically constructive and timely reviews.

Nancy N. Rabalais
R. Eugene Turner
Editors

Take'em down – to the riverside
and thro'em over the side
to be swept up by a current,
then taken to the ocean,
to be eaten by some fishes,
who were eaten by some fishes,
and swallowed by a whale,
who grew so old,
he decomposed
he died, and left his body
to the bottom of the ocean.
Now, everybody knows
that when a body decomposes
the basic elements
are given back to the ocean
and the sea does what it ought'ta.

Harry Nilsson, the song "Think About Your Troubles"
from *The Point*

1

Hypoxia in the Northern Gulf of Mexico: Description, Causes and Change

Nancy N. Rabalais and R. Eugene Turner

Abstract

Nutrient over-enrichment in many areas around the world is having pervasive ecological effects on coastal ecosystems. These effects include reduced dissolved oxygen in aquatic systems and subsequent impacts on living resources. The largest zone of oxygen-depleted coastal waters in the United States, and the entire western Atlantic Ocean, is found in the northern Gulf of Mexico on the Louisiana/Texas continental shelf influenced by the freshwater discharge and nutrient load of the Mississippi River system. The mid-summer bottom areal extent of hypoxic waters (≤ 2 mg l^{-1} O_2) in 1985-1992 averaged 8,000 to 9,000 km^2 but increased to up to 16,000 to 20,000 km^2 in 1993-2000. Hypoxic waters are most prevalent from late spring through late summer, and hypoxia is more widespread and persistent in some years than in others. Hypoxic waters are distributed from shallow depths near shore (4 to 5 m) to as deep as 60 m water depth but more typically between 5 and 30 m. Hypoxia occurs mostly in the lower water column but encompasses as much as the lower half to two-thirds of the water column. The Mississippi River system is the dominant source of fresh water and nutrients to the northern Gulf of Mexico. Mississippi River nutrient concentrations and loading to the adjacent continental shelf have changed in the last half of the 20th century. The average annual nitrate concentration doubled, and the mean silicate concentration was reduced by 50%. There is no doubt that the average concentration and flux of nitrogen (per unit volume discharge) increased from the 1950s to 1980s, especially in the spring. There is considerable evidence that nutrient enhanced primary production in the northern Gulf of Mexico is causally related to the oxygen depletion in the lower water column. Evidence from long-term data sets and the sedimentary record demonstrate that historic increases in

Coastal Hypoxia: Consequences for Living Resources and Ecosystems
Coastal and Estuarine Studies, Pages 1-36
Copyright 2001 by the American Geophysical Union

riverine dissolved inorganic nitrogen concentration and loads over the last 50 years are highly correlated with indicators of increased productivity in the overlying water column, i.e., eutrophication of the continental shelf waters, and subsequent worsening of oxygen stress in the bottom waters. Evidence associates increased coastal ocean productivity and worsening oxygen depletion with changes in landscape use and nutrient management that resulted in nutrient enrichment of receiving waters. Thus, nutrient flux to coastal systems has increased over time due to anthropogenic activities and has led to broad-scale degradation of the marine environment.

Introduction

There is increasing concern in many areas around the world that an oversupply of nutrients from multiple sources is having pervasive ecological effects on shallow coastal and estuarine areas. Marine plants provide essential habitat, and there are well-established positive relationships between dissolved inorganic nitrogen flux and phytoplankton primary production (e.g., Nixon et al. [1996], Lohrenz et al. [1997]). In addition, data from 36 marine systems show a relationship between fisheries yield and primary production [Nixon, 1988]. There are thresholds, however, where the load of nutrients to a marine system causes water quality degradation and detrimental changes to fisheries [Caddy, 1993].

While a variety of changes may result in the increased accumulation of organic matter in a marine system (= eutrophication, as defined by Nixon [1995]), the most common factor is an increase in the amount of nitrogen and phosphorus marine waters receive. With an increase in the world population, a focusing of that population in coastal regions and agricultural expansion in major river basins, eutrophication is becoming a major environmental problem in coastal waters throughout the world. Humans have altered the global cycles of nitrogen and phosphorus over large regions and increased the mobility and availability of these nutrients to marine ecosystems [Peierls et al., 1991; Howarth et al., 1995, 1996; Vitousek et al., 1997; Howarth, 1998; Caraco and Cole, 1999]. These human-controlled inputs are the result of human populations and their activities, particularly the application of nitrogen and phosphorus fertilizers, nitrogen fixation by leguminous crops, and atmospheric deposition of oxidized nitrogen from fossil-fuel combustion. Changes in the relative proportions of these nutrients may exacerbate eutrophication, favor noxious algal blooms and aggravate conditions of oxygen depletion [Officer and Ryther, 1980; Smayda, 1990; Conley et al., 1993; Justic' et al. 1995a,b; Turner et al., 1998].

The impairment of waters from nutrient over-enrichment goes well beyond scummy-looking water to threatening the suitability of water for human consumption and impairing the sustained production of useful forms of aquatic life. Excess nutrients lead to degraded water quality through increased phytoplankton or filamentous algal growth. Increasing nutrient loads are the cause of some noxious or harmful algal blooms (HABs), including some toxic forms. Secondary effects include increased turbidity or oxygen-depleted waters (= hypoxia) and eventually loss of habitat with consequences to marine biodiversity and changes in ecosystem structure and function. Over the last two decades it has become increasingly apparent that the effects of eutrophication, including oxygen depletion, are not minor and localized, but have large-scale implications and are spreading rapidly [Rosenberg, 1985; Diaz and Rosenberg, 1995; Anderson, 1995; Nixon, 1995; Paerl, 1995, 1997].

Water with less than 2 mg l^{-1} dissolved oxygen is considered hypoxic. Hypoxia occurs naturally in many parts of the world's marine environments, such as fjords, deep basins, open ocean oxygen minimum zones, and oxygen minimum zones associated with western boundary upwelling systems [Kamykowski and Zentara, 1990]. Hypoxic and anoxic (no oxygen) waters have existed throughout geologic time, but their occurrence in shallow coastal and estuarine areas appears to be increasing [Diaz and Rosenberg, 1995]. The largest zone of oxygen-depleted coastal waters in the United States, and the entire western Atlantic Ocean, is in the northern Gulf of Mexico on the Louisiana/Texas continental shelf at the terminus of the Mississippi River system (Plate 1). The size of the Gulf of Mexico hypoxic zone reaches 20,000 km^2 in mid-summer [Rabalais, 1999], and ranks third in area behind similar coastal hypoxic zones on the northwestern shelf of the Black Sea and in the Baltic basins. The hypoxic zone in the northern Gulf of Mexico (average for 1993-1999) is about the size of the state of New Jersey or the states of Rhode Island and Connecticut combined. Its extent on the bottom is twice the total surface area of the whole Chesapeake Bay, and its volume is several orders of magnitude greater than the hypoxic water mass of Chesapeake Bay [Rabalais, 1998].

The watershed that drains through the Mississippi and Atchafalaya Rivers to the Gulf of Mexico is also immense (Plate 1). The Mississippi River system ranks among the world's top ten rivers in length, freshwater discharge and sediment delivery and drains 41% of the lower forty-eight United States [Milliman and Meade, 1983]. Thus, the dimensions of the problem and the drainage system that affect it are of much greater magnitude than most nutrient-driven eutrophication problems elsewhere.

The linked Mississippi River system and the northern Gulf of Mexico is an example of the worldwide trend of increasing riverborne nutrients and worsening coastal water quality. Model simulations, research studies, empirical relationships and retrospective analyses of the sedimentary record have produced considerable evidence that nutrient loading from the Mississippi River system is the dominant factor in controlling the extent and degree of hypoxia and its worsening in the last century [Rabalais et al., 1996, 1999].

Despite recent advances in identifying links between Mississippi River system discharge and nutrient loads and coastal hypoxia in the Gulf of Mexico, defining the ecological and economic consequences of hypoxia on the living resources of the northern Gulf of Mexico has proven difficult. Long-term fisheries data are lacking, as are data specific to present-day distribution and abundance of living resources. Ecosystem level changes have occurred, however, consistent with changes in Mississippi River system discharge and nutrient loads. In this chapter, we describe the phenomenon of hypoxia in the Gulf of Mexico, its physical and biological causes, the close coupling of hypoxia with Mississippi River effluents, and the historical changes in river constituents and hypoxia that parallel each other. Against this background of watershed landscape changes, human activities, and worsening hypoxic conditions in the Gulf of Mexico, the subsequent chapters detail the state of knowledge of hypoxia on living resources of the northern Gulf within the broader context of patterns already demonstrated elsewhere in the world's coastal waters.

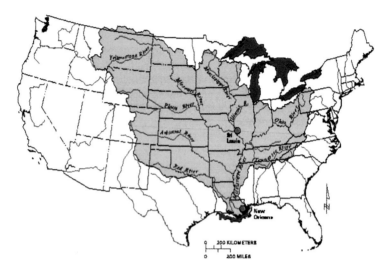

Plate 1. Mississippi River drainage basin and major tributaries, and general location of the 1999 midsummer hypoxic zone [Rabalais, 1999]. (From Goolsby [2000], used with permission of the author).

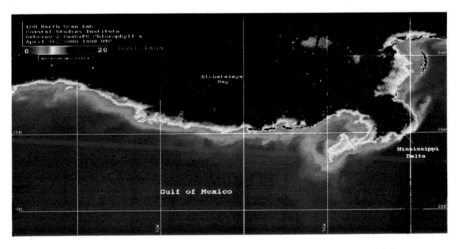

Plate 2. Orbview-2 SeaWiFS satellite image from April 26, 2000, during below average discharge of the Mississippi River, showing estimated chlorophyll along the Louisiana coast. Image supplied by Nan D. Walker, used with permission of the Earth Scan Laboratory, Louisiana State University.

Definition

Oxygen is necessary to sustain the life of most higher organisms, including the fish and invertebrates living in aquatic habitats. The normal condition is for surface water dissolved oxygen to be mixed or diffused into the lower water column where oxygen has been consumed by organisms, particularly by the micro-organisms. When the supply of oxygen to the bottom is cut off due to stratification or the consumption rate of oxygen during the decomposition of organic matter exceeds supply, oxygen concentrations become depleted.

The point at which various animals are affected by low oxygen concentration varies, but generally effects start to appear when oxygen drops below 2 or 3 mg l^{-1} (ppm) [Tyson and Pearson, 1991; Diaz and Rosenberg, 1995]. For seawater, this concentration is only about 20 to 30% of full saturation and is insufficient to support most larger aerobic organisms. The operational definition for hypoxia in the northern Gulf of Mexico is < 2 mg l^{-1} (2.8 ml l^{-1}), because trawlers seldom capture any shrimp or demersal fish in their nets below that value [Pavela et al., 1983; Leming and Stuntz, 1984; Renaud, 1986]. The oxygen concentration of surface waters is typically > 8 mg l^{-1} if they are 100% saturated with oxygen at summertime temperature and salinity conditions.

For consistency, most dissolved oxygen concentrations in this book are expressed in units of mg l^{-1}, but some are also converted to ml l^{-1}. An oxygen concentration expressed as % saturation is the % of air saturation at the ambient temperature and salinity. Physiologically relevant units are often given in oxygen tension with the unit of torr (mm mercury).

Causes

Two principal factors lead to the development and maintenance of hypoxia. First, the water column must be stratified so that the bottom layer is isolated from the surface layer and the normal resupply of oxygen. The physical structure is dictated by water masses that differ in temperature or salinity or both. Fresher waters derived from rivers and seasonally-warmed surface waters are less dense and reside above the saltier, cooler and more dense water masses near the bottom. Second, there is decomposition of organic matter that reduces the oxygen levels in the bottom waters. The source of this organic matter is mostly the result of phytoplankton growth stimulated by nutrients delivered to the coastal ocean with the riverine freshwater supply. The concentrations and total loads of nitrogen, phosphorus and silica to the coastal ocean influence the productivity of the phytoplankton community as well as the types of phytoplankton that are most likely to grow. The carbon that is produced by phytoplankton is the base of the marine food web that supports further production by multi-celled organisms including zooplankton and fish. Not all of the carbon produced in the surface waters becomes incorporated into the food web. Some of the algal cells die and sink to the bottom; others are grazed by zooplankton and are incorporated into fecal pellets that also sink to the bottom. Many algal cells and fecal pellets sink to the bottom as aggregates, or marine snow. Thus, a high percentage of the organic matter produced in coastal waters reaches the bottom and becomes the source for aerobic decomposition and causes hypoxia.

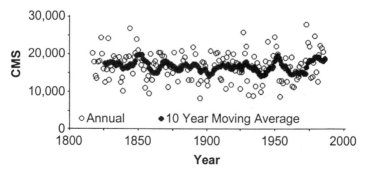

Figure 1. The annual discharge of the Mississippi River at Vicksburg, Mississippi ($m^3 s^{-1}$) with a 10-y moving average superimposed (data from U.S. Army Corps of Engineers). (From Rabalais et al. [1999].)

The relative importance of both physical structure and biological productivity in the development of hypoxia varies among environments and over an annual cycle. In the northern Gulf of Mexico the two factors are complexly inter-related and directly linked with the dynamics of the Mississippi and Atchafalaya river discharges. The Mississippi and Atchafalaya rivers are the primary riverine sources of fresh water to the Louisiana continental shelf [Dinnel and Wiseman, 1986] and to the Gulf of Mexico (80% of freshwater inflow from U.S. rivers to the Gulf [Dunn, 1996]). The discharge of the Mississippi River system is controlled so that 30% flows seaward through the Atchafalaya River delta and 70% flows through the Mississippi River birdfoot delta. The former enters through two outlets into Atchafalaya Bay, a broad shallow embayment; the latter enters the Gulf through multiple outlets, some in deep water and some in shallow water. Approximately 53% of the Mississippi River Delta discharge flows westward onto the Louisiana shelf [U.S. Army Corps of Engineers, 1974; Dinnel and Wiseman, 1986], and the general flow of the Atchafalaya River effluent is to the west.

The variability in freshwater discharge on seasonal, annual, decadal and longer scales underlies many important physical and biological processes affecting coastal productivity and food webs. There is significant interannual variability in the annual discharge with the peak in March-May and low discharge in late summer-fall (Figs. 1 and 2). The 1900-1992 average discharge rate (decadal time scale) for the lower Mississippi River is remarkably stable at about 14,000 $m^3 s^{-1}$. There was a decrease in flow during the 1950s and 1960s, and the 1990s have been a period of higher discharge. The discharge of the Mississippi River increased from 1935 to 1995 at 0.3% y^{-1}, or by 20%. The stage height, however, did not increase over the same period. There is some question as to the existence of a trend in discharge from the system, with the reported differences likely attributable to the period of record examined.

The discharge of the Atchafalaya increased during the course of the most complete record (1930-1997) [Bratkovich et al., 1994], as the U.S. Army Corps of Engineers allowed more Mississippi River water to enter the Atchafalaya basin at a diversion above St. Francisville, Louisiana. Less obvious is an increasing trend in the Mississippi River discharge as measured at Tarbert Landing. This trend is also statistically significant and increasing. It appears to be due to a tendency for increasing discharge in September through December. This period, however, is least important in the timing of important biological processes that lead to the development of hypoxia or the physical processes important in its maintenance. If a longer period of annual discharge were considered

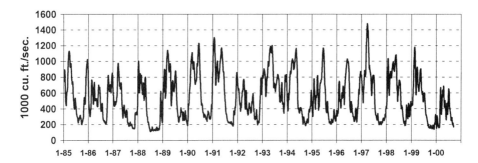

Figure 2. Daily discharge of the Mississippi River at Tarbert Landing. (Data from the U.S. Army Corps of Engineers.) (Modified from Rabalais et al. [1999].)

(e.g., from Turner and Rabalais [1991] for the early 1800s to present), the trends since the 1950s are obvious but are concealed within high interannual variability and no long-term change over a century and a half (Fig. 1).

Freshwater discharge and seasonal atmospheric warming control the strength of stratification necessary for the development and maintenance of hypoxia. The depth of the main pycnocline (depth of greatest change in density) does not always track the depth of the oxycline (Fig. 3). The existence of a strong near-surface pycnocline, usually controlled by salinity differences, is a necessary condition for the occurrence of hypoxia, while a weaker, seasonal pycnocline, influenced by temperature differences, guides the morphology of the hypoxic domain [Wiseman et al., 1997]. Stratification goes through a well-defined seasonal cycle that generally exhibits maximum stratification during summer and weakest stratification during winter months (Fig. 4). This is due to the strength and phasing of river discharge, wind mixing, regional circulation and air-sea heat exchange processes.

Dimensions and Variability of Hypoxia

Historical Occurrence and Geographic Extent

Accounts of low oxygen from the Gulf of Mexico for the mid-1930s [Conseil Permanent International pour l'Expoloration de la Mer, 1936] were not about continental shelf hypoxia, but described the oxygen minimum layer, an oceanic feature at 400-700 m depth. Coastal hypoxia was first reported in the northern Gulf of Mexico in the early 1970s off Barataria and Terrebonne/Timbalier Bays as part of environmental assessments of oil production [Ward et al., 1979] and transportation studies [Hanifen et al., 1997]. Following the initial discovery of hypoxia in 1972-1974, Ragan et al. [1978] and Turner and Allen [1982a] surveyed the shelf in 1975 and 1976 and found low oxygen in the warmer months west of the Mississippi and Atchafalaya River discharges. Environmental assessments and studies of oil and gas production revealed low oxygen conditions in most inner shelf areas of Louisiana and Texas studied in mid-summer for the period 1978-1984 (summarized in Rabalais [1992], Rabalais et al. [1999]).

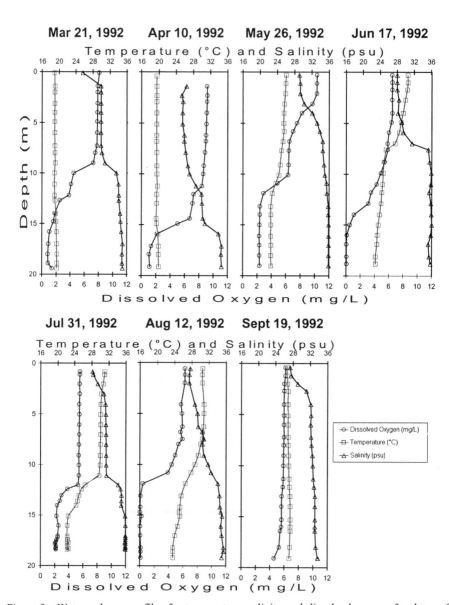

Figure 3. Water column profiles for temperature, salinity and dissolved oxygen for dates of monthly sampling at station C6B off Terrebonne Bay on the southeastern Louisiana shelf in 1992 [derived from Rabalais, Turner and Wiseman, unpublished data]. Station location is identified in Figure 6.

Hypoxia on the upper Texas coast is usually an extension of the larger hypoxic zone off Louisiana, although isolated areas may be found farther to the south (e.g., off Galveston and Freeport, Texas) [Harper et al., 1991; Pokryfki and Randall, 1987]. Isolated areas may be an artifact of the sampling, and very few systematic surveys have been conducted in this area with the exception of the summer SEAMAP cruises [Gulf States Fisheries Commission, 1982, et seq.]. Mid-summer SEAMAP cruises documented hypoxia on the Texas coast in small, isolated areas in 1983, none in 1984-1985, and again in localized areas in most years between 1991-1997 [J. K. Craig, unpublished data]. Most instances of hypoxia along the Texas coast are infrequent, short-lived, and limited in extent [Rabalais, 1992].

Hypoxia has been documented off Mississippi Sound during high stages of the Mississippi River and off Mobile Bay in bathymetric low areas [Rabalais, 1992]. There are usually more reports in flood years or when more Mississippi River water moves to the east of the birdfoot delta. Hypoxia east of the Mississippi River is infrequent, short-lived, and limited in extent [Rabalais, 1992]. From limited data where both sides of the delta were surveyed for hydrographic conditions including dissolved oxygen [Turner and Allen, 1982a], there was no evidence that the area of low oxygen formed a continuous band around the delta.

Mid-Summer Extent and Variability

The distribution of hypoxia on the Louisiana shelf has been mapped in mid-summer (usually late July to early August) over a standard 60- to 80-station grid since 1985 (representative maps in Fig. 4). Hypoxic waters are distributed from shallow depths near shore (4 to 5 m) to as deep as 60 m water depth [Rabalais et al., 1991, 1998, 1999], but more typically between 5 and 30 m.

For the period from 1985 to 1992, the zone of hypoxia was usually in a configuration of disjunct areas to the west of the deltas of the Mississippi and Atchafalaya Rivers, and the bottom area averaged 7,000 to 9,000 km^2 (1986 and 1990 are illustrated in Fig. 4, areas for all years are in Fig. 5). Hypoxia in mid-summer 1988 was confined to a single inshore station off Terrebonne Bay on Transect C. A reduced grid was mapped in 1989, and was, therefore, not comparable to data from other years. Bottom water hypoxia was continuous across the Louisiana shelf in mid-summer of 1993-1997, and the area (16,000 to 18,000 km^2) was twice as large as the 1985-1992 average (1996 is illustrated in Fig. 4). The somewhat smaller size of the hypoxic area in July 1997 was likely due to the passage of Hurricane Danny that either caused wind mixing and reaeration or forced the hypoxic water mass closer to shore. The 1998 hypoxia was concentrated on the eastern and central Louisiana shelf from the Mississippi River delta to Marsh Island near Atchafalaya Bay and in deeper water than usual. The largest area of bottom-water hypoxia to date (20,000 km^2) was mapped in July of 1999 [Rabalais, 1999]. Time or other logistical constraints often prevent the complete mapping of the extent of hypoxia, either in the offshore direction or to the west. Thus, the areal extent of bottom-water hypoxia generated from these surveys is a minimal estimate. The area estimations vary within a summer, and they should not be over-interpreted in making year-to-year comparisons or identifying trends.

There were extensive areas of hypoxia during multiple July cruises in 1993 and 1994, three and two, respectively. The multiple cruises demonstrate that the large area of hypoxia is persistent over two to three weeks, at least, although changing in configuration.

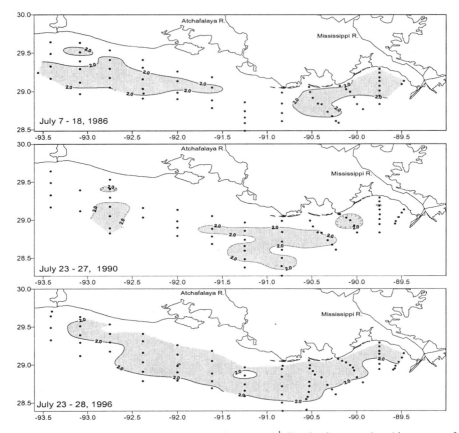

Figure 4. Distribution of bottom water less than 2 mg l^{-1} dissolved oxygen in mid-summer of the years indicated (from Rabalais et al. [1991, 1999]).

A compilation of fifteen mid-summer shelfwide surveys (1985-1999) (Fig. 6) illustrates that the frequency of occurrence of hypoxia is higher to the west of the Mississippi and Atchafalaya Rivers in a down-current direction from the freshwater discharge and nutrient load. Other gradients in biological parameters and processes are also evident in a decreasing gradient away from the river discharges [Rabalais et al., 1996; Rabalais and Turner, 1998].

Hypoxia in Flood and Drought Conditions

Conditions during extreme events such as the 1993 flood or the 1988 or 2000 droughts emphasize the importance of river discharge and nutrient load in defining the mid-summer extent of hypoxia [Rabalais et al., 1991, 1998]. The influence of the Mississippi River system was magnified during the 1993 flood. Above-normal freshwater inflow and nutrient flux from the Mississippi and Atchafalaya Rivers from late spring into mid-summer and early fall [Dowgiallo, 1994] were clearly related in time and

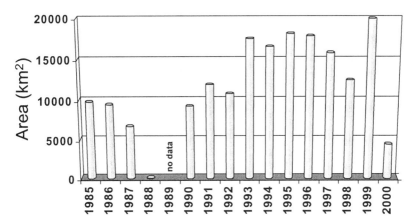

Figure 5. Estimated areal extent of bottom water hypoxia (≤ 2 mg l^{-1}) for mid-summer cruises in 1985-2000 (modified from Rabalais et al. [1998, 1999]).

space to the seasonal progression of hypoxic water formation and maintenance and its increased severity and areal extent on the Louisiana-Texas shelf in 1993 [Rabalais et al., 1998]. Flood conditions resulted in a higher flux of nutrients to the Gulf, higher concentrations of dissolved nutrients in Gulf surface waters, lower surface water salinity, higher surface water chlorophyll a biomass, increased phytoplankton abundance, modeled greater carbon export from the surface waters, increased bottom water phaeopigment concentrations (an indicator of fluxed degraded surface water chlorophyll a biomass), lower bottom water oxygen concentrations compared to the long-term averages for 1985-1992 and a doubled size of the zone [Dortch, 1994; Goolsby, 1994; Justic' et al., 1997; Rabalais et al., 1998].

A 52-yr low-river flow of the Mississippi River occurred in 1988. Discharge began at normal levels in 1988 and dropped to some of the lowest levels on record during the summer months (Fig. 2). In early June 1988, hydrographic conditions on the southeastern Louisiana shelf were similar to those observed in previous years, i.e., a stratified water column and some areas of oxygen-deficient bottom waters [Rabalais et al., 1991]. By mid-July, few areas of lower surface salinity were apparent, there was little density stratification, and low oxygen conditions were virtually absent. Reduced summer flows in 1988 also resulted in reduced suspended sediment loads and nutrient flux and subsequent increased water clarity across the continental shelf. The critical depth for photosynthesis was greater than the depth of the seabed, and there likely was photosynthetic production of oxygen in bottom waters [Rabalais et al., 1991]. A typical seasonal sequence of nutrient-enhanced primary production and flux of organic matter progressed in the spring and led to the formation of hypoxia, but hypoxia was not maintained because of weak stratification and improved light conditions in bottom waters that allowed for some photosynthesis.

Drought conditions in spring 2000 throughout the watershed resulted in decreased discharge (Fig. 2) and nutrient flux so that the typical spring sequence of increased productivity, phytoplankton biomass and carbon flux was diminished in importance. Less productivity, coupled with windy spring weather that prevented the usual development of stratified waters, resulted in little development of hypoxia in the spring. Discharge reached mean levels in the summer and hypoxia eventually developed

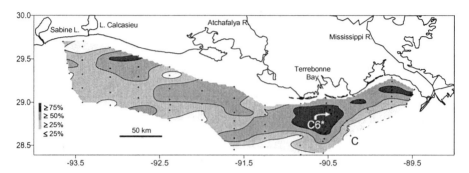

Figure 6. Distribution of frequency of occurrence of mid-summer bottom-water hypoxia over the 60- to 80-station grid from 1985-1999 [derived from Rabalais, Turner and Wiseman, various published and unpublished data]. Station C6*, which incorporates data from C6, C6A and C6B, is identified.

although over a smaller area (4,400 km²). While the total discharge in 1988 and 2000 between the months of January-May was similar, the resulting response of the system and distribution of hypoxia were quite different in the two spring-summer sequences.

Temporal Variability

Critically depressed dissolved oxygen concentrations occur below the pycnocline from as early as late February through early October and nearly continuously from mid-May through mid-September. In March, April and May, hypoxia tends to be patchy and ephemeral; it is most widespread, persistent, and severe in June, July and August [Rabalais et al., 1991, 1999]. The low oxygen water mass on the bottom during peak development in the summer changes configuration in response to winds, currents and tidal advection. The persistence of extensive and severe hypoxia into September and October depends primarily on the breakdown of stratification by winds from either tropical storm activity or passage of cold fronts. Hypoxia is rare in the late fall and winter.

Hypoxia occurs not only at the bottom near the sediments, but well up into the water column (Fig. 3). Depending on the depth of the water and the location of the pycnocline(s), hypoxia may encompass from 10% to over 80% of the total water column, but normally only 20 to 50%. Hypoxia may sometimes reach to within 2 m of the surface in a 10-m water column, or to within 6 m of the surface in a 20-m water column. Anoxic bottom waters can occur, along with the release of toxic hydrogen sulfide from the sediments.

Continuously recording oxygen meters were deployed near the bottom at a 20-m station off Terrebonne Bay (example for 1993 in Fig. 7). There was variability within the year and between years, but the pattern generally depicted is for (1) a gradual decline of bottom oxygen concentrations through the spring with reoxygenation from wind-mixing events, (2) persistent hypoxia and often anoxia for extended parts of the record in May-September, (3) isolated wind-mixing events in mid-summer that reaerate the water column followed by a decline in oxygen similar to that seen in the spring, (4) isolated intrusion of higher oxygen content waters from deeper water during upwelling favorable

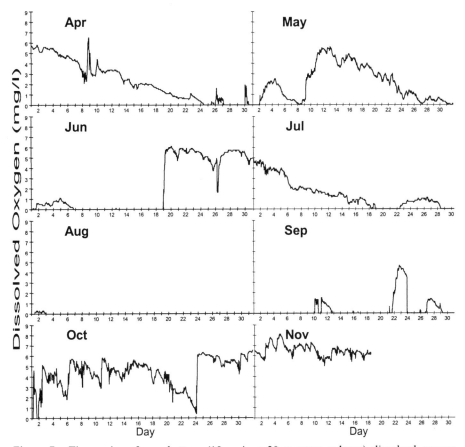

Figure 7. Time-series of near-bottom (19 m in a 20-m water column) dissolved oxygen concentration (mg l^{-1} in 1-h intervals) at station C6B (see Fig. 6) for 1993 (modified from Rabalais et al. [1999]).

wind conditions, then a relaxation of the winds and a movement of the low oxygen water mass back across the bottom at the site of the oxygen meter, and (5) wind mixing events, either tropical storms or hurricanes or cold fronts in the late summer and fall that mix the water column sufficiently to prevent prolonged instances of dissolved oxygen concentrations less than 2 mg l^{-1}.

Mississippi River Discharge, Nutrient Load and Hypoxia

Although the Mississippi River discharges organic matter, whose decomposition could consume oxygen, the principal source of organic matter reaching the hypoxic bottom waters is from *in situ* phytoplankton production [Turner and Allen, 1982b; Rabalais et al., 1992; Turner and Rabalais, 1994a; Eadie et al., 1994; Justic' et al., 1996, 1997]. The variability in primary production in the northern Gulf impacted by the

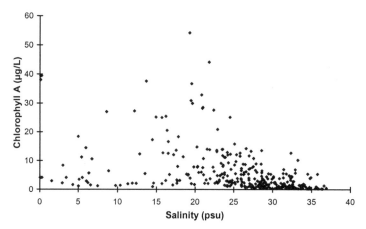

Figure 8. Relationship of surface chlorophyll *a* and salinity for six cruises (April 1992, October 1992, April 1993, July 1993, April 1994, July 1994) from 89.5°W to 97°W, 5 m to 50 m depth. One value (209 µg l⁻¹ at 13.7 psu) was deleted from the plot. (From Rabalais and Turner [1998].)

Mississippi River is quite high, due to the dynamic and heterogeneous conditions found in the river/ocean mixing zone. The highest values of primary production and chlorophyll *a* biomass (Fig. 8) are typically observed at intermediate salinities within the Mississippi River plume and across the broad region of the Louisiana shelf influenced by the river and coincide with non-conservative decreases in nutrients along the salinity gradient (Fig. 8) [Rabalais and Turner, 1998; Lohrenz et al., 1999]. Salinities of 20 to 30 psu are typical across broad regions of the Louisiana inner shelf for much of the year, and high production and chlorophyll biomass occur over broad areas (Plate 2). The rates of primary production along a salinity gradient of the Mississippi River plume are constrained by low irradiance, mixing in the more turbid, low salinity regions of the plume, and by nutrient limitation outside the plume—a pattern typical of other major rivers including the Amazon [DeMaster et al., 1986; Smith and DeMaster, 1996], Huanghe [Turner et al., 1990] and Changjiang [Xiuren et al., 1988].

The high rates of primary production in the inner shelf of the northern Gulf of Mexico can be attributed to nutrient loading from the Mississippi and Atchafalaya rivers and buoyancy flux that keeps recycled nutrients in the photic zone [Riley, 1937; Thomas and Simmons, 1960, Sklar and Turner, 1981; Lohrenz et al., 1990]. In the continental shelf waters of the northern Gulf of Mexico, there is a strong relationship between river-borne nutrient flux, nutrient concentration, primary production and net production. Primary production near the river delta and on the southeastern Louisiana shelf is significantly correlated with nitrate + nitrite concentrations and fluxes [Lohrenz et al., 1997]. Even stronger correlations were observed between the concentration of orthophosphate and primary production, but these were not significant (smaller sample size). Peak nutrient inputs generally occur in the spring with peak river discharge in March-May, although there is a temporal lag (1 mo) in nitrate load behind peak discharge [Justic' et al., 1997]. Net production at station C6* off Terrebonne Bay lags 1 mo from the peak nitrate load (Fig. 9).

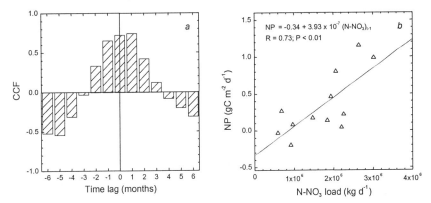

Figure 9. Left panel: Cross correlation function (CCF) for Mississippi River nitrate flux at Tarbert Landing and net production of the upper water column (1-10 m at station C6* in 20-m depth off Terrebonne Bay, see Fig. 6). Right panel: best-fit time-delayed linear model for the regression of net production (NP) on nitrate load. The model is $NP_t = -0.34 + 3.93 \times 10^{-7}$ nitrate$_{t-1}$ where t and t-1 denote values for the current and preceding months, respectively. Symbols denote monthly averages for the period 1985-1992. (From Justic' et al. [1997], used with permission of the author.)

The spring delivery of nutrients initiates a seasonal progression of biological processes that ultimately lead to the depletion of oxygen in the bottom waters. The rates of primary production in the surface waters of the Mississippi River-influenced continental shelf are high (290 to > 300 g C m^{-2} y^{-1}) [Sklar and Turner, 1981; Lohrenz et al., 1990]. One might expect that the vertical export of particulate organic carbon (POC) would also be high and be roughly proportional to the quantity of carbon fixed in the surface waters. Although the overall flux of POC on the continental shelf influenced by the Mississippi River is high, the relationship between POC export and primary production is quite variable in time and space. Redalje et al. [1994] examined the relationship between primary production and the export of POC from the euphotic zone determined with free-floating sediment traps. Productivity and POC exports exhibited similar trends in spring and fall, but were uncoupled in summer. The lowest ratio of export to production coincided with the time when production was greatest, and the highest ratios occurred when production was the lowest. Export ranged from low values of 3-9% during July-August 1990 to high values during March 1991, when export exceeded measured water column-integrated primary production by a factor of two.

In another study of the vertical flux of particulate material, particle traps were deployed on an instrument mooring within the zone of recurring hypoxia at depths of 5 m and 15 m in a 20-m water column in spring, summer and fall of 1991 and 1992 (station C6B off Terrebonne Bay, Fig. 6) [Qureshi, 1995]. Carbon flux was approximately 500 to 600 mg C m^{-2} d^{-1} in 15-m water depth [Qureshi, 1995]. A rough estimate of the fraction of production exported from the surface waters (compared to seasonal primary production data of Sklar and Turner [1981]) was highly variable and ranged from 10 to 200% with higher percentages in spring [Qureshi, 1995]. A large proportion of the POC that reached the bottom was incorporated in zooplankton fecal pellets (55%), but also as individual cells or in aggregates. Both phytoplankton and zooplankton fecal carbon flux were greater in the spring and the fall than in the summer.

The oxygen consumption rates in near-bottom waters of the seasonally oxygen-deficient continental shelf were measured during spring and summer cruises in several years [Turner et al., 1998; Turner and Rabalais, 1998]. Respiration rates varied between 0.0008 to 0.29 mg O_2 l^{-1} h^{-1}, and were sufficient to reduce the *in situ* oxygen concentration to zero in less than four weeks. The rates were inversely related to depth and decreased westward of the Mississippi River delta, consistent with the decrease in nutrients, chlorophyll *a* and total pigment concentrations, and the relative proportion of surface-to-bottom pigments. The amount of phytoplankton biomass in the bottom waters across the Louisiana inner and middle continental shelf is high, often exceeding 30 μg l^{-1}, and a high percentage is composed of phaeopigments [Rabalais and Turner, 1998]. The respiration rate is proportional to phytoplankton pigment concentration [Turner and Allen, 1982b], and, thus, higher rates of oxygen consumption would be expected where higher flux of material reaches the lower water column and sediments. Respiration rates per unit phytoplankton pigment were highest in the spring, in shallower waters, and also closest to the Mississippi River delta.

The high particulate organic carbon flux to the 15-m moored trap was sufficient to fuel hypoxia in the bottom waters below the seasonal pycnocline [Qureshi, 1995; Justic' et al., 1996]. The flux of organic material in summer, while it sustained hypoxia, was incremental to the majority flux of particulates in the spring [Qureshi, 1995]. Because the moored sediment traps were serviced by divers, they were not deployed from late fall through early spring when high fluxes might have occurred that also fueled the consumption of oxygen.

There is a time lag between nutrient delivery and production in the surface waters, and a subsequent lag in flux of carbon to the lower water column and oxygen uptake in the lower water column and sediments. The spatial and temporal variability in the distribution of hypoxia is, at least partially, related to the amplitude and phasing of the Mississippi River discharge and nutrient fluxes [Pokryfki and Randall, 1987; Justic' et al., 1993]. An annual sequence of salinity, surface chlorophyll *a*, bottom oxygen and bottom total pigments are illustrated in Figure 10. Net productivity (a surrogate for excess carbon available for export) of the upper water column appears to be an important factor controlling the accumulation of organic matter in coastal sediments and development of hypoxia in the lower water column. The seasonal dynamics of net productivity in the northern Gulf of Mexico are coherent with the dynamics of freshwater discharge [Justic' et al., 1993]. The surface layer (0 to 0.5 m at station C6*) shows an oxygen surplus relative to the saturation values during February-July; the maximum occurs during April and May and coincides with the maximum flow of the Mississippi River. Peak chlorophyll *a* biomass in the surface waters, a surrogate for net productivity, occurs in spring at station C6* in the core of the hypoxic zone (Fig. 10). The bottom layer (approximately 20 m), in contrast, exhibits an oxygen deficit throughout the year, but reaches its highest value in July (Fig. 10). Bottom hypoxia in the northern Gulf is most pronounced during periods of high water column stability when surface-to-bottom density differences are greatest (demonstrated by reduced surface salinity in spring and summer in Fig. 10) [Rabalais et al., 1991]. The correlation between Mississippi River flow and surface oxygen surplus peaks at a time-lag of one month, and the highest correlation between discharge and bottom oxygen deficit is for a time-lag of two months [Justic' et al., 1993]. A similar cross-correlation analysis verified that the seasonal maximum in net production lags riverine nitrate flux by one month (Fig. 9) [Justic' et al., 1997]. These findings suggest that the oxygen surplus in the surface layer following high

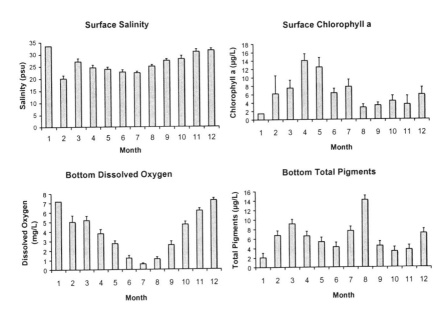

Figure 10. Surface and bottom water quality for station C6* (composite data for stations C6A, C6B and C6, Fig. 6) for 1985-1997 average conditions (± s.e.). *n* ranges between 1-10 for winter, 10-20 for spring and fall, and 20-40 for summer. (Modified from Rabalais et al. [1998].)

flow depends on nutrients ultimately coming from the river that are regenerated many times. An oxygen surplus also means that there is an excess of organic matter derived from primary production that can be redistributed within the system; much of this will eventually reach the sediments.

Similar relationships with freshwater discharge and oxygen depletion in bottom waters at stations west of the Atchafalaya River delta and expected direction of materials and freshwater flux were identified by Pokryfki and Randall [1987]. Time lags were apparent between values of river discharge, bottom dissolved oxygen and salinity. The highest cross-correlation coefficient between bottom water dissolved oxygen (in the area off the Calcasieu estuary) and river discharge (from the Atchafalaya) was −0.51 at a lag of two months. Their linear regression model did not include any factors for biological processes, and the accuracy would have been improved by "incorporating a biological component into the time series" [Pokryfki and Randall, 1987].

Nutrient Sources and Changes

The Mississippi River is the largest source of freshwater and nutrients to the northern Gulf of Mexico. This watershed, like others, has undergone major changes affecting water quality since the Native American culture was displaced by mostly European immigrants in the early 1800s. Major alterations in the morphology of the main river channel and widespread landuse patterns in the watershed, along with anthropogenic additions of nitrogen and phosphorus, have resulted in dramatic water quality changes

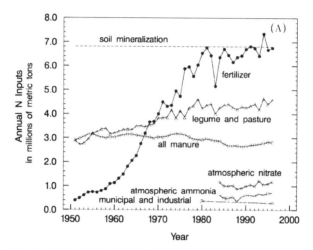

Figure 11. Annual nitrogen inputs from major sources in the Mississippi River Basin, 1951-96, (from Goolsby [2000], modified from Goolsby et al. [1999], used with permission of the author).

this century [Turner and Rabalais, 1991]. The river has been shortened by 229 km in an effort to improve navigation, and has a flood-control system of earthwork levees, revetments, weirs, and dredged channels for much of its length. These modifications have left adjacent lands drier and more susceptible to massive conversion to farmland [Abernethy and Turner, 1987]. More than half of the original wetlands in the United States have been lost to drainage practices [Zucker and Brown, 1998]. Much of this wetland loss is related to agricultural expansion.

Water quality in streams, rivers, lakes and coastal waters may change when watersheds are modified by alterations in vegetation, sediment balance, conversion of forests and grasslands to farms and cities, and increased anthropogenic activities that accompany increased population density, e.g., fertilizer application, sewage disposal or atmospheric deposition [Peierls et al., 1991; Turner and Rabalais, 1991; Howarth et al., 1996; Caraco and Cole, 1999]. The estimate of current river nitrogen export from the Mississippi River is 2.5- to 7.4-fold higher than from the watershed during pre-agricultural and pre-industrial or "pristine" conditions [Howarth et al., 1996].

In an average year the Mississippi River discharges nearly 1.6 million mt of nitrogen to the Gulf of Mexico, of which 0.95 million mt is nitrate and 0.58 million mt is organic nitrogen [Goolsby et al., 1999]. The principle sources of inputs of nitrogen to the Mississippi River system are soil mineralization, fertilizer application, legume crops, animal manure, atmospheric deposition, and municipal and industrial point discharges (Fig. 11). The highest inputs within the watershed are above the confluence of the Mississippi and Ohio Rivers (Plate 3, upper panel), and not surprisingly the yields are from sub-basins where inputs are the greatest (Plate 3, lower panel). High inputs and yields are characteristic of sub-basins were precipitation is high and agricultural drainage is extensive, resulting in the high rates of transport of soluble nitrate into streams, the Mississippi River and the Gulf of Mexico.

Large-scale industrial production and use of nitrogen and phosphorus fertilizer in the United States began in the mid-1930s and climbed to a peak in the 1980s [Turner and

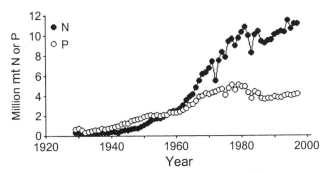

Figure 12. Nitrogen (as N) and phosphorus (as phosphate) fertilizer use this century in the United States up to 1997 (from USDA annual agriculture statistical summaries). (From Rabalais et al. [1999].)

Rabalais, 1991]. Phosphorus fertilizer use in the United States reached a plateau around 1980, whereas nitrogen fertilizer use is still increasing (Fig. 12). Forty-two percent of the nitrogen fertilizer and 37% of the phosphorus fertilizer used annually in the United States from 1981 to 1985 was applied in states that are partially or completely in the Mississippi River watershed, where it equaled 4.2 million mt of nitrogen (as N) and 0.53 million mt of phosphorus (as P). Turner and Rabalais [1991] estimated that a maximum of 44% of the applied nitrogen and 28% of the applied phosphorus may have made its way to the Gulf of Mexico. Subtracting a natural loading estimate (riverine fluxes prior to World War II), they estimated that the maximal loading from fertilizer sources probably represents no more than 22% of the applied fertilizer.

Mississippi River nutrient concentrations and loading to the adjacent continental shelf have changed dramatically this century, with an acceleration of these changes since the 1950s [Turner and Rabalais, 1991, 1994a; Rabalais et al., 1996]. Turner and Rabalais [1991] examined water quality data for four lower Mississippi River stations for dissolved inorganic nitrogen (as nitrate), phosphorus (as total phosphorus) and silicon (as silicate). The mean annual concentration of nitrate was approximately the same in 1905-1906 and 1933-1934 as in the 1950s, but it has doubled in the last 40 years (Fig. 13). The increase in total nitrogen is almost entirely due to changes in nitrate concentration. The mean annual concentration of silicate was approximately the same in 1905-1906 as in the early 1950s, then it declined by 50%. Concentrations of nitrate and silicate appear to have stabilized, but trends are masked by increased variability in the 1980s and early 1990s data. Although the concentration of total phosphorus appears to have increased since 1972, variations among years are large.

The silicate:nitrate ratios have changed as the concentrations varied. The Si:N atomic ratio was approximately 4:1 at the beginning of this century, dropped to 3:1 in 1950 and then rose to approximately 4.5:1 during the next ten years, before plummeting to 1:1 in the 1980s. The ratio appears stable at 1:1 through 1997 with little variation. The average atomic ratios of N:Si, N:P and Si:P are currently 1.1, 15 and 14, respectively, and closely approximate those of Redfield [1958] of 16:16:1, N:Si:P [Justic' et al., 1995a, b].

The seasonal patterns in nitrate and silicate concentration have also changed during this century. There was no pronounced peak in nitrate concentration earlier this century, whereas there was a spring peak from 1975 to 1985, presumably related to seasonal agricultural activities, timed with long-term peak river flow [Turner and Rabalais, 1991].

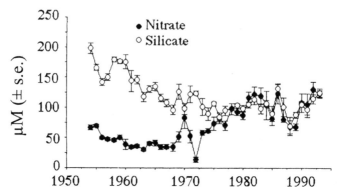

Figure 13. The average annual concentration (μM $\pm$ 1 S.E.) of nitrate and silicate in the Mississippi River at New Orleans (modified from Turner et al. [1998]).

A seasonal summer-fall maximum in silicate concentration, in contrast, is no longer evident. Consequently the seasonal signal of Si:N atomic ratio has also changed. The seasonal shifts in nutrient concentrations and ratios become increasingly relevant in light of the close temporal coupling of river flow to surface water net productivity (1-mo lag) and subsequent bottom water oxygen deficiency (2-mo lag) [Justic' et al., 1993].

Justic' et al. [1995a] compared data for two periods: 1960-1962 and 1981-1987 (Table 1). Substantial increases in N (300%) and P (200%) concentrations occurred over several decades, and Si decreased (50%). [No data on total P concentration in the Mississippi River were reported prior to 1973; however, total P in the river showed a moderate increase between 1973 and 1987. By applying a linear least-squares regression on the 1973-1987 data, they estimated (p<0.01) that the total P concentration increased two fold between 1960-1962 and 1981-1987.] Accordingly, the Si:N ratio decreased from 4.3 to 0.9, the Si:P ratio decreased from 40 to 14, and the N:P ratio increased from 9 to 15. By applying the Redfield ratio as a criterion for stoichiometric nutrient balance, one can distinguish between P-deficient, N-deficient, and Si-deficient rivers, and those having a well balanced nutrient composition. The nutrient ratios for the Mississippi River (1981-1987 data base) show an almost perfect coincidence with the Redfield ratio. The proportions of Si, N and P have changed over time in such a way that they now suggest a balanced nutrient composition.

Despite being balanced on an annual basis, seasonal variations in nutrient inputs can affect nutrient availability. In particular, there is nearly a two-fold difference in nitrate supply over the course of the year [Turner and Rabalais, 1991], but only small annual variations in the silicate and total phosphorus supply. Consequently, the nutrient supply ratios vary around the Redfield ratios on a seasonal basis, with silicate and phosphorus in the shortest supply during the spring and nitrogen more likely to be limiting (based on ratios) during the rest of the year. With nutrient concentrations so closely balanced, Justic' et al. [1995b] proposed that any nutrient can become limiting, perhaps in response to small differences in nutrient supply ratios such as these, or conversely that no single nutrient is more limiting than others. These seasonal differences in nutrient ratios co-occur with seasonal variation in river flow, so that the riverine supply of all nutrients is least in low flow periods. Fluctuations in the Si:N ratio within the major riverine effluents and differences in Si:N ratios between the effluents of the two rivers are

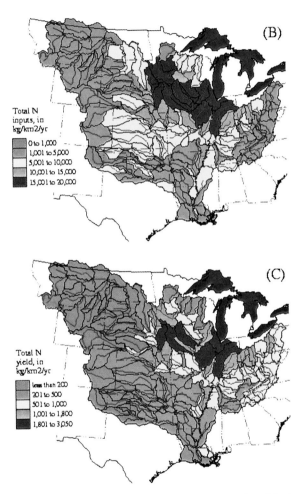

Plate 3. Upper panel: Nitrogen inputs in hydrologic accounting units of the basin during 1992, based on data from the 1992 Census of Agriculture. Lower panel: Average annual nitrogen yields of streams in the basin for 1980-96. (From Goolsby [2000], modified from Goolsby et al. [1999], used with permission of the author).

TABLE 1. Changes in concentrations and atomic ratios of nitrogen (N), phosphorus (P) and silica (Si) in the lower Mississippi River and the northern Gulf of Mexico; x - mean value, n - number of data, S - standard error, p < 0.001 - highly significant difference in nutrient concentrations between the two periods, based on a two-sample t-test. (Modified from Justic' et al. [1995a].)

		Mississippi River		Northern Gulf of Mexico	
		1960-62[d]	1981-87	1960[e]	1981-87
Nutrient concentration (μM):					
	X	36.5	114	2.23	8.13
N[a]	n	72	200	219	219
	S	2.9	6.0	0.16	0.60
			(p < 0.001)		
	x	3.9	7.7	0.14	0.34
P[b]	n	-	234	231	231
	S	-	0.4	0.01	0.02
			(p < 0.001)		
	x	155.1	108	8.97	5.34
Si[c]	n	72	71	235	235
	S	7.5	4.3	0.55	0.33
			(p < 0.001)		
Average atomic ratios:					
Si:N		4.2	0.9	4.0	0.7
N:P		9	15	16	24
Si:P		39.8	14	64	16

[a]N-NO_3 for the Mississippi, dissolved inorganic nitrogen (DIN=NO_3+NH_4+NO_2) for the Gulf of Mexico
[b]total P for the Mississippi, reactive P for the Gulf of Mexico
[c]reactive Si
[d]Turner and Rabalais [1991], for N and Si, reconstructed for P
[e]reconstructed data

believed to be major determinants in estuarine and coastal food web structure on a seasonal and annual basis, with major implications to the cycling of oxygen and carbon [Turner et al., 1998]. As the Si:N ratio falls from above 1:1 to below 1:1, there may be shifts in the phytoplankton community from diatoms to an increasing flagellated algal community, including those that are potentially harmful, an altered marine food web by reducing the diatom-to-zooplankton-to higher trophic level connection, altered carbon flux to the lower water column and sediments as prey items change and grazing influences shift, and subsequent changes in the severity and expanse of hypoxia.

Historical Trends in Productivity and Hypoxia

One might expect a propensity for high productivity and development of hypoxia, given the high volume of fresh water and associated nutrients delivered by the Mississippi River into a stratified coastal system. Unfortunately, the long-term data sets that demonstrate changes in surface water production and bottom water dissolved oxygen, such as available for the northern Adriatic Sea and areas of the Baltic and

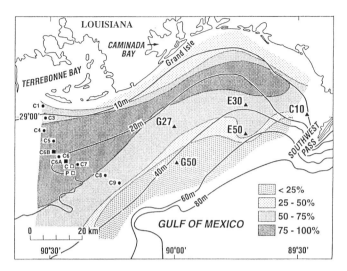

Figure 14. Station locations within the Mississippi River bight for hypoxia monitoring on transect C (closed circles), mooring locations (C6A and C6B) (closed squares), coring stations (closed triangles, those referred to in text are labeled). Stippling corresponds to frequency of occurrence of mid-summer hypoxia at monitoring stations (1985-1987, 1990-1993 [Rabalais, Turner and Wiseman, unpublished data]). (Modified from Rabalais et al. [1996].)

northwestern European coast, are few for the northern Gulf of Mexico. Therefore, biological, mineral or chemical indicators of surface water production and hypoxia preserved in sediments that accumulate under the plume of the Mississippi River (Fig. 14) provide clues to prior hydrographic and biological conditions. Sediment cores analyzed for different constituents [Turner and Rabalais, 1994a; Eadie et al., 1994] document eutrophication and increased organic sedimentation in bottom waters, with the changes being more apparent in areas of chronic hypoxia and coincident with the increasing nitrogen loads from the Mississippi River system

Indicators of Productivity

Although the marine ecosystem influenced by the Mississippi River discharge exhibits signs of incipient Si limitation, the overall silicate-based productivity of the ecosystem on the southeastern Louisiana shelf influenced by Mississippi River discharge appears to have increased in response to the increased nitrogen load. This is evidenced by (1) equal or greater diatom community uptake of silica in the mixing zone, compared to the 1950s [Turner and Rabalais, 1994b], and (2) greater accumulation rates of biogenic silica (BSi) in sediments beneath the plume (Fig. 15) [Turner and Rabalais, 1994a].

Turner and Rabalais [1994a] quantified the silica in the remains of diatoms sequestered as biologically bound silica (BSi) in dated sediment cores from the Mississippi River bight. Relative changes in the % BSi reflect changes in *in situ* production [Conley et al., 1993]. The pattern in % BSi in dated sediment cores parallels the documented increases in nitrogen loading in the lower Mississippi River, over the same period that silicate concentrations have been decreasing [Turner and Rabalais,

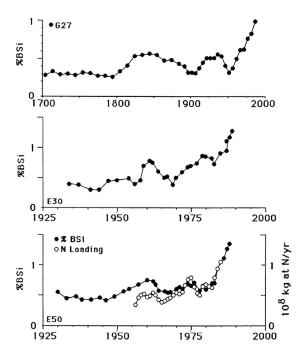

Figure 15. The average concentration of biologically bound silica (BSi) in sediments in each section of three dated sediment cores from stations in the Mississippi River Bight in depths of 27 to 50 m (stations in Fig. 14). A 3-y running average is plotted by time determined from Pb-210 dating. The figure for station E50 is superimposed with a 3-y average nitrogen loading from the Mississippi River. (Modified from Turner and Rabalais [1994a].)

1994a] (Fig. 15). The increased % BSi in Mississippi River bight sediments is direct evidence for the increase in flux of diatoms from surface to bottom waters beneath the Mississippi River plume. The highest concentrations of BSi were in sediments deposited in 25 to 50 m water depth in the middle of the sampling area. The % BSi in sediments from deeper waters (110 and 200 m) was generally stable through time, but rose in the shallower stations (10 and 20 m) around 1900 (not illustrated in Fig. 15). At the intermediate depths (27 to 50 m), where both the % BSi concentration and accumulation rates were highest, coincidental changes in the % BSi with time were evident, especially in the 1955 to 1965 period (a rise and fall) and a post 1975 (1980?) rise that was sustained to the sampling date (1989) (Fig. 15).

The increase in % BSi in sediments from the mid 1850s to the early 1900s support the hypothesis of Mayer et al. [1998] that organic nitrogen associated with the suspended sediment load may be a relatively large proportion of the total nitrogen load in river-dominated coastal regions. The % BSi peaks and declines around the mid-1800s and later around 1925 are causally related to the expansion of land clearing and land drainage efforts within the Mississippi River basin [Turner and Rabalais, unpublished data]. The trough in the 1930s to 1945 era is coincidental with the Great Depression and World War II, and the accelerated accumulation since the 1950s is associated with the tripling in the dissolved inorganic nitrogen flux. Mayer et al. [1998] predicted that the relative importance of the organic nitrogen associated with the suspended load would become less important in the Mississippi River as the suspended load decreased and as the anthropogenic dissolved inorganic nitrogen load increased.

The organic accumulation in the middle of the Mississippi River bight during the 1980s was 90 g C m^{-2} y^{-1}, based on sedimentation rates and % carbon of the sediments [Turner and Rabalais, 1994a]. This is approximately 30% of the estimated annual phytoplankton production [Sklar and Turner, 1981; Lohrenz et al., 1990]. If the assumption is made that the BSi:C ratio at the time of deposition remained constant this century, then the increased BSi deposition represents a significant change in carbon deposition rates (up to 43% higher in cores dated after 1980 than those dated between 1900 to 1960). These results are corroborated by the same rate of increase in marine-origin carbon in sediment cores also collected within the Mississippi River bight (near station E30 of Turner and Rabalais [1994a]) [Eadie et al., 1994]. They estimated accumulation rates of about 30 g C m^{-2} y^{-1} in the 1950s to 50-70 g C m^{-2} y^{-1} at present. The rate of burial was significantly higher at a station within the area of chronic hypoxia (approximately 70 g C m^{-2} y^{-1}), in comparison with another site at which hypoxia had not been documented (approximately 50 g C m^{-2} y^{-1}). The $\delta^{13}C$ partitioning of organic carbon into terrestrial and marine fractions further indicated that the increase in accumulation of carbon in both cores was exclusively in the marine fraction. The accumulation of carbon in the sediments analyzed by Eadie et al. [1994] was strongly correlated with Mississippi River nitrate flux.

Indicators of Hypoxia

The surrogates in the dated sediment cores for oxygen conditions indicate an overall increase in oxygen stress (in intensity or duration) in the last 100 years, which seems especially severe since the 1950s and coincident with the onset of increases in riverine nitrogen loading. The average glauconite abundance, a sediment mineral indicative of reducing environments and geologic anoxic settings, is ~5.8% of the coarse fraction of sediments from 1900 (oldest date in core collected near station E30, Fig. 14), rises to a transition point in the early 1940s and is ~13.4% afterwards [Nelsen et al., 1994]. These data suggest that hypoxia may have existed before the 1940 time horizon (at least to 1900) and that subsequent anthropogenic influences have exacerbated the problem.

Benthic foraminiferans, protozoans with calcium carbonate embedded cell walls, are useful indicators of reduced oxygen levels or carbon-enriched sediments or both [Sen Gupta et al., 1981; Sen Gupta and Machain-Castillo, 1993]. Benthic foraminiferal density and diversity were generally low in the Mississippi River bight, but a comparison of assemblages in surficial sediments from areas differentially affected by oxygen depletion indicated that the dominance of Ammonia parkinsoniana over Elphidium spp. (A-E index) was much more pronounced under hypoxia than in well-oxygenated waters (Fig. 16). The relative abundance of A. parkinsoniana was correlated with % BSi (i.e., a food source indicator) in sediments [Sen Gupta et al., 1996]. The A-E index also correlated strongly with the percentage of total organic carbon in surficial sediments. Thus, the index is affected by seasonal hypoxia produced by phytoplankton blooms that are recorded in the sediments in % BSi and carbon content. In the context of modern hypoxia, species distribution in dated sediment cores revealed stratigraphic trends in the

Ammonia/Elphidium ratio that indicate an overall increase in oxygen stress (in intensity or duration) in the last 100 years (Fig. 16). In particular, the stress seems especially severe since the 1950s. It is notable that there is no trend in the A-E index for station G50 outside the zone of persistent hypoxia and that the index in 1988 for station C10 fell off the trend line (i.e., no low oxygen during the mid-summer 1988 cruise). In the last 100 years, both *Ammonia* and *Elphidium* become less important components of the assemblage, while *Buliminella morgani* shows an unusual dominance (also see Blackwelder et al. [1996]). *B. morgani*, a hypoxia-tolerant species, is known only from the Gulf of Mexico and dominates the population (> 50%) within the area of chronic seasonal hypoxia [Blackwelder et al., 1996]. It increased markedly upcore in the sediments analyzed by Blackwelder et al. [1996] and for station G27 of the Sen Gupta et al. [1996] study. *Quinqueloculina* (a significant component of the modern assemblage only in well-oxygenated waters) has been absent from the record of the G27 core since the early 1900s, but was a conspicuous element of the fauna in the previous 200 years. The historical absence of *Quinqueloculina* since 1900 at station G27 matches the presence of glauconite at station 10 since 1900. The occurrence of *Quinqueloculina* prior to 1870, however, indicates that oxygen stress was not a problem before then.

Global Patterns

There is a general consensus that the eutrophication of estuaries and enclosed coastal seas worldwide has increased over the last several decades [Nixon, 1995]. Evidence from many coastal seas suggests a long-term increase in frequency of phytoplankton blooms, including noxious forms [Smayda, 1990; Hallegraeff, 1993; Anderson, 1995]. Also, an increase in the areal extent and/or severity of hypoxia was observed, for example, in Chesapeake Bay [Officer et al., 1984], the northern Adriatic Sea [Justic´ et al., 1987], some areas of the Baltic Sea (e.g., Andersson and Rydberg [1987]) and many other areas in the world's coastal ocean [Diaz and Rosenberg, 1995]. Diaz and Rosenberg [1995] documented that many systems are hypoxic now that were not historically, and others have expanded the geographic extent or increased in severity, either in lower dissolved oxygen concentrations or prolonged periods of exposure or both.

Long-term increases in nutrient concentrations in coastal waters along with increased primary production have been documented elsewhere in the world, e.g., the Baltic Sea [Larsson et al., 1985; Rosenberg, 1986; Wulff and Rahm, 1988], the Kattegat and Skaggerak [Rosenberg, 1986; Andersson and Rydberg, 1987], the sounds separating Sweden from Denmark [Rosenberg, 1986], the northwestern shelf of the Black Sea [Tolmazin, 1985], the northern Adriatic Sea [Faganeli et al., 1985; Justic´ et al., 1987] and the Dutch coast of the North Sea (Fransz and Verhagen, 1985). In the opinion of Diaz and Rosenberg [1995], no other environmental stressor has changed to the degree that oxygen depletion has in the last several decades.

Smaller and less frequent zones of hypoxia than that of the northern Gulf of Mexico occur in U.S. coastal and estuarine areas (e.g., New York Bight [Garside and Malone, 1978; Swanson and Sindermann, 1979; Falkowski et al., 1980; Swanson and Parker, 1988], Chesapeake Bay [Officer et al., 1984; Malone, 1991, 1992; Boynton et al., 1995], Long Island Sound [Welsh and Eller, 1991; Welsh et al., 1994; Parker and O'Reilly,

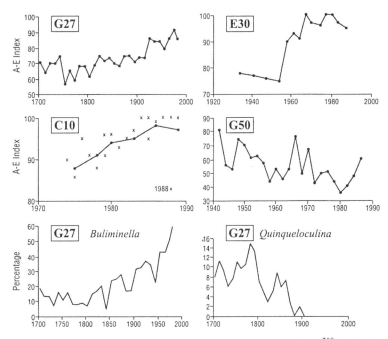

Figure 16. Changes in benthic foraminiferans with stratigraphic depth in ^{210}Pb-dated sediment cores from the Mississippi River bight. A line connecting 3-y averages is superimposed on the data for C10; the 1988 outlier reflects the absence of summer hypoxia. Foraminiferans that indicate changes in oxygen stress (*Buliminella morgani* and *Quinqueloculina* sp.) are shown for G27. Note: the time scale is variable among plots. (Modified from Rabalais et al. [1996], Sen Gupta et al. [1996].)

1991], Mobile Bay [Loesch, 1960; May, 1973; Turner et al., 1987], and the Neuse River estuary [Paerl et al., 1998]. Where sufficient long-term data exist, e.g., Chesapeake Bay, there is clear evidence for increases in nutrient flux, increased primary production, and worsening hypoxia. Thorough analyses of multiple indicators in sediment cores from the Chesapeake Bay indicate that sedimentation rates and eutrophication of the waters of the Bay have increased dramatically since the time of European settlement of the watershed [Cooper and Brush, 1991, 1993; Cooper, 1995; Karlsen et al., 2000]. In addition, results indicate that hypoxia and anoxia may have been more severe and of longer duration in the last 50 years, particularly since the 1970s. The sediment core findings corroborate long-term changes in Chesapeake Bay water column chlorophyll biomass since the 1950s [Harding and Perry, 1997]. The parallels of the Chesapeake Bay eutrophication and hypoxia to those of the Mississippi River watershed and Gulf of Mexico hypoxia are striking, in particular those of the last half century.

Consequences to Living Resources

Most marine systems respond to an increase in nutrient inputs with an increase in primary production. Shifts in the relative proportion of essential nutrients, as one or two increase and others remain the same or decrease, however, may result in altered phytoplankton communities and trophic links. There are examples of excessive nutrients and phytoplankton production leading to a shift in zooplankton communities from copepod-based to gelatinous zooplankton-based (i.e., jelly fish and ctenophores) [Zaitsev, 1993] with devastating effects on fisheries because of increased predation by the gelatinous zooplankton on fish larvae and other zooplankton. If surface productivity is enhanced in prey species that are preferred by the community of zooplankton grazers, then there will likely be increased productivity in pelagic and demersal populations that depend on either the living cells or the detrital material that sinks to the seabed, respectively. There are thresholds, however, where the load of nutrients to a marine system and the carbon produced exceeds the capacity for assimilation, and water quality degradation occurs with detrimental effects on components of the ecosystem and on ecosystem functioning.

When the depletion of oxygen worsens, the ability of organisms to reside either at the bottom or within the water column or even their survivability, is affected. When oxygen levels fall below critical values, those organisms capable of swimming (e.g., demersal fish, portunid crabs and shrimp) evacuate the area. The stress on less motile fauna varies, but they also experience stress or die as oxygen concentrations fall to zero. Important fishery resources are variably affected by direct mortality, forced migration, reduction in suitable habitat, increased susceptibility to predation, changes in food resources and disruption of life cycles. Prolonged oxygen depletion can cause mass mortalities in aquatic life, disrupt aquatic communities, cause declines in biological diversity, impact the capacity of aquatic systems to support biological populations, and disrupt the natural cycling of elements.

The effects of eutrophication, including hypoxia, are well known for some systems and include the loss of commercially important fisheries. The multi-level impacts of increased nutrient inputs and worsening hypoxia are not known for many components of productivity in the Gulf of Mexico, including pelagic and benthic, primary and secondary, food web linkages, and ultimately fisheries yield. Comparisons of ecosystems along a gradient of increasing nutrient enrichment and eutrophication or changes of a specific ecosystem over time through a gradient towards increasing eutrophication, provide information on how nutrient enrichment affects coastal communities. Work by Caddy [1993] in semi-enclosed seas demonstrates a continuum of fishery yield in response to increasing eutrophication. In waters with low nutrients, the fishery yield is low. As the quantity of nutrients increases, the fishery yield increases. As the ecosystem becomes increasingly eutrophied, there is a drop in fishery yield but the decreases are variable. The benthos are the first resources to be reduced by increasing frequency of seasonal hypoxia and eventually anoxia; bottom-feeding fishes then decline. The loss of a planktivorous fishery follows as eutrophication increases, with eventually a change in the zooplankton community composition. Where the current Gulf of Mexico fisheries lie along the continuum of increasing eutrophication is part of the discussions found in this book.

As more and more of the Unite States' and world's coastal waters become hypoxic or as hypoxia increases in severity where it exists now—a trajectory proposed by many researchers and resource managers—what will happen to the habitats, the resource base, the food webs, and ultimately resources of importance for human consumption? The northern Gulf of Mexico is not unique among the world's coastal waters, nor immune to negative impacts, as hypoxia worsens. While there have been no catastrophic losses in fisheries resources in the northern Gulf of Mexico and, in fact, increases in the abundance of some components, the potential impacts of worsening hypoxic conditions are likely given the experience in other systems (e.g., Baltic and Black Seas) where there was a precipitous decline of ecologically and commercially important species.

Reducing excess nutrient delivery to estuarine and marine waters for the improvement of coastal water quality, including the alleviation of hypoxia, requires individual, societal and political will. Proposed solutions are often controversial and have societal and economic costs in a narrow and short-term sense. Yet, multiple, cost-effective methods of reducing nutrient use and delivery can be integrated into a management plan that results in improved habitat and water quality, both within the watershed and the receiving waters [National Research Council, 2000]. Successful plans with successful implementation and often with successful results span geopolitical boundaries, for example the Chesapeake Bay Agreement, the Comprehensive Conservation and Management Plans developed under the U.S. National Estuary Program for many of the nation's estuaries, a Long Island Sound agreement, the efforts of Denmark, Holland and Sweden, and international cooperation among the nations fringing the Baltic Sea as part of the Helsinki Commission [Boesch and Brinsfield, 2000]. These efforts are usually more successful in reducing point sources of nitrogen and phosphorus than with the multiple nonpoint sources of high solubility and growing atmospheric inputs of nitrogen. But success it is for examples such as coral recovery in Kaneohe Bay and for the improved water clarity and recovery of seagrass beds in Tampa and Sarasota Bays [Smith, 1981; Johansson and Lewis, 1992; Sarasota Bay National Estuary Program, 1995]. The growing decline of coastal water quality, and also the proven successes of reducing nutrients, are reasons enough for continued and expanded efforts to reduce nutrient overenrichment and the detrimental effects of hypoxia.

References

Abernethy, Y. and R. E. Turner, US forested wetlands: 1940-1980, *BioScience, 37*, 721-727, 1987.

Anderson, D. M. (ed.), *ECOHAB, The Ecology and Oceanography of Harmful Algal Blooms: A National Research Agenda,* Woods Hole Oceanographic Institution, Woods Hole, Massachusetts, 1995.

Andersson, L. and L. Rydberg, Trends in nutrient and oxygen conditions within the Kattegat: Effects of local nutrient supply, *Estuar. Coast. Shelf Sci., 26*, 559-579, 1987.

Blackwelder, P., T. Hood, C. Alvarez-Zarikian, T. A. Nelsen and B. McKee, Benthic foraminifera from the NECOP study area impacted by the Mississippi River plume and seasonal hypoxia, *Quaternary Intl., 31*, 19-36, 1996.

Boesch, D. F. and R. B. Brinsfield, Coastal eutrophication and agriculture: contributions and solutions, in *Biological Resource Management: Connecting Science and Policy*, edited by E. Balázs, E. Galante, J. M. Lynch, J. S. Schepers, J.-P. Toutant, E. Werner, and P. A. Th. J. Werry, pp. 93-115, Springer, Berlin, 2000.

Boynton, W. R., J. H. Garber, R. Summers and W. M. Kemp, Inputs, transformations, and transport of nitrogen and phosphorus in Chesapeake Bay and selected tributaries, *Estuaries, 18*, 285-314, 1995.

Bratkovich, A, S. P. Dinnel, and D. A. Goolsby, Variability and prediction of freshwater and nitrate fluxes for the Louisiana-Texas shelf: Mississippi and Atchafalaya River source functions, *Estuaries, 17*, 766-778, 1994.

Caddy, J. F., Toward a comparative evaluation of human impacts on fishery ecosystems of enclosed and semi-enclosed seas. *Rev. Fisher. Sci., 1*, 57-95, 1993.

Caraco, N. F. and J. J. Cole, Human impact on nitrate export: An analysis using major world rivers, *Ambio, 28*, 167-170, 1999.

Conley, D. J., C. L. Schelske, and E. F. Stoermer, Modification of the biogeochemical cycle of silica with eutrophication, *Mar. Ecol. Prog. Ser., 101*, 179-192, 1993.

Conseil Permanent International pour l'Exporation de la Mer, *Bulletin Hydrographique pour l'Année 1935*, Series B1, Le Bureau du Conseil, Service Hydrographique, Charlottelund Slot, Danemark, 1936.

Cooper, S. R., Chesapeake Bay watershed historical land use: Impact on water quality and diatom communities, *Ecol. Appl., 5*, 703-723, 1995.

Cooper, S. R. and G. S. Brush, Long-term history of Chesapeake Bay anoxia, *Science, 254*, 992-996, 1991.

Cooper, S. R. and G. S. Brush, A 2500 year history of anoxia and eutrophication in Chesapeake Bay, *Estuaries, 16*, 617-626, 1993.

DeMaster, D. J., G. B. Knapp and C. A. Nittrouer, Effect of suspended sediments on geochemical processes near the mouth of the Amazon River: examination of biogenic silica uptake and the fate of particle-reactive elements, *Cont. Shelf Res., 6*, 107-125, 1986.

Diaz, R. J. and R. Rosenberg, Marine benthic hypoxia: A review of its ecological effects and the behavioural responses of benthic macrofauna, *Oceanogr. Mar. Biol. Ann. Rev., 33*, 245-303, 1995.

Dinnel, S. P. and W. J. Wiseman, Jr., Fresh-water on the Louisiana and Texas Shelf, *Cont. Shelf Res., 6*, 765-784, 1986.

Dortch, Q., Changes in phytoplankton numbers and species composition, in Coastal Oceanographic Effects of Summer 1993 Mississippi River Flooding, Special National Oceanic and Atmospheric Administration Report, edited by M. J. Dowgiallo, pp. 46-49, NOAA Coastal Ocean Office/National Weather Service, Silver Spring, Maryland, 1994.

Dowgiallo, M. J. (ed.), Coastal Oceanographic Effects of Summer 1993 Mississippi River Flooding, Special NOAA Report, NOAA Coastal Ocean Office/National Weather Service, Silver Spring, Maryland, 1994.

Dunn, D. D., Trends in Nutrient Inflows to the Gulf of Mexico from Streams Draining the Conterminous United States 1972 – 1993. U.S. Geological Survey, Water-Resources Investigations Report 96—4113, Prepared in cooperation with the U.S. Environmental Protection Agency, Gulf of Mexico Program, Nutrient Enrichment Issue Committee, U.S. Geological Survey, Austin, Texas, 1996.

Eadie, B. J., B. A. McKee, M. B. Lansing, J. A. Robbins, S. Metz, and J. H. Trefry, Records of nutrient-enhanced coastal productivity in sediments from the Louisiana continental shelf, *Estuaries, 17*, 754-765, 1994.

Faganeli , J., A. Avcin, N. Fanuko, A. Malej, V. Turk, P. Tusnik, B. Vriser, and A. Vukovic, Bottom layer anoxia in the central part of the Gulf of Trieste in the late summer of 1983, *Mar. Pollut. Bull., 16*, 75-78, 1985.

Falkowski, P. G., T. S. Hopkins, and J. J. Walsh, An analysis of factors affecting oxygen depletion in the New York Bight, *J. Mar. Res., 38*, 479-506, 1980.

Fransz, H. G. and J. H. G. Verhagen, Modeling research on the production cycle of phytoplankton in the southern bight of the North Sea in relation to riverborne nutrient loads, *Netherlands J. Sea Res.,* 19, 241-250, 1985.

Garside, C. and T. C. Malone, Monthly oxygen and carbon budgets of the New York Bight apex, *Estuar. Coast. Shelf Sci., 6*, 93-104, 1978.

Goolsby, D. A., Flux of herbicides and nitrate from the Mississippi River to the Gulf of Mexico, in Coastal Oceanographic Effects of Summer 1993 Mississippi River Flooding, Special National Oceanic and Atmospheric Administration Report, edited by M. J. Dowgiallo, pp. 32-35, NOAA Coastal Ocean Office/National Weather Service, Silver Spring, Maryland, 1994.

Goolsby, D. A., W. A. Battaglin, G. B. Lawrence, R. S. Artz, B. T. Aulenbach, R. P. Hooper, D. R. Keeney, and G. J. Stensland, Flux and Sources of Nutrients in the Mississippi-Atchafalaya River Basin, Topic 3 Report for the Integrated Assessment of Hypoxia in the Gulf of Mexico. NOAA Coastal Ocean Program Decision Analysis Series No. 17, NOAA Coastal Ocean Program, Silver Springs, Maryland, 1999.

Goolsby, D. A., Mississippi Basin nitrogen flux believed to cause Gulf hypoxia, *Eos, Trans. Amer. Geophys. Union, 81*, 325-327, 2000.

Gulf States Marine Fisheries Commission, SEAMAP Environmental and Biological Atlas of the Gulf of Mexico, 1982, Gulf States Marine Fisheries Commission, Ocean Springs, Mississippi, 1982 et seq.

Hallegraeff, G. M., A review of harmful algal blooms and their apparent global increase, *Phycologia, 32*, 79-99, 1993.

Hanifen, J. G., W. S. Perret, R. P. Allemand, and T. L. Romaire, Potential impacts of hypoxia on fisheries: Louisiana's fishery-independent data, in Proc., First Gulf of Mexico Hypoxia Management Conference, December 1995, New Orleans, Louisiana, pp. 87-100, Publ. No. EPA-55-R-97-001, Gulf of Mexico Program Office, Stennis Space Center, Mississippi, 1997.

Harding, Jr., L. W. and E. S. Perry, Long-term increase of phytoplankton biomass in Chesapeake Bay, 1950-1994, *Mar. Ecol. Prog. Ser., 157*, 39-52, 1997.

Harper, D. E., Jr., L. D. McKinney, J. M. Nance, and R. R. Salzer, Recovery responses of two benthic assemblages following an acute hypoxic event on the Texas continental shelf, northwestern Gulf of Mexico, in *Modern and Ancient Continental Shelf Anoxia*, edited by R. V. Tyson and T. H. Pearson, pp. 49-64, Geological Society Special Publ., 58, 1991.

Howarth, R. W., An assessment of human influences on fluxes of nitrogen from the terrestrial landscape to the estuaries and continental shelves of the North Atlantic Ocean, *Nutrient Cycling in Agroecosystems, 52*, 213-223, 1998.

Howarth, R. W., H. S. Jensen, R. Marino, and H. Postma, Transport to and processing of P in near-shore and oceanic waters, in *Phosphorus in the Global Environment*, edited by H. Tiessen, SCOPE 54, pp. 323-356, John Wiley & Sons Ltd., Chichester, 1995.

Howarth, R. W., G. Billen, D. Swaney, A Townsend, N. Jaworski, K. Lajtha, J. A. Downing, R. E. Elmgren, N. Caraco, T. Jordan, F. Berendse, J. Freney, V. Kudeyarov, P. Murdoch and Z.-L. Zhu, Regional nitrogen budgets and riverine N & P fluxes for the drainages to the North Atlantic Ocean: Natural and human influences, *Biogeochemistry, 35*, 75-139, 1996.

Johansson, J. O. R. and R. R. Lewis, III. Recent improvements of water quality and biological indicators in Hillsborough Bay, a highly impacted subdivision of Tampa Bay, Florida, USA, in *Marine Coastal Eutrophication*, edited by R. A. Vollenweider, R. Marchetti, and R. Viviani, pp. 1199-1215, *Sci. Total Environ.*, suppl. no. 0048-9697, 1992..

Justic', D., T. Legovic', and L. Rottini-Sandrini, Trends in oxygen content 1911-1984 and occurrence of benthic mortality in the northern Adriatic Sea, *Estuar. Coast. Shelf Sci., 24*, 435-445, 1987.

Justic', D., N. N. Rabalais, R. E. Turner, and W. J. Wiseman, Jr., Seasonal coupling between riverborne nutrients, net productivity and hypoxia, *Mar. Pollut. Bull., 26*, 184-189, 1993.

Justic', D., N. N. Rabalais, and R. E. Turner, Stoichiometric nutrient balance and origin of coastal eutrophication, *Mar. Pollut. Bull., 30*, 41-46, 1995a.

Justic', D., N. N. Rabalais, R. E. Turner, and Q. Dortch, Changes in nutrient structure of river-dominated coastal waters: Stoichiometric nutrient balance and its consequences, *Estuar. Coast. Shelf Sci., 40*, 339-356, 1995b.

Justic', D., N. N. Rabalais, and R. E. Turner, Effects of climate change on hypoxia in coastal waters: A doubled CO_2 scenario for the northern Gulf of Mexico, *Limnol. Oceanogr., 41*, 992-1003, 1996.

Justic', D., N. N. Rabalais, and R. E. Turner, Impacts of climate change on net productivity of coastal waters: Implications for carbon budget and hypoxia, *Climate Res., 8*, 225-237, 1997.

Kamykowski, D. and S. J. Zentara, Hypoxia in the world ocean as recorded in the historical data set, *Deep-Sea Res., 37*, 1861-1874, 1990.

Karlsen, A.W., T. M. Cronin, S.E. Ishman, D. A. Willard, R. Kerhin, C. W. Holmes, and M. Marot, Historical trends in Chesapeake Bay dissolved oxygen based on benthic Foraminifera from sediment cores, *Estuaries, 23*, 488-508, 2000.

Larsson, U. R., R. Elmgren, and F. Wulff, Eutrophication and the Baltic Sea: Causes and consequences, *Ambio, 14*, 9-14, 1985.

Leming, T. D. and W. E. Stuntz, Zones of coastal hypoxia revealed by satellite scanning have implications for strategic fishing, *Nature, 310*, 136-138, 1984.

Loesch, H., Sporadic mass shoreward migrations of demersal fish and crustaceans in Mobile Bay, Alabama, *Ecology, 41*, 292-298, 1960.

Lohrenz, S. E., M. J. Dagg and T. E. Whitledge, Enhanced primary production at the plume/oceanic interface of the Mississippi River, *Cont. Shelf Res., 10*, 639-664, 1990.

Lohrenz, S. E., G. L. Fahnenstiel, D. G. Redalje, G. A. Lang, X. Chen, and M. J. Dagg, Variations in primary production of northern Gulf of Mexico continental shelf waters linked to nutrient inputs from the Mississippi River, *Mar. Ecol. Prog. Ser., 155*, 435-454, 1997.

Lohrenz, S. E., G. L. Fahnenstiel, D. G. Redalje, G. A. Lang, M. J. Dagg, T. E. Whitledge, and Q. Dortch, The interplay of nutrients, irradiance and mixing as factors regulating primary production in coastal waters impacted by the Mississippi River plume, *Cont. Shelf Res., 19*, 1113-1141, 1999a.

Lohrenz, S. E., D. A. Wiesenburg, R. A. Arnone, and X. Chen, What controls primary production in the Gulf of Mexico? in *The Gulf of Mexico Large Marine Ecosystem, Assessment, Sustainability, and Management*, edited by K. Sherman, H. Kumpf and K. Steidinger, pp. 151-170, Blackwell Science, Malden, Massachusetts, 1999b.

Malone, T.C., River flow, phytoplankton production and oxygen depletion in Chesapeake Bay, in *Modern and Ancient Continental Shelf Anoxia*, edited by R. V. Tyson and T. H. Pearson, pp. 83-93, Geological Society Special Publ., 58, 1991.

Malone, T. C., Effects of water column processes on dissolved oxygen, nutrients, phytoplankton and zooplankton, in *Oxygen Dynamics in Chesapeake Bay. A Synthesis of Recent Research*, edited by D. E. Smith, M. Leffler, and G. Makiernan, pp. 61-112 Maryland Sea Grant Program, College Park, Maryland, 1992.

May, E. B., Extensive oxygen depletion in Mobile Bay, Alabama, *Limnol. Oceanogr., 18*, 353-366, 1973.

Mayer, L. M., R. G. Keil, S. A. Macko, S. B. Joye, K. C. Ruttenberg, and R. C. Aller, Importance of suspended particulates in riverine delivery of bioavailable nitrogen to coastal zones, *Global Biogeochemical Cycles, 12*, 573-579, 1998.

Milliman, J. D. and R. H. Meade, World-wide delivery of river sediment to the ocean, *J. Geol., 91*, 1-21, 1983.

National Research Council, *Clean Coastal Waters: Understanding and Reducing the Effects of Nutrient Pollution,* Committee on Causes and Management of Coastal Eutrophication, Ocean Studies Board and Water Science and Technology Board, Commission on Geosciences, Environment, and Resources, National Research Council, National Academy Press, Washington, D.C., 2000.

Nelsen, T. A., P. Blackwelder, T. Hood, B. McKee, N. Romer, C. Alvarez-Zarikian, and S. Metz, Time-based correlation of biogenic, lithogenic and authigenic sediment components with anthropogenic inputs in the Gulf of Mexico NECOP study area, *Estuaries, 17*, 873-885, 1994.

Nixon, S. W., Physical energy inputs and comparative ecology of lake and marine ecosystems, *Limnol. Oceanogr., 33(4, part 2)*, 1005-1025, 1988.

Nixon, S. W., Coastal marine eutrophication: A definition, social causes, and future concerns, *Ophelia, 41*, 199-219, 1995.

Nixon, S. W., J. W. Ammerman, L. P. Atkinson, V. M. Berounsky, G. Billen, W. C. Boicourt, W. R. Boynton, T. M. Church, D. M. DiToro, R. Elmgren, J. H. Garber, A. E. Giblin, R. A. Jahnke, N. J. P. Owens, M. E. Q. Pilson, and S. P. Seitzinger, The fate of nitrogen and phosphorus at the land-sea margin of the North Atlantic Ocean, *Biogeochemistry, 35*, 141-180, 1996.

Officer, C. B. and J. H. Ryther, The possible importance of silicon in marine eutrophication, *Mar. Ecol. Prog. Ser., 3*, 83-91, 1980.

Officer, C. B., R. B. Biggs, J. L. Taft, L. E. Cronin, M. A. Tyler, and W. R. Boynton, Chesapeake Bay anoxia. Origin, development and significance, *Science, 223*, 22-27, 1984.

Paerl, H. W., Emerging role of anthropogenic nitrogen deposition in coastal eutrophication: Biogeochemical and trophic perspectives, *Canad. J. Fish. Aquat. Sci., 50*, 2254-2269, 1995.

Paerl, H. W., A comparison of cyanobacterial bloom dynamics in freshwater, estuarine and marine environments, *Phycologia, 35(suppl. 6)*, 25-35, 1996.

Paerl, H., Coastal eutrophication and harmful algal blooms: Importance of atmospheric deposition and groundwater as "new" nitrogen and other nutrient sources, *Limnol. Oceanogr., 42*, 1154-1165, 1997.

Paerl, H., J. Pinckney, J. Fear, and B. Peierls, Ecosystem responses to internal and watershed organic matter loading: Consequences for hypoxia in the eutrophying Neuse River Estuary, NC, USA, *Mar. Ecol. Prog. Ser., 166*, 17-25, 1998.

Parker, C. A. and J. E. O'Reilly, Oxygen depletion in Long Island Sound: A historical perspective, *Estuaries, 14*, 248-264, 1991.

Pavela, J. S., J. L. Ross, and M. E. Chittenden, Sharp reductions in abundance of fishes and benthic macroinvertebrates in the Gulf of Mexico off Texas associated with hypoxia. *Northeast Gulf Sci., 6*, 167-173, 1983.

Peierls, B. L., N. Caraco, M. Pace, and J. Cole, Human influence on river nitrogen, *Nature, 350*, 386-387, 1991.

Pokryfki, L. and R. E. Randall, Nearshore hypoxia in the bottom water of the northwestern Gulf of Mexico from 1981 to 1984, *Mar. Envtl. Res., 22*, 75-90, 1987.

Qureshi, N. A., The role of fecal pellets in the flux of carbon to the sea floor on a river-influenced continental shelf subject to hypoxia, Ph.D. Dissertation, Department of Oceanography & Coastal Sciences, Louisiana State University, Baton Rouge, 1995.

Rabalais, N. N., An Updated Summary of Status and Trends in Indicators of Nutrient Enrichment in the Gulf of Mexico, Environmental Protection Agency Publ. No. EPA/800-R-92-004, Gulf of Mexico Program, Nutrient Enrichment Subcommittee, Stennis Space Center, Mississippi, 1992.

Rabalais, N. N., Oxygen Depletion in Coastal Waters, in NOAA's State of the Coast Report. National Oceanic and Atmospheric Administration, Silver Spring, Maryland, http:// state_of_coast.noaa.gov/bulletins/html/hyp_09/hyp.html, 1998 (on-line).

Rabalais, N. N., Press release dated July 29, 1999, Louisiana Universities Marine Consortium, Chauvin, Louisiana, 1999.

Rabalais, N. N. and R. E. Turner, Pigment and nutrient distributions, in An Observational Study of the Mississippi-Atchafalaya Coastal Plume, Final Report, edited by S. P. Murray,

OCS Study MMS 98-0040, pp. 208-230, U.S. Dept. of Interior, Minerals Management Service, Gulf of Mexico OCS Region, New Orleans, Louisiana, 1998.

Rabalais, N. N., R. E. Turner, W. J. Wiseman, Jr., and D. F. Boesch, A brief summary of hypoxia on the northern Gulf of Mexico continental shelf: 1985-1988, in *Modern and Ancient Continental Shelf Hypoxia,* edited by R. V. Tyson and T. H. Pearson, pp. 35-47, Geological Society Special Publ. No. 58, 1991.

Rabalais, N.N., R. E. Turner, and Q. Dortch, Louisiana continental shelf sediments: Indicators of riverine influence, in *Nutrient-Enhanced Coastal Ocean Productivity Workshop Proceedings*, pp. 77-81, TAMU-SG-92-109 Technical Report, Texas A&M University Sea Grant Program, Galveston, Texas, 1992.

Rabalais, N. N., R. E. Turner, D. Justic, Q. Dortch, W. J. Wiseman, Jr., and B. K. Sen Gupta, Nutrient changes in the Mississippi River and system responses on the adjacent continental shelf, *Estuaries, 19*, 386-407, 1996.

Rabalais, N. N., R. E. Turner, W. J. Wiseman, Jr., and Q. Dortch, Consequences of the 1993 Mississippi River flood in the Gulf of Mexico, *Regulated Rivers: Research & Management, 14*, 161-177, 1998.

Rabalais, N. N., R. E. Turner, D. Justic', Q. Dortch, and W. J. Wiseman, Jr., Characterization of hypoxia: Topic 1 Report for the Integrated Assessment of Hypoxia in the Gulf of Mexico. NOAA Coastal Ocean Program Decision Analysis Series No. 15, NOAA Coastal Ocean Program, Silver Springs, Maryland, 1999.

Ragan, J. G., A. H. Harris, and J. H. Green, Temperature, salinity and oxygen measurements of surface and bottom waters on the continental shelf off Louisiana during portions of 1975 and 1976, *Professional Papers Series (Biology) (Nicholls State Univ., Thibodaux, Louisiana), 3*, 1-29, 1978.

Redalje, D. G., S. E. Lohrenz, and G. L. Fahnenstiel, The relationship between primary production and the vertical export of particulate organic matter in a river-impacted coastal ecosystem, *Estuaries, 17*, 829-838, 1994.

Redfield, A. C., The biological control of chemical factors in the environment, *American Scientist, 46*, 205-222, 1958.

Renaud, M., Hypoxia in Louisiana coastal waters during 1983: implications for fisheries, *Fishery Bull., 84*, 19-26, 1986.

Riley, G. A., The significance of the Mississippi River drainage for biological conditions in the northern Gulf of Mexico, *J. Mar. Res., 1*, 60-74, 1937.

Rosenberg, R., Eutrophication—The future marine coastal nuisance?, *Mar. Pollut. Bull.,16*, 227-231, 1985.

Rosenberg, R. (ed.), A Review. Eutrophication in Marine Waters Surrounding Sweden, National Swedish Environmental Protection Board Report, 3054, [translation of SNV PM 1808 (1984)], Soina, Sweden, 1986.

Sarasota Bay National Estuary Program, *Sarasota Bay: The Voyage to Paradise Reclaimed,* Southwest Florida Water Management District, Brooksville, Florida, 1995.

Sen Gupta, B. K. and M. L. Machain-Castillo, Benthic formainifera in oxygen-poor habitats, *Mar. Micropaleontology, 20*, 183-201, 1993.

Sen Gupta, B. K., R. F. Lee, and M. S. May, Upwelling and an unusual assemblage of benthic foraminifera on the northern Florida continental slope, *J. Paleontology, 55*, 853-857, 1981.

Sen Gupta, B. K., R. E. Turner, and N. N. Rabalais, Seasonal oxygen depletion in continental-shelf waters of Louisiana: Historical record of benthic foraminifers, *Geology, 24*, 227-230, 1996.

Sklar, F. H. and R. E. Turner, Characteristics of phytoplankton production off Barataria Bay in an area influenced by the Mississippi River, *Contr. Mar. Sci., 24*, 93-106, 1981.

Smayda, T. J., Novel and nuisance phytoplankton blooms in the sea: Evidence for global epidemic, in *Toxic Marine Phytoplankton*, edited by E. Graneli, B. Sundstrom, R. Edler, and D. M. Anderson (eds.), pp. 29-40, Elsevier, New York, 1990.

Smith, S. V., Responses of Kaneohe Bay, Hawaii, to relaxation of sewage stress, in *Estuaries and Nutrients* edited by B. J. Neilson and L. E. Cronin, pp. 391-410, Humana Press, Inc., Clifton, New Jersey, 1981.

Smith, W. O. and D. J. Demaster, Phytoplankton biomass and productivity in the Amazon River plume: correlation with seasonal river discharge, *Cont. Shelf Res., 16*, 291-319, 1996,

Swanson, R. L. and C. A. Parker, Physical environmental factors contributing to recurring hypoxia in the New York Bight, *Trans. Amer. Fisher. Soc., 117*, 37-47, 1988.

Swanson, R. L. and C. J. Sindermann (eds.), Oxygen Depletion and Associated Benthic Mortalities in New York Bight, 1976, National Oceanic and Atmospheric Administration Professional Paper 11, 1979.

Thomas, W. H. and E. G. Simmons, Phytoplankton production in the Mississippi River Delta, in *Recent Sediments, Northwest Gulf of Mexico*, edited by F. P. Shepard, pp. 103-116, American Association of Petroleum Geologists, Tulsa, Oklahoma, 1960

Tolmazin, R., Changing coastal oceanography of the Black Sea. I. Northwestern shelf, *Progr. Oceanogr.*, 15, 2127-276, 1985.

Turner, R. E. and R. L. Allen, Bottom water oxygen concentration in the Mississippi River Delta Bight, *Contr. Mar. Sci., 25*, 161-172, 1982a.

Turner, R. E. and R. L. Allen, Plankton respiration in the bottom waters of the Mississippi River Delta Bight, *Contr. Mar. Sci., 25*, 173-179, 1982b.

Turner, R. E. and N. N. Rabalais, Changes in Mississippi River water quality this century. Implications for coastal food webs, *BioScience, 41*, 140-148, 1991.

Turner, R. E. and N. N. Rabalais, Coastal eutrophication near the Mississippi river delta, *Nature,* 368, 619-621, 1994a.

Turner, R. E. and N. N. Rabalais, Changes in the Mississippi River nutrient supply and offshore silicate-based phytoplankton community responses, in *Changes in Fluxes in Estuaries: Implications from Science to Management*, edited by K. R. Dyer and R. J. Orth, pp. 147-150, Proceedings of ECSA22/ERF Symposium, International Symposium Series, Olsen & Olsen, Fredensborg, Denmark, 1994b.

Turner, R. E. and N. N. Rabalais, Bottom water respiration rates in the hypoxia zone within the Louisiana Coastal Current, in An Observational Study of the Mississippi-Atchafalaya Coastal Plume, Final Report, edited by S. P. Murray, OCS Study MMS 98-0040, pp. 354-364, U.S. Dept. of Interior, Minerals Management Service, Gulf of Mexico OCS Region, New Orleans, Louisiana, 1998.

Turner, R. E., W. W. Schroeder, and W. J. Wiseman, Jr., The role of stratification in the deoxygenation of Mobile Bay and adjacent shelf bottom waters, *Estuaries,* 10, 13-19, 1987.

Turner, R. E., N. N. Rabalais, and Z.-N. Zhang, Phytoplankton biomass, production and growth limitations on the Huanghe (Yellow River) continental shelf, *Cont. Shelf Res., 10*, 545-571, 1990.

Turner, R. E., N. Qureshi, N. N. Rabalais, Q. Dortch, D. Justic', R. F. Shaw, and J. Cope, Fluctuating silicate:nitrate ratios and coastal plankton food webs, *Proc. Natl. Acad. Sci., USA, 95*, 13048-13051, 1998.

Tyson, R. V. and T. H. Pearson, Modern and ancient continental shelf anoxia: an overview in *Modern and Ancient Continental Shelf Anoxia,* edited by R. V. Tyson and T. H. Pearson, pp. 1-24, *Geological Society Special Pub., 58,* 1991.

U. S. Army Corps of Engineers, Deep draft access to the ports of New Orleans and Baton Rouge, Draft Environmental Statement, U. S. Army Corps of Engineers, New Orleans District, New Orleans, Louisiana, 1974.

Vitousek, P. M., J. D. Aber, R. W. Howarth, G. E. Likens, P. A. Matson, D. W. Schindler, W. H.

Schlesinger, and D. G. Tilman, Human alterations of the global nitrogen cycle: Sources and consequences, *Ecol. Applic., 7*, 737-750, 1997.

Ward, C. H., M. E. Bender, and D. J. Reish (eds.), The Offshore Ecology Investigation. Effects of oil drilling and production in a coastal environment, *Rice Univ. Stud., 65*, 1-589, 1979.

Welsh, B. L. and F. C. Eller, Mechanisms controlling summertime oxygen depletion in western Long Island Sound, *Estuaries, 14*, 265-278, 1991.

Welsh, B. L., R. I. Welsh, and M. L. DiGiacomo-Cohen, Quantifying hypoxia and anoxia in Long Island Sound, in *Changes in Fluxes in Estuaries: Implications from Science to Management*, edited by K. R. Dyer and R. J. Orth, pp. 131-137, Proceedings of ECSA22/ERF Symposium, International Symposium Series, Olsen & Olsen, Fredensborg, Denmark, 1994.

Wiseman, Jr., W. J., N. N. Rabalais, R. E. Turner, S. P. Dinnel, and A. MacNaughton, Seasonal and interannual variability within the Louisiana Coastal Current: Stratification and hypoxia, *J. Mar. Systems, 12*, 237-248, 1997.

Wulff, F. and L. Rahm, Long-term, seasonal and spatial variations of nitrogen, phosphorus silicate in the Baltic: An overview, *Mar. Envtl. Res., 26*, 19-37, 1988.

Xiuren, N., D. Vaulot, L. Zhensheng, and L. Zilin, Standing stock and production of phytoplankton in the estuary of the Changjiang (Yangtse) River and the adjacent East China Sea, *Mar. Ecol. Prog. Ser., 49*, 141-150, 1988.

Zaitsev, Y. P., Recent changes in the trophic structure of the Black Sea, *Fisher. Oceanogr., 1*, 180-189, 1992.

Zucker, L. A. and L. C. Brown, Agriculture Drainage, Water Quality Impacts and Subsurface Drainage Studies in the Midwest, Ohio State University Extension Bulletin 871, The Ohio State University, Columbus, 1998.

2

Impacts of Changing Si/N Ratios and Phytoplankton Species Composition

Quay Dortch, Nancy N. Rabalais, R. Eugene Turner, and Naureen A. Qureshi

Abstract

While nitrogen (N) and phosphorus (P) inputs from the Mississippi River have increased since the 1950s concomitantly with increasing productivity and hypoxia, silicate (Si) inputs have decreased. As a result, nutrient ratios have changed so that Si can now be limiting, especially in the spring. Si limitation controls the size and species composition of the diatom bloom by selecting species with lower Si requirements. Evidence from the Louisiana shelf indicates that phytoplankton sinking, especially of diatoms in the spring, contributes to the vertical carbon flux that causes hypoxia. Most of the sinking phytoplankton are diatoms that are moderately to heavily silicified. Similar results have been obtained in other eutrophic areas. Consequently, the Si/N input ratio may influence the environmental impacts of increasing nutrient inputs through control of phytoplankton species composition. Nutrient control strategies to reduce hypoxia need to consider the consequences of changing nutrient ratios as well as changing nutrient concentrations.

Introduction

A large area of hypoxic bottom water occurs annually on the Louisiana shelf, caused by high degradation rates of organic matter deposited as a result of high nutrient stimulated primary production [Rabalais et al., 1996, 1999]. Since the 1950s nitrogen and phosphorus inputs to the Louisiana coastal zone from the Mississippi River system have increased substantially, whereas silicon inputs have decreased [Turner and Rabalais, 1991]. Although there are few historical measurements of primary production or hypoxia, indirect indicators preserved in cores, suggest that both primary productivity and hypoxia have increased, concomitantly with the increase in nutrients [Eadie et al., 1994; Sen Gupta et al., 1996; Turner and Rabalais, 1994].

At present, levels of primary productivity are extremely high and proportional to N inputs from the Mississippi River [Lohrenz et al., 1997]. Ratios of nutrient inputs, however, have changed due to the increasing N and P and decreasing Si, so that nutrient

Coastal Hypoxia: Consequences for Living Resources and Ecosystems
Coastal and Estuarine Studies, Pages 37-48
Copyright 2001 by the American Geophysical Union

availability is now balanced with regard to phytoplankton requirements [Justic' et al., 1995]. As a consequence, depending on the season or location, N, P, or Si availability can limit phytoplankton growth [Dortch and Whitledge, 1992; Justic' et al., 1995; Lohrenz et al., 1999; Nelson and Dortch, 1996; Rabalais et al., 1999] with Si limitation observed most frequently [Dortch and Whitledge, 1992, Justic' et al., 1995; Rabalais et al., 1999].

While increasing nutrient inputs stimulate increased productivity, resource competition theory predicts that changes in nutrient ratios will result in changes in phytoplankton species composition [Sommer, 1989, 1993, 1994, 1995; Tilman, 1977; Tilman et al., 1986]. In marine systems the Si/N and N/P ratios are most important, due to overall N limitation. This chapter will focus on how changes in Si/N ratios determine the impact of increased productivity through changes in phytoplankton species composition. There is sufficient evidence from the northern Gulf of Mexico and other eutrophic areas to provide some insights about the role of Si/N ratios. P limitation does occur in this region and N/P input ratios have changed over time [Justic' et al., 1995; Rabalais et al., 1999]. It is likely that these changes may also influence phytoplankton species composition and the effects of eutrophication. There are, however, little data from the Louisiana shelf and most of the pertinent data from other areas concerns species not commonly found on the Louisiana shelf, so it is not possible to address questions concerning changing N/P ratios in the northern Gulf of Mexico.

Hypotheses about Impacts of Changing Si/N Ratios

In the most extreme case, severe Si limitation prevents the growth of diatoms, which, unlike other phytoplankton, have an absolute growth requirement for Si. It also allows the growth of non-diatoms, including toxic and noxious phytoplankton [Officer and Ryther, 1980; Smayda, 1989, 1990]. Culture studies show that less severe Si limitation may cause shifts within the diatom species composition [Sommer, 1994, 1995]. Both major and more subtle changes in phytoplankton species composition will affect the size structure of the plankton, transfer of carbon (C) to higher trophic levels, nutrient cycling, vertical flux of organic matter [Conley et al., 1993] and, possibly the extent and severity of hypoxia [Dortch et al., 1992]. Thus, changes in phytoplankton composition may affect how the entire ecosystem functions and the environmental impact of the increased productivity.

If N inputs are high and P is not limiting, a series of hypotheses can be made relating the Si/N input ratio, phytoplankton species composition, and the environmental consequences of increased productivity (Table 1). When Si/N inputs are extremely high, diatoms would be the most abundant phytoplankton, be heavily silicified, and have high direct sinking rates. Further, they would be consumed by zooplankton that produce fecal pellets with high sinking rates. At Si/N ratios that are so low they approach the ability of diatoms to grow, the abundance of diatoms would be less and lightly silicified diatoms would be expected to predominate. These would have lower direct sinking rates, but still be grazed, although the resulting fecal pellets might not have as great a density and sinking rate. Finally, when Si/N ratios are very low, non-diatoms would predominate. Many of these are motile or small species with very low sinking rates. They are also sometimes less favorable food for large zooplankton, which produce rapidly sinking fecal pellets. As a consequence, carbon flux to the bottom would be greatest when heavily silicified diatoms are present, resulting in persistent hypoxia. Vertical carbon flux would be least

TABLE 1. Potential consequences of high N input and variable Si/N ratios (based on dissolved inorganic nutrient concentrations) in the absence of P limitation.

Si/N	Phytoplankton	Consequence
> 1	Sinking Diatoms	Persistent Hypoxia
= 1	Non-sinking Diatoms Sinking Diatoms Non-diatoms	Seasonal Hypoxia Sporadic Harmful Algal Blooms
< 1	Non-diatoms Non-sinking Diatoms	Frequent Harmful Algal Blooms Algal Blooms

when non-diatoms are present, although some large dinoflagellate blooms can sink out rapidly and cause sporadic hypoxia [e.g., Graneli et al., 1989; Falkowski et al., 1980]. This chapter will focus on the effect of variations in Si/N availability on phytoplankton species composition and direct sinking. The next chapter will consider the effects on zooplankton and fecal pellet flux.

Evidence from the Louisiana Shelf

Phytoplankton Species Composition

At present the ratio of Si/N inputs averages 1.1 for the Mississippi River and 1.7 for the Atchafalaya River [Rabalais et al., 1999] over the course of a year. It varies, however, so that there are times when Si is more likely to be limiting [late spring and early summer) and times when N is more likely to be limiting (late summer and early fall) (Fig. 1). Thus, within any given year, all the phytoplankton stages listed in Table 1 may be encountered, resulting in seasonal hypoxia and sporadic Harmful Algal Blooms (HABs) on the Louisiana Shelf.

Diatoms bloom every spring along the Louisiana shelf, just after nutrient inputs are maximal due to high river flow when Si/N input ratios are decreasing (Fig. 1A). During the peak of the diatom bloom abundances average 1×10^7 cells l^{-1}, which is extremely high in comparison with other coastal areas. The other major phytoplankton group is picocyanobacteria (Fig. 1B). The remainder are primarily cryptomonads and other, small flagellates, but occasionally larger dinoflagellates (Fig. 1B). While picocyanobacteria dominate numerically during the summer and late fall, they rarely dominate the biomass, because of their small size relative to diatoms. Their high abundance in late summer and fall may be related to temperature rather than nutrient availability, since a very similar pattern is observed in local estuarine waters where nutrients are rarely limiting [Dortch and Soniat, unpublished]. Other phytoplankton are numerically abundant in the winter and early spring, but do not dominate the biomass because of their small size. Thus, diatoms dominate numerically during the spring, but probably dominate the biomass most of the year.

Annual relative diatom abundance varies considerably from year to year (Fig. 2). For all years except 1992, there was an inverse linear relationship between the average annual

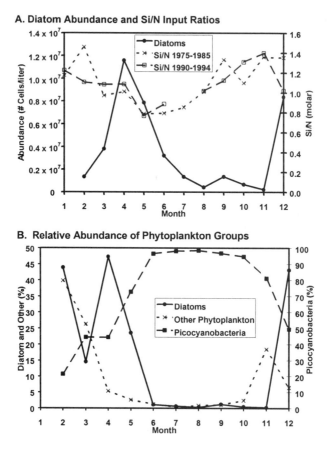

Figure 1. Average monthly diatom abundance (A) and relative abundance of phytoplankton groups (B) from 1990 to 1995 at C6A/B and Si/N input ratios (A) from 1975-1985 (from Turner and Rabalais [1991]) and 1990-1994 (NASQAN data, 0 to 6 points per month).

diatom abundance and the % of samples showing Si limitation, calculated according to criteria developed in Dortch and Whitledge [1992] and Nelson and Dortch [1996]. Thus, on an annual basis diatom abundance is determined by Si availability. Analyses on shorter time scales do not result in a relationship between diatoms and Si concentration because, for any given water sample, the highest diatom abundances are often associated with very Si depleted water.

The fact that maximum diatom abundance occurs in the spring, when Si/N input ratios are lowest and Si limitation is more frequent [Dortch and Whitledge, 1992] and minimum diatom abundance occurs in the summer, when ratios are much higher, suggests a more complicated scenario. A study was undertaken comparing the effect of Si limitation on Si uptake and phytoplankton species composition in a spring, high flow period and a summer low flow period [Dortch et al., 1995; Nelson and Dortch, 1996]. As expected, diatoms dominated in the spring and picocyanobacteria dominated during the summer (Table 2). Both nutrient concentrations and ratios and Si uptake kinetics indicated that Si was limiting in approximately half the samples tested in the spring, but in very few in the

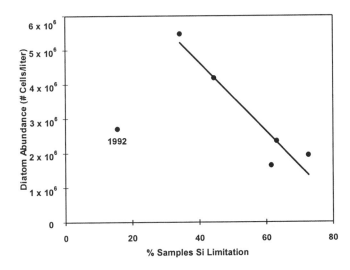

Figure 2. Annual average diatom abundance versus Si limitation in the surface layer (< 10 m) at C6A/B from 1990 to 1995. % Si Limitation = (# Samples [Si] < 5 μM & [Si]/[N] < 1/Total # Samples) x 100. The line for all data except 1992 (n=6) is described by Diatom = -1.01 x 10^5 x % Limit + 8.67 x 10^6, r^2 = 0.90.

summer. In fact some of the lowest Si concentrations and half saturation constants for silicate uptake ever observed were measured during the spring. It was suspected, but has not been tested directly, that in the summer diatom abundance may be limited by N availability and that picocyanobacteria are better competitors for N. Finally, in both spring and summer the diatom species composition differed markedly between samples with and without Si limitation. *Skeletonema costatum* dominated when Si was not limiting, whereas *Chaetoceros* spp., especially *C. socialis* in the spring, dominated when Si was limiting. Thus, in this area, where nutrient inputs are balanced with regard to phytoplankton growth requirements, Si limitation determines the type of diatom which dominates rather than a switch from diatoms to non-diatoms. The data from this study are insufficient to definitively relate the types of diatoms to the degree of Si limitation, although it is possible to hypothesize that Si limitation will result in more lightly silicified diatoms with a lower Si requirement.

Phytoplankton Direct Sinking

Phytoplankton sinking into sediment traps was measured in 1990 to 1992 at a station in the core of the hypoxic region, using several different approaches. In 1990 and 1991 phytoplankton cells counts [Dortch et al., 1992], supplemented in 1992 with volume estimates, were made on trap material. In 1991 and 1992 chlorophyll and particulate carbon measurements were made [Qureshi, 1995]. The traps were located just below the pycnocline (surface) and in the hypoxic layer several meters above the bottom in 20 m water depth at a station (C6A/B) in the core of the hypoxic zone.

Phytoplankton direct sinking was greatest in the spring when diatom abundance was at a maximum (Fig. 3A; note log scale). Phytoplankton carbon comprised a very large

TABLE 2. Indicators of Si limitation at locations across the Louisiana shelf, during two 10-day periods in April/May and July/August (adapted from Dortch et al. [1995]). n = number of samples. Si limitation based on Si concentrations and ratios as in Fig. 2 and based on uptake as in Nelson and Dortch [1996]. Adapted from Dortch et al. [1995].

	Spring 1993	Summer 1992
Phytoplankton (# Cells/liter)		
Diatoms	1.1×10^7	2.5×10^5
Cyanobacteria	5.3×10^6	3.3×10^8
n	40	36
% Si Limitation		
Si Concentration & Si/DIN Ratio	58	4
n	67	50
^{30}Si Uptake Kinetics	47	9
n	15	11

portion of the total vertical carbon flux in the spring, less in the rest of the year, with a small increase in the fall (Fig. 3B, note different years for A and B). Unfortunately, trap data are not available for the late winter because the traps were serviced by divers and poor weather precluded diving. The high % Phytoplankton/Total Carbon flux in 1992 (Fig. 3B), the only year with early spring sampling, suggests that phytoplankton direct sinking could be high in late winter as well.

Diatoms comprised a larger fraction of the phytoplankton cells sinking into traps than they represented in the water in three out of four cases (Table 3), suggesting selective sinking of diatoms. In the fourth case, during high flow in 1990, the remainder of the sinking phytoplankton was picocyanobacteria. Their flux was proportional to the diatom flux, so that the peak in picocyanobacteria flux occurred in the spring, when their numbers were low, rather than in the summer/fall when their numbers were high (not shown), suggesting they sank as part of diatom aggregates. In contrast, in 1991 diatoms dominated the cell flux and no other group dominated the other cell flux (not shown). Some species that dominated the flux, such as *Skeletonema costatum*, formed aggregates more often than others, such as *Thalassionema nitzschioides* or *Pseudo-nitzschia* spp. Thus, diatom chains, diatom aggregates or diatom/picocyanobacteria aggregates are the major component of direct sinking phytoplankton.

Besides the selective sinking of diatoms, there is selective sinking of some diatom groups (Table 3). Only some diatom species occurring in the water sink into traps, including in approximate order of abundance, *Thalassionema nitzschioides, Skeletonema costatum, Pseudo-nitzschia* spp., *Thalassiosira* spp., *Cyclotella* spp., *Coscinodiscus* spp., some *Chaetoceros* spp. (not *socialis* or *debilis*, which dominate when Si is limiting), *Rhizosolenia setigera*, and *Probiscia alata*. Based on comparison with data in Conley et al. [1989], the diatoms in the traps were generally more heavily silicified than those rarely or never found in traps. Thus, there is a selective loss of more heavily silicified diatoms (defined as Sinking Diatoms, based on list above) due to direct sinking (Table 3).

In summary, Si limitation occurs on the Louisiana shelf, especially in the spring when Si/N inputs are lowest and diatom blooms occur. Si availability appears to regulate the abundance of diatoms on an annual basis and determine the diatom species composition.

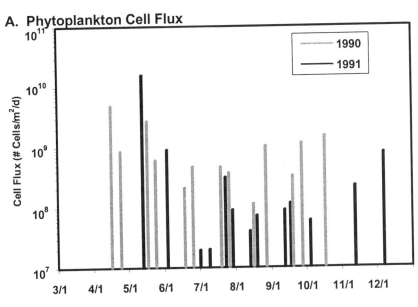

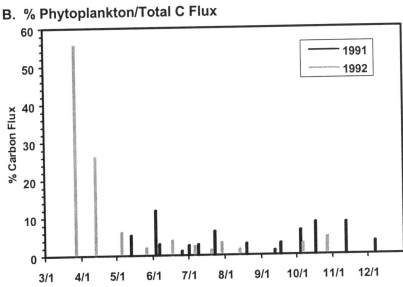

Figure 3. Phytoplankton cell flux, determined from direct counts [Dortch et al., 1992], in 1990 and 1991 and % Phytoplankton/total Carbon Flux (B), determined from chlorophyll and carbon flux measurements [Qureshi, 1995] in 1991 and 1992 at C6A/B. Chlorophyll flux was converted to C flux, using the C/Chl ratio derived from cell volume to C conversions [Rabalais et al., 1999]. Flux is mean of top and bottom traps.

Phytoplankton direct sinking is a substantial part of the total vertical flux and is maximal in the spring. It is dominated by diatoms, especially more heavily silicified diatoms, and organisms which sediment with the diatoms. Blooms of non-diatoms, especially pico-cyanobacteria and, occasionally, HABs [Dortch et al., 1999] do occur in this situation

TABLE 3. Selective sinking of diatoms, compared with total phytoplankton, and sinking diatoms (see text for list), compared to total diatoms, into sediment traps at C6A/B. Counts described in Dortch et al. [1992]; traps described in Qureshi [1995]. na = not available.

	% Diatoms/Total Cells		% Sinking Diatoms/Total Diatoms	
	1990	1991	1990	1991
High River Flow				
Water < 10 m	48	12	78	na
Top Trap	30	70	98	na
Bottom Trap	39	49	98	na
Low River Flow				
Water < 10 m	1	<1	45	na
Top Trap	32	74	99	na
Bottom Trap	44	76	99	na

where nutrient inputs are nearly balanced. The relationship between Si/N input ratios and hypoxia can only be hypothesized in general terms because key information is lacking. Interannual differences in the data presented here complicate interpretation in terms of Si limitation. More generally, phytoplankton nutrient requirements with regard to Si and how this affects phytoplankton species composition are not well known. Further the effect of changing nutrient availability on direct sinking has been studied in only a few species (see next section). A more thorough field test, with supplemental culture studies, is required to provide a rigorous test of hypotheses linking Si availability, diatom species composition, and cell and carbon flux.

Evidence from Other Eutrophic Coastal Areas

Phytoplankton Species Composition

A shift from diatoms to non-diatoms in response to Si depletion has been observed under a number of circumstances. As mentioned above, it has been documented in chemostat culture experiments with mixed diatoms and flagellates [Sommer, 1994, 1995]. In northern Europe, where blooms of the noxious *Phaeocystis pouchettii* and *Emiliana huxleyi* have increased at the expense of diatoms, mesocosm experiments have examined the effect of changing nutrient ratios on natural phytoplankton species composition [Egge and Aksnes, 1992; Egge and Jacobsen, 1997: Escravage et al., 1996]. In general diatom abundance and productivity was significantly higher in mesocosms with excess Si, N and P, compared with those with only excess N and P. Egge and Aksnes [1992] showed that below a threshold of approximately 2 µM Si, the relative abundance of diatoms to total phytoplankton dropped precipitously, which is consistent with what is known about the kinetics of Si limited growth [Nelson and Dortch, 1996]. Similar changes in phytoplankton species composition have been measured in field studies in a variety of eutrophic environments, including the North Sea [Cadée and Hegeman, 1991; Reigman et al.,

1992], the Baltic Sea [Maestrini and Graneli, 1991], and the Chesapeake Bay [Conley et al., 1992]. Changes within the diatom community in response to Si availability have also been observed. In mixed cultures, some diatoms could persist at much lower Si/N supply ratios than others, so that flagellate dominance did not occur. The low Si adapted species were *Asterionella glacialis*, *Nitzschia closterioides*, and *Chaetoceros socialis* [Sommer, 1995]. In tanks with natural phytoplankton populations, Si availability determined which of two *Chaetoceros* spp., one with a heavily silicified, rapidly sinking, resting cyst and the other without, predominated [Kuosa et al., 1997]. In the field it is more difficult to conduct the time series measurements necessary to relate species changes to subtle changes in nutrient availability. In the Bay of Brest, France, the diatom succession as Si became depleted was almost identical to what would be predicted from data from the northern Gulf of Mexico, terminating in *Chaetoceros socialis* dominance when Si was most limiting [Ragueneau et al., 1994]. Finally, in Oslofjord, *C. socialis* and *C. debilis* dominated when Si was limiting, but *Skeletonema costatum* dominated during a later bloom which was not Si limited [Paasche and Erga, 1988]. These data suggest that there is a group of diatoms, which may be adapted to growth when Si availability is low and others with much higher Si requirements.

Carbon Flux

Vertical C flux is theorized to be greater when large phytoplankton predominate, especially diatoms, due either to greater direct sinking or fecal pellet sinking [Michaels and Silver, 1988]. Direct sinking should result in a larger proportion of primary production being lost to the bottom than grazing and fecal pellet production because there is less respiratory loss. In several eutrophic areas C flux is greatest during the diatom bloom and when Si concentrations are above a threshold value [Conley and Malone, 1992; Heiskanen and Kononen, 1994; Ragueneau et al., 1994]. In experimental mesocosms with variable additions of N, P, and Si, excess Si addition resulted in much greater carbon flux [Egge and Jacobsen, 1997; Wassmann et al., 1996]. This led Wassmann et al. [1996] to propose increasing Si inputs as a management strategy for reducing nutrients in surface water to prevent HABs, although it was noted that this might exacerbate problems with low oxygen bottom water.

There is not enough data on flux of individual diatom species from other eutrophic areas to assess the hypothesis that only some species of, usually more heavily silicified, diatoms contribute to vertical flux. Both field and culture studies suggest, however, that Si limitation triggers diatom sinking [Bienfang et al., 1982] and that some species, such as some *Thalassiosira* spp. or *Skeletonema costatum*, are more vulnerable to nutrient limitation (Si and other nutrients) induced increases in sedimentation rate than others, such as some *Chaetoceros* spp. vegetative cells [Harrison et al., 1986; Heiskanen and Kononen, 1994; Waite et al., 1992].

Conclusions and Management Implications

We have shown, both for the northern Gulf of Mexico and other eutrophic areas, that Si/N ratios influence the phytoplankton species composition and the vertical flux of carbon, which in turn determines the extent and severity of hypoxia. It also will influence the frequency of occurrence of HABs, a growing problem in many eutrophic coastal areas [Hallegraeff, 1993]. Changes in the size and species composition of phytoplankton will also effect trophodynamics at many levels [Turner et al., 1998]. The impacts of variable Si supply are less well understood than that of other nutrients because the occurrence and importance of Si limitation have been underestimated.

Future research on eutrophication should include studies of the role of silicate. This is critical from a management point of view because changing ratios of nutrient inputs may influence the effectiveness of management strategies that reduce nutrient loads [Rabalais et al., 1996]. This is especially true if only small reductions of one nutrient can be accomplished. For example, at present N is the overall limiting nutrient, but spring Si limitation may reduce the magnitude or duration of the diatom bloom, lower C flux and reduce the extent and severity of hypoxia in comparison with what would occur if Si were not limiting in the spring. Reductions in P inputs to freshwater systems can be accomplished more readily than N reductions, but this will increase Si availability downstream to marine ecosystems [Rabalais et al., 1996: Turner and Rabalais, 1991]. Unless the P decrease is large enough to cause overall P limitation in marine waters, the increased Si inputs will decrease Si limitation, resulting in larger and more persistent diatom blooms and possibly exacerbating hypoxia. Thus, a small decrease in P could have an opposite effect to that which was intended. More substantial decreases in nutrient inputs may be needed to decrease hypoxia.

Acknowledgments. This work was funded by NOAA Coastal Ocean Program Office, Nutrient Enhanced Coastal Ocean Productivity Grant No. NA 90AA-D-SG691 to the Louisiana Sea Grant Program (award Nos. MAR02 and MAR92-02 to Q. Dortch and Nos. MAR31A and MAR 92-07A to N. N. Rabalais, D. E. Harper, Jr., and R. E. Turner. We also thank D. Milsted, S. Pool, R. Robichaux, B. Cole, L. Smith and T. Oswald for technical assistance.

References

Bienfang, P. K., P. J. Harrison, and L. M. Quarmby, Sinking rate response to depletion of nitrate, phosphate and silicate in four marine diatoms, *Mar. Biol.*, *67*, 295-302, 1982.

Cadée, G. C. and J. Hegeman, Historical phytoplankton data for the Marsdiep, *Hydrobiol. Bull.*, *24*, 111-118, 1991.

Conley, D. J., S. S. Kilham, and E. Theriot, Differences in silica content between marine and freshwater diatoms, *Limnol. Oceanogr.*, *34*, 205-213, 1989.

Conley, D. J., and T. C. Malone, Annual cycle of dissolved silicate in Chesapeake Bay: implications for the production and fate of phytoplankton biomass, *Mar. Ecol. Prog. Ser.*, *81*, 121-128, 1992.

Conley, D. J., C. L. Schelske, and E. F. Stoermer, Modification of the biogeochemical cycle of silica with eutrophication, *Mar. Ecol. Prog. Ser.*, *101*, 179-192, 1993.

Dortch, Q., D. Milsted, N. N. Rabalais, S. E. Lohrenz, D. G. Redalje, M. J. Dagg, R. E. Turner, and T. E. Whitledge, Role of silicate availability in phytoplankton species composition and the fate of carbon, in *Nutrient-Enhanced Coastal Ocean Productivity Workshop Proceedings*, pp. 76-83, TAMU-SG-92-109 Technical Report, Texas A&M University Sea Grant Program, Galveston, Texas, 1992.

Dortch, Q., D. M. Nelson, R. E. Turner, and N. N. Rabalais, Silicate limitation on the Louisiana continental shelf, in *Proceedings of 1994 Synthesis Workshop, Nutrient-Enhanced Coastal Ocean Productivity Program, Baton, Rouge LA*, pp. 34-39, Louisiana Sea Grant

College Program, Louisiana State University, Baton Rouge, Louisiana, 1995.

Dortch, Q., M. L. Parsons, N. N. Rabalais, and R. E. Turner, What is the threat of Harmful Algal Blooms in Louisiana coastal waters?, in Symposium Proceedings, Recent Research in Coastal Louisiana: Natural System Function and Response to Human Influences, Lafayette, Louisiana, February 1998, edited by L. P. Rozas, J. A. Nyman, C. E. Profitt, N. N. Rabalais, D. J. Reed, and R. E. Turner, pp. 1-11, Louisiana Sea Grant College Program, Louisiana State University, Baton Rouge, Louisiana, 1999.

Dortch, Q. and T. E. Whitledge, Does nitrogen or silicon limit phytoplankton production in the Mississippi River plume and nearby regions?, *Cont. Shelf Res., 12*, 1293-1309, 1992.

Eadie, B. J., B. A. McKee, M. B. Lansing, J. A. Robbins, S. Metz, and J. H. Trefry, Records of nutrient-enhanced coastal ocean productivity in sediments from the Louisiana continental shelf, *Estuaries, 17*, 754-766, 1994.

Egge, J. K. and D. L. Aksnes, Silicate as regulating nutrient in phytoplankton competition, *Mar. Ecol. Prog. Ser., 83*, 281-289, 1992.

Egge, J. K. and A. Jacobsen, Influence of silicate on particulate carbon production in phytoplankton, *Mar. Ecol. Prog. Ser., 147*, 219-230, 1997.

Escravage, V., T. C. Prins, A. C. Smaal, and J. C. H. Peeters, The response of phytoplankton communities to phosphorus input reduction in mesocosm experiments, *J. Exp. Mar. Biol. Ecol., 198*, 55-79, 1996.

Falkowski, P. G., T. S. Hopkins, and J. J. Walsh, An analysis of factors affecting oxygen depletion in the New York Bight, *J. Mar. Sci., 38*, 479-506, 1980.

Graneli, E., P. Carlsson, P. Olsson, B. Sundstrom, W. Graneli, and O. Lindahl, From anoxia to fish poisoning: the last ten years of phytoplankton blooms in Swedish marine waters, in *Novel Phytoplankton Blooms*, edited by E. M. Cosper, V. M. Bricelj, and E. J. Carpenter, pp. 407-427, Springer-Verlag, Berlin, 1989.

Hallegraeff, G. M., A review of harmful algal blooms and their apparent global increase, *Phycologia, 32*, 79-99, 1993.

Harrison, P. J., D. H. Turpin, P. K. Bienfang, and C. O. Davis, Sinking as a factor affecting phytoplankton species succession: the use of selective loss semi-continuous cultures, *J. Exp. Mar. Biol. Ecol., 99*, 19-30, 1986.

Heiskanen, A. S. and K. Kononen, Sedimentation of vernal and late summer phytoplankton communities in the coastal Baltic Sea, *Arch. Hydrobiol., 131*, 175-198, 1994.

Justic', D., N. N. Rabalais, R. E. Turner, and Q. Dortch, Changes in nutrient structure of river-dominated coastal waters: stoichiometric nutrient balance and its consequences, *Estuar. Coast. Shelf Sci., 40*, 339-356, 1995.

Kuosa, H., R. Autio, P. Kuuppo, O. Setälä, and S. Tanskanen, Nitrogen, silicon and zooplankton controlling the Baltic spring bloom: an experimental study, *Estuar. Coast. Shelf Sci., 45*, 813-821, 1997.

Lohrenz, S. E., G. L. Fahnenstiel, D. G. Redalje, G. A. Lang, X. Chen, and M. J. Dagg, Variations in primary production of northern Gulf of Mexico continental shelf waters linked to nutrient inputs from the Mississippi river, *Mar. Ecol. Prog. Ser., 155*, 45-54, 1997.

Lohrenz, S. E., G. L. Fahnenstiel, D. G. Redalje, G. A. Lang, M. J. Dagg, T. E. Whitledge, and Q. Dortch, The interplay of nutrients, irradiance, and mixing as factors regulating primary production in coastal waters impacted by the Mississippi River plume, *Cont. Shelf Res., 19*, 1113-1141, 1999.

Maestrini, S. Y. and E. Graneli, Environmental conditions and ecophysiological mechanisms which led to the 1988 *Chrysochromulina polylepsis* bloom: an hypothesis, *Oceanologia Acta, 14*, 397-413, 1991.

Michaels, A. F. and M. W. Silver, Primary production, sinking fluxes and the microbial food web, *Deep-Sea Res., 35*, 473-490, 1988.

Nelson, D. M. and Q. Dortch, Silicic acid depletion and silicon limitation in the plume of the Mississippi River: evidence from kinetic studies in spring and summer, *Mar. Ecol. Prog. Ser., 136*, 163-178, 1996.

Officer, C. B., and J. R. Ryther, The possible importance of silicon in marine eutrophication, *Mar. Ecol. Prog. Ser., 3*, 83-91, 1980.

Paasche, E. and S. R. Erga, Phosphorus and nitrogen limitation of phytoplankton in the inner Oslofjord (Norway), *Sarsia, 73*, 229-243, 1988.

Qureshi, N. A., The role of fecal pellets in the flux of carbon to the sea floor on a river-influenced continental shelf subject to hypoxia, Ph.D. Dissertation, Department of Oceanography and Coastal Sciences, Louisiana State University, Baton Rouge, Louisiana, 1995.

Rabalais, N. N., R. E. Turner, D. Justic', Q. Dortch, and W. J. Wiseman, Jr., Characterization of Hypoxia: Topic 1 Report for the Integrated Assessment of Hypoxia in the Gulf of Mexico, NOAA Coastal Ocean Program Decision Analysis Series No. 15, NOAA Coastal Ocean Program, Silver Spring, Maryland, 1999.

Rabalais, N. N., R. E. Turner, D. Justic', Q. Dortch, W. J. Wiseman, Jr., and B. K. Sen Gupta, Nutrient changes in the Mississippi River and system responses on the adjacent continental shelf, *Estuaries, 19*, 386-407, 1996.

Ragueneau, O., E. De B. Varela, P. Tréguer, B. Quéguiner, and Y. Del Amo, Phytoplankton dynamics in relation to the biogeochemical cycle of silicon in a coastal ecosystem of western Europe, *Mar. Ecol. Prog. Ser., 106*, 157-172, 1994.

Reigman, R., A. A. M. Noordeloos, and G. C. Cadée, *Phaeocystis* blooms and eutrophication of the continental coastal zones of the North Sea, *Mar. Biol., 112*, 479-484, 1992.

Sen Gupta, B. K., R. E. Turner, and N. N. Rabalais, Seasonal oxygen depletion in continental shelf waters of Louisiana: Historical record of benthic formanifers, *Geology, 24*, 227-230, 1996.

Smayda, T. J., Primary production and the global epidemic of phytoplankton blooms in the sea: a linkage?, in *Novel Phytoplankton Blooms*, edited by E. M. Cosper, V. M. Bricelj and E. J. Carpenter, pp. 449-483, Springer-Verlag, Berlin, 1989.

Smayda, T. J., Novel and nuisance phytoplankton blooms in the sea: Evidence for a global epidemic, in *Toxic Marine Phytoplankton*, edited by E. Granelli, B. Sundstrom, R. Edler and D. M. Anderson, pp. 29-40, Elsevier, New York, 1990.

Sommer, U., Nutrient status and nutrient competition of phytoplankton in a shallow, hypertrophic lake, *Limnol. Oceanogr., 34*, 1162-1173, 1989

Sommer, U., Phytoplankton competition in Pluss-see: a field test of the resource-ratio hypothesis, *Limnol. Oceanogr., 38*, 838-845, 1993.

Sommer, U., The impact of light intensity and day length on silicate and nitrate competition among marine phytoplankton, *Limnol. Oceanogr.*, 39, 1680-1688, 1994.

Sommer, U., Eutrophication related changes in phytoplankton species composition: is there a role of nutrient competition?, ICES-CM-1995/T:7, 1995.

Tilman, D., Resource competition between planktonic algae, An experimental and theoretical approach, *Ecology, 68*, 338-348, 1977.

Tilman, D., R. L. Kriesling, R. W. Sterner, S. S. Kilham, and F. A. Johnson, Green, bluegreen, and diatom algae: taxonomic differences in competitive availability for phosphorus, silicon, and nitrogen, *Arch. Hydrobiol., 106*, 473-485, 1986.

Turner, R. E., N. Qureshi, N. N. Rabalais, Q. Dortch, D. Justic', R. F. Shaw, and J. Cope, Fluctuating silicate:nitrate ratios and coastal plankton food webs, *Proc. Natl. Acad. Sci. USA, 95*, 13048-13051, 1998.

Turner, R. E. and N. N. Rabalais, Changes in the Mississippi River water quality this century. - Implications for coastal food webs, *BioScience, 41*, 140-147, 1991.

Turner, R. E. and N. N. Rabalais, Coastal eutrophication near the Mississippi River delta, *Nature, 368*, 619-621, 1994.

Waite, A., P. K. Bienfang, and P. J. Harrison, Spring bloom sedimentation in a subarctic ecosystem 1. Nutrient sensitivity, *Mar. Biol., 112*, 119-130, 1992.

Wassman, P., J. K. Egge, M. Regstad, and D. L. Aksnes, Influence of dissolved silicate on vertical flux of particulate biogenic matter, *Mar. Pollut. Bull., 33*, 10-21, 1996.

3

Zooplankton: Responses to and Consequences of Hypoxia

Nancy H. Marcus

Abstract

Zones of reduced dissolved oxygen (< 2.86 mg l^{-1}) in the water column occur throughout the world's oceans. In association with the permanent oxygen minimum zones found in the open-ocean, most macrozooplankton are generally not found where dissolved oxygen concentrations are < 0.29 mg l^{-1}, though some species are able to survive at oxygen concentrations as low as 0.07 mg l^{-1}. Physiological studies indicate that survival of these more tolerant species is facilitated by anaerobic respiration in the oxygen-depleted waters followed by upward vertical migration into oxygenated waters to repay the oxygen debt. Microzooplankton, which tend to be more common than macrozooplankton in regions with extremely reduced dissolved oxygen concentrations, are also capable of anaerobic respiration since some have been found in anoxic waters. Episodic periods of hypoxia/anoxia are becoming more common in coastal waters. The impact of hypoxia/anoxia, however, on estuarine and coastal zooplankton has received little attention yet micro- and mesozooplankton are important elements in marine food webs. Copepods in particular constitute a critical component of the diet of larval fish in these regions. Based on a limited number of studies, coastal zooplankton appear to be more sensitive to hypoxia/anoxia than zooplankton associated with permanent oxygen minimum zones. Mortality increases markedly at dissolved oxygen concentrations < 1.43 to 2.0 mg l^{-1}. The potential consequences of the increasing occurrence of hypoxic/anoxic zones in coastal waters on zooplankton community structure, predator-prey interactions and energy flow in these systems are discussed.

Introduction

Zones of reduced dissolved oxygen (< 2.86 mg l^{-1}) in the water column occur throughout the world's oceans. In many regions, e.g., the Oxygen Minimum Zones

Coastal Hypoxia: Consequences for Living Resources and Ecosystems
Coastal and Estuarine Studies, Pages 49-60
Copyright 2001 by the American Geophysical Union

(OMZs) of the Eastern Equatorial Pacific, Arabian Sea and Black Sea the reduction/absence of dissolved oxygen is permanent. Temporary but typically recurrent hypoxia/anoxia occurs in coastal regions during summer when increased temperatures and/or riverine flow lead to seasonal stratification of the water column. Stratification can prevent the resupply of oxygen to bottom waters and thus promotes the development of hypoxic/anoxic conditions if the biochemical oxygen demand of the system is sufficiently high. The frequency, duration and spatial coverage of hypoxic/anoxic bottom waters in the coastal zone appear to be increasing [Diaz and Rosenberg, 1995].

From a biological perspective, reduced dissolved oxygen is a major problem for the constituent biota because few metazoans can withstand long periods without oxygen. Concern for the increasing occurrence of coastal zone hypoxia/anoxia has focused on the direct, short-term impact of reduced dissolved oxygen concentrations on the survival of commercially important demersal invertebrate and fish species, e.g., blue crabs, shrimp, oysters, flounder [see Pihl et al., 1991; Diaz and Rosenberg, 1995]. The results of these studies have shown that severe mortality usually occurs when dissolved oxygen concentrations fall below 2.86 mg l^{-1}. Few studies, however, have considered the potential impact of indirect effects on these important taxa due to the influence of hypoxia/anoxia on other water column biota. For example, zooplankton are key elements of marine food webs because they serve as mediators of energy flow from lower to higher trophic levels. Copepods, especially the naupliar stages, are important prey items in the diets of many larval marine fish. Shifts in zooplankton community structure due to hypoxia/anoxia could alter predator-prey interactions and affect the population dynamics of fish on long time scales. In addition, changes in the zooplankton community could alter the flux of organic matter to the seabed, due to a shift in grazing pressure on the phytoplankton community.

Zooplankton Responses to Hypoxia/Anoxia

Several approaches have been used to gain insight into the response of zooplankton to hypoxia/anoxia. Descriptive studies have examined the distribution and abundance of zooplankton in the field in relation to the occurrence of reduced dissolved oxygen concentrations. Experimental and analytical studies of zooplankton physiology, biochemistry, behavior and life history traits have yielded an understanding of the mechanisms underlying the patterns observed in the field.

Distribution and Abundance

Persistent oxygen minimum zones

Several studies have documented the distribution and abundance of zooplankton in relation to the changes in dissolved oxygen concentrations that are associated with persistent OMZs [e.g., Vinogradov and Voronina, 1961; Longhurst, 1967; Judkins, 1980; Sameoto, 1986; Saltzman and Wishner, 1997a; Wishner et al., 1998]. Like most other studies that have examined the vertical distribution of fauna in the ocean, these studies found that zooplankton biomass generally declined with depth in the ocean. More importantly, a marked decrease of biomass occurred in the oxycline, and biomass levels often increased again below the OMZ. Some species had a disjunct distribution occurring above and below the OMZ, but not within the layer of reduced oxygen. Most macrozooplankton were generally not found where dissolved oxygen concentrations fell below < 0.29 mg l^{-1} [e.g., Judkins, 1980; Saltzman and Wishner, 1997a; Wishner et al., 1998]. However, some species occurred at extremely low dissolved oxygen levels giving rise to biomass peaks within or at the lower boundaries of OMZs [Wishner et al., 1998]. The euphausids *Euphausia diomedeae, E. distinguenda* and *Nematoscelis gracilis* have been found at depths where the dissolved oxygen concentration was < 0.14 mg l^{-1} [Brinton, 1979; Sameoto, 1986]. Boyd and Smith [1980] reported that the copepod, *Eucalanus enermis,* was abundant off Peru where dissolved oxygen concentrations ranged from 0 to 0.14 mg l^{-1}. Saltzman and Wishner [1997b] also reported that *E. inermis* was generally associated with waters containing dissolved oxygen concentrations < 0.14 mg l^{-1}. The copepod, *Rhincalanus nasutus* was found in the eastern tropical Pacific and Arabian Sea at dissolved oxygen concentrations in the range of 0.19 to 0.01 mg l^{-1} [Vinogradov and Voronina, 1961; Sameoto, 1986]. Diapausing fifth stage copepodites of *Calanus* were found in the OMZ of the Santa Barbara basin [Longhurst, 1967; Alldredge et al., 1984]. The dissolved oxygen concentration in the surrounding waters was 0.29 mg l^{-1}. An ostracod species co-existed with the copepod aggregation, but unlike *Calanus* this organism was quite active. Several species, e.g., *Pleuromamma robusta, P. indica* and *P. gracilis,* have been reported to migrate in and out of zones of reduced dissolved oxygen concentration on a diel basis [Vinogradov and Voronina, 1961; Longhurst, 1967; Saltzman and Wishner, 1997b]. These species appear to tolerate a wide range of dissolved oxygen concentrations (0.14 to 7.14 mg l^{-1}). Evidence of diel migration into waters of extremely low oxygen concentration (0.07 to 1.43 mg l^{-1}) was also observed for macrozooplankton in the Arabian Sea OMZ [Wishner et al., 1998]. As noted by Wishner et al. [1998], differences among studies, in estimates of oxygen tolerance in the field, could reflect differences in the methods used to determine dissolved oxygen concentration as well as differences in the thickness of the depth interval over which zooplankton were collected.

Due to differences in the overall metabolic requirements of small and large sized organisms, it is reasonable to expect that the distribution and abundance of these two groups would be different. Although smaller organisms tend to have higher metabolic rates, the surface to volume ratio is greater than larger organisms and this may enable their survival at lower oxygen concentrations. In the Black Sea, Vinogradov et al. [1984] did not find mesozooplankton, e.g., *Calanus helgolandicus, Pleurobrachia, Sagitta setosa,* or *Pseudocalanus* at depths where the dissolved oxygen concentration dropped below 0.5 to 0.4 mg l^{-1}. However, microflagellates were found below these layers at even lower dissolved oxygen levels. More recent studies in the Red Sea and Arabian Sea indicated that some micrometazoan taxa occurred within the core of the OMZ [Bottger-Schnack, 1994, 1996]. Fenchel et al. [1990] examined the distribution of protozoans in relation to oxygen in two fjords and noted that several species survived in anoxic waters. Moreover, distinct assemblages were found above, within and below the oxycline suggesting that different taxa have very distinct oxygen requirements.

Since OMZs are permanent features of the marine system, it is likely that species have adapted to the low dissolved oxygen concentrations. This is supported by the occurrence of many species at oxygen concentrations (< 0.29 mg l^{-1}) that are generally thought to be below the tolerance limits of most metazoans. Since the occurrence of hypoxia/anoxia in coastal zones is a relatively recent phenomenon, it would not be surprising to find few species able to tolerate the low levels of dissolved oxygen common in permanent OMZs. On the other hand, it is reasonable to expect that over time some coastal species might adapt to the low dissolved oxygen concentrations associated with temporary but recurrent zones of hypoxia/anoxia.

Episodic hypoxic/anoxic coastal zones

In the coastal zone, episodic periods of hypoxia and anoxia lasting days to months are becoming more and more common in both frequency and duration [Diaz and Rosenberg, 1995]. In some regions these events re-occur on an annual basis. Only a few studies, however, have reported on the distribution and abundance of zooplankton in relation to episodic hypoxic/anoxic events in the coastal zone. Roman et al. [1993] examined the distribution of the copepods, *Acartia tonsa* and *Oithona colcarva*, in association with anoxic bottom waters in Chesapeake Bay. *A. tonsa* normally undergoes diel vertical migration during summer. Stages that normally migrated, occurring in the bottom waters during the day, were not found at depth when oxygen levels were < 1.0 mg l^{-1}. The highest concentrations of zooplankton were found at the pycnocline. Uye [1994] examined the distribution of zooplankton in Tokyo and Osaka Bays and reported an increased abundance of small sized copepods in areas subject to intense eutrophication. Oxygen values were not reported, but dissolved oxygen concentrations may have been low due to increased biochemical oxygen demand associated with the eutrophic conditions. Cervetto et al. [1995] examined the distribution of *Acartia tonsa* in a shallow stratified lagoon in France and found that individuals did not occur in anoxic bottom waters. The lower limit of the species distribution was set by the oxycline, but the actual depth varied with time as the oxycline moved up and down. Protozoans appear to be more tolerant. Fenchel et al. [1990] examined the distribution of protozoans in fjords that experienced episodic bouts of anoxia and found distinct species assemblages associated with the upper oxygenated layer, the oxycline and even the zone of no oxygen.

Physiological Responses

Since metazoans require oxygen for their development, growth and reproduction, various structural (e.g., smallness, gills) and biochemical (e.g., oxygen binding proteins) mechanisms have evolved for extracting oxygen from the environment to meet metabolic demands.   The metabolic response of metazoans to reduced dissolved oxygen concentrations varies.   Some species are metabolic regulators, whereas others are metabolic conformers.   Metabolic regulators maintain a constant rate of oxygen consumption over a range of dissolved oxygen concentration, whereas metabolic conformers reduce their demands in association with a decline in dissolved oxygen concentration.   In the case of metabolic regulators there is typically a critical oxygen level below which the organism cannot meet its metabolic needs.   At this level oxygen consumption can no longer be regulated, consumption drops rapidly, and mortality is likely.
Most of the research on metabolic responses of marine zooplankton to reduced dissolved oxygen concentrations has focused on bathypelagic animals from the open-

ocean OMZs [e.g., Childress, 1971; Thuesun et al., 1998]. Childress [1968] showed that the bathypelagic mysid, *Gnathophausia ingens*, regulated its respiratory rate when exposed to reduced dissolved oxygen concentrations. The critical oxygen threshold was at or below what the species normally encountered in the OMZ, i.e. 0.20 to 0.37 mg l^{-1}. On the other hand, the midwater copepod, *Gaussia princeps*, a vertical migrator, was shown to regulate its oxygen consumption rate under reduced dissolved oxygen concentrations, but the critical oxygen threshold was higher than the concentrations the species encountered in the environment [Childress, 1977]. Experiments with male *G. princeps* showed they could survive 10 to 14 hours without oxygen. These results led to the conclusion that *G. princeps* respires anaerobically while at depth in midwater low oxygen conditions. Moreover, it was suggested that individuals migrated upwards at night into well oxygenated waters to undergo aerobic respiration and thereby repay the oxygen debt generated while at depth.

Many lakes and ponds stratify during summer creating a hypolimnion with reduced oxygen levels. Various daphnids occur in the hypolimnion even when dissolved oxygen concentrations are greatly reduced. Their survival under these conditions has been related to their ability to produce the oxygen binding protein hemoglobin [Kobayashi, 1982; Hanazato and Dodson, 1995]. The increased production of hemoglobin in response to decreasing dissolved oxygen concentrations enables individuals to fulfill their oxygen demands via aerobic respiration. Similarly, the tadpole shrimp, *Triops longicaudatus*, that inhabits ephemeral pools that undergo wide fluctuations in dissolved oxygen concentration, produces hemoglobin when dissolved oxygen concentrations decline. This response enables the shrimp to sustain a high growth rate [Scholnick and Snyder, 1996]. There is no evidence that freshwater copepods produce hemoglobin or other oxygen binding proteins, and typically copepods are not found in the hypolimnion when dissolved oxygen concentrations are reduced. One freshwater copepod species, *Cyclops varicans*, was able to tolerate several hours of exposure to anoxic conditions [Chaston, 1969]. This study also showed that as the duration of exposure to reduced oxygen increased, treated animals increased their rate of oxygen consumption following their return to oxygenated waters. It was suggested that individuals rely on anaerobic metabolism for a limited period of time to meet their energy requirements. In nature diel vertical migration between hypolimnetic layers with reduced oxygen and epilimnetic layers with higher oxygen concentrations enables individuals to complete the metabolic steps needed to breakdown the waste products generated by anaerobic metabolism. Childress [1977] used this same explanation to account for the distribution and migratory behavior of *Gaussia princeps* in the open ocean OMZ.

Determining survival rates in controlled laboratory experiments is one approach to assessing the capacity of coastal marine zooplankton to cope with the increasing occurrence of reduced dissolved oxygen concentrations. Vargo and Sastry [1977] reported that 2-h LD_{50} values for *Acartia tonsa* and *Eurytemora affinis* adults collected from the Pettaquamscutt River Basin, Rhode Island ranged from dissolved oxygen concentrations of 0.36 to 1.40 mg l^{-1} and 0.57 to 1.40 mg l^{-1}, respectively. Roman et al. [1993] tested the oxygen tolerance of adults of *Acartia tonsa* and *Oithona colcarva* from Chesapeake Bay. Survival was considerably less after 24 h in < 2.0 mg l^{-1} oxygenated water. *Oithona* showed slightly better survival than *Acartia*. Experiments with eggs of *Acartia* revealed that none hatched at concentrations < 0.57 mg l^{-1} within 30 h. This threshold differs slightly from earlier studies [Uye and Fleminger, 1976; Ambler, 1985]. Lutz et al. [1992] suggested that the differences observed in the various studies could have been due to differences in the lengths of incubation, methods for determining oxygen, or the condition of the eggs.

Stalder and Marcus [1997] examined the 24-h survival of three coastal copepod species in response to low oxygen. *Acartia tonsa* showed excellent survival at concentrations as low as 1.43 mg l^{-1}. Between 1.29 and 0.86 mg l^{-1} survival declined markedly and at 0.71 mg l^{-1} mortality was 100%. *Labidocera aestiva* and *Centropages hamatus* were more sensitive to reduced dissolved oxygen concentrations. The survival of these species was significantly lower at 1.43 mg l^{-1}. The survival of nauplii of *Labidocera aestiva* and *Acartia tonsa* at low dissolved oxygen concentrations was generally better than adult survival.

No studies have examined the role of biochemical adaptations in enabling copepods to survive under anoxic conditions. Moreover, oxygen-binding proteins that could enhance survival at low oxygen conditions have not been identified in copepods [Thuesen et al., 1998].

Life History Features

Behavioral avoidance

An alternative response of zooplankton that are confronted with reduced dissolved oxygen levels is to actively avoid the zone of reduced oxygen. For example, in a comparative study of two deep fjords, Devol [1981] showed that the distribution of zooplankton was different and concluded that the zooplankton altered the depth of their vertical migration to avoid reduced oxygen levels. While this interpretation may be correct, it is generally difficult to determine from field studies whether differences in vertical distribution of zooplankton are due to avoidance of reduced dissolved oxygen concentrations or differential mortality. A few laboratory studies have addressed this issue.

Using artificial microcosms, Tinson and Laybourn-Parry [1985] showed that freshwater copepods moved upward in a stratified water column where the lower layer was anoxic. When the lower layer was hypoxic (20 to 35% saturation) the distribution of the copepods was no different than that in the fully oxygenated control columns. The threshold dissolved oxygen concentration for this movement was not established. They also showed that animals could not survive extended exposure (> 5 hours) to low dissolved oxygen concentrations. Survival was possible at 25% saturation.

Stalder and Marcus [1997] also examined the behavioral avoidance response of three common inshore copepods, *Acartia tonsa*, *Labidocera aestiva* and *Centropages hamatus*, to reduced dissolved oxygen concentrations. Their results showed that none of the species avoided either severely hypoxic < 0.71 mg l^{-1} or moderately hypoxic 1.43 mg l^{-1} waters. It was suggested that in the field these species would not avoid hypoxic or anoxic regions. If this interpretation is correct, then the patterns of reduced abundance of *Acartia tonsa* in bottom waters reported by Roman et al. [1993] and Cervetto et al. [1995] reflected mortality rather than avoidance. On the other hand it is possible that the results obtained by Stalder and Marcus [1997] are not applicable to all regions. The animals used in their experiments were obtained from an area that does not experience hypoxic conditions. Thus, in areas like Chesapeake Bay that have been experiencing recurring hypoxia/anoxia for a number of years, it is possible that the members of the zooplankton community have adapted to these environmental conditions by evolving avoidance responses to minimize mortality.

Dormancy

Dormancy is defined as a state of reduced development. While the occurrence of dormancy may not be a response to reduced dissolved oxygen conditions, dormant organisms may have a survival advantage because metabolic activity is typically reduced in such individuals and oxygen consumption rates are extremely low or undetectable.

In copepods, dormancy has been shown to occur at different life stages, but most work has focused on eggs and copepodites. Two principal types of eggs have been identified, non-diapause (subitaneous) which typically hatch within a few days of being spawned and diapause eggs that must complete a refractory period before they become competent to hatch [Grice and Marcus, 1981]. The refractory period may last days to weeks to months. Eggs that are spawned into the water column may sink to the seabed before hatching and once in the seabed may become quiescent due to a lack of oxygen [Uye and Fleminger, 1976]. Large accumulations of copepod eggs have been documented in the seabed of coastal waters. Since the seabed is typically anoxic below a depth of a several millimeters, considerable research in recent years has focused on the impact of hypoxia/anoxia on the survival and hatching of copepod eggs. The results of this research indicate that copepod eggs can hatch at extremely low dissolved oxygen concentrations. For example, the eggs of *Centropages hamatus* hatched at concentrations as low as 0.03 mg l^{-1} after 11 days of incubation [Lutz et al., 1992]. Generally hatching occurs within a few days at oxygen concentrations > 0.21 mg l^{-1} [Kasahara et al., 1975; Uye and Fleminger 1976; Uye et al., 1979; Ambler, 1985; Lutz et al., 1994]. Non-diapause eggs are less tolerant of long-term exposure to anoxia than diapause eggs. For example, non-diapause eggs of *Acartia tonsa*, *Labidocera aestiva* and *Centropages hamatus* survived days to a few weeks without oxygen [Marcus and Lutz, 1994], but diapause eggs of *Centropages hamatus* were able to withstand many months without oxygen [Marcus and Lutz, 1998].

Reproductive mode

Not all copepods release their eggs into the water column. Some produce egg sacs that remain attached to the body until the nauplii hatch. As long as the females avoid anoxic bottom waters, the eggs should be protected from reduced oxygen levels. Uye [1994] attributed the success of *Oithona davisae* in Tokyo Bay to its ability to tolerate low dissolved oxygen conditions as well as its strategy of producing egg sacs. Eutrophic conditions in the bay since the 1970s have seen a rise in the dominance of *Oithona* and a decline in the occurrence of *Acartia omorii* and *Paracalanus*, both of which release their eggs into the water column. Uye [1994] also noted that the shift to the smaller *Oithona* species was associated with a decline in fish production within the bay and suggested that this may have been due to their small size which makes them difficult for visual predators to locate [Uye, 1994].

Consequences

Predator-Prey Interactions

Judkins [1980] suggested that the presence of OMZs made zooplankton more susceptible to predation by compressing their distribution into the well-oxygenated surface layers and limiting vertical migration. This would be beneficial to the predators of the zooplankton.   It could also ultimately lead to reduced grazing pressure on the phytoplankton.

On the other hand, zones of reduced dissolved oxygen may also provide a refuge from predation. For example, while there was some predation on an aggregation of *Calanus* in the Santa Barbara Basin OMZ, Alldredge at al. [1984] reported that there were many more predator taxa, e.g., siphonophores, medusae and ctenophores in the water column above the aggregation. They assumed that the location of the aggregation served to limit predation. A similar layering of predators, the ctenophore, *Pleurobrachia* sp. and its prey (the copepod, *Calanus helgolandicus*), as well as the chaetognath, *Sagitta setosa* and its prey (the copepod, *Pseudocalanus* sp.) in association with an oxygen gradient was observed in the Black Sea [Vinogradov et al., 1984]. In freshwater ponds and lakes there is also evidence that the reduced dissolved oxygen concentrations in the hypolimnion provides a refuge from predation. Hanazato et al. [1989] found that predation on *Daphnia longispina* was reduced during summer because fish did not penetrate into the low oxygenated bottom waters. The cladoceran found refuge in the bottom waters during the day. At night the *Daphnia* migrated into the epilimnion to feed. In the fall, however, stratification disappeared, oxygen levels increased in the hypolimnion, and the cladoceran population declined.   Apparently the bottom waters no longer provided a refuge from fish predation. Similarly Naess [1996] suggested that the deeper anoxic layers of the seabed provide a refuge from predation for resting eggs since fewer predators occupy those layers. The ability to survive for extended periods in the seabed is important as resting eggs represent a pool of potential recruits for the planktonic population.

Life History Trade-Offs

Studies of freshwater zooplankton indicate that reduced dissolved oxygen concentrations impact several life history traits. For example, the survival of *Daphnia* was > 85% when exposed to levels of oxygen > 1.86 mg l^{-1}. However, reproduction was impaired below 2.71 mg l^{-1} and growth below 3.71 mg l^{-1} [Homer and Waller, 1983]. Nebeker et al. [1992] reported reduced reproductive effort in *D. pulex* at oxygen concentrations $\leq$ 1.57 mg l^{-1}. Hanazato [1996] found that the growth rate of *Daphnia pulex* that were exposed to reduced dissolved oxygen concentrations declined and maturation time was delayed. Hanazato and Dodson [1995] suggested that these changes could have been due to a reduction in the metabolic rate and/or a shift in energy allocation due to the cost of producing hemoglobin.   In another study, Weider and Lampert [1985] found that clones of *Daphnia* varied in their response to low dissolved oxygen concentrations. While this variation was related to differences in hemoglobin concentration, the results suggested that organisms adjusted to reductions in dissolved oxygen concentrations through acclimation and adaptation.   Thus, spatial and temporal fluctuations in dissolved oxygen concentrations may play an important selective role influencing the species composition of natural planktonic populations. It is likely that species that inhabit regions where reduced dissolved oxygen concentration is a persistent or predictable aspect of the environment can adapt over time to this feature of the habitat. The adaptive responses might be expressed at the physiological level resulting in the ability to survive at reduced dissolved oxygen concentrations or at the behavioral level by altering migration patterns. Thus, Kerambrun et al. [1993] attributed differences in the respiration rates of two populations of *Acartia clausi*, one from a eutrophic environment and one from a more oligotrophic system, to genetic differentiation arising from directional selection.

Community Structure and Energy Flow

Due to differences in metabolic requirements, it is apparent that microzooplankton, including microcopepods and protozoans, may become the chief mediators of organic matter transfer instead of larger macrozooplankton, when dissolved oxygen levels decline. These changes in community structure may alter the flux of organic matter to the sediments since smaller zooplankton tend to enhance the rate at which organic matter is recycled in the water column. On the other hand, if organic matter is not intercepted by microzooplankton the deposition of material on the sea floor may increase due to an overall decrease in grazing within hypoxic/anoxic zones by macrozooplankton. Unlike the open ocean, where zones of reduced oxygen are sandwiched between strata of higher oxygen concentration, hypoxic/anoxic zones in coastal waters typically extend upwards from the seafloor into the water column. Thus, the phenomenon of enhanced grazing by macrozooplankton at the lower boundary of OMZs that has been reported [Gowing and Wishner, 1998] probably does not occur in association with hypoxic/anoxic zones in coastal waters.

Changes in zooplankton population dynamics and community structure may also arise due to sublethal effects of reduced dissolved oxygen concentrations on development, growth and fecundity. Such changes could also alter the flux of organic material through the food web due to shifts in grazing pressure and predator-prey interactions.

Finally, many studies have shown that in coastal waters, especially shallow bays and estuaries, the concentration of copepod eggs in the seabed can be quite high. Under normal conditions, the physical resuspension of bottom sediment returns buried eggs to the water column where they can hatch. If the bottom water becomes depleted of oxygen because of stratification, then the supply of naupliar recruits from the seabed is cut off. Moreover, eggs produced by the planktonic population may sink into the anoxic layer before having sufficient time to hatch. As a result, the naupliar recruitment of these species will be reduced which could have an impact on their population growth and thereby alter zooplankton community structure.

Recommendations

Because the lethal limits of zooplankton under reduced dissolved oxygen concentrations are lower than fish and larger invertebrates, it is likely that oxygen concentrations that are suitable for the survival of larger taxa will also be suitable for the survival of zooplankton. While survival may not be obviously affected by reduced dissolved oxygen concentrations, longer-term chronic effects may impact the population dynamics of zooplankton and this in turn may impact other elements of the food web. Research that distinguishes the sublethal impacts of hypoxia on coastal zooplankton is needed. In addition research must consider how extended periods of hypoxic/anoxic bottom water may decouple the benthic and pelagic environments and lead to shifts in plankton community structure due to the disruption of life cycles.

References

Alldredge, A. L., B. H. Robison, A. Fleminger, J. J. Torres, J. M. King, and W. M. Hamner, Direct sampling and in situ observation of a persistent copepod aggregation in the mesopelagic zone of the Santa Barbara basin, *Mar. Biol., 80*, 75-81, 1984.

Ambler, J. W., Seasonal factors affecting egg production and viability of eggs of *Acartia tonsa* Dana, from East Lagoon, Galveston, Texas, *Estuar. Coastal Shelf Sci., 20*, 743-760, 1985.

Bottger-Schnack, R., The microcopepod fauna in the Eastern Mediterranean and Arabian Seas: a comparison with Red Sea fauna. *Hydrobiologia, 292/293*, 271-282, 1994.

Bottger-Schnack, R., Vertical structure of small metazoan plankton, especially non-calanoid copepods. I. Deep Arabian Sea, *J. Plankton Res.,18*, 1073-1101, 1996.

Boyd, C. M. and S. L. Smith, Grazing patterns of copepods in the upwelling system off Peru, *Limnol. Oceanogr., 25*, 583-596, 1980.

Brinton, E., Parameters relating to the distributions of plankton organisms, especially euphausids in the eastern tropical Pacific, *Prog. Oceanogr., 8*, 1-64, 1979.

Cervetto, G., M. Pagano, and R. Gaudy, Feeding behavior and migrations in a natural population of the copepod *Acartia tonsa, Hydrobiologia , 300/301*, 237-248, 1995.

Chaston, I., Anaerobiosis in *Cyclops varicans, Limnol.Oceanogr.,14*, 298-301, 1969.

Childress, J. J., Oxygen minimum layer: vertical distribution and respiration of the mysid *Gnathophausia ingens, Science, 160*, 1242-1243, 1968.

Childress, J. J., Respiratory adaptations to the oxygen minimum layer in the bathypelagic mysid *Gnathophausia ingens, Biol. Bull., 141*, 109-121, 1971.

Childress, J. J., Effects of pressure, temperature, and oxygen on the oxygen-consumption rate of the midwater copepod, *Gaussia princeps, Mar. Biol., 39*, 19-24, 1977.

Devol, A., Vertical distribution of zooplankton respiration in relation to the intense oxygen minimum zones in two British Columbia fjords, *J. Plankton Res., 3*, 593-601, 1981.

Diaz, R. and R. Rosenberg, Marine benthic hypoxia: a review of its ecological effects and the behavioural responses of benthic macrofauna, *Oceanogr. Mar. Biol. Ann. Rev., 33*, 245-303, 1995.

Fenchel, T., L. D. Kristensen, and L. Rasmussen, Water column anoxia: vertical zonation of planktonic protozoa, *Mar. Ecol. Prog. Ser., 62*, 1-10, 1990.

Gowing , M. M. and K. F. Wishner, Feeding ecology of the copepod *Lucicutia aff. L. grandis* near the lower interface of the Arabian Sea oxygen minimum zone. *Deep-Sea Res. II., 45*, 2433-2459, 1998.

Grice, G. and N. H. Marcus, Dormant eggs of marine copepods, *Oceanogr. Mar. Biol. Ann. Rev., 19*, 125-140, 1981.

Hanazato, T., Combined effects of food shortage and oxygen deficiency on life history characteristics and filter screens of *Daphnia, J. Plankton Res., 18*, 757-765, 1996.

Hanazato, T. and S. I. Dodson, Synergistic effects of low oxygen concentration, predator kairomone, and a pesticide on the cladoceran *Daphnia pulex*, *Limnol. Oceanogr., 40*, 700-709, 1995.

Hanazato, T., M. Yasuno, and M. Hosomi, Significance of a low oxygen layer for a *Daphnia* population in Lake Yunoko, Japan, *Hydrobiologia, 185*, 19-27, 1989.

Homer, D. H. and W. T. Waller, Chronic effects of reduced dissolved oxygen on *Daphnia magna*, *Water, Air, and Soil Pollution, 20*, 23-28, 1983.

Judkins, D., Vertical distribution of zooplankton in relation to the oxygen minimum off Peru, *Deep- Sea Res., 27A*, 475-487, 1980.

Kasahara, S., T. Onbe, and M. Kamigaki, Calanoid copepod eggs in sea bottom muds. III. Effects of temperature, salinity, and other factors on the hatching of resting eggs of *Tortanus forcipatus*, *Mar. Biol., 31*, 31-35, 1975.

Kerambrun, P., M. Thessalou-Legaki, and G. Verriopoulos, Comparative effects of environmental conditions, in eutrophic polluted and oligotrophic non-polluted areas of the Saronikos Gulf (Greece), on the physiology of the copepod, *Acartia clausi*, *Comp. Biochem. Physiol., 105C*, 415-420, 1993.

Kobayashi, M., Influence of body size on haemoglobin concentration and resistance to oxygen deficiency in *Daphnia magna*, *Comp. Biochem. Physiol., 72A*, 599-602, 1982.

Longhurst, A., Vertical distribution of zooplankton in relation to the eastern Pacific oxygen minimum, *Deep-Sea Res., 14*, 51-63, 1967.

Lutz, R. V., N. H. Marcus, and J. P. Chanton, Effects of low oxygen concentrations on the hatching and viability of eggs of marine calanoid copepods, *Mar. Biol., 114*, 241-247, 1992.

Lutz, R. V., N. H. Marcus, and J. P. Chanton, Hatching and viability of copepod eggs at stages of embryological development: anoxic/hypoxic effect, *Mar. Biol., 119*, 199-204, 1994.

Marcus, N. H. and R. V. Lutz, Effects of anoxia on the viability of subitaneous eggs of planktonic copepods, *Mar. Biol., 121*, 83-87, 1994.

Marcus, N. H. and R. V. Lutz, Longevity of subitaneous and diapause eggs of *Centropages hamatus* (Copepoda: Calanoida) from the northern Gulf of Mexico, *Mar. Biol., 131*, 249-257, 1998.

Naess, T., Benthic resting eggs of calanoid copepods in Norwegian enclosures used in mariculture: abundance, species, composition, and hatching, *Hydrobiologia, 320*, 161-168, 1996.

Nebeker, A. V., S. T. Onjukka, D. G. Stevens, G. A. Chapman, and S. E. Dominguez, Effects of low dissolved oxygen on survival, growth and reproduction of *Daphnia, Hyalella*, and *Gammarus*, *Environmental Toxicol. Chem., 11*, 373-379, 1992.

Pihl, L., S. P. Baden, and R. J. Diaz, Effects of periodic hypoxia on distribution of demersal fish and crustaceans, *Mar. Biol., 108*, 349-360, 1991.

Roman, M., A. L. Gauzens, W. K. Rhinehart, and J. R. White, Effects of low oxygen water on Chesapeake Bay zooplankton, *Limnol. Oceanogr., 38*, 1603-1614, 1993.

Saltzman, J. and K. Wishner, Zooplankton ecology in the eastern tropical Pacific oxygen minimum zone above a seamount: 1. General trends, *Deep-Sea Res., 44*, 907-931, 1997a.

Saltzman, J. and K. Wishner, Zooplankton ecology in the eastern tropical Pacific oxygen minimum zone above a seamount: 2. Vertical distribution of copepods, *Deep-Sea Res., 44*, 931-954, 1997b.

Sameoto, D. D., Influence of the biological and physical environment on the vertical distribution of mesozooplankton and micronekton in the eastern tropical Pacific, *Mar. Biol., 93*, 263-279, 1986.

Scholnick, D. A. and G. K. Snyder, Response of the tadpole shrimp *Triops longicaudatus* to hypoxia, *Crustaceana, 69*, 937-948, 1996.

Stalder, L. C. and N. H. Marcus, Zooplankton responses to hypoxia: behavioral patterns and survival of three species of calanoid copepods, *Mar. Biol. 127*, 599-607, 1997.

Thuesen, E. V., C. B. Miller, and J. J. Childress, Ecophysiological interpretation of oxygen consumption rates and enzymatic activities of deep-sea copepods, *Mar. Ecol. Prog. Ser., 168*, 95-107, 1998.

Tinson, S. and J. Laybourn-Parry, The behavioral responses and tolerance of freshwater benthic cyclopoid copepods to hypoxia and anoxia, *Hydrobiologia, 127*, 257-263, 1985.

Uye, S., Replacement of large copepods by small ones with eutrophication of embayments: cause and consequence, *Hydrobiologia, 292/293*, 513-519, 1994.

Uye, S. and A. Fleminger, Effects of various environmental factors on egg development of several species of *Acartia* on southern California, *Mar. Biol., 38*, 253-262, 1976.

Uye, S., S. Kasahara, and T. Onbe, Calanoid copepod eggs in sea bottom muds. IV. Effects of some environmental factors on the hatching of resting eggs, *Mar. Biol., 51*, 151-156, 1979.

Vargo, S. L. and A. N. Sastry, Interspecific differences in tolerance of *Eurytemora affinis* and *Acartia tonsa* from an estuarine anoxic basin to low dissolved oxygen and hydrogen sulfide. *Physiology and Behavior of Marine Organisms*. D. S. McCluskey and A. J. Berry (Eds.), Pergamon Press. 12th European Marine Biology Symposium, 219-226, 1977.

Vinogradov, M. Y., E. A. Shuskina, M. V. Flint, and N. I. Tumantsev, Plankton in the lower layers of the oxygen zone in the Black Sea, *Oceanology, 26*, 222-228, 1984.

Vinogradov, M. E. and N. M. Voronina, Influence of the oxygen deficit on the distribution of plankton in the Arabian Sea, *Okeanologiya, 1*, 670-678, 1961. Eng. transl. *in Deep-Sea Res., 9*, 523-530, 1962.

Weider, L. and W. Lampert, Differential response of *Daphnia* genotypes to oxygen stress: respiration rates, hemoglobin content and low-oxygen tolerance, *Oecologia, 65*, 487-491, 1985.

Wishner, K. F., M. M. Gowing, and C. Gelfman, Mesozooplankton biomass in the upper 1000 m in the Arabian Sea: overall seasonal and geographic patterns, and relationship to oxygen gradients, *Deep-Sea Res. II., 45*, 2405-2432, 1998.

4

Distribution of Zooplankton on a Seasonally Hypoxic Continental Shelf

Naureen A. Qureshi and Nancy N. Rabalais

Abstract

The vertical distribution of zooplankton was documented for a station in 20-m water depth through a seasonal decline of bottom-water dissolved oxygen concentration, and across a broad area of hypoxic bottom-water in mid-summer of two years. There was a seasonal progression of zooplankton abundance with a spring peak and summer decline and a change in the relative proportion of taxa through the year. Copepods (adults and copepedites) were more abundant in the lower water column than the upper water column (daytime samples) across all monthly samples at the 20-m station that experienced severe hypoxia for extended parts of the summer. Copepods were present at normal or negligible densities for the two sampling dates when the oxygen concentration was below 1 mg l^{-1}. Copepod nauplii, on the other hand, were reduced in abundance in the bottom water when the oxygen was less than 1 mg l^{-1}. Across the broad area of hypoxia in the two summer surveys, copepods and copepod nauplii were concentrated below the pycno-oxycline but above the bottom water where they were reduced when the bottom-water oxygen concentration was less than 1 mg l^{-1}. Meroplankton were concentrated above oxygen-deficient bottom waters in summer and were either delaying metamorphosis or were unable to recruit to the seabed. Bottom-water oxygen concentrations less than 1 mg l^{-1} may have disrupted the daytime migration of copepods and copepod nauplii into that layer. The potential for indirect effects of altered zooplankton distributions and behavior on zooplankton food webs, energy transfer, trophic interactions, and secondary production, both pelagic and benthic, exist but are not known.

Coastal Hypoxia: Consequences for Living Resources and Ecosystems
Coastal and Estuarine Studies, Pages 61-76
Copyright 2001 by the American Geophysical Union

Introduction

The inner to mid continental shelf of Louisiana is influenced by the freshwater discharge and high nutrient loads of the Mississippi-Atchafalaya river system. Stratified and highly productive waters exist for much of the year. Degradation of the fluxed organic matter from the surface waters (either via sinking phytoplankton, fecal pellets or aggregates) to the lower water column and seabed contributes to the annual depletion of dissolved oxygen in the water column beneath the pycnocline [Qureshi, 1995; Justic' et al., 1996]. Hypoxia (defined as < 2 mg O_2 l^{-1}) forms as early as February and persists through early October, but is most widespread and severe from May through September [Rabalais et al., 1999; this volume]. The hypoxic zone covers as much as 20,000 km^2 of the seabed in mid-summer. Hypoxic waters are distributed from shallow depths near shore (4 to 5 m) to as deep as 60 m water depth, but more typically between 5 and 30 m. Hypoxia occurs mostly in the lower water column but may encompass as much as the lower half to two-thirds.

The presence of hypoxic or anoxic (0 mg O_2 l^{-1}) water below the pycnocline may affect, either directly or indirectly, the abundance and distribution of zooplankton. Copepods and copepod nauplii were in low abundance or were absent from the hypoxic (< 1 mg l^{-1}) mesohaline region of Chesapeake Bay [Roman et al., 1993]. Conversely, copepod abundances were highest during the day in normoxic bottom waters during spring and summer. Anoxia in the bottom waters also disrupted the normal diel migration of copepods in Chesapeake Bay. Hypoxia, and not predation by gelatinous zooplankton, was proposed by Roman et al. [1993] as a possible explanation for a decrease in the abundance of copepod nauplii in Chesapeake Bay. In fact, Purcell et al. [1994] demonstrated that there was no limitation on the abundance of *Acartia tonsa* due to predation by gelatinous zooplankton. In the Patuxent subestuary of Chesapeake Bay, all zooplankton were proportionally less common in the bottom layer when the bottom-water dissolved oxygen was less than 2 mg l^{-1} [Keister et al., in press]. Other examples of hypoxia-affected zooplankton distribution are provided by Marcus [this volume]. The potential exists, therefore, for widespread seasonally severe hypoxia on the Louisiana shelf to affect zooplankton abundance, distribution and composition.

Low dissolved oxygen results in copepod mortality in laboratory experiments. A value of 1 to 1.43 mg l^{-1} (= 0.7 to 1 ml l^{-1}) is a typical value below which mortality for *Acartia tonsa* increases significantly [Roman et al., 1993; Stalder and Marcus, 1997]. *Labidocera aestiva* and *Centropages hamatus* suffered significant mortality as the dissolved oxygen dropped to 1.43 mg l^{-1} [Stalder and Marcus, 1997]. Copepod nauplii appear to be more tolerant to hypoxia than adults with nauplii of *Acartia tonsa* withstanding 24-h exposure to dissolved oxygen as low as 0.3 to 0.7 mg l^{-1} [Stalder and Marcus, 1997]. The inhibition of copepod egg hatching in less than 1 mg l^{-1} dissolved oxygen [Roman et al., 1993], or in near anoxic waters (0.08 mg l^{-1}, Marcus and Lutz [1994]) depending on the stage of development, could affect recruitment and eventually population levels.

If hypoxia affects mortality of zooplankton individuals or eggs, grazing patterns or diel migratory behavior, there may be cascading consequences to the transfer of carbon and the flux of carbon to the seabed. Copepods graze as much as 14 to 62% of the daily primary production in plume-influenced waters, but only 4 to 5% within the plume [Dagg, 1995], as do other mesozooplankton and microzooplankton [Dagg et al., 1996; Strom and Strom, 1996]. An estimated 55% of surface productivity is transferred to the

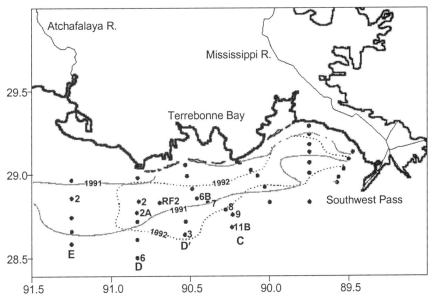

Figure 1. Distribution of hypoxia monitoring stations, stations for zooplankton collections and summertime bottom-water hypoxia during 24-28 July 1991 and 3-6 August 1992.

lower water column via zooplankton fecal pellets at a station in 20-m water depth within the core of the Louisiana hypoxic zone off Terrebonne Bay (station C6B, Fig. 1) [Qureshi, 1995]. Changes in the zooplankton community composition and distribution as a result of hypoxia may affect the flux of organic matter from productive surface waters to the bottom, trophic interactions, transfer of energy within the food web, and overall secondary production (both within the water column and the benthos).

Zooplankton distribution studies for the Louisiana continental shelf or within the influence of the Mississippi River plume have focused on copepods and copepod nauplii [e.g., Dagg, 1988; Ortner et al., 1989; Dagg and Whitledge, 1991], but none have addressed zooplankton distribution in relation to hypoxia. Our objectives were to examine the vertical distribution of zooplankton at a station located within the core of the seasonally severe hypoxic area on a monthly basis through the development and dissipation of hypoxia, and at multiple stations across a broad area during the peak of hypoxia development in mid-summer.

Methods

Monthly sampling was conducted in 1992 at station C6B in 20-m depth off Terrebonne Bay within the core of hypoxic waters that form annually on the southeastern Louisiana shelf (Fig. 1). A 2500-km^2 area, including station C6B, was sampled during five days in late July 1991 and in early August 1992. All monthly samples were taken between 0930 and 1730 h, but stations for the broader spatial survey were sampled during both day and night.

Water column vertical profiles for temperature, salinity, density and dissolved oxygen were recorded with a Hydrolab Surveyor 3 or a SeaBird CTD unit. Continuous dissolved oxygen data (15-min intervals) were obtained with an Endeco 1184 on a mooring at 19 m in a 20-m water column [Rabalais et al., 1994]. Biological data, including chlorophyll a and phaeopigments, were determined for water collected at the same depths as the zooplankton samples.

Water (10-15 l) was collected with a bucket from the surface, and with multiple 5-l Niskin bottles at 6.5 m, 14 m, and 19 m (bottom). The mid-depths at station C6B were chosen to correspond with the openings of moored particle traps within the upper mixed layer above the usual 10-m pycnocline and below the pycnocline in the lower water column. Samples for the upper water column above the pycnocline and lower water column below the pycnocline were at variable depths during the spatial surveys because of varying total and pycnocline depths.

The mesozooplankton size fraction (153 μm) sampling with Niskin bottles provided consistent temporal and spatial sampling, although a towed net would have provided better estimates of total abundance. For limited dates on which surface towed net samples (234 μm) for mesozooplankton were made in conjunction with surface discrete samples [Qureshi, 1995], the total abundance with discrete sampling was approximately 30% of the total abundance for a towed sample. The mesozooplankton data were used only to identify relative differences across time and space. Copepod nauplii data from the smaller size fractions also provided the same ability to identify relative differences, but could also be compared quantitatively to other studies for which the same methods, suitable for collection of the smaller size fraction zooplankters, were employed.

Water samples were transferred to spigotted carboys from which water was dispensed to a nested series of 153 μm, 63 μm and 20 μm Nitex screens. Water in the carboys was continuously agitated to ensure collection of a mixed sample. A 2- to 3-l volume was filtered for the 20 μm size fraction, and 10- to 15-l samples (average 13.5 l) were filtered for the two larger size fractions. The zooplankton collected on each screen was backwashed with filtered sea water (0.2 μm) into 60-ml bottles, preserved in 2.5% glutaraldehyde and refrigerated until further analysis.

Samples collected on the 153 μm size fraction were counted under a dissection scope, and the sample was split with a Folsom plankton splitter, if necessary. The > 63 μm and > 20 μm size fractions were stained with proflavin and filtered through an 8 μm cellulose filter which was mounted on a slide for enumeration on an epifluorescence microscope (primarily to enumerate the fecal pellets, not reported here). Only the copepod nauplii of the microzooplankton component were counted.

The zooplankton data (no. l^{-1}) were categorized into total zooplankton (all three size fractions), copepods (adults and copepodites of the 153 μm sample) and copepod nauplii (63 μm and 20 μm size fractions combined). Temporal variability was tested at the monthly scale for station C6B and spatially among stations with varying degrees of bottom water dissolved oxygen ranging from 0.05 to 3.53 mg l^{-1} (Table 1). Each data set was tested for variability in the three main effects month/station, depth and size fraction using a completely randomized design ANOVA with three factorial (4*4*3) treatment arrangements. The assumptions for normally distributed data and homogeneity of variance were tested for all dependent variables. Data were transformed to $\log_{10}(x+1)$ values to obtain normality and to stabilize variances. The Type III sum of squares was used to compute mean square errors, because of the unbalanced data [Steele and Torrie, 1980]. Tukey's studentized range test, a pair-wise comparison of means, was employed to specify variability within a treatment, when main effects were significant. The abundance data were correlated with environmental parameters of the water column including temperature, salinity, dissolved oxygen, chlorophyll *a* and phaeopigments. All statistical analyses were conducted with PC SAS version 6.04 [SAS Institute, 1985].

Results

Size Fraction Differences

Most of the organisms were copepod nauplii, and most organisms were collected in the 63 and 20 μm fractions. There was a strong positive correlation for most samples between total zooplankton and copepod nauplii. The abundance of organisms for the 63 and 20 μm size fractions were similar to each other and higher than the 153 μm fraction (Appendices 1-4).

Monthly Variability

Dissolved oxygen patterns at station C6B

The water column at station C6B from March through September was stratified during most sampling trips, and the density structure was controlled primarily by salinity differences but strengthened by temperature differences in the summer (Fig. 2). The pycnocline was usually located between 8 and 12 m. Continuous bottom water dissolved oxygen data for station C6B indicated fluctuations above and below 2 mg l^{-1} from March through April, but the monthly samples were taken when the oxygen was less than 2 mg l^{-1} (Fig. 3). The May sample was collected when the bottom waters were near 2 mg l^{-1} but following a severely hypoxic period. The June sample was in the middle of an extended period of severely hypoxic/anoxic bottom waters. Bottom water during the July sample fluctuated around 2 mg l^{-1}. The reoxygenation in late July resulted from the intrusion of colder, saltier water from offshore, but was followed by another severely hypoxic period during which the August sample was taken. Hurricane Andrew moved through the study area on August 24-25 and mixed the water column, a condition that persisted through September because of the passage of several cold fronts.

Mesozooplankton composition

The zooplankton in the > 153 μm size fraction were composed of a high percentage of copepods and copepod nauplii at all depths, but the percent composition varied with month and depth (Fig. 4). Copepods comprised an average of 55% of the total sample (by number), with values from 15 to 85%, and were generally a higher percentage of the total zooplankton in the spring and the single fall sample, than in the summer. Mesozooplankton other than copepods were larvaceans, chaetognaths, meroplankton and miscellaneous others (e.g., decapod larvae, doliolids, medusae, pteropods, ostracods, amphipods, and fish eggs and larvae). Meroplankton (predominantly larvae of benthos, such as polychaete larvae, *Balanoglossus* tornaria, pilidium larvae of nemerteans and

Table 1. Sample stations, dates, times, total water column depth and bottom water
dissolved oxygen (DO).

Station	Date	Time	Day/Night	Depth (m)	Bottom DO (mg l^{-1})
Monthly					
C6B	3.21.92	09:25*	D	19.0	0.86-1.39
C6B	4.10.92	16:15	D	19.0	1.14
C6B	5.26.92	11:02	D	19.0	2.36
C6B	6.17.92	17:25	D	19.0	0.07
C6B	7.31.92	16:30	D	19.0	2.09
C6B	8.12.92	11:03	D	19.0	0.12
C6B	9.18.92	11:36	D	19.0	4.56
Spatial					
RF2	8.3.92	16:55	D	16.3	0.05
D2A-2	7.30.91	09:30	D	16.5	0.06
C8	7.29.91	09:40	D	23.0	0.07
D2-1	7.28.91	21:10	N	12.5	0.07
C6B-4	7.27.91	16:20	D	19.0	0.14
C6B-5	7.31.91	00:01	N	19.0	0.54
C9	7.26.91	08:30	D	27.6	0.68
D2A-3	7.30.91	16:00	D	16.5	0.69
D6	7.30.91	00:35	N	41.7	1.80
C7	8.6.92	03:13	N	20.8	1.98
C6B-1	7.31.92	16:30	D	19.0	2.09
C11B	7.28.91	03:28	N	47.6	2.31
D'3	7.27.91	01:40	N	17.6	2.89
E2	8.2.92	11:09	D	8.3	3.53

*all times except this one (central standard) are central daylight savings time

ophiopleuteus larvae of brittle stars) comprised nearly 80 to 90% of the total community
in the lower water column (14 m and bottom) when oxygen was severely depleted.
Meroplankton comprised less of the total community in the lower water column when the
oxygen concentration was higher.

Zooplankton seasonal and vertical distributions

The concentration of total zooplankton was significantly different among months
across all months, and was highest in March and April and lowest in June (Appendices 1
and 2). Total zooplankton were evenly distributed at all depths across all months
(Appendix 2).

Copepod concentrations varied among months and with depth (Appendices 1 and 2,
Fig. 5). The highest mean concentrations of copepods occurred in March and May and
were lowest in April and June. More copepods were found in the lower water column
(14 m and bottom) than in the upper water column (surface and 6.5 m) for all months
combined. There was no clear pattern of effect of bottom-water dissolved oxygen less
than 1 mg l^{-1}. Copepod abundances were very low in June (dissolved oxygen 0.07 mg l^{-1})
but similar to other months in August (dissolved oxygen 0.12 mg l^{-1}). The number of
copepods at 6.5 m were positively correlated with higher dissolved oxygen

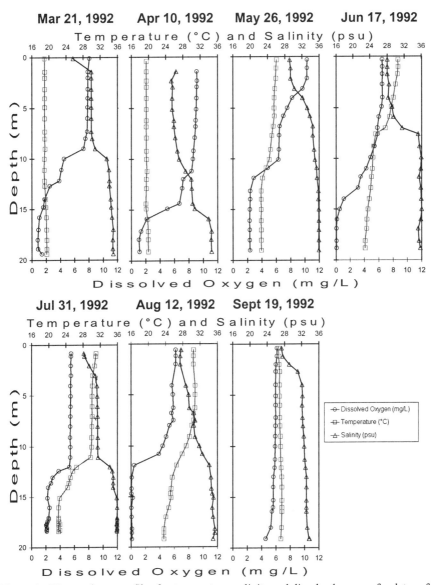

Figure 2. Water column profiles for temperature, salinity and dissolved oxygen for dates of monthly sampling at station C6B.

concentrations and negatively related to lower water temperatures, both conditions more likely in the spring ($r^2 = 0.81$ and 0.92, respectively, $P < 0.05$).

The abundances of copepod nauplii were significantly different among months, but not by depth (Appendices 1 and 2, Fig. 5). Concentrations of copepod nauplii were significantly higher in March and April, decreased from May to June and increased again in July through September. While the copepods were more concentrated in the lower

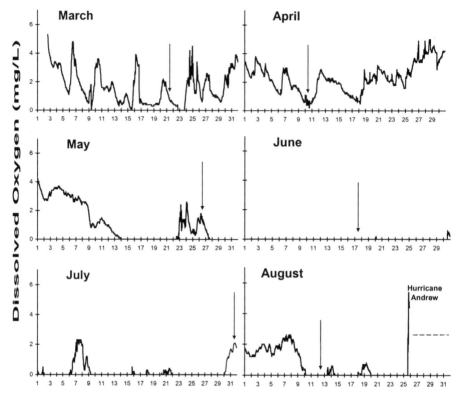

Figure 3. Continuous dissolved oxygen (mg l^{-1}) time series for station C6B in 1992 for the period of sample collections; arrows indicate times of samples. The oxygen meter malfunctioned following Hurricane Andrew on August 24, but discrete bottom-water measurements in September indicated a well-oxygenated bottom layer.

water column during daytime, the nauplii were evenly distributed across all month/depth combinations. The number of copepod nauplii in bottom waters during the severely hypoxic June and August periods were negligible. Within the lower water column (14 m and bottom), copepod nauplii were positively correlated with dissolved oxygen concentration ($r^2 = 0.79$ and 0.90, respectively, $P < 0.05$), i.e., the lower the dissolved oxygen concentration the lower the copepod nauplii density.

Spatial Variability

Oxygen and depth conditions

Collections across a larger area in July 1991 and August 1992 (Fig. 1) necessitated a broad range of depths to find some stations that were not hypoxic at the bottom (Table 1). The water column was strongly stratified at most stations (exception at the shallowest station, E2) with a halocline located at 6 to 12 m, depending on total depth. It was also necessary to take collections during both day and night in order to collect the samples

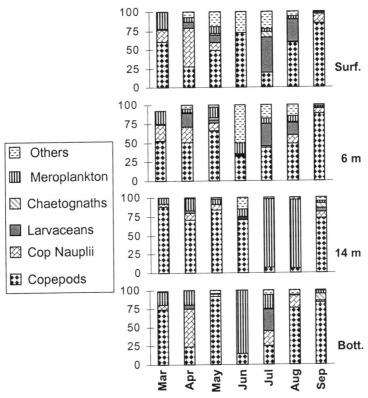

Figure 4. Percent composition of zooplankton in the > 153 μm by depth and date at station C6B from March – September 1992.

during the cruise time limitations. Thus, results for a range of bottom-water oxygen concentrations are complicated by depth and day/night differences (Table 1).

Mesozooplankton composition

Zooplankton in the > 153 μm fraction were composed of an average of 45% copepods (by number) and ranged from 5 to 80% [Qureshi, 1995]. A greater proportion of meroplankton was present in the bottom waters at all stations with the exception of stations D6 and C11B; neither of these stations was severely hypoxic and both were in deeper water than the others (Table 1). Meroplankton comprised 80 to 90% of the total zooplankton in the lower water column (below the pycno-oxycline and in the bottom) at station D2A, a station with severe hypoxia.

Zooplankton spatial and vertical distributions

The concentration of total zooplankton was significantly different for both station and depth (Appendices 3 and 4). The highest concentration of total zooplankton was at station D2 (Fig. 6). The concentration of total zooplankton was lowest in the bottom waters, and did not appear to be related to day/night differences, but was possibly

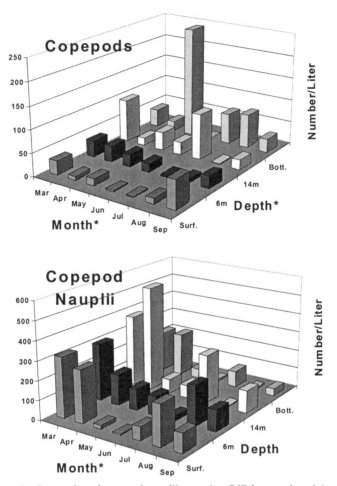

Figure 5. Copepods and copepod nauplii at station C6B by month and depth.

affected by the severity of hypoxia. For example, there were several daytime samples with few organisms in hypoxic bottom waters, whereas in oxygenated bottom waters there were zooplankton in the lower water column during daytime.

The copepod concentrations differed among stations and with depth (Appendices 3 and 4). Copepod concentrations were significantly higher at station D2 than other stations. Copepods were mostly present in the upper water column (surface and above pycno-oxycline) with a few exceptions (stations C6B-4, C6B-1, C9 and C8), where more copepods were present in the lower water column (below the pycno-oxycline and bottom) (Fig. 6). The stations that were the exceptions were sampled during day, and migration into the lower water column by the copepods may have caused the depth differences. In the severely hypoxic stations, the copepods were concentrated below the pycno-oxycline and not at the bottom, but were in the upper water column at night, i.e., an indication that diel migration may have been disrupted. In more oxygenated waters, there were more copepods in the lower water column, especially during daytime, but also at night.

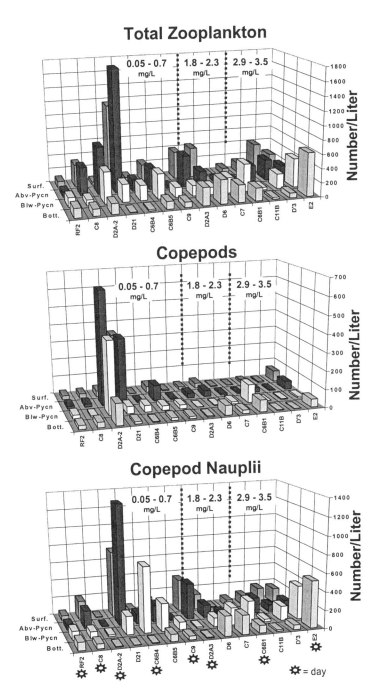

Figure 6. Total zooplankton, copepods and copepod nauplii at several stations in late July 1991 and early August 1992.

The copepod nauplii concentrations were significantly different among stations and by depth (Appendices 3 and 4). Highest concentrations of copepod nauplii occurred at station D2. Most copepod nauplii were present in the upper water column (surface and above pycno-oxycline) (Fig. 6). Copepod nauplii were less abundant in bottom samples of stations where the bottom-water dissolved oxygen concentration was less than 1 mg l^{-1} (Fig. 6). Like the copepod adults the nauplii seemed to concentrate below the pycno-oxycline at stations where the bottom-water oxygen was less than 1 mg l^{-1}. They did occur into the upper water column at night, but did not occur in the bottom during the day, another potential disruption of diel behavior (assuming that these broadly spaced samples taken on different days but at day/night differences can be assumed to show a pattern of diel behavior).

Discussion

The study area was stratified primarily by salinity from March to August and well-mixed in September following a hurricane and series of cold fronts. Hypoxic (< 2 mg l^{-1}) and extremely low oxygen (below 1 mg l^{-1} and near 0 mg l^{-1}) existed at C6B for much of May-August 1992 and across most of the larger area surveyed in July 1991 and August 1992. The overall abundance of total zooplankton, copepods, and nauplii followed a typical seasonal sequence of higher values in spring with a decline into the summer that followed the seasonal trend of food availability (see monthly chlorophyll *a* concentrations at station C6B in Rabalais and Turner [this volume]). Copepods were more abundant in the lower water column (daytime samples) at Station C6B across all months sampled than in the upper water column, whereas copepod nauplii were evenly distributed across all depths. There were inconsistent results of the effect of dissolved oxygen less than 1 mg l^{-1} on copepods; copepod density was low in one sample but normal in another. Copepod nauplii abundance was correlated with dissolved oxygen in the lower water column, i.e., fewer nauplii in lower oxygen conditions, which could indicate an association with that parameter or a similar seasonal pattern in the decline of food availability. There were no significant correlations, however, of copepod nauplii abundance with chlorophyll *a* and phaeopigment concentrations at any depth of the water column. Both copepods and copepod nauplii were concentrated below the pycno-oxycline and were low in abundance in bottom samples for stations in the broad spatial survey where bottom-water dissolved oxygen concentrations were less than 1 mg l^{-1}.

Despite the seasonal decline in overall abundance of copepod nauplii and low densities when the oxygen was below 1 mg l^{-1} (both the monthly samples and the spatial survey), the numbers of copepod nauplii that we documented were similar to, or exceeded, those previously reported from Louisiana shelf areas influenced by discharge from the Mississippi River [Dagg et al., 1987; Dagg, 1988; Dagg and Whitledge, 1991]. Copepod nauplii concentrations averaged 135 l^{-1} over all depths in the monthly samples at station C6B. Dagg and Whitledge [1991] reported high naupliar concentrations (> 800 l^{-1}) in the Mississippi River plume in July 1987, but otherwise they were < 100 l^{-1} at depths greater than 10 m. Copepod naupliar concentrations at stations farther from the Mississippi River plume were 100 to 300 l^{-1} in the upper 10 m and < 100 l^{-1} below 10 m in July 1987, and generally 50 l^{-1} in April 1988. At station C6B, nauplii density was 200 to 500 l^{-1} at most depths in March and April 1992, then decreased to < 100 up to 300 l^{-1} in May through September. In our broad spatial surveys of mid-summer, copepod naupliar concentrations peaked at > 700 l^{-1} in the surface water and exceeded 1200 l^{-1} above the pycno-oxycline at station D2.

Meroplankton (including mostly larvae of benthos) comprised nearly 80 to 90% of the total community at station C6B in the lower water column when oxygen was severely hypoxic, but decreased to approximately 20% in July when the mesozooplankton community structure was more diverse. In the spatial survey, meroplankton comprised a large percentage of the mesozooplankton community, and more were present in the bottom waters at all stations with the exception of stations D6 and C11B; neither station was severely hypoxic and both were in deeper water than the others. The concentration of meroplankton above oxygen-depleted bottom waters is consistent with the results of Powers et al. [this volume] that indicate either delayed metamorphosis by benthic larvae or an avoidance of settlement.

The diel vertical migration of zooplankton from bottom layers during the day to surface waters at night facilitates the biological pump of organic matter as surface grazers defecate at depth, accelerating fecal pellet flux to the benthos (e.g., Ohman [1990]). Disruption of typical zooplankton migratory patterns when oxygen concentrations in bottom waters fall below 1 mg l^{-1} (e.g., in Chesapeake Bay [Roman et al., 1993]), would have the potential to alter this normal pattern. The exclusion of zooplankters from the bottom waters might also negate any predatory escape advantages facilitated by daytime movements to the bottom. Differences in vertical distribution displayed during the day/night sampling across the large area of hypoxia, however, indicated that most zooplankters were excluded from severely hypoxic bottom waters even during daytime. Zooplankton were found in well-oxygenated bottom waters (2.9 to 3.5 mg l^{-1}) during the day, and their density was relatively higher in the surface waters at night at those stations.

We attribute the lack of zooplankters in the hypoxic lower water column in the spatial survey to avoidance rather than mortality. Laboratory experiments with *Acartia tonsa*, however, indicate that they do not avoid hypoxia but die when subjected to lethal concentrations of dissolved oxygen [Stalder and Marcus, 1997]. This would indicate that the absence of copepods in severely hypoxic or anoxic bottom waters, therefore, could be due to mortality rather than avoidance. We have acquired numerous video tapes from remotely operated vehicles (ROVs) combined with CTD units that show high concentrations of fish, darting zooplankters and particles as the depth increases towards the interface of the pycno-oxycline [N. Rabalais and D. Harper, personal observations]. Below a steep oxycline, visible particles and certainly living organisms darting about are absent. If mortality were the cause of their absence, then some evidence of sinking zooplankton bodies seems reasonable. Exclusion may also be due to strong stratification. Concentrations of *Calanus ponticus* (V-VI copepodite) were confined above the depth of the maximal gradient of the main pycnocline under an unusually sharp oxycline in the Black Sea, but this was attributed to the effects of density rather than the concentration of oxygen [Vinogradov et al., 1992], a consideration that cannot be excluded for the Louisiana shelf hypoxic zone. However, there was often strong salinity stratification at C6B that did not preclude zooplankters (cf. CTD casts in Fig. 2 and zooplankton in Figs. 5 and 6). Thus, we suggest that exclusion due to low oxygen is important.

The effects of low dissolved oxygen (< 1 mg l^{-1}) were inconsistent for copepods in the monthly study. Copepods maintained average densities in bottom waters when the oxygen was 0.12 mg l^{-1}, but were reduced in bottom water oxygen of 0.07 mg l^{-1}. Copepod nauplii, however, were less abundant in bottom waters in these same two months than other months sampled. In the spatial survey, both copepods and copepod nauplii densities were reduced in bottom waters where the dissolved oxygen concentration was less than 1 mg l^{-1}. Despite some evidence of exclusion from suitable and preferred habitats, the numbers of microzooplankton (as copepod nauplii) were consistent with similar numbers for coastal Louisiana, if not higher. The distribution of zooplankton on the seasonally hypoxia Louisiana shelf clearly requires more study. Potential implications of low oxygen to zooplankton communities and subsequent trophic interactions that impact zooplankton food webs, energy transfer, pelagic and benthic secondary production, and eventually fisheries, are not known and also deserve study.

Acknowledgements. Funding for this research was provided by grants from the National Oceanic and Atmospheric Administration, Nutrient Enhanced Coastal Ocean Productivity (NECOP) program, the NOAA National Undersea Research Center, a U.S. Minerals Management Service University Research Initiative, and the Louisiana Universities Marine Consortium Foundation, Inc. and financial support of U.S. AID (Agency for International Development). Ben Cole assisted with figure preparation.

Appendix 1. Three factorial completely randomized design ANOVA (P > F) for concentration (no. l^{-1}) of total zooplankton, copepods and copepod nauplii collected from station C6B from March – September 1992.

	Df	Total Zooplankton P > F	Copepods** P > F	Copepod Nauplii** P > F
Pr < W		0.6602	0.7329	0.8823
Model	47	0.0011*	0.0007*	0.0001*
Months	6	0.0018*	0.0003*	0.0001*
Depth	3	0.1825	0.0113*	0.3453
Size Fraction (SF)	2	0.0001*	0.3564	0.0001*
Month*Depth	18	0.1425	0.0012*	0.0479*
Month *SF	12	0.0088*	0.6731	0.8756
Depth*SF	6	0.4241	0.0602	0.6265

*significant at alpha = 0.05
**$\log_{10}(x+1)$ transformed data

Appendix 2. Tukey's studentized range test for concentration (no. l^{-1}) for total zoopalnkton, copepods and copepod nauplii collected from station C6B during March – September 1992. Means with the same letter are not significantly different (alpha = 0.05)

	Number	Total Zooplankton	Copepods*	Copepod Nauplii*
Month				
March	12	125.92 A	25.09 A	99.93 AB
April	10	99.90 AB	8.01 BDC	100.22 A
May	12	56.58 ABC	40.75 AB	23.26 BC
June	12	24.26 C	6.89 D	20.34 C
July	10	58.66 ABC	14.40 BCD	35.99 ABC
August	12	51.79 BC	12.19 CD	43.74 B
September	11	44.61 BC	17.40 ABC	30.17 ABC
Depth				
Surface	21	64.91 A	8.25 B	55.29 A
6.5 m	19	61.44 A	9.69 B	48.35 A
14 m	20	88.38 A	20.04 A	71.77 A
Bottom (19 m)	19	51.44 A	34.07 A	32.96 A
Size Fraction				
> 153 μm	24	20.27 B	12.30 A	3.15 A

Appendix 2. Continued.

> 63 μm	27	77.62 A	25.78 A	50.33 A
> 20 μm	28	104.56 A	4.00 A	101.32 A

*analysis was performed on $\log_{10}(x+1)$ transformed data, but means are presented as untransformed data.

Appendix 3. Three factorial completely randomized design ANOVA (P > F) for concentration (no. l^{-1}) of total zooplankton, copepods and copepod nauplii collected from stations with varying degrees of bottom water dissolved oxygen during July and August of 1991 and 1992.

	Df	Total Zooplankton P > F	Copepods** P > F	Copepod Nauplii P > F
Pr < W		0.3334	0.9213	0.9223
Model	83	0.0001*	0.0001*	0.0001*
Station	12	0.0066*	0.0001*	0.0002*
Depth	3	0.0002*	0.0275*	0.0001*
Size Fraction (SF)	2	0.0001*	0.0001*	0.0001*
Station*Depth	36	0.0001*	0.0001*	0.0001*
Station*SF	24	0.0638	0.0065*	0.0050*
Depth*SF	6	0.9064	0.8334	0.0001*

*significant at alpha = 0.05
** $\log_{10}(x+1)$ transformed data

Appendix 4. Tukey's studentized range test for concentration (no. l^{-1}) for total zooplankton, copepods and copepod nauplii collected from stations with varying degrees of bottom water dissolved oxygen during July and August of 1991 and 1992. Means with the same letter are not significantly different (alpha = 0.05)

	Time	Number	Total Zooplankton	Copepods*	Copepod Nauplii*
Stations					
RF2	D	12	36.53 ABC	9.89 BC	24.57 BC
C8	D	12	75.30 ABC	10.02 BCD	47.63 AB
D2	N	12	264.90 AB	109.50 A	206.88 AB
C6B-4	D	12	38.83 ABC	5.70 E	74.88 ABC
C6B-5	N	12	84.85 ABC	20.24 B	68.68 A
C9	D	8	31.03 C	7.43 BCDE	22.46 C
D2A	D	12	104.61 ABC	6.04 CDE	77.54 ABC
D6	N	12	90.55 ABC	14.97 B	57.06 AB
C7	N	12	46.20 ABC	5.66 BCDE	45.36 ABC
C6B-1	D	12	58.66 ABC	14.40 BC	35.99 ABC
C11B	N	12	35.88 BC	5.27 CDE	29.63 BC
D'3	N	12	82.79 ABC	2.41 DE	38.87 ABC
E2	D	12	122.72 A	20.21 B	96.73 AB
Depth					
Surface		37	98.35 A	21.88 AB	65.84 A
Above pycno-oxycline		37	113.11 A	24.59 A	78.90 A
Below pycno-oxycline		39	69.97 A	10.50 B	72.06 AB
Bottom		39	56.33 B	11.73 AB	40.76 B
Size Fraction					
> 153 μm		52	21.20 B	11.56 B	2.66 B
> 63 μm		50	105.60 A	24.66 A	70.50 A
> 20 μm		50	127.35 A	0.00	111.17 A

*analysis was performed on $\log_{10}(x+1)$ transformed data, but means are presented as untransformed data.
Time: D= day, N = night.

References

Dagg, M. J., Physical and biological responses to the passage of a winter storm in the coastal and inner shelf waters of the northern Gulf of Mexico, *Cont. Shelf Res., 8,* 167-178, 1988.

Dagg, M. J. and T. E. Whitledge, Concentrations of copepod nauplii associated with the nutrient-rich plume of the Mississippi River, *Cont. Shelf Res., 11,* 1409-1423, 1991.

Dagg, M. J., E. P. Green, B. A. McKee, and P. B. Ortner, Biological removal of fine-grained lithogenic particles from a large river plume, *J. Mar. Res., 54,* 149-160, 1996.

Dagg, M. J., P. B. Ortner, and F. Al-Yamani, Winter-time distribution and abundance of copepod nauplii in the northern Gulf of Mexico, *Fishery Bull., 86,* 319-330, 1987.

Justic', D., N. N. Rabalais, and R. E. Turner, Effects of climate change on hypoxia in coastal waters: A doubled CO_2 scenario for the northern Gulf of Mexico. *Limnol. Oceanogr.,* 41, 992-1003, 1996.

Keister, J. E., E. D. Houde, and D. L. Breitburg, Effects of bottom-layer hypoxia on abundances and depth distributions of organisms in Patuxent River, Chesapeake Bay, *Mar. Ecol. Prog. Ser.,* in press.

Marcus, N. H. and R. V. Lutz, Effects of anoxia on the viability of subitaneous eggs of planktonic copepods, *Mar. Biol., 121,* 83-87, 1994.

Ohman, M. D., The demographic benefits of diel vertical migration by zooplankton, *Ecol. Monogr., 60,* 257-281, 1990.

Ortner, P. B., L.C. Hill, and S. R. Cummings, Zooplankton community structure and copepod species composition in the northern Gulf of Mexico, *Cont. Shelf Res., 9,* 387-402, 1989.

Purcell, J. E., J. R. White, and M. R. Roman, Predation by gelatinous zooplankton and resource limitation as potential controls of *Acartia tonsa* copepod populations in Chesapeake Bay, *Limnol. Oceanogr., 39,* 263-278, 1994.

Qureshi, N. A., The role of fecal pellets in the flux of carbon to the sea floor on a river-influenced continental shelf subject to hypoxia, Ph.D. Dissertation, Department of Oceanography and Coastal Sciences, Louisiana State University, Baton Rouge, Louisiana, 1995.

Rabalais, N. N., W. J. Wiseman, Jr., and R. E. Turner, Comparison of continuous records of near-bottom dissolved oxygen from the hypoxia zone of Louisiana, *Estuaries, 17,* 850-861, 1994.

Rabalais, N. N., R. E. Turner, D. Justic', Q. Dortch, and W. J. Wiseman, Jr., Characterization of hypoxia: Topic 1 Report for the Integrated Assessment of Hypoxia in the Gulf of Mexico. NOAA Coastal Ocean Program Decision Analysis Series No. 16, NOAA Coastal Ocean Program, Silver Springs, Maryland, 167 pp., 1999.

Roman, M. R., A. L. Gauzens, W. Kr. Rhinehart, and J. R. White, Effects of low oxygen waters on Chesapeake Bay zooplankton, *Limnol. Oceanogr., 38,* 1603-1614, 1993.

SAS Institute, Inc., SAS User's Guide: Statistics, Vers. 5 edition, SAS Institute, Inc., Cary North Carolina, 1985.

Stalder, L. C. and N. H. Marcus, Zooplankton responses to hypoxia: behavioral patterns and survival of three species of calanoid copepods, *Mar. Biol., 127,* 599-607, 1997.

Steele, R. G. and J. H. Torrie, *Principles and Procedures of Statistics,* McGraw Hill, New York, 1980.

Strom, S. L. and M. W. Strom, Microplankton growth, grazing, and community structure in the northern Gulf of Mexico, *Mar. Ecol. Prog. Ser., 130,* 229-240, 1996.

Vinogradov, M. E., E. G. Arashkevic, and S. V. Ilchenko, The ecology of *Calanus ponticus* population in the deeper layer of its concentration in the Black Sea, *J. Plankton Res., 14,* 447-458, 1992.

5

Pelagic Cnidarians and Ctenophores in Low Dissolved Oxygen Environments: A Review

Jennifer E. Purcell, Denise L. Breitburg, Mary Beth Decker, William M. Graham, Marsh J. Youngbluth, and Kevin A. Raskoff

Abstract

Low dissolved oxygen concentrations occur at depth in many marine and estuarine environments. Human effects on coastal ecosystems, especially nutrient enrichment and depletion of commercial fish and shellfish species, may promote outbreaks of jellyfish populations. We review existing information on the vertical distribution of pelagic cnidarians and ctenophores (referred to subsequently as "jellyfish") relative to dissolved oxygen concentrations in environments that experience seasonal hypoxia (Chesapeake Bay, the northern Gulf of Mexico) and in environments having a permanent oxycline (the Black Sea, marine lakes and fjords, and oxygen minimum layers). Most species of jellyfish do not live in hypoxic waters, however, some species occur in high densities at very low dissolved oxygen concentrations. Some jellyfish accumulate at the pycnocline/oxycline just above severely hypoxic waters. Experiments on the scyphomedusan, *Chrysaora quinquecirrha*, and the ctenophore, *Mnemiopsis leidyi*, in Chesapeake Bay have shown prolonged survival in dissolved oxygen concentrations < 2 mg l^{-1}. Hypoxic conditions alter trophic interactions in complex ways. For example, low dissolved oxygen concentrations have been shown to reduce the escape abilities of fish larvae, thereby increasing their vulnerability to predation. Jellyfish, in general, may be more tolerant of low dissolved oxygen than fishes, which may give jellyfish an adaptive advantage over fishes in eutrophic coastal environments.

Introduction

The episodic or persistent occurrence of low dissolved oxygen concentrations (hypoxia) is an important physical feature of many aquatic habitats including estuaries, lakes, nearshore coastal waters, fjords, and oxygen minimum layers of the deep sea. In shallow marine systems, hypoxia generally occurs during summer when density stratification of the water

Coastal Hypoxia: Consequences for Living Resources and Ecosystems
Coastal and Estuarine Studies, Pages 77-100

77

column limits reaeration of bottom waters [Renaud, 1986; Turner et al., 1987; Swanson and Parker, 1988; Rabalais et al., 1991]. Oxygen depletion in estuaries and coastal waters is often a consequence of excess nutrient loadings [Officer et al., 1984; Rosenberg et al., 1990], and is predicted to become more common in the event of global warming [Kennedy, 1990].

Low dissolved oxygen concentrations are major threats to the ecology and fisheries of estuaries and other coastal waters worldwide [Caddy, 1993]. In addition to mortality directly attributable to lethal low dissolved oxygen concentrations [e.g., Breitburg, 1994; Stalder and Marcus, 1997], major ecological effects may occur depending on how low dissolved oxygen influences the vertical distributions and behaviors of key species. Studies in freshwater and coastal systems indicate that predators and their prey can change vertical or horizontal distributions under hypoxic conditions [Rudstam and Magnuson, 1985; Kolar and Rahel, 1993; Roman et al., 1993; Breitburg, 1994; Howell and Simpson, 1994; Keister, 1996], potentially altering encounter rates between predator and prey [Breitburg et al., 1999]. Low dissolved oxygen concentrations can also affect prey capture rates by modifying the percent of time spent foraging [Bejda et al., 1987], the attack rates of predators [Breitburg et al., 1994], and the behaviors of prey that influence their susceptibility to predation [Poulin et al., 1987; Rahel and Kolar, 1990; Pihl et al., 1992; Kolar and Rahel, 1993]. Compromised escape abilities of prey species under hypoxic conditions may influence susceptibility of zooplankton and ichthyoplankton to predation [Roman et al., 1993; Breitburg et al., 1994, 1997; Keister, 1996; Stalder and Marcus, 1997].

Because outbreaks of jellyfish pose public health risks and can impact tourism and fisheries, considerable attention has been directed towards understanding the causes and effects of such outbreaks. There is some evidence, and much speculation, that jellyfish populations are increasing [Mills, 1995; Arai, 1997; Anderson and Piatt, 1999; Brodeur et al., 1999]. Ironically, human activities may promote outbreaks in jellyfish populations. Legović [1987] concluded that jellyfish populations increase as a consequence of nutrient enrichment, increased mortality rates of competitors (e.g., from harvesting of zooplanktivorous fishes) and predators, and increased survival rates at warmer winter sea water temperatures. Some combination of these changes has occurred in nearly every coastal ecosystem.

Trophic analyses have indicated that jellyfish, when numerous, can regulate plankton dynamics and reduce planktivorous fish populations during summer in Chesapeake Bay [Baird and Ulanowicz, 1989]. Jellyfish potentially have far-reaching effects on fish populations because jellyfish consume the eggs and larvae of fishes as well as the prey of fish larvae and zooplanktivorous fish species. For example, predation by the ctenophore, Mnemiopsis leidyi, has been blamed for the demise of fisheries in the Black and Azov Seas [e.g., Volovik et al., 1993]. Alternatively, over-exploitation of fish stocks such as anchovies and herrings, which has occurred there and in many other areas, may increase foods available to zooplanktivorous gelatinous species and result in outbreaks of their populations [Shiganova, 1998]. Increases in jellyfish populations may inhibit recovery of damaged fish stocks.

Jellyfish biomasses are highest in coastal environments, some of which historically may have experienced hypoxia. The summertime occurrence of vertical stratification and hypoxia coincides with the peak production of many coastal jellyfish species. Many species aggregate along density discontinuities in the water column [e.g., Arai, 1992; Graham, 1994], where primary and secondary production is enhanced [Franks, 1992; Tiselius, 1994]. Therefore, some jellyfish species might be expected to tolerate and exploit hypoxic waters. We review existing evidence that many predaceous jellyfish species appear to be tolerant of low dissolved oxygen. All of the preceding characteristics of jellyfish may lead to their predominance over fishes in eutrophic coastal waters.

Distributions of Jellyfish Relative to Dissolved Oxygen

Chesapeake Bay

During summer, dissolved oxygen concentrations in sub-pycnocline waters in the mainstem Chesapeake Bay and its tributaries vary spatially and temporally, ranging from 0 mg l^{-1} (anoxic) to near saturation levels depending on water column depth, local wind conditions, and proximity to freshwater inflows and the influence of the Atlantic Ocean [e.g., Breitburg, 1990; Sanford et al., 1990; Boicourt, 1992]. The most severe and seasonally persistent areas of low oxygen occur in the mesohaline reaches of the mainstem bay and tributaries. The deepest areas of the mesohaline mainstem Chesapeake Bay can remain severely hypoxic or anoxic for extended periods of time [e.g., Sanford et al., 1990; Boicourt, 1992]. Shallower sub-pycnocline and pycnocline depths in the mainstem and tributaries exhibit moderate to severe oxygen depletion, which is periodically disrupted by winds and tides, and also varies among years. In some areas, the oxygen-depleted bottom layers occasionally intrude into nearshore shallow waters because of winds and lateral internal tides [Breitburg, 1990; Sanford et al., 1990]. Annual variation in the extent and severity of hypoxia is directly related to the volume of freshwater entering the bay in the spring [Boicourt et al., 1999]. The severity and extent of hypoxia has increased due to a combination of excess nutrient loading from natural and anthropogenic sources, deforestation, and, possibly, the substantial decrease in top-down control of primary production due to over harvesting and disease mortality of oysters [Taft et al., 1980; Officer et al., 1984; Newell, 1988; Cooper and Brush, 1993].

Sampling of the surface, bottom and pycnocline layers of the water column by Tucker trawl indicates that the vertical distributions of ctenophores and medusae in the mesohaline Patuxent River, a tributary of Chesapeake Bay, vary with bottom dissolved oxygen concentrations [Keister et al., in review; Kolesar et al., unpublished data]. The lobate ctenophore, Mnemiopsis leidyi, was rare or absent in samples from bottom waters with 0.2 mg l^{-1} dissolved oxygen, but was found at higher densities in the hypoxic bottom layer than in surface or pycnocline waters when bottom dissolved oxygen concentrations were 1.3 and 2.3 mg l^{-1} (Fig. 1A). Ctenophore density averaged 54 ind. m^{-3} in the six bottom layer samples taken at dissolved oxygen concentrations of 1.3 to 2.9 mg l^{-1}. By contrast, Chrysaora quinquecirrha medusae showed no preference for hypoxic bottom waters. The proportional density (i.e., the proportion of the population that would be found in a particular layer, given equal volumes of surface, pycnocline, and bottom waters) of the C. quinquecirrha medusa population in the bottom layer increased from 0% at 0.0 mg l^{-1} to 40% when bottom dissolved oxygen concentrations reached 4 mg l^{-1} (Fig. 1B). At dissolved oxygen concentrations $\geq$ 4 mg l^{-1}, medusa densities were similar throughout the water column. Similarities between medusae and copepod distributions suggest that the vertical distribution of C. quinquecirrha may more closely reflect that of their zooplankton prey than the sensitivity of the medusae to low dissolved oxygen concentrations. Because of the low abundance of C. quinquecirrha in hypoxic bottom waters, hypoxia in tributaries may provide a refuge for M. leidyi from its most important summer predator in the Chesapeake Bay system. In contrast to the Patuxent River study, Purcell et al. [1994a] found more medusae and ctenophores above the

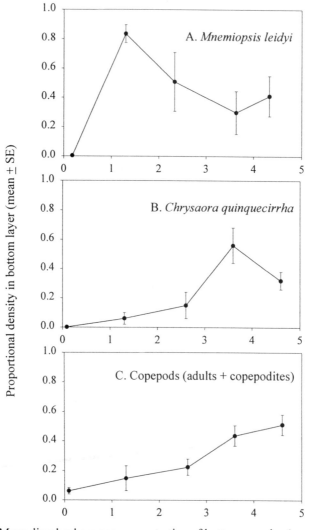

Figure 1. Vertical distributions of (A) *Mnemiopsis leidyi*, (B) *Chrysaora quinquecirrha*, and (C) copepods, relative to dissolved oxygen concentrations in bottom waters of the Patuxent River. Data are presented as proportional densities in the bottom layer, i.e., the proportion of individuals that would be found in the bottom layer given equal volumes of surface, pycnocline, and bottom water. Data included consist of all 1992-1993 samples [Keister et al., in review], and 1998 samples [Kolesar et al., unpublished data] in which at least 10 individuals were collected in a matched set of surface, pycnocline, and bottom tows with a 1 m² Tucker trawl. Both day and night samples are included. Position along the x-axis indicates the mean bottom dissolved oxygen concentration at which samples were taken. Between 2 and 14 samples met the inclusion criteria at each dissolved oxygen interval.

pycnocline than below it in the mainstem Chesapeake Bay regardless of bottom dissolved oxygen concentrations, which suggests that both species may avoid the deeper hypoxic waters where density stratification is stronger and more persistent or where the surface layer is thicker.

Recent laboratory experiments have shown direct evidence of a behavioral response to low oxygen by jellyfish. The positions of *Mnemiopsis leidyi* ctenophores relative to low dissolved oxygen concentrations in 1.2 m-deep cylindrical containers with stratified water columns revealed strong, statistically significant, avoidance of both low but non-lethal dissolved oxygen levels (0.5 mg l^{-1}) and high surface temperatures ($\geq 29°C$) [E. McLaughlin, Breitburg, Decker, and Purcell, unpublished data]. Preliminary results also suggested that the ctenophores avoided moderately hypoxic bottom dissolved oxygen concentrations (1.5 mg l^{-1}) in the absence of their predators and prey. Breitburg has suggested that vertical distributions of *M. leidyi* in nature may result from their avoidance of hypoxic bottom waters and high surface temperatures, and predation by *Chrysaora quinquecirrha* medusae.

Northern Gulf of Mexico

Hypoxia in the northern Gulf of Mexico is strongly related to seasonal stratification induced by freshwater outflow, high algal and zooplankton production in near-surface waters, and deposition of organic particles to sub-pycnocline depth. Extensive hypoxia in the northern Gulf occurs from May to September [Rabalais et al., 1991]. Although predictable from year to year, northern Gulf of Mexico hypoxia varies widely in magnitude and extent as a function of river discharge. The greatest extent of hypoxia is near the Mississippi and Atchafalaya Rivers, however, seasonal hypoxia also often develops proximally to the numerous, river-dominated estuaries along the northern gulf. The region of most concern for ecological change is the so-called "Dead Zone" where dissolved oxygen concentrations < 2 mg l^{-1} cover 8,000 to 18,000 km^2 of the continental shelf off the Louisiana and northeastern Texas coasts [Rabalais et al., 1998]. The broad, shallow continental shelf of the northern Gulf of Mexico restricts extensive hypoxia to bottom waters of 5 to 30 m depth [Rabalais et al., 1994]. Hypoxia may at times encompass as much as 80% of the water column in affected areas [Rabalais et al., 1994]. Human activities (e.g., increased agrochemical usage) in the Mississippi River watershed over the past 50 years may be major factors contributing to hypoxia [Rabalais et al., 1996; other citations in this volume].

We examined trends in jellyfish distribution around hypoxic areas in the northern Gulf of Mexico by using trawl by-catch data. Since 1982, systematic living resource surveys by the National Marine Fisheries Service (NMFS) Southeast Area Monitoring and Assessment Program (SEAMAP) have been conducted twice annually, in summer and fall. SEAMAP has occupied 250 to 700 trawl and environmental stations during 4 to 6 week periods. Large jellyfish are recovered intact from these samples and have been counted as part of the by-catch since 1982. The 1-m high trawl net opening remains open during deployment and recovery, therefore the whole water column has been sampled at each station. Details of the trawling and hydrographic sampling protocols are presented in Craig et al. [this volume]. By use of SEAMAP records since 1987, when nearshore stations were added, we compared the presence of *Chrysaora quinquecirrha* medusae with the extent of hypoxia in the northern Gulf of Mexico, as measured at the time of the trawl by a dissolved oxygen sensor. Despite strong historical evidence that *C. quinquecirrha* was limited to very nearshore estuarine areas before the 1980s [Gunter, 1950; Reid, 1955; Phillips, 1972], the recent surveys show a range extension of *C. quinquecirrha* over the Texas-Louisiana continental shelf where hypoxic conditions are prevalent during the summer (Fig. 2).

The summertime peak abundance of *Chrysaora quinquecirrha* medusae coincides with peak seasonal hypoxia in the northern Gulf of Mexico. The frequency of *C. quinquecirrha* occurrence, measured as presence in a trawl, has increased from about 5% of all SEAMAP stations sampled between 1987 and 1991 to 13 to 18% since 1992. This increased catch frequency coincides with increased numbers of stations with hypoxic bottom waters (Fig. 3). For the 11-year period, the frequency of hypoxia explains almost 60% of the variation in the catch frequency of *C. quinquecirrha* (data not shown; $r^2 = 0.58$; $p < 0.005$). Although we do not fully understand the role of hypoxia in determining jellyfish distributions, it is apparent that the range expansion of *C. quinquecirrha* away from shore has not been restricted by hypoxic bottom water in the northern Gulf of Mexico. The percentages of stations with *C. quinquecirrha* and hypoxic bottom waters have increased from 0% in 1987 to 25% in 1997. In the entire SEAMAP region, mean dissolved oxygen concentrations at the bottom of the water column containing *C. quinquecirrha* have dropped significantly from 6.3 mg l^{-1} in 1987 to 3.4 mg l^{-1} in 1997 (ANOVA; $p = 0.05$).

A study of fine-scale vertical distribution of jellyfish using *in situ* profiling videography constitutes the only information on coastal jellyfish distribution relative to oxygen in the Gulf of Mexico to date [Graham and Cinkovich, unpublished data]. During summer 1998, five vertical profiles were made in a region of locally persistent hypoxic bottom water southeast of Mobile Bay, Alabama. Hypoxic conditions encompassed 13 to 30% of the shallow (7 to 13 m), strongly stratified water column. Most jellyfish observed in these profiles were *Chrysaora quinquecirrha* medusae and *Mnemiopsis leidyi* ctenophores. Fewer than 6% of the total observations of the medusae and ctenophores were in the hypoxic bottom layer. Of those occurring in hypoxic water, the frequency of *M. leidyi* (0.072) was double the frequency of *C. quinquecirrha* (0.036). *Mnemiopsis leidyi* occurred significantly deeper than *C. quinquecirrha* in these profiles (Student's t-test, $p < 0.001$; Fig. 4). The mean dissolved oxygen concentration where these species were observed was significantly lower for *M. leidyi* (4.4 mg l^{-1}) than for *C. quinquecirrha* (5.0 mg l^{-1}, Student's t-test, $p < 0.05$). The vertical distributions of *M. leidyi* relative to their predator *C. quinquecirrha* are similar in the Gulf of Mexico and Chesapeake Bay.

Black Sea

The Black Sea is a strongly stratified system, characterized by 5 layers, and is permanently anoxic below 200 m depth [Vinogradov et al., 1985]. The oxygenated surface layer has lower salinity, and varies in depth from 80 to 120 m in the central portion to 160 to 200 m near shore. Below the surface oxygenated layer is the oxycline, where oxygen decreases sharply to 1.2 to 0.8 mg l^{-1}. Between the oxycline and the upper boundary of the hydrogen sulfide zone is an oxygen-deficient layer averaging 30 to 35 m in thickness with dissolved oxygen concentrations decreasing to 0.6 to 0.4 mg l^{-1}. In the fourth layer, the "C zone", which is $\geq$ 60 m thick, < 0.4 mg l^{-1} dissolved oxygen is found with hydrogen sulfide. Below this zone, there is no detectable oxygen and high levels of hydrogen sulfide persist.

Vinogradov et al. [1985] reported the vertical distribution of zooplankton relative to dissolved oxygen concentrations from water bottle and net samples, as well as visual counts from the *Argus* submersible in April to May, 1984. Where dissolved oxygen concentrations were 0.7 to 1.0 mg l^{-1}, mesozooplankton were almost absent, except for polychaete larvae and

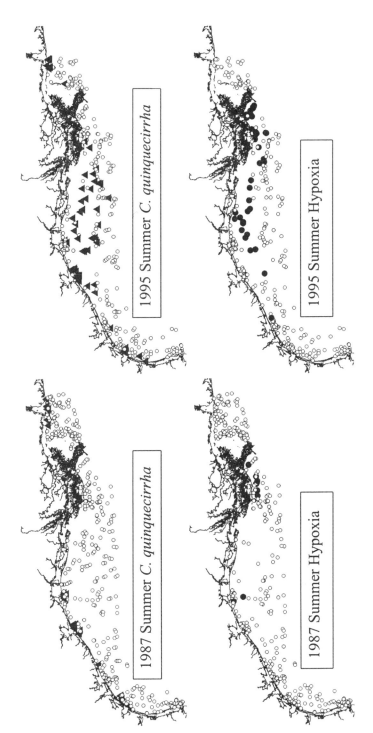

Figure 2. Comparison of summertime distribution of *Chrysaora quinquecirrha* medusae and bottom dissolved oxygen < 2 mg l⁻¹ between 1987 and 1995 in the Gulf of Mexico. Locations of *C. quinquecirrha* in SEAMAP trawls are marked by solid triangles in the upper panels, open circles indicate stations where they were absent. Presence of hypoxia is indicated by solid circles in the lower panels; open circles indicate non-hypoxic bottom waters.

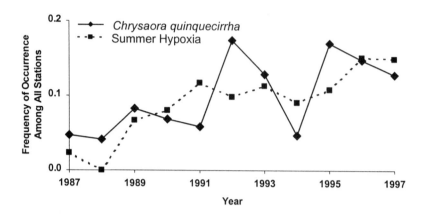

Figure 3. An 11 year trend in the occurrence of jellyfish, *Chrysaora quinquecirrha*, and hypoxic waters (< 2 mg l⁻¹) in the northern Gulf of Mexico since 1986. Frequencies are the fraction of all summertime SEAMAP trawls with *C. quinquecirrha* medusae in the catch. Frequencies of hypoxia in bottom waters are reported similarly.

planulae of the scyphomedusan, *Aurelia aurita*, which occurred at densities of "dozens m^{-3}" in this layer [Vinogradov et al., 1985]. High densities of vertically-migrating mesozooplankton (especially stages V and VI of the copepod, *Calanus helgolandicus*, the ctenophore, *Pleurobrachia pileus*, and the chaetognath, *Sagitta setosa*, were layered just above the 0.7 to 1.0 mg l⁻¹ layer during the daytime. Within the oxycline at dissolved oxygen concentrations of 2 to 4 mg l⁻¹, the abundances of large (1.0 to 1.6 cm) *P. pileus* greatly increased. This ctenophore layer was 2 to 6 m thick, and densities reached 70 ind. m^{-3}. Densities of *C. helgolandicus*, which were eaten by the ctenophores, increased in this layer. Maximum densities of *S. setosa* (360 ind. m^{-3}) occurred several meters deeper. The high-density layers of copepods and chaetognaths dispersed toward the surface at night, but the vertical distribution of *P. pileus* did not change significantly by night.

Counter to the results of Vinogradov et al. [1985], Mutlu and Bingel [1999], who used an opening-closing net at 15 m depth intervals, reported that in August, 1993 and March, 1995, *Pleurobrachia pileus* migrated vertically at night into waters of higher dissolved oxygen concentrations. By day, *P. pileus* ctenophores were mostly in the lower levels of the pycnocline down to the anoxic hydrogen sulfide zone, with maximum abundances at 90 to 120 m depth (Fig. 5). By contrast, peak densities of the ctenophore, *Mnemiopsis leidyi*, which was introduced into the Black Sea in the early 1980s, were mostly in the upper 20 m, well above the oxycline/pycnocline on the same dates [Mutlu, 1999]. In Chesapeake Bay and the Gulf of Mexico, *M. leidyi* encounter hypoxic water at similar depths.

Fjords and Marine Lakes

Many fjords and other partially enclosed coastal inlets have restricted water exchange with the ocean, which leads to persistent stratification and the development of hypoxia in the lower water column. A few studies document jellyfish in those hypoxic waters. In Oslofjord, Norway, the hydromedusae *Aglantha digitale* and *Rathkea octopunctata* occurred in greatest

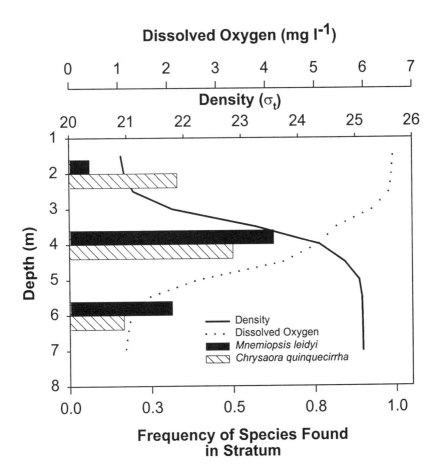

Figure 4. Water column density and oxygen profiles superimposed on depth distributions of the ctenophore, *Mnemiopsis leidyi*, and its scyphomedusan predator, *Chrysaora quinquecirrha*, in the northern Gulf of Mexico. Three sets of bars represent 10 min horizontal transects with an *in situ* underwater video camera at 2, 4, and 6 m depth. Data are from a typical station.

numbers in the most polluted areas, where bottom waters were anoxic [Beyer, 1968]. Smedstadt [1972] illustrated that *A. digitale* medusae were concentrated in the sampling intervals just above dissolved oxygen concentrations of 0.5 mg l⁻¹. In Lindåspollene, another Norwegian fjord, the highest concentration of > 1000 μm zooplankton, which was comprised mainly of *A. digitale* and the chaetognath, *Sagitta elegans*, occurred in waters with < 0.2 mg l⁻¹ dissolved oxygen [Magnesen, 1989]. Hoos [in Davis, 1975] found four species of hydromedusae (*A. digitale, Aequorea* sp., *Phialidium gregarium*, and *Aegina* sp.) and two species of ctenophores (*Pleurobrachia* sp., *Beroe cucumis*) in dissolved oxygen concentrations of 0.6 mg l⁻¹ in Saanich Inlet, a fjord in British Columbia, Canada. By contrast, observations from the *PISCES IV* submersible indicated that *A. digitale, Aegina citrea* and the ctenophore, *Bolinopsis infundibulum*, were concentrated at 100 to 150 m depth, just above water with < 2 mg l⁻¹ dissolved oxygen, and were not seen in the hypoxic waters

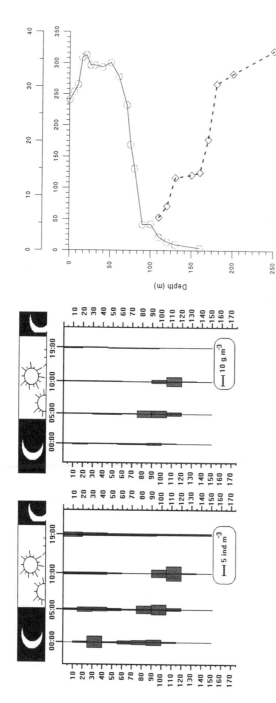

Figure 5. Day and nighttime vertical distributions (in ind. m^{-3} and g m^3) of the ctenophore, *Pleurobrachia pileus* in the Black Sea relative to dissolved oxygen (circles in µm l^{-1}, lower axis) and hydrogen sulfide (diamonds, upper axis) concentrations in August, 1993 [from Mutlu and Bingel, 1999].

of Saanich Inlet [Mackie and Mills, 1983]. In a restricted Irish inlet, two species of hydromedusae, *Bougainvillia britannica* and *B. principis*, which had preferred depths below 20 m when bottom waters were not anoxic, showed clear avoidance of anoxic waters below 30 m depth, changing their distributions to become concentrated at the oxycline where dissolved oxygen concentrations were 0.2 to 1.2 mg l⁻¹ [Ballard and Myers, 1996].

Some medusae apparently make excursions into anoxic waters in marine lakes. In Brooks Lake, Antarctica, hydromedusae, *Rathkea lizzoides*, occurred in high densities (mean 230 m⁻³) in the oxylimnion beginning about 35 cm above the sharp oxycline at 8.15 m [Bayly, 1986]. Medusae also occurred 10 cm into the anoxic zone, but may have spent only short periods of time there. The authors speculated that the medusae were feeding on flagellates that were abundant in that layer. In Eil Malk Jellyfish Lake, Palau, populations of *Aurelia aurita* and *Mastigias* sp. medusae were observed by SCUBA divers above the chemocline at 15 m depth, which separated the oxygenated surface layer from the anoxic hydrogen sulfide zone [Hamner et al., 1982]. *Aurelia aurita* performed a typical diel vertical migration in the lake, rising close to the surface at night, but *Mastigias* sp. exhibited a reverse migration pattern, dispersing at night from near surface to throughout the surface layer and 1 m into the anoxic zone. The authors speculated that this behavior enabled uptake of high concentrations of dissolved nutrients that occurred only in the anoxic zone. The nutrients presumably would promote growth of the symbiotic dinoflagellates (zooxanthellae) that are present in *Mastigias* sp., but not in *A. aurita*.

Oxygen Minimum Layers

Oxygen minimum layers, defined here as having dissolved oxygen concentrations < 1.0 mg l⁻¹, occur as permanent features in mesopelagic waters of all the major oceans [Kamykowski and Zentara, 1990]. The depletion of oxygen results from an interplay of biological activity and water circulation. As organic material produced in surface waters sinks beneath the photic zone, it is oxidized by animal respiration and bacterial decomposition. Weak physical mixing processes at intermediate depths limit the renewal of oxygen. The vertical extent of oxygen minimum layers, from about 60 m to nearly 1500 m, differs on basin and regional scales. In this midwater regime, temperatures vary from 4 to 20°C and salinities of 34.5 to 36 are common. The most expansive oxygen minimum layer is in the eastern tropical Pacific, between 27° N to 27° S and extends westward at 10° N and 10° S to 170° W and 130° E, respectively [Reid, 1965]. Other extensive oxygen minimum layers are in the central Arabian Sea [Swallow, 1984] and in the Atlantic Ocean off western Africa.

Knowledge about the biodiversity, relative abundance, vertical distribution, and trophic roles of gelatinous zooplankton within oxygen minimum layers has been obtained primarily with nets. Unfortunately, most of the specimens collected with such gear are ignored or discarded because their soft bodies are easily damaged and preserve poorly. In most publications, gelatinous species have been lumped into major taxa (e.g., ctenophores, siphonophores, medusae, and salps) and not identified further. However, it is clear from *in situ* research with undersea vehicles that many midwater gelatinous species are undescribed and that some species are often numerous within oxygen minimum layers [e.g., Madin and Harbison, 1978; Pugh and Youngbluth, 1988; Mills et al., 1996].

In the eastern Pacific, ctenophores, siphonophores, and medusae have been observed from *Mir* submersible dives throughout the oxygen minimum layer in the region of the Costa Rica dome [Vinogradov et al., 1991]. Wishner et al. [1995] reported an abundant stock of medusae at 775 to 880 m near the Volcano 7 seamount. Medusae and siphonophores occur throughout the oxygen minimum layer in the Arabian Sea, but collections with open-closing nets indicate that these taxa are most abundant in the uppermost 150 m of the oxygen minimum layer [Madhupratap and Haridas, 1990; Bottger-Schnack, 1996; Herring et al., 1998]. Jellyfish and salps constituted 4 to 75% of the biomass of macrozooplankton and micronekton collected in trawls taken in that mesopelagic realm [Gjosaeter, 1984]. The vertical distributions of gelatinous zooplankton in the oxygen minimum layers off western Africa are unknown [F. Pages, personal communication].

   The *in situ* mesopelagic research program at the Monterey Bay Aquarium Research Institute (MBARI) is providing a wealth of new information about the biodiversity and ecology of midwater macrozooplankton and micronekton in the oxygen minimum layer (typically from 500 to 850 m depth, but occasionally deeper) of Monterey Bay, California. The dissolved oxygen sensor on the *ROV Ventana* has allowed fine scale oxygen measurements to be linked with observations of pelagic fauna over the past decade [Robison, 1993].

   Pelagic cnidarians are diverse and numerous in Monterey Bay, with many taxa found in the oxygen minimum layer. Of the siphonophores, *Apolemia* sp., which is at least 20 m in length, occurs mainly from 500 to 750 m at dissolved oxygen concentrations of 0.2 to 0.9 mg l^{-1} [Youngbluth et al., 1994]. Ongoing studies indicate that this species eats various crustaceans, medusae, siphonophores, ctenophores, and fishes. The occurrence of *Apolemia* sp. in low oxygen waters at 400 to 480 m depths off southern California where diapausing stages of *Calanus pacificus californicus* are concentrated, suggests that this siphonophore may be an important predator of this copepod [Alldredge et al., 1984]. Other siphonophore species have not been observed often in the oxygen minimum layer, however, the second highest abundance (65 ind. 100 m^{-3}) of the siphonophore, *Nanomia bijuga*, quantified along midwater transects was at 600 m [Robison et al., 1998]. Thus, species that are not regular inhabitants of hypoxic environments occur occasionally in high abundance within the oxygen minimum layer. Conversely, mass abundances of typical oxygen minimum layer residents, *Apolemia* sp. and the medusa *Periphylla periphylla,* have appeared in coastal oxygenated surface waters in Norway [Fosså, 1992; Båmstedt et al., 1998].

   Hydromedusae have been the numerically predominant medusae observed in the oxygen minimum layer of Monterey Bay. During 1989 to 1995, 37% of 13,700 individuals in the eleven most numerous species of hydromedusae occupied depths that extended into the oxygen minimum layer. All occurrence data presented here are normalized by the time the ROV spent at each depth, giving numbers h^{-1} as a proxy of abundance. The narcomedusan, *Aegina citrea*, has been the most numerous medusa, sometimes occurring at 120 ind. h^{-1} [Raskoff, 1998]. Over 66% of *A. citrea* were observed in the oxygen minimum layer (Fig. 6). Most of the hydromedusae observed in the oxygen minimum layer, including *A. citrea, Halicreas minimum, Haliscera conica, Solmissus incisa, Solmissus marshalli,* and *Solmundella bitentaculata*, are very active swimmers, rarely seen drifting passively [Raskoff, unpublished data]. The lights and mechanical disturbances from the ROV did not appear to influence the behavior of most jellyfish in the oxygen minimum layer.

   In Monterey Bay, most mesopelagic scyphomedusae observed with the ROV from 1989 to 1995 were in the oxygen minimum layer. Of the species encountered (coronates, *Atolla vanhoeffeni, A. wyvillei, Periphylla periphylla,* and *Paraphyllina* sp., and semaeostomes, *Deepstaria enigmatica* and *Poralia* sp.), 63% of the individuals were in the oxygen minimum layer. Over half of the coronates, *Atolla* spp. and *P. periphylla,* have been observed in waters with dissolved oxygen concentrations < 1.0 mg l^{-1} [D. Murray, personal communication].

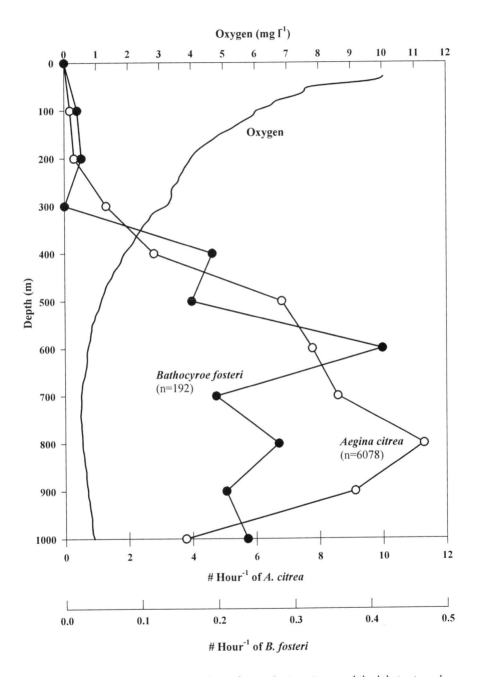

Figure 6. Vertical distribution of the hydromedusan, *Aegina citrea*, and the lobate ctenophore, *Bathocyroe fosteri*, relative to dissolved oxygen concentrations during a three year period in Monterey Bay, California, which has a distinct midwater oxygen minimum layer. Oxygen data were averaged from 12 typical ROV dives at the MBARI midwater site [Robison et al., 1998].

Many of the ctenophores in Monterey Bay have been observed in the oxygen minimum layer. Over 32 % of the ctenophores observed with the ROV between 1991 and 1995 occurred in waters with dissolved oxygen concentrations < 1.0 mg l^{-1}. Among the ctenophores found in the oxygen minimum layer, 72% were lobates and 27% were cydippids. *Bolinopsis infundibulum* accounted for 65% of the lobate ctenophores in the oxygen minimum layer. *Bathocyroe fosteri* is another lobate species that has a distribution centered in the oxygen minimum layer, with over 72% of those observed found between 500 and 900 m depth (Fig. 6) [Raskoff, unpublished data]. Beroid and cestid ctenophores also occur in the bay, but are uncommon in the oxygen minimum layer.

Effects of Low Dissolved Oxygen on Survival of Jellyfish

To our knowledge, the only pelagic cnidarians and ctenophores whose tolerance to low oxygen has been tested experimentally are *Chrysaora quinquecirrha* medusae and *Mnemiopsis leidyi* ctenophores from Chesapeake Bay. In 96-h laboratory experiments, medusa survival was 100% at dissolved oxygen concentrations ≥ 1.0 mg l^{-1} [Houde and Zastrow, unpublished data]. At 0.5 mg l^{-1}, all *C. quinquecirrha* survived to 48 h, but died by 72 h. Even at dissolved oxygen concentrations of 0.3 mg l^{-1}, 2 of 3 medusae tested survived 24 h. *Mnemiopsis leidyi* appears to be even more tolerant of exposure to low dissolved oxygen concentrations than *C. quinquecirrha*. In similar experiments, 96-h survival of ctenophores was 100% at dissolved oxygen concentrations ≥ 0.5 mg l^{-1} [Breitburg, Decker, and Purcell, unpublished data]. Ctenophores have not been tested at dissolved oxygen concentrations below 0.5 mg l^{-1}.

Low dissolved oxygen concentrations in bottom waters may limit populations of scyphomedusan and hydromedusan species that exhibit alternation of an asexually-multiplying benthic generation (scyphistoma or hydroid) with the sexual planktonic medusa generation, as suggested by Benović et al. [1987]. Unlike the swimming medusae, the attached perennial polyps would be unable to escape from hypoxic bottom waters. In 1910, forty species of hydromedusae were reported from the Northern Adriatic, which has become increasingly eutrophic with marked decreases in near-bottom dissolved oxygen concentrations since 1965; however, by 1984-1985, only 25% of the species remained [Benović et al., 1987]. The losses were of species with benthic hydroids. The lowest numbers of hydromedusan species were adjacent to the Po River plume, where high concentrations of nutrients enter the Adriatic Sea.

By contrast, taxa that are holoplanktonic with direct development of the fertilized eggs (siphonophores, ctenophores, and some medusae, such as *Pelagia noctiluca*) may not be as vulnerable to low dissolved oxygen concentrations as are taxa with benthic stages. For example, the ability to reproduce in the water column may have allowed two species of hydromedusae to survive where bottom waters were depleted of oxygen by domestic sewage in Oslo Harbor; *Rathkea octopunctata* asexually buds medusae from the manubrium, and *Aglantha digitale* is holoplanktonic [Beyer, 1968; Smedstad, 1972].

The tolerance of benthic scyphistomae to low oxygen levels is poorly documented. Cargo and Schultz [1966] reported that the polyps of *Chrysaora quinquecirrha* occurred above 11 m depth in Chesapeake Bay, which corresponds with the depth of persistent seasonal hypoxia as well as the maximum depth of available hard substrate (oyster reefs). Prolonged exposure to low oxygen may kill the benthic stages, but this has not been tested. The polyps encyst in response to temperature and salinity shocks [Cargo and Schultz, 1966, 1967], but their response to oxygen stress has not been tested. The presence of the enzyme phosphoenolpyruvate carboxykinase in scyphistomae of *C. quinquecirrha* may indicate the presence of an anaerobic metabolic pathway [Lin and Zubkoff, 1977]. The planula larvae of the scyphomedusan, *Cyanea capillata*, showed abnormal attachment orientation in reduced dissolved oxygen concentrations and high mortality at 0% oxygen [Brewer, 1976].

Effects of Low Dissolved Oxygen on Trophic Interactions of Jellyfish

The effects of low dissolved oxygen concentrations on the prey of jellyfish, on the jellyfish themselves, and on their predators, may have pronounced effects on trophic interactions in the water column. We discuss the effects of low dissolved oxygen concentrations on two of the main prey of jellyfish, copepods and ichthyoplankton. The only data on the effects of low dissolved oxygen concentrations on trophic interactions between jellyfish and their prey, of which we are aware, are from Chesapeake Bay [Breitburg et al., 1994, 1997, 1999].

Copepods are the main prey items of most species of jellyfish, with varying proportions of other zooplankton taxa and ichthyoplankton being eaten depending on their availability [reviewed by Purcell, 1997]. The calanoid copepod, *Acartia tonsa*, is the predominant zooplankton species in the mesohaline Chesapeake Bay during the summer, reaching densities of 200 nauplii and copepodites l^{-1}. It represents 50 to 90% of the prey items in gut contents of resident medusae and ctenophores [Purcell, 1992; Purcell et al., 1994a,b]. In the mesohaline portion of Chesapeake Bay during 1987 and 1988, *Chrysaora quinquecirrha* medusae consumed up to 94% d^{-1} of the copepods in the Broad Creek and Tred Avon tributaries, but predation by medusae and ctenophores did not appear to control copepod populations in the mainstem bay in those years [Purcell, 1992; Purcell et al., 1994b]. Copepods are more sensitive to low dissolved oxygen concentrations than are medusae or ctenophores in Chesapeake Bay. Low dissolved oxygen concentrations affect the survival and distribution of copepods like *A. tonsa*, which are rare in severely hypoxic bottom waters [Roman et al., 1993; Keister et al., in review]. *Acartia tonsa* produces negatively buoyant eggs that can sink into hypoxic waters [Roman et al., 1988]. In experiments with *A. tonsa* in Chesapeake Bay, survival of copepods and the hatching rate of their eggs were low at dissolved oxygen concentrations < 2 mg l^{-1} [Roman et al., 1993].

Consumption of ichthyoplankton by gelatinous zooplankton also can be very important. Some species show strongest selection for fish eggs and larvae above all other prey types [reviewed by Purcell, 1997]. Estimates of predation on ichthyoplankton are very high, with scyphomedusae and ctenophores consuming ≤ 54% d^{-1} of the fish eggs and larvae in Chesapeake Bay in July, 1991 [Cowan and Houde, 1993; Purcell et al., 1994a; reviewed by Purcell, 1997]. Naked goby (*Gobiosoma bosc*) and bay anchovy (*Anchoa mitchilli*) are the most abundant fish larvae in mesohaline Chesapeake Bay and its tributaries during summer when low dissolved oxygen concentrations occur [e.g., Olney, 1983; Shenker et al., 1983; Keister et al., in review]. Pelagic fish eggs and larvae are eaten by *Chrysaora quinquicirrha* and *Mnemiopsis leidyi* [Cowan and Houde, 1993; Purcell et al. 1994a; Breitburg et al., 1997]. Larvae of both fish species are rare in the bottom waters where dissolved oxygen concentrations were < 2 mg l^{-1} [Keister et al., in review].

Laboratory and mesocosm experiments that measured predation rates of *Chrysaora quinquecirrha* showed that low dissolved oxygen concentrations can dramatically alter trophic interactions [Breitburg et al., 1994, 1997]. Predation on naked goby larvae is approximately 4 times higher at 1.5 mg l^{-1} than under saturated dissolved oxygen conditions.

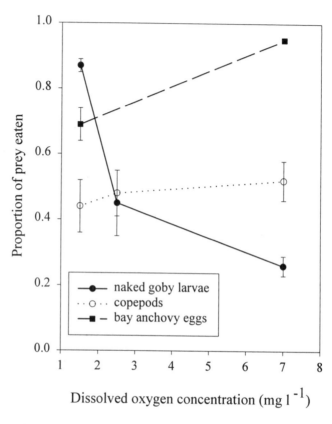

Figure 7. Effect of low dissolved oxygen on feeding by *Chrysaora quinquecirrha* medusae on fish eggs and larvae, and copepods. Experiments using fish larvae as prey were conducted in 1 m³ mesocosms, and experiments using fish eggs and copepods were conducted in 80 l cylindrical containers [from Breitburg et al., 1997].

By contrast, predation on copepods (primarily *Acartia tonsa*), is unaffected by dissolved oxygen concentrations ≥ 1.5 mg l⁻¹ and predation on fish eggs (*Anchoa mitchilli*) decreases at low dissolved oxygen concentrations (Fig. 7). Furthermore, the oxygen concentrations that result in increased predation on fish larvae by *C. quinquecirrha* decrease predation on fish larvae by juvenile and adult fishes (*Morone saxatilis* and *Gobiosoma bosc*, respectively) [Breitburg et al., 1994, 1997].

Breitburg et al. [1997] suggested that differences in physiological tolerance and behavioral responses between among predators and prey, and the ability of prey to escape from predator encounters under high dissolved oxygen concentrations, are likely to determine the net effect of low dissolved oxygen concentrations on trophic interactions in coastal systems. In the above experiments, fish larvae were more sensitive to hypoxia than were the scyphomedusae, and the escape swimming speed of larvae was reduced under hypoxic conditions. As a result, escape frequency of larvae from medusae probably was reduced, resulting in the observed increase in predation rates. By contrast, fish eggs have no escape behavior, and are unaffected by low dissolved oxygen exposure. Consequently, decreased predation on fish eggs at low dissolved oxygen concentrations may be due solely to reduced predator activity or capabilities.

Discussion and Conclusions: Ecosystem Implications

Evidence indicates that many species of pelagic cnidarians and ctenophores live in hypoxic waters. Although most jellyfish species occur in waters where dissolved oxygen concentrations are > 2 mg l^{-1}, some species accumulate at hypoxic interfaces above anoxic waters, in hypoxic bottom waters, in oxygen minimum zones, or migrate into hypoxic/anoxic waters on a diel basis. In addition, populations of coastal jellyfish have increased, rather than diminished, with increasing frequency of hypoxia in sub-pycnocline waters. Some species have been shown experimentally to be tolerant of low dissolved oxygen concentrations.

Survival in hypoxic waters appears to have ecological costs and advantages. Jellyfish feeding rates may be low in hypoxic bottom waters, as suggested by the finding that non-gelatinous prey species in Chesapeake Bay are less tolerant of low dissolved oxygen concentrations than their jellyfish predators. Keister et al. [in review] found that when bottom dissolved oxygen concentrations were < 2 mg l^{-1}, the prey of ctenophores (copepods and fish larvae) were more abundant within and above the pycnocline than below it. This cost may be offset by reduced predation on jellyfish in hypoxic environments, as suggested by Breitburg et al. [1999] and Keister et al. [in review]. For example, fishes are much less tolerant of low dissolved oxygen concentrations than at least some jellyfish species, and actively avoid hypoxic waters [Breitburg et al., this volume]. Several fish species eat jellyfish [e.g., Harbison, 1993], and juvenile fish (e.g., *Peprilus alepidotus*) associate with medusae and consume parts of them [Mansueti, 1963]. *Peprilus alepidotus* also eats ctenophores [Oviatt and Kremer, 1977]. Often other jellyfish species are important predators of other jellyfish [reviewed by Purcell, 1991]. For example, *Chrysaora quinquecirrha* medusae eat *Mnemiopsis leidyi*, and at times eliminate them from tributaries of Chesapeake Bay [Purcell and Cowan, 1995]. When bottom waters are hypoxic, *C. quinquecirrha* medusae generally occur at shallower depths and higher dissolved oxygen concentrations than *M. leidyi*. The ctenophores, therefore, may find a refuge from predation at dissolved oxygen concentrations lower than those at which the medusae are abundant [Keister et al., in review; Graham, unpublished data].

In coastal ecosystems impacted by human activities, many factors interact that may favor populations of jellyfish over fishes. Such factors include nutrient enrichment and resulting increases in available prey [Parsons et al., 1977] and in the frequency of hypoxic events [Boicourt et al., 1999], reduction of commercial species and resulting increases in zooplankton prey [Legović, 1987; Parsons, 1995; Pauly et al., 1998; Shiganova, 1998; Houde et al., 1999; Newell and Ott, 1999], and reduced water clarity, which could favor non-visual predators like jellyfish [Eiane et al., 1999]. Direct connections between human effects on estuarine systems and changes in jellyfish populations are difficult to document. For example, no data on *Chrysaora quinquecirrha* abundances exist before 1960 when Chesapeake Bay was already nutrient enriched, and there has been no general increase in medusa numbers since then [Cargo and King, 1990]. Similarly, there was general agreement by researchers that outbreaks of *Pelagia noctiluca* in the Mediterranean were not directly related to increases in nutrient loading, especially because the episodic phenomenon has been described several times over the last 200 years [reviewed in Purcell et al., 1999]. Nevertheless, seasonally high plankton biomass in eutrophic areas may result in greater jellyfish biomass. The lack of long-term data hinders evaluation of possible relationships between these various factors and jellyfish abundance.

Jellyfish may be superior competitors to fish in many ways. Unlike fish, they feed continuously and do not satiate at natural food concentrations [e.g., Purcell, 1992]; capture both small and large zooplankton [e.g., Purcell, 1992]; do not decrease feeding in turbid waters; eat their competitors, including ichthyoplankton and other jellies [e.g., Purcell, 1991; Purcell et al., 1994a]; and shrink rather than die when food is withheld [Hamner and Jenssen, 1974; Purcell, unpublished observations]. Additionally, their populations can increase rapidly, because most cnidarian species have both asexual and sexual reproduction, most ctenophores are hermaphroditic, and ctenophores and some medusae are holoplanktonic and have short generation times. We have shown here that many jellyfish are tolerant of low dissolved oxygen concentrations. By contrast, most fishes are impaired at the same dissolved oxygen concentrations [Davis, 1975; Breitburg et al., 1994, 1997, this volume].

Coronate scyphomedusae (e.g., *Atolla* spp. and *Periphylla periphylla*) from the oxygen minimum zone off California have high levels of a glycolytic enzyme lactate dehydrogenase, indicating the potential to function anaerobically [Thuesen and Childress, 1994]. Oxygen consumption rates of medusae did not decline dramatically with increasing depth of occurrence, in contrast with results for fishes and crustaceans [Bailey et al., 1994; Childress, 1995]. Metabolism of fishes and crustaceans are believed to decrease with depth because of reduced reliance on visual predation. Mesopelagic species living in hypoxic waters compensate for low dissolved oxygen concentrations by increased surface areas of gills, increased ventilation and circulation, and efficient extraction of DO [Childress, 1995]. Pelagic cnidarians and ctenophores from all depths respire across their entire surface, which is only one cell-layer thick, providing a large surface for oxygen exchange that may predispose them to tolerate low dissolved oxygen concentrations.

Tolerance of low dissolved oxygen concentrations may enable gelatinous species to inhabit extensive volumes of hypoxic water that exclude fish with high respiratory demands. This may lead to the predominance of gelatinous species over fish in regions affected by hypoxia. Comparisons of the effects of low dissolved oxygen concentrations on predation rates of *Chrysaora quinquecirrha* medusae and fish, as well as individual-based models that incorporated changes in predator capture rate, vertical distribution and larval growth rate, indicate the potential for low dissolved oxygen concentrations in bottom waters of stratified coastal systems to increase the importance of tolerant gelatinous predators relative to fish predators [Breitburg et al., 1994, 1997, 1999]. Thus, low dissolved oxygen concentrations could potentially alter carbon pathways in these systems.

Acknowledgments. Some data presented here for the first time are from research supported by the National Science Foundation (Grant Nos. OCE-9633607 and OCE-973341), Maryland Sea Grant College (Grant No. R/P-98-PD), NOAA Coastal Ocean Program funding to the COASTES project, NOAA West Coast National Undersea Research Center (Grant No. 97-0032), and the David and Lucile Packard Foundation through MBARI's Midwater Ecology Group. We thank Dr. V. S. Kennedy for editorial comments. Contribution numbers 235 (HPL) and 1318 (HBOI).

References

Alldredge, A. L., B. H. Robison, A. Fleminger, J. J. Torres, J. M. King, and W. M. Hamner, Direct sampling and *in situ* observation of a persistent copepod aggregation in the mesopelagic zone of the Santa Barbara Basin, *Mar. Biol., 80*, 75-81, 1984.

Anderson, P. J., and J. F. Piatt, Trophic reorganization in the Gulf of Alaska, *Mar. Ecol. Prog. Ser.*, in press, 1999.

Arai, M. N., Active and passive factors affecting aggregations of hydromedusae: a review, *Sci. Mar., 56*, 99-108, 1992.

Arai, M. N., *A Functional Biology of Scyphozoa*, Chapman and Hall, London, 1997.

Bailey, T. G., J. J. Torres, M. J. Youngbluth and G. P. Owen, Effect of decompression on mesopelagic gelatinous zooplankton: A comparison of *in situ* and shipboard measurements of metabolism. *Mar. Ecol. Prog. Ser., 113*, 13-27, 1994.

Baird, D., and R. E. Ulanowicz, The seasonal dynamics of the Chesapeake Bay ecosystem, *Ecol. Monogr., 59*, 329-364, 1989.

Ballard, L., and A. Myers, Seasonal changes in the vertical distribution of five species of the family Bougainvilliidae (Cnidaria: Anthomedusae) at Lough Hyne, south-west Ireland, *Sci. Mar., 60*, 69-74, 1996.

Båmstedt, U., J. H. Fosså, M. B. Martinussen, and A. Fosshagen, Mass occurrence of the physonect siphonophore *Apolemia uvaria* (Lesueur) in Norwegian waters, *Sarsia, 83*, 79-85, 1998.

Bayly, A. E., Ecology of the zooplankton of a meromictic antarctic lagoon with special reference to *Drepanopus bispinosus* (Copepods: Calanoida), *Hydrobiologia, 140*, 199-231, 1986.

Bejda, A. J., A. L. Studholme, and B. L. Olla, Behavioral responses of red hake, *Urophycis chuss*, to decreasing concentrations of dissolved oxygen, *Env. Biol. Fishes, 19*, 261-8, 1987.

Benović, A., D. Justić, and A. Bender, Enigmatic changes in the hydromedusan fauna of the northern Adriatic Sea, *Nature, 326*, 597-599, 1987.

Beyer, F., Zooplankton, zoobenthos, and bottom sediments as related to pollution and water exchange in Oslofjord, *Helgol. Wiss. Meeresunters, 17*, 496-509, 1968.

Boicourt, W. C., Influences of circulation processes on dissolved oxygen in the Chesapeake Bay, in *Oxygen Dynamics in the Chesapeake Bay: A Synthesis of Recent Research*, edited by D. E. Smith, M. Leffler, and G. Mackiernan, pp. 7-10, Maryland Sea Grant, College Park, 1992.

Boicourt, W. C., M. Kuzmić, and T. S. Hopkins, The inland sea: circulation of Chesapeake Bay and the northern Adriatic, in *Ecosystems at the Land-Sea Margin: Drainage Basin to Coastal Sea*, edited by T. C. Malone, A. Malej, L. W. Harding, Jr., N. Smodlaka, and R. E. Turner, pp. 81-129, American Geophysical Union, *Coastal and Estuarine Studies, 55*, 1999.

Bottger-Schnack, R., Vertical structure of small metazoan plankton, especially non-calanoid copepods. 1. Deep Arabian Sea, *J. Plankton Res., 18*, 1073-1101, 1996.

Breitburg, D. L., Nearshore hypoxia in the Chesapeake Bay, Patterns and relationships among physical factors, *Estuar. Coastal Shelf Sci., 30*, 593-609, 1990.

Breitburg, D. L., Behavioral responses of fish larvae to low oxygen risk in a stratified water column., *Mar. Biol., 120*, 615-625, 1994.

Breitburg, D. L., T. Loher, C. A. Pacey, and A. Gerstein, Varying effects of low dissolved oxygen on trophic interactions in an estuarine food web, *Ecol. Monogr., 67*, 489-507, 1997.

Breitburg, D. L., L. Pihl, and S. E. Kolesar, Effects of low dissolved oxygen on the behavior, ecology and harvest of fishes: a comparison of the Chesapeake and Baltic-Kattegatt systems. This volume.

Breitburg, D. L., K. Rose, and J. H. Cowan, Jr., Linking water quality to larval survival: predation mortality of fish larvae in an oxygen-stratified water column. *Mar. Ecol. Prog. Ser. 178*, 39-54 , 1999.

Breitburg, D. L., N. Steinberg, S. DuBeau, C. Cooksey, and E. D. Houde, Effects of low dissolved oxygen on predation on estuarine fish larvae, *Mar. Ecol. Prog. Ser., 104*, 235-246, 1994.

Brewer, R. H., Some microenvironmental influences on attachment behavior of the planula of *Cyanea capillata* (Cnidaria: Scyphozoa), in *Coelenterate Ecology and Behavior*, edited by G. O. Mackie, pp. 347-354, Plenum Publ. Corp., New York, 1976.

Brodeur, R. D., C. E. Mills, J. E. Overland, and G. E. Walters, Recent increase in jellyfish biomass in the Bering Sea: possible links to climate change, *Fish. Oceanogr.*, in press, 1999.

Caddy, J. F., Toward a comparative evaluation of human impacts on fishery ecosystems of enclosed and semi-enclosed seas, *Rev. Fish. Sci., 1*, 57-95, 1993.

Cargo, D. G., and D. R. King, Forecasting the abundance of the sea nettle, *Chrysaora quinquecirrha*, in the Chesapeake Bay, *Estuaries, 13*, 486-491, 1990.

Cargo, D. G., and L. P. Schultz, Notes on the biology of the sea nettle, *Chrysaora quinquecirrha*, in Chesapeake Bay, *Chesapeake Sci., 7*, 95-100, 1966.

Cargo, D. G., and L. P. Schultz, Further observations on the biology of the sea nettle and jellyfishes in the Chesapeake Bay, *Chesapeake Sci., 8*, 209-220, 1967.

Childress, J. J., Are there physiological and biochemical adaptations of metabolism in deep-sea animals?, *Trends Ecol. Evol., 10*, 30-36, 1995.

Cooper, S. R., and G. S. Brush, Long-term history of Chesapeake Bay anoxia, *Science, 254*, 992-996, 1993.

Cowan, J. H., Jr., and E. D. Houde, Relative predation potentials of scyphomedusae, ctenophores and planktivorous fish on ichthyoplankton in Chesapeake Bay, *Mar. Ecol. Prog. Ser., 95*, 55-65, 1993.

Craig, K., L. B. Crowder, C. D. Gray, C. J. McDaniel, T. A. Henwood, and J. G. Hanifer, Ecological effects of hypoxia on fish, sea turtles, and marine mammals in the northwestern Gulf of Mexico, this volume.

Davis, J. C., Minimal dissolved oxygen requirements of aquatic life with emphasis on Canadian species: a review, *J. Fish. Res. Board Can., 32*, 2295-2332, 1975.

Eiane, K, D. L. Aksnes, E. Bagøien, and S. Kaartvedt, Fish or jellies -- a question of visibility?, *Limnol. Oceanogr., 44*, 1352-1357, 1999.

Fosså, J. H., Mass occurrence of *Periphylla periphylla* (Scyphozoa, Coronatae) in a Norwegian Fjord, *Sarsia, 77*, 237-251, 1992.

Franks, P. J. S., Phytoplankton blooms at fronts: Patterns, scales, and physical forcing mechanisms, *Reviews in Aquatic Sciences 6*, 121-137, 1992.

Gjosaeter, J., Mesopelagic fish, a large potential resource in the Arabian Sea, *Deep-Sea Res., 31*, 1019-1035, 1984.

Graham, W. M., The physical oceanography and ecology of upwelling shadows. Ph.D. thesis, Dept. of Biology, University of California, Santa Cruz, 1994.

Gunter, G., Seasonal population changes and distributions as related to salinity, of certain invertebrates of the Texas coast, including the commercial shrimp, *Publ. Inst. Mar. Sci., Univ. of Texas. 1*, 7-51, 1950.

Hamner, W. M., R. W. Gilmer, and P. P. Hamner, The physical, chemical, and biological characteristics of a stratified, saline, sulfide lake in Palau, *Limnol. Oceanogr., 27*, 896-909, 1982.

Hamner, W. M., and R. M. Jenssen, Growth, degrowth, and irreversible cell differentiation in *Aurelia aurita, Am. Zool., 14*, 833-849, 1974.

Harbison, G. R., The potential of fishes for the control of gelatinous zooplankton, *ICES* C.M. 1993/L, 74, 1993.

Herring, P. J., M. J. R. Fasham, A. R. Weeks, J. C. P. Hemmings, H. S. J. Roe, P. R. Pugh, S.

Holley, N. A. Crisp, and M. V. Angel, Across-slope relations between the biological populations, the euphotic zone and the oxygen minimum layer off the coast of Oman during the southwest monsoon (August, 1994), *Prog. Oceanogr., 41*, 69-109, 1998.

Houde, E. D., S. Jukiê-Peladiê, S. B. Brandt, and S. D. Leach, Fisheries: Trends in catches, abundance and management, in *Ecosystems at the Land-Sea Margin: Drainage Basin to Coastal Sea*, edited by T. C. Malone, A. Malej, L. W. Harding, Jr., N. Smodlaka, and R. E. Turner, pp. 341-366, American Geophysical Union, *Coastal and Estuarine Studies, 55*, 1999.

Howell, P., and D. Simpson, Abundance of marine resources in relation to dissolved oxygen in Long Island Sound, *Estuaries, 17*, 384-402, 1994.

Kamykowski, D., and S. Zentara, Hypoxia in the world ocean as recorded in the historical data set, *Deep-Sea Res., 37*, 1861-1874, 1990.

Keister, J. E., Habitat selection and predation risk: effects of low dissolved oxygen on zooplankton and fish larvae in Chesapeake Bay, M.S. Thesis, University of Maryland, College Park, 1996.

Keister, J. E., E.D. Houde, and D.L. Breitburg, Effects of bottom-layer hypoxia on abundance and depth distributions of organisms in Patuxent River, Chesapeake Bay, *Mar. Ecol. Prog. Ser.*, in review.

Kennedy, V. S., Anticipated effects of climate change on estuaries and coastal fisheries. *Fisheries, 16*, 16-24, 1990.

Kolar, C. S., and F. J. Rahel, Interaction of a biotic factor (predator presence) and an abiotic factor (low oxygen) as an influence on benthic invertebrate communities, *Oecologia, 95*, 210-219, 1993.

Legović, T., A recent increase in jellyfish populations: a predator-prey model and its implications, *Ecol. Modelling, 38*, 243-256, 1987.

Lin, A. L., and R. L. Zubkoff, Enzymes associated with carbohydrate metabolism of scyphistomae of *Aurelia aurita* and *Chrysaora quinquecirrha* (Scyphozoa: Semaeostomae), *Comp. Biochem. Physiol., 57B*, 303-308, 1977.

Mackie, G. O., and C. E. Mills, Use of the *Pisces IV* submersible for zooplankton studies in coastal waters of British Columbia, *Can. J. Fish. Aquat. Sci., 40*, 763-776, 1983.

Madin, L. P. and G. R. Harbison, *Bathocyroe fosteri* gen. nov., sp. nov.: A mesopelagic ctenophore observed and collected from a submersible, *J. Mar. Biol. Ass. U.K., 58*, 559-564, 1978.

Madhupratap, M., and P. Haridas, Zooplankton, especially calanoid copepods, in the upper 1000 m of the south-east Arabian Sea, *J. Plankton Res., 12*, 305-321, 1990.

Magnesen, T., Vertical distribution of size-fracions in the zooplankton community in Lindåspollene, western Norway. 1. Seasonal variations, *Sarsia, 74*, 59-68, 1989.

Mansueti, R., Symbiotic behavior between small fishes and jellyfishes, with new data on that between the stromateid, *Peprilus alepidotus*, and the scyphomedusa, *Chrysaora quinquecirrha*, *Copeia, 1963*, 40-80, 1963.

Mills, C. E.., Medusae, siphonophores, and ctenophores as planktivorous predators in changing global ecosystems, *ICES J. Mar. Sci., 52*, 575-581, 1995.

Mills, C. E., P. R. Pugh, G. R. Harbison, and S. H. D. Haddock, Medusae, siphonophores and ctenophores of the Alboran Sea, south western Mediterranean, *Sci. Mar., 60*, 145-163, 1996.

Mutlu, E., Distribution and abundance of ctenophores and their zooplankton food in the Black Sea. II. *Mnemiopsis leidyi, Mar. Biol., 135*, 603-613, 1999.

Mutlu, E., and F. Bingel, Distribution and abundance of ctenophores and their zooplankton food in the Black Sea. I. *Pleurobrachia pileus, Mar. Biol., 135*, 589-601, 1999.

Newell, R. E. I., Ecological changes in Chesapeake Bay: Are they the result of overharvesting the American oyster, *Crassostrea virginica? CRC Publ., 129*, 29-31, 1988.

Newell, R. E., and J. Ott, Macrobenthic communities and eutrophication, in *Ecosystems at the Land-Sea Margin: Drainage Basin to Coastal Sea*, edited by T. C. Malone, A. Malej, L. W.

Harding, Jr., N. Smodlaka, and R. E. Turner, pp. 265-293, American Geophysical Union, *Coastal and Estuarine Studies, 55*, 1999.

Officer, C. B., R. B. Biggs, J. L. Taft, L. E. Cronin, M. A. Tyler, and W. R. Boynton, Chesapeake Bay anoxia: origin, development and significance, *Science, 223*, 22-27, 1984.

Olney, J. E., Eggs and early larvae of the bay anchovy, *Anchoa mitchilli*, and the weakfish, *Cynoscion regalis*, in lower Chesapeake Bay with notes on associated ichthyoplankton, *Estuaries, 6*, 20-35, 1983.

Oviatt, C. A., and P. M. Kremer, Predation on the ctenophore, *Mnemiopsis leidyi*, by butterfish, *Peprilus triacanthus*, in Narragansett Bay, Rhode Island, *Chesapeake Sci., 18*, 236-240, 1977.

Parsons, T. R., The impact of industrial fisheries on the trophic structure of marine ecosystems, in *Food Webs: Integration of Patterns and Dynamics*, edited by G. A. Polis and K. O. Winemiller, pp. 352-357, Chapman and Hall, New York, 1995.

Parsons, T. R., K. von Brockel, P. Koeller, and M. Takahashi, The distribution of organic carbon in a marine planktonic food web following nutrient enrichment, *J. Exp. Mar. Biol. Ecol., 26*, 235-247, 1977.

Pauly, D., V. Christensen, J. Dalsgaard, R. Froese, and F. Torres, Jr., Fishing down marine food webs, *Science, 279*, 860-863, 1998.

Phillips, P. J., The Pelagic Cnidaria of the Gulf of Mexico: Zoogeography, Ecology and Systematics. Ph.D. thesis, Dept. of Oceanography, Texas A&M University, College Station, 1972.

Pihl, L., S. P. Baden, R. J. Diaz, and L. C. Schaffner, Hypoxia-induced structural changes in the diet of bottom-feeding fish and crustacea, *Mar. Biol., 112*, 349-361, 1992.

Poulin, R., N. G. Wolf, and D. L. Kramer, The effect of hypoxia on the vulnerability of guppies (*Poecilia reticulata*, Poeciliidae) to an aquatic predator (*Astronotus ocellatus*, Cichlidae), *Env. Biol. Fish., 20*, 285-292, 1987.

Pugh, P. R., and M. J. Youngbluth, Two new species of prayine siphonophore (Calycophorae, Prayidae) collected by the submersibles 'Johnson-Sea-Link' I and II, *J. Plankton Res., 10*, 637-657, 1988.

Purcell, J. E., A review of cnidarians and ctenophores feeding on competitors in the plankton, *Hydrobiologia, 216/217*, 335-342, 1991.

Purcell, J. E., Effects of predation by the scyphomedusan *Chrysaora quinquecirrha* on zooplankton populations in Chesapeake Bay, *Mar. Ecol. Prog. Ser., 87*, 65-76, 1992.

Purcell, J. E., Pelagic cnidarians and ctenophores as predators: Selective predation, feeding rates and effects on prey populations, *Ann. Inst. Oceanogr., Paris, 73*, 125-137, 1997.

Purcell, J. E., and J. H. Cowan, Jr., Predation by the scyphomedusan *Chrysaora quinquecirrha* on *Mnemiopsis leidyi* ctenophores, *Mar. Ecol. Prog. Ser., 128*, 63-70, 1995.

Purcell, J. E., A. Malej, and A. Benović, Potential links of jellyfish to eutrophication and fisheries, in *Ecosystems at the Land-Sea Margin: Drainage Basin to Coastal Sea*, edited by T. C. Malone, A. Malej, L. W. Harding, Jr., N. Smodlaka, and R. E. Turner, pp. 241-263, American Geophysical Union, *Coastal and Estuarine Studies, 55*, 1999.

Purcell, J. E., D. A. Nemazie, S. E. Dorsey, E. D. Houde, and J. C. Gamble, Predation mortality of bay anchovy *Anchoa mitchilli* eggs and larvae due to scyphomedusae and ctenophores in Chesapeake Bay, *Mar. Ecol. Prog. Ser., 114*, 47-58, 1994a.

Purcell, J. E., J. R. White, and M. R. Roman, Predation by gelatinous zooplankton and resource limitation as potential controls of *Acartia tonsa* copepod populations in Chesapeake Bay, *Limnol. Oceanogr., 39*, 263-278, 1994b.

Rabalais, N. N., R. E. Turner, W. J. Wiseman, Jr., and D. F. Boesch, A brief summary of hypoxia on the northern Gulf of Mexico continental shelf: 1985-1988, in *Modern and Ancient Continental Shelf Anoxia*, edited by R. V. Tyson and T. H. Person, pp. 35-47, *Geological Society Special Pub., 58*, 1991.

Rabalais, N. N., R. E. Turner, D. Justić, Q. Dortch, W. J. Wiseman, Jr., and B. K. Sen Gupta, Nutrient changes in the Mississippi River and system responses on the adjacent continental shelf, *Estuaries, 19*, 386-407, 1996.

Rabalais, N. N., R. E. Turner, W. J. Wiseman, Jr., and Q. Dortch, Consequences of the 1993 Mississippi River flood in the Gulf of Mexico, *Regulated Rivers: Research & Management, 14*, 161-177, 1998.

Rabalais, N. N., W. J. Wiseman, Jr., and R. E. Turner, Comparison of continuous records of near-bottom dissolved oxygen from the hypoxia zone along the Louisiana coast, *Estuaries, 17*, 850-861, 1994.

Rahel, F. J., and C. S. Kolar, Trade-offs in the response of mayflies to low oxygen and fish predation, *Oecologia, 84*, 39-44, 1990.

Raskoff, K. A., Distributions and trophic interactions of mesopelagic hydromedusae in Monterey Bay, CA: *In situ* studies with the MBARI ROVs *Ventana* and *Tiburon, Supplement to EOS, Transactions, American Geophysical Union, 79, American Society of Limnology and Oceanography, San Diego, CA*, 1998.

Reid, G. K., Jr., A summer study of the biology and ecology of East Bay, Texas. I. *Texas J. Sci., 7*, 316-343, 1955.

Reid, J. L., Jr., *Intermediate Waters of the Pacific*, John Hopkins Press, Baltimore, MD, 1965.

Renaud, M. L., Hypoxia in Louisiana coastal waters during 1983: implication for fisheries, *Fish. Bull. U.S., 84*, 19-26, 1986.

Robison, B. H., Midwater research methods with MBARI's ROV, *MTS J., 26*, 32-39, 1993.

Robison, B. H., K. R. Reisenbichler, R. E. Sherlock, J. M. B. Silguero and F. P. Chavez, Seasonal abundance of the siphonophore, *Nanomia bijuga*, in Monterey Bay, *Deep-Sea Res. II, 45*, 1741-1751, 1998.

Roman, M. R., K. A. Ashton, and A. L. Gauzens, Day/night differences in the grazing impact of marine copeopds, *Hydrobiologia, 167/168*, 21-30, 1988.

Roman, M. R., A. L. Gauzens, W. K. Rhinehart, and J. R. White, Effects of low oxygen waters on Chesapeake Bay zooplankton, *Limnol. Oceanogr., 38*, 1603-1614, 1993.

Rosenberg, R., R. Elmgren, S. Fleischer, P. Jonsson, G. Presson, and H. Dahlin, Marine eutrophication case studies in Sweden: a synopsis, *Ambio, 19*, 102-108, 1990.

Rudstam, L. G., and J. J. Magnuson, Predicting the vertical distribution of fish populations: analysis of cisco, *Coregonus artedii*, and yellow perch, *Perca flavescens, Can. J. Fish. Aquat. Sci., 42*, 1178-1188, 1985.

Sanford, L. P., K. G. Sellner, and D. L. Breitburg, Covariability of dissolved oxygen with physical processes in the summertime Chesapeake Bay, *J. Mar. Res., 48*, 567-590, 1990.

Shenker, J. M., D. J. Hepner, P. E. Frere, L. E. Currence, and W. W. Wakefield, Upriver migration and abundance of naked goby (*Gobiosoma bosc*) larvae in the Patuxent River Estuary, Maryland, *Estuaries, 5*, 36-42, 1983.

Shiganova, T. A., Invasion of ctenophore *Mnemiopsis leidyi* in the Black Sea and recent changes in its pelagic community structure, *Fisheries Oceanogr., 7*, 305-310, 1998.

Smedstad, O. M., On the biology of *Aglantha digitale rosea* (Forbes) (Coelenterata: Trachymedusae) in the inner Oslofjord, *Norw. J. Zool., 20*, 111-135, 1972.

Stalder, L. C., and N. H. Marcus, Zooplankton responses to hypoxia: behavioral patterns and survival of three species of calanoid copepods, *Mar. Biol., 127*, 599-607, 1997.

Swallow, J. C., Some aspects of the physical oceanography of the Indian Ocean, *Deep-Sea Res., 31*, 639-650, 1984.

Swanson, R. L., and C. A Parker, Physical environmental factors contributing to recurring hypoxia in the New York Bight, *Trans. Am. Fish. Soc., 117*, 37-47, 1988.

Taft, J. L., W. R. Taylor, E. O. Hartwig, and R. Loftus, Seasonal oxygen depletion in Chesapeake Bay, *Estuaries, 4*, 242-247, 1980.

Thuesen, E. V., and J. J. Childress, Oxygen consumption rates ad metabolic enzyme activities of oceanic California medusae in relation to body size and habitat depth, *Biol. Bull., 187*, 84-98, 1994.

Tisileus, P., G. Nielsen, and T. Nielsen, Microscale patchiness of plankton within a sharp pycnocline, *J. Plankton Res., 16*, 543-554, 1994.

Turner, R. E., W. E. Shroeder, and W. J. Wiseman, Jr., The role of stratification in the deoxygenation of Mobile Bay and adjacent shelf bottom waters, *Estuaries, 19*, 13-20, 1987.

Vinogradov, M. Ye., M. V. Flint, and E. A. Shushkina, Vertical distribution of mesoplankton in the open area of the Black Sea, *Mar. Biol., 89*, 95-107, 1985.

Vinogradov, M. Ye., E. A. Shushkina, A. Ye. Gorbunov, and N. L. Shashkov, Vertical distribution of the macro- and mesoplankton in the region of the Costa Rica dome, *Oceanology, 31*, 559-565, 1991.

Volovik, S. P., Z. A. Myrzoyan, and G. S. Volovik, *Mnemiopsis leidyi*: Biology, population dynamics, impact to the ecosystem and fisheries, ICES Statutory Meeting C.M. 1993/L:69, Dublin, 1993.

Wishner, K. F., C. J. Ashjian, C. Gelfman, M. M. Gowing, L. Kann, L. Levin, L. S. Mullineau and J. Saltzman, Pelagic and benthic ecology of the lower interface of the Eastern Tropical Pacific oxygen minimum zone, *Deep-Sea Res., 42*, 93-115, 1995.

Youngbluth, M .J., B. H. Robison, and K. R. Reisenbichler, Predation by large physonect siphonophores in an hypoxic midwater environment, *Abstract, 7th Symposium on Deep Sea Biology*, Hersonisus, Crete, 1994.

6

Physiological Responses to Hypoxia

Louis E. Burnett and William B. Stickle

Abstract

Hypoxia can have profound effects on individual organisms. This chapter focuses on the mechanisms different kinds of animals possess to avoid, tolerate, and adapt to low levels of oxygen in water; selected examples illustrate these mechanisms. While some organisms can detect and avoid hypoxic water, avoidance is not always possible, especially in the case of sessile organisms. When an organism cannot avoid hypoxia, its response may depend on the intensity and the duration of the bout of low oxygen. Examples of responses to hypoxia include a depression in feeding as well as a decrease in molting and growth rates. During acute exposures to hypoxia some organisms can maintain aerobic metabolism by making effective use of a respiratory pigment, or increasing ventilation rates, or increasing the flow of blood past the respiratory surfaces or combinations of all three. Responses to chronic hypoxia are different and include the production of greater quantities of respiratory pigment and changing the structure of the pigment to one with an adaptive higher oxygen affinity. Many organisms respond to hypoxia by switching from aerobic to anaerobic metabolism and some simply reduce their overall metabolism. Hypoxia is often accompanied by hypercapnia (an elevation in water CO_2), which produces an acidification of the body tissues, including the blood, and has physiological implications that can also be profound and separate from the effects of low oxygen. Finally, there is evidence that hypoxia can inhibit immune responses, causing greater mortality than would otherwise occur when organisms are challenged with a pathogen.

Introduction

An obvious and dramatic effect of low ambient oxygen on an organism is a lethal response. The general public is well aware of the results of hypoxia when large "fish kills" are reported. However, organisms can be affected by a lack of oxygen in other ways.

Coastal Hypoxia: Consequences for Living Resources and Ecosystems
Coastal and Estuarine Studies, Pages 101-114
Copyright 2001 by the American Geophysical Union

Hypoxia may limit the energy budget or scope for growth and activity of an organism, it may cause an organism to alter its behavior, and/or it may limit the tolerance of an organism to other environmental challenges. The manifestations of these effects may be seen as changes in the population structure within a species, changes in the range of distribution, or a decrease in the population density of an organism. In this manner individual organismal effects are transferred to the population and ecosystem levels of organization. From a population perspective, there is a big difference between the ability of an organism simply to survive and its ability to thrive. This difference can be evidenced in the structure of food webs and the size distribution of organisms within a population. In this chapter, we explore some of the individual organismal responses to hypoxia and the effects of hypoxia on basic physiological mechanisms and behavior. The examples we use are selected from the literature based on their relevance to the kinds of organisms living in the Gulf of Mexico and are summarized in Table 1. In addition, we discuss some new information that suggests that resistance to disease is compromised in hypoxic environments.

Environmental hypoxia can be moderate (e.g., half air saturation) or severe (e.g., less than 20 to 30% air saturation), designations not used consistently in the literature. Obviously, the more severe the hypoxia, the greater the physiological challenge to the organism. More severe hypoxia may, for example, require an organism to utilize anaerobic metabolism to sustain its energy production. Another water quality variable that nearly always accompanies hypoxia is elevated carbon dioxide or hypercapnia [Cochran and Burnett, 1996; Burnett, 1997]. The biological demand for oxygen responsible for lowering oxygen partial pressures produces carbon dioxide as the main product of metabolism and even slight hypercapnia lowers water pH dramatically. Even though marine systems are considered to be well buffered against changes in pH, significant fluctuations in carbon dioxide occur in coastal waters resulting in highly variable water pH [Christmas and Jordan, 1987; Cochran and Burnett, 1996; Burnett, 1997]. Thus, aquatic organisms facing a drop in dissolved oxygen also face an acidification that causes a concomitant acidosis in the blood and the tissues. Below, we examine some of the consequences of hypercapnic hypoxia.

The duration of exposure to environmental hypoxia may be relatively short and diurnal or tidal [Summers et al., 1997; Spicer et al., 1999; Das and Stickle, unpublished observations] or it may be long term exposure for weeks to years [Stickle et al., 1989; Das and Stickle, 1993]. Adaptations to hypoxia likewise vary depending on the duration of exposure. Different response patterns to diurnal hypoxia compared to chronic hypoxia of the estuarine crab *Callinectes sapidus* and the offshore crab *C. similis* highlight these metabolic, feeding, and growth rate adaptations [Das and Stickle, 1993, unpublished observations] and are discussed below.

Thus, in this chapter we explore examples of the responses of organisms to hypoxia that illustrate the state of our knowledge and the problems organisms face. In addition, we suggest some areas for future research.

Physiological and Behavioral Responses

Many organisms living in coastal environments are well adapted to endure and even thrive for short durations in hypoxic water. Organisms are able to exist for short durations in hypoxia because they possess respiratory mechanisms to take up oxygen from the

TABLE 1. Behavioral and physiological responses of different organisms to hypoxia.

Organism	Response to Hypoxia	Reference
Shrimp		
Penaeus aztecus	detect and avoid	Renaud, 1986
Penaeus setiferus	detect and avoid	Renaud, 1986
Penaeus monodon	decrease hemocyte phagocytosis	Direkbusarakom & Danayadol, 1998
Penaeus stylirostris	decrease total hemocyte count	Le Moullac et al., 1998
	increased mortality induced by *Vibrio alginolyticus*	Le Moullac et al., 1998
Crabs		
Callinectes sapidus	detect and avoid	Das & Stickle, 1994
	decrease feeding	Das & Stickle, 1993
	reduce growth rate	Das & Stickle, 1993
	Acute Hypoxia	
	increase ventilation rate	Batterton & Cameron, 1978
	increase heart rate	deFur & Pease, 1988
	slight increase in cardiac output	deFur & Pease, 1988
	Chronic Hypoxia	Das & Stickle, 1993
	decrease oxygen consumption	deFur & Pease, 1988
	no change in ventilation	deFur & Pease, 1988
	no change in heart rate	deFur et al., 1990
	increase hemocyanin O_2 affinity and concentration	
Callinectes similis	detect and avoid	Das & Stickle, 1994
	increase oxygen consumption	Das & Stickle, 1993
	decrease feeding	Das & Stickle, 1993
Gastropod Molluscs		
Stramonita haemastoma	reduce growth rate	Das & Stickle, 1993
	large reduction in metabolism	Liu et al., 1990
	decrease oxygen consumption	Das & Stickle, 1993
Bivalved Molluscs		
Crassostrea virginica	switch to anaerobic metabolism	Stickle et al., 1989
	small reduction in metabolism	Stickle et al., 1989
	decrease production of reactive oxygen species	Boyd & Burnett, 1999

environment when it is scarce or because they sustain energy production by switching to anaerobic biochemical pathways. Many organisms can do both. Yet another option is to lower the overall rate of metabolism. However, if an organism is to sustain its "normal" rate of activity, it must maintain its energy production. This may be especially important if the levels of dissolved oxygen fluctuate on a tidal and a diurnal basis as they often do [Cochran and Burnett, 1996; Das and Stickle, unpublished observations]. Finally, some organisms appear to make physiological adjustments when confronted with hypoxia on a chronic basis. Examples of these responses are presented below.

Avoidance Behavior

Perhaps the first line of defense of a mobile organism against hypoxia is to avoid it. A number of coastal species appear to be able to detect and avoid hypoxia. Investigations in the field have documented migration of fishes, crustaceans and annelids in response to low oxygen [Loesch, 1960; May, 1973; Garlo et al., 1979; Pavela et al., 1983; Pihl et al., 1991; Nestlerode and Diaz, 1998]. Differential predator and prey activity in hypoxic water can enhance feeding by the predator species [Nestlerode and Diaz, 1998]. Careful laboratory studies have supported these observations. Two such cases are mentioned below.

Juvenile brown shrimp, *Penaeus aztecus*, and juvenile white shrimp, *Penaeus setiferus*, can detect and avoid hypoxia [Renaud, 1986]. Furthermore, a number of behavioral responses are associated with hypoxia. These responses include an initial increase in activity, retreat from the hypoxia by walking or swimming, rapid eye-stalk movements, and flexing of antennal scales. The non-homogeneous patterns of hypoxia and well oxygenated water along the Gulf coast are thought to cause brown and white shrimp to aggregate in areas that are less hypoxic. Renaud [1986] suggests that this crowding could lead to increased animal stress and greater predation.

Juvenile blue crabs (both *Callinectes sapidus* and the lesser blue crab *Callinectes similis*) are also able to detect hypoxia [Das and Stickle, 1994]. Both species of blue crabs prefer water greater than two-thirds air saturation, and both species are more active in water of higher oxygen pressures than in hypoxic water [Das and Stickle, 1994]. Interestingly, unlike *C. similis*, juvenile *C. sapidus* do not avoid hypoxic water. Upon exposure to hypoxia, blue crabs demonstrate behaviors similar to that of shrimp (see above). They exhibit "restless and erratic movements" when exposed to hypoxia and move their eye-stalks with a greater frequency [Das and Stickle, 1994]. They also rapidly move their antennae. In addition, *C. sapidus* and *C. similis* exhibit decreased feeding rates in hypoxic water. There was a dramatic decline in growth rate as a function of the severity of hypoxia in *C. sapidus*, but not in *C. similis*. The reduced growth of *C. sapidus* was seen even with mild hypoxia, i.e., 74% air saturation [Das and Stickle, 1994].

Obviously, hypoxia has profound effects on the behavior of shrimp and crabs, even though they have physiological and biochemical mechanisms that enable them to cope with it (see below). That these responses include a decrease in feeding and growth rates suggests why populations of organisms, although they may survive hypoxia, do not thrive.

A behavioral pattern that is stronger than hypoxia avoidance in crustaceans is the diel vertical migration of Nordic krill, *Meganyctiphanes norvegica*, into depths that are severely hypoxic (> 70 m depth) in Gullmarsfjord, Sweden during daylight [Spicer et al., 1999]. Krill trawled at dusk had hemolymph lactate concentrations that were similar to lactate concentrations of krill caged at 70 m depth for the day, indicating that migrating krill had also undergone anaerobic metabolism during the day.

Effects of Hypoxia on Physiological Functions

Little is known of how acute or chronic hypoxia affects activity and growth in organisms. Chronic hypoxia (28 days exposure) results in a reduction in growth rate, as measured by Scope for Growth in the blue crab, *Callinectes sapidus*, the lesser blue crab, *C. similis* and the southern oyster drill, *Stramonita haemastoma* [Das and Stickle, 1993]. Feeding rate, the primary determinant of Scope for Growth [Stickle, 1985], was reduced in blue crabs exposed to severe hypoxia and varied directly with the severity of hypoxia in the southern oyster drill. Oxygen consumption rates of blue crabs and southern oyster drills exposed to severe hypoxia were significantly lower than in animals exposed to normoxia. In contrast, the oxygen consumption rate of *C. similis* was higher in hypoxia than in lesser blue crabs exposed to normoxia, indicating overcompensation in the oxygen delivery system. Growth and molting rates in both species of crabs were significantly higher in the normoxic exposure than in crabs exposed to hypoxia.

Exposure of these species to 28 days of a 100-16-100% diurnal pattern of air saturation produced bioenergetic results that differed from the earlier chronic exposure to various levels of dissolved oxygen [Das and Stickle, unpublished observations]. There was 3% mortality in *C. sapidus*, 11% in *C. similis*, and 0% in *S. haemastoma* exposed to this fluctuating pattern of diurnal variation in oxygen tension [Das and Stickle, unpublished observations]. The feeding rate of *C. sapidus* exposed to a fluctuating diurnal pattern of hypoxia was significantly higher than that of blue crabs exposed to constant 100 or 16% saturation. However, the feeding rate of *C. similis* exposed to diurnal hypoxia was lower than that of lesser blue crabs exposed to 100% oxygen saturation or 16% oxygen saturation. Both species of crabs exposed to the diurnal pattern of oxygen variation increased in weight during the experimental period at a rate faster than those exposed to 16% oxygen saturation, but not significantly different than crabs exposed to normoxia. Growth rate of *C. sapidus* exposed to the diurnal variation of oxygen tension was significantly higher (95% increase in wet weight) than that of *C. similis* (73% increase in wet weight).

A Case Study of Mechanisms Organisms Use to Handle Hypoxia

Even though organisms may have the ability to detect and avoid hypoxia, they may not always escape it. In this case, organisms must rely on physiological mechanisms to extract as much oxygen as possible from the water and transport it to the tissues or switch to anaerobic metabolic pathways to supply energy, or both. A number of organisms are able to maintain oxygen uptake when oxygen supplies in the ambient environment are limiting. The term "oxygen regulation" is often applied to this phenomenon. The ability to maintain or regulate oxygen uptake is obviously important in allowing an organism to sustain its normal scope of activity.

One of the best known examples of organisms that tolerate and respond to hypoxia is found in the blue crab, *Callinectes sapidus*. Blue crabs are highly active animals that are known to tolerate a wide range of temperatures, salinities, and oxygen conditions. This tolerance allows them to exploit a wide range of habitats. Adult *C. sapidus* are able to tolerate moderate hypoxia (i.e., 32% air saturation) quite well for 25 days with only 20% mortality [deFur et al., 1990]. In contrast, juvenile *C. sapidus* exhibit 50% mortality at 68%

air saturation after 28 days exposure [Das and Stickle, 1993]. Juvenile *C. sapidus* are also poor regulators of oxygen uptake when exposed to severe hypoxia, but *C. similis* exhibit elevated oxygen consumption rates relative to normoxia at all hypoxia levels analyzed [Das and Stickle, 1993].

The responses to short-term hypoxia exposure in adult blue crabs are similar to those found in other crabs. Many crabs respond to short-term hypoxia by increasing the ventilatory flow of water past the gills [Truchot, 1975; Burnett, 1979; Burnett and Johansen, 1981; Lallier and Truchot, 1989], which favors the diffusion of oxygen into the blood. Blue crabs acutely exposed to hypoxia (one-third air saturation) hyperventilate [Batterton and Cameron, 1978]. During chronic exposure to hypoxia, hyperventilation persists for five days and then returns to the normoxic baseline [deFur and Pease, 1988]. Hyperventilation often produces a respiratory alkalosis in crabs [Burnett and Johansen, 1981], elevating the hemolymph pH and thus giving rise to an adaptive increase in the oxygen affinity of hemocyanins with normal Bohr shifts. This alkalosis associated with hypoxia has been observed in *C. sapidus* [Pease et al., 1986]. Heart rate increases by as much as 30%, remains elevated for five days and returns to the normoxic baseline despite the persistence of hypoxia [deFur and Pease, 1988]. While cardiac output increases during hypoxia in the lobster *Homarus americanus* [McMahon and Wilkens, 1975] and the spider crab *Libinia emarginata* [Burnett, 1979], it increases by only a small amount in blue crabs, despite the increase in heart rate [deFur and Pease, 1988].

Long-term exposure to moderate hypoxia has been well studied in the blue crab [Stickle et al., 1989; deFur et al., 1990; Das and Stickle, 1993]. After 25 days of exposure of adult blue crabs to moderate hypoxia (Po_2 = 50-55 torr or 33% air saturation), oxygen uptake is no different from that of crabs living in well-oxygenated water [deFur and Pease, 1988]. Furthermore, cardiac output was only slightly elevated. Most of the adjustments that account for sustaining oxygen uptake during chronic hypoxia occur with hemocyanin. Small increases in hemocyanin oxygen affinity (i.e., decreases in P_{50}) occur as a result of changes in hemolymph lactate, urate, and calcium ion concentrations.

The elevated water CO_2 that accompanies hypoxia [Cochran and Burnett, 1996] results in elevated hemolymph CO_2 in blue crabs [Cameron, 1978]. Carbon dioxide, independent of pH, increases oxygen affinity [Mangum and Burnett, 1986], contributing to the adaptive response. These changes comprise extrinsic factors that interact with the hemocyanin molecule to modify its oxygen affinity. Extrinsic factors generally include ions such as hydrogen, calcium, magnesium and sodium, or organic molecules such as lactate and urate. Taken together, the extrinsic factors described above result in an increase in oxygen affinity of approximately 3 torr.

Chronic hypoxia stimulates significant changes in the concentration and the structure of the hemocyanin molecules of *C. sapidus* [Mangum, 1997]. Hemocyanin concentration increases by about 40%, enhancing the capacity of the hemolymph to carry oxygen. Hemocyanin levels also increase in the shrimp *Crangon crangon* during prolonged exposure to hypoxia [Hagerman, 1986]. But perhaps more interesting is the intrinsic adaptation of hemocyanin oxygen affinity to low oxygen. deFur et al. [1990] postulated that the net synthesis or degradation of hemocyanin during hypoxia produces replacement molecules that differ from those in normoxic crabs. This possibility becomes even greater given the net synthesis of hemocyanin that occurs during long-term hypoxia [Senkbeil and Wriston, 1981].

Intrinsic changes in the structure of hemocyanin can alter the oxygen affinity in response to chronic changes in environmental variables in crayfish [Rutledge, 1981] and crabs [Mauro and Mangum, 1982; Mason et al., 1983; Mangum and Rainer, 1988]. There are six different kinds of subunits that make up the large hemocyanin molecule in *C. sapidus* and these can be distinguished electrophoretically [Mangum and Rainer, 1988]. Three of the six subunits are known to be variable in different populations of blue crabs [Mangum and Rainer, 1988]. Subunits 3, 5, and 6, the variable subunits, decrease in concentration in relation to the other subunits in response to chronic hypoxia. The net result of the changes in subunit composition of hemocyanin is an overall increase in oxygen affinity (decrease in P_{50}) of 5 torr. It appears that the electrophoretic patterns observed in both the field and the laboratory bring about a higher hemocyanin oxygen affinity by favoring the more primitive subunits of hemocyanin [Mangum, 1997].

The Role of Anaerobic Metabolism

For many organisms avoidance of hypoxia is not an option. The eastern oyster, *Crassostrea virginica*, is obviously not able to avoid hypoxic water. The mechanisms it uses to cope with hypoxia are fundamentally different from those used by organisms such as the blue crab. *C. virginica* is a weak oxygen regulator [Galtsoff and Whipple, 1931; Shumway and Koehn, 1982; Willson and Burnett, 2000]. Like many bivalves it has well developed biochemical pathways to sustain energy production anaerobically [Gäde, 1983]. Using microcalorimetry, Stickle et al. [1989] showed that oysters maintained 75% of their normoxic energy consumption in water that was extremely hypoxic (< 5% air saturation). At this same oxygen level, oxygen uptake is approximately 10% of that in well-aerated water [Willson and Burnett, 2000]. Therefore, in severely hypoxic water most of the energy production is via anaerobic pathways with the remainder contributed by aerobic means.

Another metabolic pattern is found in the southern oyster drill, *S. haemastoma*; metabolism declines in environments with acutely declining oxygen levels. *S. haemastoma* is a poor regulator of oxygen uptake [Kapper and Stickle, 1987]; the index of oxygen regulation [Mangum and Van Winkle, 1973], B2 is -0.073×10^{-3} at 30°C and 30‰ S. Furthermore, there is no change in the response to ambient oxygen when snails are held for 28 days at $Po_2 = 53$ torr (34% air saturation) [Kapper and Stickle, 1987]. Although *S. haemastoma* is capable of switching from aerobic to anaerobic metabolism, it does not rely on this strategy during anoxia (i.e., oxygen < 5% air saturation). Rather, it reduces its overall metabolism. This has been demonstrated in calorimetric studies in which Stickle et al. [1989] showed a steady state rate of heat dissipation of 0.73 joules g dry weight^{-1} h^{-1} under anoxia, or 9% of the rate during normoxia (8.76 joules g dry weight^{-1} h^{-1}). This contrasts with a normoxic energy consumption of 75% of the normoxic rate in the oyster, *C. virginica* under similar conditions (above).

Although highly active species such as blue crabs are capable of anaerobic metabolism, and accumulate L-lactate, they do not tolerate hypoxia to the extent of organisms like *C. virginica* and *S. haemastoma* [Stickle et al., 1989; Das and Stickle, 1993]. Reduction of metabolic rate upon exposure to hypoxia is a survival strategy utilized much more effectively by annelids and molluscs than by crustaceans [Gnaiger, 1983; Stickle et al., 1989]. Thus, it is not surprising that blue crabs have well-developed mechanisms for procuring oxygen.

Hypercapnic Hypoxia

As it does in aerial environments, photosynthesis fixes carbon dioxide in the aquatic environment, removing it from the water. However, gases are roughly 7,000 times less diffusible in water than in air [Dejours, 1975] and bodies of water are rarely homogeneous with respect to dissolved oxygen or carbon dioxide. In addition, the capacity of water to hold oxygen is significantly lower than that of air (53.8 μmol O_2 l^{-1} torr^{-1} in air at 25°C as compared to 1.4 μmol O_2 l^{-1} torr^{-1} in sea water or 1.7 μmol O_2 l^{-1} torr^{-1} in fresh water). Water is able to hold more carbon dioxide in all its chemical forms than oxygen because of the hydration reactions of carbon dioxide that produce bicarbonate and carbonate ions. In estuaries the production of oxygen by photosynthesis can lead to oxygen pressures that are significantly higher than that of air [Atkinson et al., 1987]. However, during the night when photosynthesis is greatly reduced or absent, the respiratory consumption of oxygen and production of carbon dioxide results in water that is both hypoxic and hypercapnic. In the Chesapeake Bay, the bacterioplankton can account for 60 to 100% of the planktonic oxygen consumption, especially in water rich in dissolved organics [Jonas, 1997]. In shallow salt marshes, water can become hypoxic as well as hypercapnic [Cochran and Burnett, 1996].

Many investigators studying the effects of hypoxia on organisms have induced hypoxia experimentally by gassing the water with nitrogen. However, hypoxia in the field is perhaps most often accompanied by a slight, but significant elevation in water carbon dioxide pressure (hypercapnia), resulting in large decreases in water pH [Cochran and Burnett, 1996; Burnett, 1997]. Through this mechanism water pH is highly correlated with oxygen levels [Christmas and Jordan, 1987; Burnett, 1997]. Hypercapnia, independent of dissolved oxygen, can have dramatic effects on the physiology of marine organisms. An elevation of ambient CO_2 results in a concomitant elevation of CO_2 in the bodies of organisms. The direct result of hypercapnia is a decrease in the pH of tissues and body fluids, which can have profound effects on a number of functions. For example, respiratory pigments are highly sensitive to pH resulting in a decrease of oxygen affinity in most organisms, but as discussed below, some organisms have a separate CO_2-specific effect that increases oxygen affinity counteracting the pH-specific effect. In a hypoxic environment, this would be maladaptive. In addition, hemocytes circulating in the blood may have a reduced protective function as a cellular defense mechanism (see below).

Again the blue crab, *Callinectes sapidus*, offers an example of an adaptive response to hypercapnic hypoxia. The hemocyanin of *C. sapidus* has a specific CO_2 effect [Mangum and Burnett, 1986] that contributes to an increase in oxygen affinity as CO_2 pressure increases. The CO_2-specific effect on hemocyanin is opposite to the well-known pH-specific effect (the Bohr shift in which a decrease in pH reduces oxygen affinity). During hypercapnic hypoxia the production of lactate specifically increases hemocyanin oxygen affinity (a third and also separate effect) and this will also contribute to an adaptive response. Carbon dioxide also produces a large increase in hemocyanin oxygen affinity in the grass shrimp, *Palaemonetes pugio*, independent of pH. Cochran and Burnett [1996] have shown that moderate hypercapnia does not affect the ability of grass shrimp to regulate oxygen uptake during declining oxygen tensions. Thus, the presence of a CO_2-specific effect brings about an adaptive increase in oxygen affinity. This stabilization of hemocyanin function in grass shrimp is thought to result in the maintenance in hypercapnic water of a "critical oxygen pressure," the point at which regulation of oxygen uptake ceases [Cochran and Burnett, 1996].

TABLE 2. Critical Po$_2$ and oxygen uptake at Po$_2$ > 100 torr in a variety of small coastal fishes collected in Charleston Harbor or on the ocean beach of Folly Beach, South Carolina. No distinction was made between male and female fish. Fishes were held at 25‰S and 30°C and subjected to declining oxygen using the methods of Cochran and Burnett [1996]. Critical Po$_2$ and O$_2$ Uptake are reported as mean (SEM; N).

Species	CO$_2$ Treatment	Critical Po$_2$ (torr)	O$_2$ Uptake (μmol g^{-1} h^{-1})	Mean Wt. (g)
Cyprinodon variegatus sheepshead minnow	Low (< 1 torr)	17.1 (1.8; 8)	10.0 (0.7; 8)	1.67
	High (7 torr)	20.5 (4.1; 8)	11.8 (0.9; 8)	1.95
Poecilia latipina Sailfin molly	Low (< 1 torr)	37.5 (2.3; 6)	20.0 (1.6; 6)	1.01
	High (7 torr)	33.9 (4.0; 7)	12.5 (1.8; 7)[b]	1.16
Trachinotus carolinus Pompano	Low (< 1 torr)	32.5 (1.7; 8)	21.6 (2.4; 8)	8.10
	High (7 torr)	71.7 (9.2; 5)[a]	17.8 (1.4; 5)	4.86
Mugil cephalus mullet	Low (< 1 torr)	44.5 (5.4; 6)	53.5 (5.2; 6)	0.18
	High (7 torr)	36.4 (1.8; 7)	35.5 (2.8; 7)[b]	0.16

[a]significant difference (p < 0.05) from low CO$_2$ according to Mann-Whitney rank sum test.
[b]significant difference (p < 0.05) from low CO$_2$ according to t-test.

Hypercapnia can have respiratory effects on fishes. The spot, *Leiostomus xanthurus*, showed a significantly elevated rate of oxygen uptake in mild (Pco$_2$ = 7 torr) hypercapnia [Cochran and Burnett, 1996]. The authors attributed this response to a Root effect in this species of fish [Bonaventura et al., 1976] that, under acidic environmental conditions, effectively ties up a portion of its hemoglobin preventing it from transporting oxygen and forcing the fish to use ventilatory and circulatory mechanisms to extract oxygen from the water. Increasing ventilation and circulation requires more metabolic energy, thus elevating oxygen uptake. On the other hand, the mummichog, *Fundulus heteroclitus*, was not affected by similar mild hypercapnia [Cochran and Burnett, 1996].

Using methods identical to those of Cochran and Burnett [1996] the responses of small coastal fishes to hypercapnic hypoxia were investigated (Table 2). Fish were held in respirometers for several hours and the respirometers were flushed with well-oxygenated water, until oxygen uptake by fish declined to steady levels. Respirometers were then closed and the oxygen pressures allowed to decline as fish consumed oxygen. Carbon dioxide within the respirometer was regulated by monitoring its level within the respirometer and periodically flushing the chamber with water of identical oxygen pressure (i.e., when the flushing began) and either high (1%) or low (< 0.1%) carbon dioxide, depending upon the treatment. The pompano, *Trachinotus carolinus*, had a critical Po$_2$ that was significantly elevated in mild hypercapnia, a result not surprising for a species that inhabits the well oxygenated waters of the inlets and the beaches of coastal South Carolina where it rarely encounters hypercapnic hypoxia. The critical Po$_2$ of the other species was insensitive to mild hypercapnia. Only *Poecilia latipina* and *Mugil cephalus* showed a significant depression in oxygen uptake at high Po$_2$ and mild hypercapnia (Table 2). These results are unlike those of *L. xanthurus* (above) and may be due to different mechanisms that reduce metabolism. A similar depression of metabolism by hypercapnia and the concomitant decline in pH has been documented in trout hepatocytes [Walsh et al., 1988]. The mullet (*M. cephalus*) used in the study had a very high rate of oxygen uptake at low CO_2 pressures. Mullet are a very active fish and it is difficult to obtain "resting" oxygen uptake rates within a closed respirometer. *Cyprinodon variegatus*, the most hardy of the fishes listed in Table 2, is commonly found in estuarine environments where hypercapnic hypoxia occurs frequently. It would be interesting to know the responses of these fishes to more severe hypercapnia.

Disease Resistance

The effects of environmental hypoxia on the cellular defenses of aquatic organisms is largely unknown, but is an emerging field of interest. The data that exist suggest that hypoxia can have profound effects on immune systems. This area of research has much relevance to organisms that are cultured for commercial purposes, where organism densities are high and the potential for pathogenic infections is great.

Hemocytes in invertebrates are responsible for the phagocytosis of potential pathogens as a part of an innate immune response. There is evidence that phagocytic activity in shrimp is depressed when the shrimp (*Penaeus monodon*) are exposed to hypoxia [Direkbusarakom and Danayadol, 1998]. Total hemocyte numbers can also be influenced by hypoxia. Le Moullac et al. [1998] have shown that shrimp (*Penaeus stylirostris*) exposed to severe hypoxia (1 mg O_2 l^{-1} = 25 torr = 16% air saturation) had a decrease in total hemocyte count. These effects appear to be relevant, as it was shown that shrimp injected with the pathogenic *Vibrio alginolyticus* showed significantly greater mortality when exposed to hypoxia [Le Moullac et al., 1998]

Hypoxia negatively impacts the production of reactive oxygen species by hemocytes of the oyster *Crassostrea virginica* [Boyd and Burnett, 1999]. In this case, oyster hemocytes incubated at the physiological O_2 and CO_2 pressures that exist in oyster hemolymph during moderate hypercapnic hypoxia (Po_2 = 40 to 45 torr) produced only 33% of the reactive oxygen species compared with normoxic conditions. Furthermore, this result was shown to be due to the separate and independent influences of oxygen (64% of normoxia) and pH (44% of normoxia). This study points to the importance of performing *in vitro* experiments at the physiological gas pressures that components of the immune system experience *in vivo*.

Future Directions

One final point deserves mention because it is often overlooked in discussions of the effects of hypoxia on organisms. Other environmental factors can induce indirectly an "internal" hypoxia in organisms. For example, bivalved molluscs can close themselves from the aquatic environment when conditions are unfavorable, e.g., when salinity changes rapidly or when they are disturbed by a predator. A closed bivalve does not exchange gas with the environment and tissue oxygen levels decline [Crenshaw and Neff, 1969]. The same thing can happen during air exposure when the shells of a bivalve close or when the gills of a water breather collapse. Anaerobic metabolism can predominate when tissues become hypoxic. The polychaete annelid, *Arenicola marina*, enters anaerobiosis when ambient temperature is 4°C above the aerobic threshold boundary temperature [Sommer et al., 1997; Sommer and Pörtner, 1999]. The sea urchin, *Strongylocentrotus droebachiensis*, switches to anaerobic metabolism and produces lactic acid when the coelomic cavity is full with ripe gonads [Bookbinder and Shick, 1986]; lactic acid concentrations also increase in the coelomic cavity of ripe sea urchins as environmental salinity decreases [Roller and Stickle, 1994]. These types of hypoxia may be indistinguishable, from a physiological perspective, from that induced by ambient environmental hypoxia and they should not be ignored.

As hypoxic water becomes a greater threat to coastal habitats [Diaz and Rosenburg, 1995], the need to know how low oxygen affects organisms becomes more important. There is a need for greater understanding of the sublethal effects of hypoxia on individual organisms and how these effects influence population densities and distributions. Studies done carefully on the influence of oxygen on the growth, the feeding and the predation behavior of organisms will help us to understand better the results of hypoxia at the population level. It is also important to distinguish between the acute responses and the chronic responses of an organism to hypoxia. Diurnal fluctuations in oxygen can produce responses different from chronic exposures. Investigators should also be aware that the hypercapnic and low pH water that accompanies hypoxia can have separate and profound negative effects on organisms. Finally, the immune responses of organisms appear to be sensitive to low oxygen. Studies of the components of the immune systems of aquatic organisms are needed that take into account the gas pressures and the physiological conditions that occur *in vivo*.

Acknowledgements. This work was supported in part by a grant to L. Burnett from the Charleston Harbor Project and SC Sea Grant R/ER-14. Contribution 163 of the Grice Marine Laboratory.

References

Atkinson, M. J., T. Berman, B. R. Allanson, and J. Imberger, Fine-scale oxygen availability in a stratified estuary: Patchiness in aquatic environments, *Mar. Ecol. Prog. Ser., 36*, 1-10, 1987.

Batterton, C. V. and J. N. Cameron, Characteristics of resting ventilation and response to hypoxia, hypercapnia, and emersion in the blue crab *Callinectes sapidus* (Rathbun), *Amer. J. Physiol., 203*, 403-418, 1978.

Bonaventura, C., B. Sullivan, and J. Bonaventura, Spot hemoglobin; studies on the root effect hemoglobin of a marine teleost, *J. Biol. Chem., 7*, 1871-1876, 1976.

Bookbinder, L. H. and J. M. Shick, Anaerobic and aerobic energy metabolism in ovaries of the sea urchin *Strongylocentrotus droebachiensis. Mar. Biol., 93*, 103-110, 1986.

Boyd, J. N. and L. E. Burnett, Reactive oxygen intermediate production by oyster hemocytes exposed to hypoxia, *J. Exp. Biol., 202*, 3135-3143, 1999.

Burnett, L. E., The effects of environmental oxygen levels on the respiratory function of hemocyanin in the crabs, *Libinia emarginata* and *Ocypode quadrata, J. Exp. Zool., 210*, 289-300, 1979.

Burnett, L. E., The challenges of living in hypoxic and hypercapnic aquatic environments, *Amer. Zool., 37*, 633-640, 1997.

Burnett, L. E. and K. Johansen, The role of branchial ventilation in hemolymph acid-base changes in the shore crab *Carcinus maenas* during hypoxia, *J. Comp. Physiol., 141*, 489-494, 1981.

Cameron, J. N., Effects of hypercapnia on blood acid-base status, NaCl fluxes, and trans-gill potential in freshwater blue crabs *Callinectes sapidus*, *J. Comp. Physiol., 123*, 137-141, 1978.

Christmas, J. F. and S. J. Jordan, Biological monitoring of selected oyster bars in the lower Choptank, in *Dissolved Oxygen in the Chesapeake Bay*, edited by G. B. MacKiernan, Maryland Sea Grant, College Park, Maryland, No. UM-SG-TS-87-03, pp. 125-128., 1987.

Cochran, R. E. and L. E. Burnett, Respiratory responses of the salt marsh animals, *Fundulus heteroclitus, Leiostomus xanthurus*, and *Palaemonetes pugio* to environmental hypoxia and hypercapnia and to the organophosphate pesticide, azinphosmethyl, *J. Exp. Mar. Biol. Ecol., 195*, 125-144, 1996.

Crenshaw, M. A. and J. M. Neff, Decalcification at the mantle-shell interface in molluscs, *Amer. Zool., 9*, 881-885, 1969.

Das, T. and W. B. Stickle, Sensitivity of crabs *Callinectes sapidus* and *C. similis* and the gastropod *Stramonita haemastoma* to hypoxia and anoxia, *Mar. Ecol. Prog. Ser., 98*, 263-274, 1993.

Das, T. and W. B. Stickle, Detection and avoidance of hypoxic water by juvenile *Callinectes sapidus* and *C. similis, Mar. Biol., 120*, 593-600, 1994.

deFur, P. L. and A. L. Pease, Metabolic and respiratory compensation during long term hypoxia in blue crabs, *Callinectes sapidus*, in *Understanding the Estuary: Advances in Chesapeake Bay Research*, Chesapeake Research Consortium Publication, pp. 608-616, 1988.

deFur, P. L., C. P. Mangum, and J. E. Reese, Respiratory responses of the blue crab *Callinectes sapidus* to long-term hypoxia, *Biol. Bull., 178*, 46-54, 1990.

Dejours, P., *Principles of Comparative Respiratory Physiology*, American Elsevier Publishing Company, Inc., New York, 253 pp., 1975.

Diaz, R. J. and R. Rosenburg, Marine benthic hypoxia: A review of its ecological effects and the behavioral responses of benthic macrofauna, *Oceanography and Mar. Biol.: Ann. Rev., 33*, 245-303, 1995.

Direkbusarakom, S. and Y. Danayadol, Effect of oxygen depletion on some parameters of the immune system in black tiger shrimp (*Penaeus monodon*), in *Advances in Shrimp Biotechnology*, edited by T. W. Flegel, National Center for Genetic Engineering and Biotechnology, Bangkok, pp. 147-149, 1998.

Gäde, G., Energy metabolism of arthropods and molluscs during environmental and functional anaerobiosis, *J. Exp. Zool., 228*, 415-429, 1983.

Galtsoff, P. S. and D. V. Whipple, Oxygen consumption of normal and green oysters, *Fish. Bull. U. S. N.M.F.S., 46*, 489-508, 1931.

Garlo, E. V., C. B. Milstein, and A. E. Jahn, Impact of hypoxic conditions in the vicinity of Little Egg Inlet, New Jersey in summer 1976, *Estuar. Cstl. Mar. Sci., 8*, 421-432, 1979.

Gnaiger, E., Heat dissipation and energetic efficiency in animal anoxibiosis: economy contra power, *J. Exp. Zool., 228*, 471-490, 1983.

Hagerman, L., Haemocyanin concentration in the shrimp *Crangon crangon* (L.) after exposure to moderate hypoxia, *Comp. Biochem. Physiol., 85A*, 721-724, 1986.

Jonas, R. B., Bacteria, dissolved organics and oxygen consumption in salinity stratified Chesapeake Bay, an anoxia paradigm, *Amer. Zool., 37*, 612-620, 1997.

Kapper, M. A. and W. B. Stickle, Metabolic responses of the estuarine gastropod *Thais haemastoma* to hypoxia, *Physiol. Zool., 60*, 159-173, 1987.

Lallier, F. and J.-P. Truchot, Hemolymph oxygen transport during environmental hypoxia in the shore crab, *Carcinus maenas, Respir. Physiol., 77*, 323-336, 1989.

Le Moullac, G., C. Soyez, D. Saulnier, D. Ansquer, J. C. Avarre, and P. Levy, Effect of hypoxic stress on the immune response and the resistance to vibriosis of the shrimp *Penaeus stylirostris*, *Fish and Shellfish Immunol., 8*, 621-629, 1998.

Liu, L.L., W.B. Stickle, and E. Gnaiger, Normoxic and anoxic energy metabolism of the southern oyster drill *Thais haemastoma* during salinity adaptation, *Mar. Biol., 104*, 239-245, 1990.

Loesch. H., Sporadic mass shoreward migrations of demersal fish and crustaceans in Mobile Bay, Alabama, *Ecology, 41*, 292-298, 1960.

Mangum, C. P., Adaptation of the oxygen transport system to hypoxia in the blue crab, *Callinectes sapidus, Amer. Zool., 37*, 604-611, 1997.

Mangum, C. and W. Van Winkle, Responses of aquatic invertebrates to declining oxygen conditions, *Amer. Zool., 13*, 529-541, 1973.

Mangum, C. P. and A. L. Weiland, The function of hemocyanin in respiration of the blue crab *Callinectes sapidus, J. Exp. Zool., 193*, 257-264, 1975.

Mangum, C. P. and L. E. Burnett, The CO_2 sensitivity of the hemocyanins and its relationship to Cl^- sensitivity, *Biol. Bull., 171*, 248-263, 1986.

Mangum, C. P. and J. S. Rainer, The relationship between subunit composition and oxygen binding of blue crab hemocyanin, *Biol. Bull., 174*, 77-82, 1988.

Mason, R. P., C. P. Mangum and G. Godette, The influence of inorganic ions and acclimation salinity of hemocyanin-oxygen binding in the blue crab *Callinectes sapidus, Biol. Bull., 164*, 104-123, 1983.

Mauro, N. A. and C. P. Mangum, The role of the blood in the temperature dependence of oxidative metabolism in decapod crustaceans. I. Intraspecific responses to seasonal differences in temperature, *J. Exp. Zool., 219*, 179-188, 1982.

May, E. B., Extensive oxygen depletion in Mobile Bay, Alabama, *Limnol. Oceanogr., 18*, 353-366, 1973.

McMahon, B. R. and J. L. Wilkens, Respiratory and circulatory responses to hypoxia in the lobster *Homarus americanus, J. Exp. Biol., 62*, 637-655, 1975.

Nestlerode, J. A. and R. J. Diaz, Effects of periodic environmental hypoxia on predation of a tethered polychaete, *Glycera americana*: implications for trophic dynamics, *Mar. Ecol. Prog. Ser., 172*, 185-195, 1998.

Pavela, J. S., J. L. Ross, and M. E. Chittenden, Jr., Sharp reductions in abundance of fishes and benthic macro-invertebrates in the Gulf of Mexico off Texas associated with hypoxia, *NE Gulf Sci., 6*, 167-173, 1983.

Pease, A. L., P. L. deFur, and C. Chase, Physiological compensation to long term hypoxia in the blue crab, *Callinectes sapidus, Amer. Zool., 26*, 122A, 1986.

Pihl, L., S. P. Baden, and R. J. Diaz, Effects of periodic hypoxia on distribution of demersal fish and crustacean, *Mar. Biol., 108*, 349-360, 1991.

Renaud, M. L., Detecting and avoiding oxygen deficient sea water by brown shrimp, *Penaeus aztecus* (Ives), and white shrimp, *Penaeus setiferus* (Linnaeus), *J. Exp. Mar. Biol. Ecol., 98*, 283-292, 1986.

Roller, R.A. and W. B. Stickle, Effects of adult salinity acclimation on larval survival and early development of *Strongylocentrotus droebachiensis* and *Strongylocentrotus pallidus* (Echinodermata: Echinoidea), *Can. J. Zool., 72*, 1931-1939, 1994.

Rutledge, P. S., Effects of temperature acclimation on crayfish hemocyanin oxygen binding, *Amer. J. Physiol., 240*, R93-R98, 1981.

Senkbeil, E. G. and J. C. Wriston, Jr., Hemocyanin synthesis in the American lobster, *Homarus americanus, Comp. Biochem. Phyiol., 68B*, 163-171, 1981.

Shumway, S. E. and R. K. Koehn, Oxygen consumption in the American Oyster *Crassostrea virginica, Mar. Ecol. Prog. Ser., 9*, 59-68, 1982.

Sommer, A., H. Klein, and H.-O. Pörtner, Temperature induced anaerobiosis in two populations of the polychaete worm *Arenicola marina* (L.), *J. Comp. Physiol., 167*, 25-35, 1997.

Sommer, A. and H.-O. Pörtner, Exposure of *Arenicola marina* (L.) to extreme temperatures: adaptive flexibility of a boreal and a subpolar population, *Mar. Ecol. Prog. Ser. 181*, 215-226, 1999.

Spicer, J. I., M. A. Thommasson, and J.-O. Strömberg, Possessing a poor anaerobic capacity does not prevent the diel vertical migration of Nordic krill *Meganyctiphanes norvegica* into hypoxic waters, *Mar. Ecol. Prog. Ser., 185*, 181-187, 1999.

Stickle, W. B., Effects of environmental factor gradients on scope for growth in several species of carnivorous marine invertebrates, in *Marine Biology of Polar Regions and Effects of Stress on Marine Organisms*, edited by J. S. Gray and M. E. Christiansen, John Wiley & Sons, Ltd., London, p. 601-616, 1985.

Stickle, W. B., M. A. Kapper, L.-L. Liu, E. Gnaiger, and S. Y. Wang, Metabolic adaptations of several species of crustaceans and molluscs to hypoxia: Tolerance and microcalorimetric studies, *Biol. Bull., 177*, 303-312, 1989.

Summers, J. K., S. B. Weisberg, A. F. Holland, J. Kou, V. D. Engle, D. L. Breitberg, and R. J. Diaz., Characterizing dissolved oxygen conditions in estuarine environments, *Env. Monitor. and Assess., 45*, 319-328, 1997.

Truchot, J.-P., Changements de l'etat acide-base du sang en foncton de l'oxygenation de l'eau chez le crabe, *Carcinus maenas* (L), *J. Physiol., Paris, 70*, 583-592, 1975.

Walsh, P. J., T. P. Mommsen, T. W. Moon, and S. F. Perry, Effects of acid-base variables on *in vitro* hepatic metabolism in rainbow trout, *J. Exp. Biol., 135*, 231-241, 1988.

Willson, L. L. and L. E. Burnett, Whole Animal and Gill Tissue Oxygen Uptake in the Eastern Oyster, *Crassostrea virginica*: Effects of Hypoxia, Hypercapnia, Air Exposure, and Infection with the Protozoan Parasite *Perkinsus marinus. J. Exp. Mar. Biol. Ecol.*, (in press), 2000.

7

Responses of Nekton and Demersal and Benthic Fauna to Decreasing Oxygen Concentrations

Nancy N. Rabalais, Donald E. Harper, Jr., and R. Eugene Turner

Abstract

We assembled 12 years of diver observations and five years of remotely operated vehicle video tapes on the responses of nekton and demersal and benthic fauna to decreasing concentrations of dissolved oxygen. The responses of the fauna vary, depending on the concentration of dissolved oxygen, but there is a fairly consistent pattern of progressive stress and mortality as the oxygen concentration decreases from 2 mg l^{-1} to anoxia (0 mg l^{-1}). Motile organisms (fish, portunid crabs, stomatopods, penaeid shrimp and squid) are seldom found in bottom waters with oxygen concentrations less than 2 mg l^{-1}. Below 1.5 to 1 mg l^{-1} oxygen concentration, less motile and burrowing invertebrates exhibit stress behavior, such as emergence from the sediments, and eventually die if the oxygen remains low for an extended period. At minimal concentrations just above anoxia, sulfur-oxidizing bacteria form white mats on the sediment surface, and at 0 mg l^{-1}, there is no sign of aerobic life, just black anoxic sediments.

Introduction

The persistent or episodic occurrence of low dissolved oxygen in estuarine and coastal waters occurs generally during summer when stratification of the water column limits reaeration of the bottom waters. Organic loading from surface water production stimulated by excess nutrients fuels the consumption of oxygen in the lower water column. The presence of oxygen-depleted waters affects in various ways an organism's

Coastal Hypoxia: Consequences for Living Resources and Ecosystems
Coastal and Estuarine Studies, Pages 115-128
Copyright 2001 by the American Geophysical Union

ability to feed, grow, reproduce or even survive. Effects leading to altered community structure and trophic interactions begin as dissolved oxygen concentrations approach 3 to 2 mg l^{-1} [Tyson and Pearson, 1991; Burnett and Stickle, this volume; Breitburg et al., this volume; Diaz and Rosenberg, this volume; Rabalais et al., this volume]. Many of these oxygen tolerance and behavioral modification limits have been defined by experimental work (e.g., Renaud [1986a], Breitburg et al. [1994], Johansson [1997], Nestlerode and Diaz [1998]), but others are inferred from presence/absence data in trawl and benthic samples associated with dissolved oxygen measurements from discrete or continuous samples (e.g., Leming and Stuntz [1984], Renaud [1986b], Baden et al. [1990a], Pihl et al. [1991] and Rabalais et al. [this volume]).

Very little direct observation of the effects of hypoxia on marine and estuarine organisms exists. An example is provided by Baden et al.'s [1990a] descriptions of stressed or dead benthic organisms in the southeastern Kattegat off Sweden. Perhaps the best example of the sequence of stress and then mortality of a benthic community was documented by Stackowitsch [1984, 1992] in the Gulf of Trieste, northern Adriatic Sea, as epifaunal and then infaunal organisms succumbed during a four-day anoxic event (0 mg l^{-1}). This sequence, although it followed an acute event, permitted distinction into levels of sensitivity, tolerance and mortality of organisms through a time-course of the community's demise. We accumulated, through the course of 12 years of diver observations and five years of remotely operated vehicle (ROV) video tapes, considerable information on the responses of nekton and demersal and benthic fauna to decreasing concentrations of hypoxia on the southeastern Louisiana shelf where hypoxia is seasonally severe. Here we outline the variable responses of the fauna, depending on the concentration of dissolved oxygen as it decreases from 2 mg l^{-1} to anoxia (0 mg l^{-1}).

Methods

We have five years of one-week mid-summer cruises (1989-1993) of diving observations and ROV video tapes, coupled with CTD/DO (conductivity, temperature, depth, dissolved oxygen) data from numerous stations on the southeastern Louisiana shelf (Fig. 1). The ROV was a SuperPhantom operated by personnel of the National Oceanic and Atmospheric Administration, National Undersea Research Center, University of North Carolina at Wilmington. The ROV was manipulated through a tethered cable via signals from onboard ship. CTD data were obtained prior to a dive or ROV observation with a Hydrolab Surveyor II or 3, or a SeaBird CTD unit. During 1989 and 1991-1993, a SeaCat CTD was connected to the ROV so that video images were made simultaneously while obtaining continuous dissolved oxygen data. Factory calibration and/or laboratory calibration of the oxygen sensor for the various CTD units were corroborated with Winkler titrations. We also had observations from divers from 1989 through 2000 during deployment and recovery of oxygen meters at stations C6A and C6B (Fig. 1) in 20-m water depth in the core of the hypoxic zone off Terrebonne Bay (approximately 20 dives per year from February/March through October/November). The instrument moorings were deployed next to an offshore oil platform and tethered with a cable, so that divers usually traversed the depth of the offshore platform and then across the seabed for 30 to 40 m, depending on the location of the subsurface mooring weight. Other observations were made by research cruise participants onboard ship or small outboards.

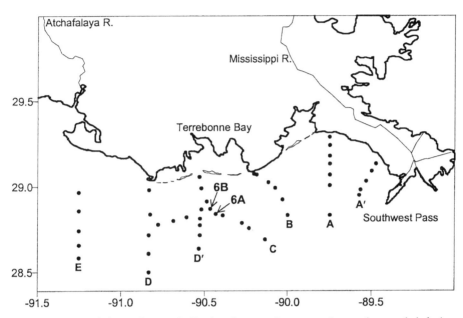

Figure 1. Map of the study area indicating those stations most frequently sampled during 1989-1993 for ROV videos and stations C6A and C6B of the instrument mooring. Depths along transects range from 5 m inshore to 60 m offshore on A' through C and to 30 m offshore on transects D' through E.

Results and Discussion

We assembled our multiple observations into a diagram that describes the responses of nekton, demersal invertebrates, and epifaunal and infaunal benthos along a gradient of decreasing dissolved oxygen concentration (Fig. 2). The value of oxygen concentration for the different behavioral responses or when mortality occurs is not absolute, because (1) individual species vary in their physiological limits to oxygen deficiency, and (2) the history of severity or persistence of low oxygen for an organism is seldom known. Dashed lines suggest approximate values for presence/absence or stress behavior, but solid lines indicate a rather precipitous decline in presence and/or dead organisms. Further explanation and details are given below.

Nekton and Demersal Fishes

Highly motile fishes (nekton) that are associated with the seabed (demersal) or more pelagic environs are seldom captured in a trawl when the oxygen level is below 2 mg l^{-1} [Pavela et al., 1983; Leming and Stuntz, 1984; Renaud, 1986b]. Fish that are brought to the surface in a trawl when the bottom-water oxygen is less than 2 mg l^{-1} are usually pelagic species that were likely caught after the trawl left the seabed on its return to the ship. Fish, squid and large mobile bottom-dwelling organisms are routinely seen in ROV tapes when the oxygen concentration is above 2 mg l^{-1} (normoxia) (Plate 1A). While

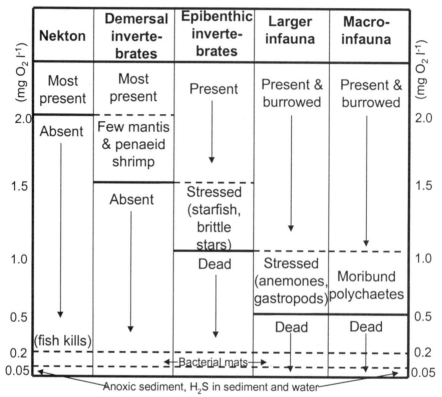

Figure 2. Progressive changes in fish and invertebrate fauna as oxygen concentration decreases from 2 mg l^{-1} to anoxia (0 mg l^{-1}).

some demersal invertebrates were seen in ROV videotapes when oxygen levels were less than 2 mg l^{-1} (see below), nekton and demersal fish usually were not. Fish were seen swimming in bottom waters of 0.95 mg O$_2$ l^{-1} on once occasion [ROV observations]. Eels, which occupy burrows in the seabed, were observed as low as 1 mg l^{-1} dissolved oxygen.

Dead fish have not been seen on the bottom by divers, but a few dead fish were observed on the sediment surface when the oxygen concentration was 0.4 mg l^{-1} during one 1993 ROV taping. The lack of fish is attributed, therefore, not to mortality, but to their avoidance of the hypoxic bottom layer by either (1) swimming upward to water above the oxycline, (2) horizontally in either an offshore/inshore direction or to the east or west of the hypoxia or (3) a combination of both upward and horizontal movement. Large schools of sting rays (*Dasyatis americana*), bottom residents and feeders, were once observed swimming at the water's surface in a shoreward direction away from a large area of hypoxic bottom water [N. Rabalais, personal observation]. Fish, such as red snapper (*Lutjanus campechanus*), mangrove snapper (*L. griseus*), red drum (*Sciaenops ocellatus*), black drum (*Pogonias cromis*), sheepshead (*Archosargus probatocephalus*), trigger fish (*Balistes capriscus*), and a variety of grouper, that normally swim into the lower water column or to the seabed to forage are concentrated above the oxycline when the waters below are less than 2 mg l[-1], and they do not move into the hypoxic waters [diver observations]. There is anecdotal evidence from recreational fishers that red snapper move offshore into deeper oxygenated bottom waters when preferred summertime habitat (15-30 m water depth) is hypoxic for extended periods, and red snapper habitat carrying capacity is thought to be reduced in the hypoxic zone of Louisiana and Texas [Gallaway et al., 1999]. Further, no fish have been observed to enter into the bottom waters to prey upon stressed or dead invertebrate fauna (see below). Fishes (spot, *Leistomus xanthurus*; pinfish, *Lagodon rhomboides*; croaker, *Micropogonias undulatus*; mullet, *Mugil cephalus*; and menhaden, *Brevoortia patronus*) and brown shrimp (*Farfantepenaeus aztecus*) all responded to low dissolved oxygen by moving away from the hypoxic area in laboratory experiments, beginning at oxygen levels above 1 mg l[-1] [Wannamaker and Rice, 2000].

Pihl et al. [1991] documented the avoidance of hypoxic waters (< 2 mg l[-1]) in the York River subestuary of Chesapeake Bay by three demersal fish species—spot, hogchoker (*Trinectes maculatus*) and croaker and two crustacean species, the mantis shrimp (*Squilla empusa*) and blue crab (*Callinectes sapidus*). The effect of periodic hypoxia on the demersal species appeared to be related to their oxygen tolerance, mobility and feeding habits. The fishes and blue crabs moved quickly into and out of the deeper parts of the river, depending on the level of dissolved oxygen. Mantis shrimp seemed to remain in the deep hypoxic strata until their tolerance was exceeded, and then they migrated to, and remained in, shallow water. All species reacted to oxygen concentrations below 2 mg l[-1] with migration from deeper to shallower waters, but differences in oxygen tolerance were found among species. When the oxygen conditions improved, all species, except for mantis shrimp, returned to the deeper areas.

On one ROV dive, numerous anchovies (later identified as *Anchoa nasuta* from a specimen recovered from the ROV frame) were observed dipping into the low oxygen bottom waters (0.07 mg l[-1]) head-first into the sediments, performing a wriggling motion sideways through the sediment, and then exiting rapidly back up into the water column and out of sight. This was presumed to be stress behavior, because anchovies are pelagic fish and do not feed on sediment organisms.

Fish kills do occur related to hypoxia in Louisiana, but not in offshore bottom waters. Kills occur rarely when hypoxic water masses, often containing toxic hydrogen sulfide, are forced towards shore along barrier island beaches following a wind from the north and upwelling favorable conditions offshore. The hypoxic waters trap fishes, shrimp and crabs that cannot escape and then die. An example was documented by personnel from the Louisiana Department of Wildlife and Fisheries and the Louisiana Department of Environmental Quality in 1990 off Grand Isle. Approximately 150,000 fish, mostly hardhead catfish (*Arius felis*), small croakers and redfish littered the beach along with dead blue crabs, shrimp (*Penaeus* spp.), eels, sheepshead, sting rays and mullet (*Mugil cephalus*) (Plate 1B) [New Orleans *Times-Picayune*, August 28, 1990]. The dissolved oxygen in shallow waters off the beach was 0.2 to 0.4 mg l[-1] the day following the fish kill [Kerry M. St. Pé, Louis. Dept. Envtl. Qual., personal communication]. Precursors to a fish kill are "jubilees" in which listless fish, crabs and shrimp are concentrated along the beach and subject to easy capture by people with nets. Jubilees are mid-summer features of the eastern Mobile Bay shoreline when hypoxic bottom-water is advected onto beach areas by tides and wind-driven baroclinic motions with similar results of concentrating fish and invertebrates [Loesch, 1960; May, 1973; Schroeder and Wiseman, 1988]. A fish kill in June 1984 on the beaches between Freeport and Galveston, Texas was attributed to low oxygen levels and/or hydrogen sulfide generation following the sinking and decomposition of a massive bloom of *Gymnodinium sanguineum* [Harper and Guillen, 1989].

Demersal Invertebrates

Demersal invertebrates have some swimming capability, but normally reside in the sediments, below the sediment-water interface, and usually feed on the sediment surface during night. Day/night differences in trawl capture clearly show that these demersal invertebrates are more active at night when commercial trawlers target the shrimp population. Examples are penaeid shrimp, including the commercially important white and brown shrimp, portunid crabs and stomatopods (= mantis shrimp).

These invertebrates are also seldom caught in trawls when the oxygen levels fall below 2 mg l^{-1} [Pavela et al., 1983; Leming and Stuntz, 1984; Renaud, 1986b], but are routinely seen in ROV tapes at oxygen above that level. Some penaeid shrimp and stomatopods were observed in ROV tapes at oxygen levels as low as 1.7 to 1.8 mg l^{-1}, but never below 1.5 mg l^{-1}. Stomatopods remaining in hypoxic waters down to 1.5 mg l^{-1} is consistent with the results of Pihl et al. [1991] who indicated that *Squilla empusa* first adapted physiologically to hypoxia and then migrated as hypoxia became more severe, which is consistent with its more stationary and territorial behavior. *Callinectes sapidus*, however, in the York River studies [Pihl et al., 1991], migrated from hypoxic waters as soon as the oxygen concentration fell below 2 mg l^{-1}, and then returned when conditions improved.

The demersal invertebrates are thought to move away from the hypoxic zone by swimming and further avoid it for its duration, but their actual behaviors over long distances and time are not known. Juvenile brown shrimp, *Farfantepenaeus aztecus*, detected and avoided water with a dissolved oxygen content of less than 2 mg l^{-1} in aquarium-scale experiments, but juvenile white shrimp, *Litopenaeus setiferus*, were more tolerant and did not show avoidance until the oxygen concentration was 1.5 mg l^{-1} [formerly *Penaeus*, Renaud, 1986a]. Juvenile blue crabs (both *Callinectes sapidus* and the lesser blue crab *C. similis*) are also able to detect hypoxia (< 2 mg O_2 l^{-1}), and *C. similis* avoided it, but *C. sapidus* did not [Das and Stickle, 1994]. *Callinectes similis* is the dominant member of this genus in the nearshore continental shelf of Louisiana most likely to be affected by hypoxia. Juvenile *C. sapidus* remain in the estuaries, and only adult female *C. sapidus* return to offshore surface waters in summer to spawn where they are not likely to be affected by the low oxygen on the bottom.

The presence of hypoxic bottom waters over an area as large as 20,000 km^2 results in the removal of a large portion of essential habitat for commercially important shrimps during part of the summer [Craig et al., this volume; Zimmerman and Nance, this volume]. Demersal invertebrates can be trapped and killed in the same manner described above for fish kills on barrier island beaches, or remain stressed and easily captured under conditions of a jubilee. No dead penaeid shrimp have been observed on the bottom in the ROV tapes or by divers. Several lines of evidence indicate that demersal invertebrates swim up out of the low oxygen bottom waters. *Penaeus* sp. were observed clinging to instrument mooring cables and on top of particle traps at 14 m depth in a 21-m water column when bottom conditions were hypoxic (0.2 to 0.3 mg l^{-1} on the bottom

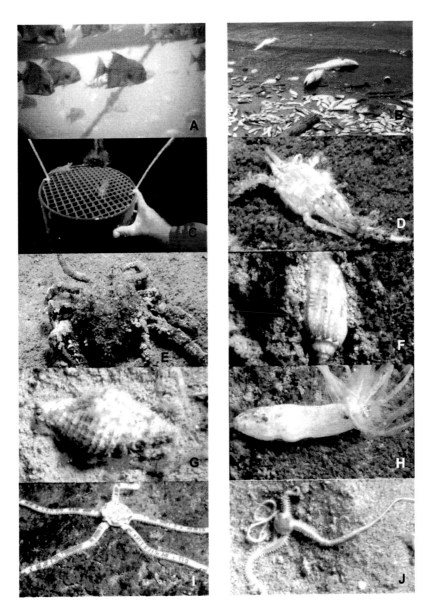

Plate 1. A, School of spadefish swimming above the oxycline at an offshore oil platform; B, Fish kill on Grand Isle barrier beach; C, *Penaeus* sp. on top of particle trap at 14 m; D, Dead portunid crab (probably *Portunus gibbesii*); E, Dead majid crab (*Libinia dubia*) , F, *Oliva sayana* plowing through surface sediments; G, *Distorsio clathrata* plowing through surface sediments; H, stressed cerianthid anemone on sediment surface; I, stressed ophiuroid with disk raised above sediment surface, note accumulation of detrital material; J, stressed ophiuroid with disk raised above sediment surface. Photo credits: A, Nancy Rabalais; B, Kerry St. Pé; C, D, E © Franklin Viola/www.violaphoto.com; F, G, H, I, J, Donald E. Harper, Jr.

and 2.1-2.3 mg l^{-1} at 14 m) (Plate 1C). Large aggregations of rock shrimp (*Sicyonia dorsalis* and *S. brevirostris*) and portunid crabs (*Portunus gibbesii* and *Callinectes similis*) were observed swimming at the surface above bottom waters that were severely oxygen depleted (0.1 to 0.2 mg l^{-1}). [Some portunid crabs succumbed to the low oxygen, however (Plate 1D).] Whether demersal invertebrates, including penaeid shrimp, are capable of moving towards normoxic waters over large distances (up to 50 km in an offshore-inshore direction, and even longer distances in an alongshore direction) is not known. Escaping the bottom and the cover of sediments, however, makes these organisms highly susceptible to pelagic predators.

We spent many hours video taping the seabed at the interface of normoxic/hypoxic water on the inshore leading edge, and never documented a "herding" effect of large numbers of escaping invertebrates (or fish) on the normoxic side or at the interface. We did document a concentration of swarming fish and darting zooplankton at mid-water column above a steep interface of > 2 mg l^{-1} at 15 m that fell to < 0.1 mg l^{-1} at 19-20 m with oxygen levels of 1.2 mg l^{-1} from 20 m to the 30-m deep seabed. Once the ROV entered the severely hypoxic water mass, no fish or zooplankton were observed.

The lobster (*Nephrops norvegicus*) fishery of the southeastern Kattegat declined as progressively worsening and more widespread hypoxia affected the area from 1985 to 1989 [Baden et al., 1990b]. These organisms, however, unlike the demersal penaeid shrimp, which can detect and avoid hypoxia, are burrowers and remain in or near their burrows until hypoxia forces them out and eventually kills them. The behavior of this commercially important crustacean makes it more susceptible to reductions in oxygen concentration than penaeid shrimp.

Although nekton were never observed foraging in bottom waters less than 2 mg l^{-1}, squid (*Lolliguncula brevis*) were observed dipping into and out of low oxygen (1 mg l^{-1}) water during ROV missions in 1989. The squid hovered above the hypoxia and dipped down into it and up without lingering on the bottom, presumably feeding on benthos. Squid were very common in the ROV tapes when the oxygen concentration was above 2 mg l^{-1}.

Less Motile Epibenthic and Infaunal Benthos

In the hypoxic zone off Louisiana where soft sediments dominate, there are no attached epifaunal organisms such as sponges or soft corals. Biofouling communities of barnacles, hydroids, bryozoans, anemones, sponges and ascidians with their associated fauna occur on offshore oil platform pilings, but not below a persistent oxycline where dissolved oxygen falls below 2 mg l^{-1}. The lack of a fouling community on the lower platform pilings is not just a feature of depth and depth-related variables such as light, food availability and temperature but instead a lack of oxygen to support the fauna. A biofouling community was present below the usual depth of the oxycline (10 to 15 m in a 20-m water column) at the Station C6B oil platform in late spring and early summer of 2000 when hypoxia had not yet developed, but disappeared following the formation of hypoxia in mid-July of that year (later than normal) [diver observations]. The lack of a biofouling community where severe seasonal hypoxia occurs on the Louisiana shelf is different from the epifauna that persisted on artificial substrates in relatively mild and short hypoxia episodes in the York River subestuary of the Chesapeake Bay [Sagasti et al., 2000].

At oxygen concentrations below 1.5 mg l⁻¹, the less motile invertebrates, burrowing invertebrates and benthic macroinfauna begin to display evidence of stress or die as the oxygen concentration declines toward 0 mg l⁻¹. The less motile larger invertebrates are not swimmers and instead reside in the sediments during day where they remain to feed, or they may move onto the sediment surface at night to feed; e.g., spider crabs, brittle stars, sea stars and gastropods. Typical burrowing invertebrates include alpheid shrimp, thalassinid shrimp, mud crabs (families Xanthidae and Goneplacidae), cerianthid anemones, gastropods and bivalves. The smaller vermiform invertebrates (e.g., polychaetes, nemerteans and sipunculids) reside within the sediments to feed on surface particles with the aid of palps or tentacles or remain below the sediment-water interface as subsurface deposit feeders.

As the oxygen level decreases from 1.5 to 1 mg l⁻¹, bottom-dwelling organisms exhibit stress behavior. Crabs (e.g., *Libinia* sp., *Persephona* sp.) and sea stars (*Astropecten* sp.) climb on top of high spots, such as burrow excavation mounds (Plate 2A). Hermit crabs (unidentified) cluster on top of shells lying on the bottom. Similar clustering and aggregations on elevated areas were observed in the sequence leading to an acute mortality event in the Gulf of Trieste [Stachowitsch, 1984, 1992]. Eventually these crustaceans die as the oxygen remains low (Plate 1E). Brittle stars emerge from the sediment and use their arms to raise their disks off the substrate (Plates 1I and 1J), similar to behavior documented for ophiuroids in the southeastern Kattegat [Baden et al., 1990a]. The ophiuroids *Amphiura filiformis* and *A. chiajei* emerged from the sediment in experimental tests when oxygen fell to 1.2 mg l⁻¹ and 0.8 ml l⁻¹, respectively [Rosenberg et al., 1991]. Burrowing shrimp (*Alpheus* sp.) emerge from their burrows (Plates 2C and 2D). Gastropods (*Oliva sayana, Terebra* sp., *Cantharus cancellarius* and *Distorsio clathrata*) (Plates 1F and 1G) move through the surface sediments with their siphons extended directly upward. Bivalves (unidentified ark shells) burrow through the sediments. Large burrowing anemones (*Cerianthus* sp.) become partly or completely extended from their tubes and lie on the substrate, in a flaccid and non-responsive condition (Plate 1H). Polychaete worms emerge from the substrate and lie motionless on the surface (e.g. *Chloeia viridis* and *Lumbrineris* sp.) (Plate 2B). These behaviors are presumed to position the organisms in higher oxygen content waters, even though moving from the safety of the sediments exposes them to greater risk of predation. The presence of large typically infaunal organisms on the sediment surface supports the idea presented earlier that bottom-feeding fish are excluded from the hypoxic lower water column.

At oxygen levels of 1 to 0.5 mg l⁻¹ even the most tolerant burrowing organisms, principally polychaetes, emerge partially or completely from their burrows and lie motionless on the bottom. Several polychaetes, one hemichordate, one ophiuroid and several cerianthid anemones that appeared lifeless on the bottom became active when they were brought to the surface in sealed containers of ambient water and placed in shallow dishes with the ambient water that naturally re-oxygenated by diffusion [Rabalais and Harper, 1992]. Jørgensen [1980] also found that many of the organisms seen lying on the bottom in hypoxic areas were moribund, not dead. If these organisms survive, they may re-enter the sediment and may partially account for the recolonization of benthos when hypoxia abates. Below oxygen concentrations of 0.5 mg l⁻¹, there is a

Plate 2. A, Bottom landscape with burrow excavation mound, brittle stars, and exposed cerianthid anemone; B, Stressed lumbrinerid polychaete; C, Live alphaeid shrimp at entrance to burrow; D, Dead alphaeid shrimp at entrance to burrow; E, Dead mud crab, Goneplacidae; F. Dead and decomposing thalassinid shrimp; G. Dead and decomposing spionid polychaete on anoxic sediments; H. Anoxic sediments, bacterial filaments, burrow tube of *Diopatra cuprea*; I, small patch of *Beggiatoa* spp.; J, large, white cottony mat of *Beggiatoa* spp. Photo credits: A, NURC ROV; B, C, D, E, F, I, Donald E. Harper, Jr.; G, H, J © Franklin Viola/www.violaphoto.com.

fairly linear decline in species richness, abundance and biomass of benthic macroinfauna [Rabalais et al., this volume]. Dead and decaying polychaetes and crustaceans were observed on the sediment surface at oxygen concentrations less than 0.25 mg l^{-1} (Plates 2E, 2F and 2G). The gaping valves of a freshly dead *Tellina* were observed at an oxygen concentration of 0.2 mg l^{-1}. Despite the anoxic appearance of sediments and detection of hydrogen sulfide in overlying waters, there usually remain some surviving fauna, typically polychaetes of the genera *Magelona, Paraprionospio* or *Sigambra* or sipunculans [Rabalais and Harper, 1992; Rabalais et al., this volume], so that the sediments are not completely azoic. Resistant organisms in the Gulf of Trieste mass mortality were two species of anemones, two species of gastropods and the sipunculan *Sipunculus nudis* [Stachowitsch, 1992].

The succession of changes along a decreasing oxygen gradient composed from widespread and temporally diverse observations on the Louisiana shelf parallels the sequence of events following a mass mortality of epifauna and infauna in the Gulf of Trieste [Stachowitsch, 1984, 1992].

Bacterial Mats and Anoxic Sediments

At oxygen values below 0.2 mg l^{-1} but above anoxia (0 mg l^{-1}) various sized patches of "cottony" mats cover the sediment surface and were identified as *Beggiatoa* spp. [Larkin and Strohl, 1990]. Filaments of the bacteria *Beggiatoa* spp. and other unidentified filamentous bacteria form on the surface of the sediments at oxygen levels as high as 1 mg l^{-1} but not higher [L. Duet, Q. Dortch, N. Rabalais, unpublished data]. The obvious *Beggiatoa* mats are observed at the extremely low oxygen concentrations (< 0.2 mg l^{-1}) (Plates 2H, 2I and 2J). These white, cottony mats are commonly seen on the sediment surface on ROV missions, and by divers traversing the seabed during deployment of oxygen meters on the instrument mooring. Many of the *Beggiatoa* mats observed in the 1993 ROV tapes were yellowish in color, and the presence of intracellular sulfur inclusions has been verified [Duet et al., unpublished data]. Similar bacterial mats were observed by divers off Freeport, Texas in June 1979 following an anoxic event [Harper et al., 1981], and are common in the Baltic where bacterial mats (*Beggiatoa* spp.) form at the sediment surface following long periods of hypoxia [Rumohr et al., 1996]. Colorless sulfur bacteria (*Beggiatoa* spp. and *Thiovulum* sp.) grow at the interface between oxygen and sulfide zones and cover the mud during periods of extremely low oxygen concentration and indicates that trace amounts of oxygen did still reach the sediments [Jørgensen, 1980]. *Beggiatoa* spp. colonies were of two types within the sediments of the Stockholm Archipelago [Rosenberg and Diaz, 1993]—either delicate lacey white-looking colonies that did not completely cover the surface sediments, and a yellowish, denser colony that blanketed the sediments. At 0 mg l^{-1} the sediment becomes almost uniformly black, and there are no obvious signs of aerobic life. When the bottom water is depleted in oxygen, hydrogen sulfide builds up in the bottom waters as anaerobic bacterial metabolism reduces sulfate to H_2S [Jørgensen, 1980]. H_2S was detected in bottom waters during the 1979 anoxic event off Texas [Harper et al., 1981] and on many occasions during sampling of bottom waters or on dives within the Louisiana shelf study area [D. E. Harper and N. N. Rabalais, personal observations].

Implications for Trophic Interactions

Pihl et al. [1992] indicate that short-lived hypoxic episodes did not appear to lessen habitat value for fisheries species and in fact may have facilitated predation upon benthos at times when the infauna were stressed from low oxygen. Stressed infauna may become suitable and easily exploited prey for demersal organisms that can enter hypoxic waters for a short period to feed. In this scenario, enhancement of energy transfer may be temporarily facilitated by hypoxia. On the other hand, if the time of hypoxia lengthens or the dissolved oxygen concentrations approach anoxia, the system may change significantly in community composition and trophic interactions. The shift in energy transfer under these conditions may be towards minimal transfer. Re-occupation of the hypoxic zone after severe hypoxia may result in continued low populations of organisms with a continued low transfer of energy [Rabalais et al., this volume].

As oxygen concentration falls on the southeastern Louisiana shelf, fish move away from the area before the less motile invertebrates and macroinfauna become stressed. This sequence is in contrast to other areas where infauna are stressed first and the fish benefit from increased ease of prey capture. Predators in the York River subestuary of Chesapeake Bay exhibited dietary evidence of optimal prey exploitation during or immediately after hypoxic events [Pihl et al., 1992]. Unlike the results of the Pihl et al. study, where predators exploited moribund or slowly recovering benthos affected by hypoxia, predators were excluded from the hypoxic bottom waters of the Louisiana shelf, and moribund and/or decomposing benthic fauna remained on the surface of the sediments.

Conclusions

The soft sediments of the southeastern Louisiana shelf are uniform, flat and non-descript even when well oxygenated. When the oxygen concentration is greater than 2 mg l^{-1}, however, the seabed is characterized by the movements of nekton and demersal fishes and invertebrates and is marked with evidence of feeding deposits, burrowing activity and trails of organisms. There are gradients of behavioral changes, stress indicators and eventually mortality in groups of nektonic, demersal and benthic organisms as oxygen concentrations decrease towards zero. In oxygen concentrations less than 1 mg l^{-1}, the sense is of a vast emptiness on the sea bottom and the absence of multi-celled organisms. There is no evidence of burrowing activity in the sedimentary landscape, and it is often characterized by bacterial mats or black anoxic muds.

The problems associated with seasonally-severe bottom-water hypoxia or anoxia worldwide have accelerated with increased input of nutrients during the last half of the twentieth century [Rosenberg, 1985; Rosenberg et al., 1990; Diaz and Rosenberg, 1995] and is expected to worsen with changing global climate scenarios [Kennedy, 1990; Justic' et al., 1996; Justic' et al., this volume]. As the persistence of hypoxia expands in estuarine and coastal regions, or the severity increases, gradients of stress and mortality similar to those observed on the Louisiana shelf will expand with potentially negative impacts on commercially important fisheries.

Acknowledgments. The National Oceanic and Atmospheric Administration, National Undersea Research Center at the University of North Carolina at Wilmington provided funding for research cruises, and ROV and diving support. The Louisiana Education Quality Support Fund and the NOAA Nutrient Enhanced Coastal Ocean Productivity Program funded the instrument mooring from which many of the diver observations were made. We thank Kerry St. Pé and Franklin Viola for the use of their photographs, and Ben Cole for converting the color slides into the photographic plates. We thank the many divers and assistants over the many years of our observations. Louisiana Sea Grant Program and the San Diego Foundation funded color plate production, and the Department of Energy provided funds for the preparation of this manuscript.

References

Baden, S. P., L.-O. Loo, L. Pihl and R. Rosenberg, Effects of eutrophication on benthic communities including fish: Swedish west coast, *Ambio, 19,* 113-122, 1990a.

Baden, S. P., L. Pihl and R. Rosenberg, Effects of oxygen depletion on the ecology, blood physiology and fishery of the Norway lobster *Nephrops norvegicus, Mar. Ecol. Prog. Ser., 67,* 141-155, 1990b.

Breitburg, D. L., N. Steinberg, S. DuBeau, C. Cooksey, and E. D. Houde, Effects of low dissolved oxygen on predation on estuarine fish larvae, *Mar. Ecol. Prog. Ser., 104,* 235-246, 1994.

Das, T. and W. B. Stickle, Detection and avoidance of hypoxic water by juvenile *Callinectes sapidus* and *C. similis, Mar. Biol., 120,* 593-600, 1994.

Diaz, R. J. and R. Rosenberg, Marine benthic hypoxia: A review of its ecological effects and the behavioural responses of benthic macrofauna, *Oceanogr. Mar. Biol. Ann. Rev., 33,* 245-303, 1995.

Gallaway, B. J., J. G. Cole, R. Meyer, and P. Roscigno, Delineation of essential habitat for juvenile red snapper in the northwestern Gulf of Mexico, *Trans. Amer. Fish. Soc., 128,* 713-726, 1999.

Harper, D. E., Jr. and G. Guillen, Occurrence of a dinoflagellate bloom associated with an influx of low salinity water at Galveston, Texas, and coincident mortalities of demersal fish and benthic invertebrates, *Contr. Mar. Sci., 31,* 147-161, 1989.

Harper, D. E., Jr., L. D. McKinney, R. R. Salzer, and R. J. Case, The occurrence of hypoxic bottom water off the upper Texas coast and its effects on the benthic biota, *Contr. Mar. Sci., 24,* 53-79, 1981.

Johansson, B., Behavioural response to gradually declining oxygen concentration by Baltic Sea macrobenthic crustaceans, *Mar. Biol., 129,* 71-78, 1997.

Jørgensen, B. B., Seasonal oxygen depletion in the bottom waters of a Danish fjord and its effect on the benthic community, *Oikos, 34,* 68-76, 1980.

Justic', D., N. N. Rabalais, and R. E. Turner, Effects of climate change on hypoxia in coastal waters: A doubled CO_2 scenario for the northern Gulf of Mexico, *Limnol. Oceanogr., 41,* 992-1003, 1996.

Kennedy, V. S., Anticipated effects of climate change on estuarine and coastal fisheries, *Fisheries, 16,* 16-24.

Larkin, J. M. and W. R. Strohl, *Beggiatoa, Thiotrix,* and *Thioplaca, Ann. Rev. Microbiol., 37,* 341-67, 1990.

Leming, T. D. and W. E. Stuntz, Zones of coastal hypoxia revealed by satellite scanning have implications for strategic fishing, *Nature, 310,* 136-138, 1984

Loesch, H., Sporadic mass shoreward migrations of demersal fish and crustaceans in Mobile Bay, Alabama, *Ecology, 41,* 292-298, 1960.

May, E. B.. Extensive oxygen depletion in Mobile Bay, Alabama, *Limnol. Oceanogr., 18,* 353-366, 1973.

Nestlerode, J. A. and R. J. Diaz, Effects of periodic environmental hypoxia on predation of a tethered polychaete, *Glycera americana*: implications for trophic dynamics, *Mar. Ecol. Prog. Ser., 172,* 185-195, 1998.

Pavela, J. S., J. L. Ross, and M. E. Chittenden, Sharp reductions in abundance of fishes and benthic macroinvertebrates in the Gulf of Mexico off Texas associated with hypoxia, *Northeast Gulf Sci., 6,* 167-173, 1983.

Pihl, L., S. P. Baden, and R. J. Diaz, Effects of periodic hypoxia on distribution of demersal fish and crustaceans, *Mar. Biol., 108,* 349-360, 1991.

Pihl, L., S. P. Baden, R. J. Diaz, and L. C. Schaffner, Hypoxia-induced structural changes in the diet of bottom-feeding fish and Crustacea, *Mar. Biol., 112,* 349-361, 1992.

Rabalais, N. N., R. E. Turner, D. Justic', Q. Dortch, and W. J. Wiseman, Jr., Characterization of hypoxia: Topic 1 Report for the Integrated Assessment of Hypoxia in the Gulf of Mexico. NOAA Coastal Ocean Program Decision Analysis Series No. 16, NOAA Coastal Ocean Program, Silver Springs, Maryland, 1999.

Rabalais, N. N. and D. E. Harper, Jr., Studies of benthic biota in areas affected by moderate and severe hypoxia, in *Nutrient Enhanced Coastal Ocean Productivity Workshop Proceedings*, pp. 150-153, Publ. No. TAMU-SG-92-109, Texas A&M University Sea Grant College Program, College Station, Texas, 1992.

Renaud, M. L., Detecting and avoiding oxygen deficient sea water by brown shrimp, *Penaeus aztecus* (Ives), and white shrimp, *Penaeus setiferus* (Linnaeus), *J. Exp. Mar. Biol. Ecol., 98,* 283-292, 1986a.

Renaud, M., Hypoxia in Louisiana coastal waters during 1983: implications for fisheries, *Fishery Bull., 84,* 19-26, 1986b.

Rosenberg, R., Eutrophication—the future marine coastal nuisance?, *Mar. Pollut. Bull., 16,* 227-231, 1985.

Rosenberg, R., R. Elmgren, S. Fleisher, P. Jonsson, G. Persson, and H. Dahlin, Marine eutrophication case studies in Sweden—a synopsis, *Ambio, 19,* 102-108, 1990.

Rosenberg, R., B. Hellman, B. Johansson, Hypoxic tolerance of marine benthic fauna, *Mar. Ecol. Prog. Ser., 79,* 127-131, 1991.

Rosenberg, R. and R. J. Diaz, Sulfur bacteria (*Beggiatoa* spp.) mats indicate hypoxic conditions in the inner Stockholm Archipelago, *Ambio, 22,* 32-36, 1993.

Rumohr, H., E. Bonsdorff, and T. H. Pearson, Zoobenthic succession in Baltic sedimentary habitats, *Arch. Fish. Mar. Res. 44,* 170-214, 1996.

Sagasti, Al., L. C. Schaffner, and J. E. Duffy, Epifaunal communities thrive in an estuary with hypoxic conditions, *Estuaries, 23,* 474-487, 2000.

Schroeder, W. W. and W. J. Wiseman, Jr., The Mobile Bay estuary: stratification, oxygen depletions, and jubilees, in *Hydrodynamics of Estuaries, Volume II, Case Studies*, edited by B. J. Kjerfve, pp. 41-52, CRC Press, Boca Raton, Florida, 1988.

Stachowitsch, M., Mass mortality in the Gulf of Trieste: The course of community destruction, *P.S.Z.N.I: Mar. Ecol., 5,* 243-264, 1992.

Stachowitsch, M., Benthic communities: eutrophication's "memory mode", in *Marine Coastal Eutrophication*, edited by R. A. Vollenweider, R. Marchetti, and R. Viviani, pp. 1017-1028, *Sci. Total Environ.*, suppl. no. 0048-9697, 1992.

Tyson, R. V. and T. H. Pearson, Modern and ancient continental shelf anoxia: an overview in *Modern and Ancient Continental Shelf Anoxia*, edited by R. V. Tyson and T. H. Pearson, pp. 1-24, *Geological Society Special Pub., 58,* 1991.

Wannamaker, C. M. and J. A. Rice, Effects of hypoxia on movements and behavior of selected estuarine organisms from the southeastern United States, *J. Exper. Mar. Biol. Ecol., 249,* 145-163, 2000.

8

Overview of Anthropogenically-Induced Hypoxic Effects on Marine Benthic Fauna

Robert J. Diaz and Rutger Rosenberg

Abstract

While hypoxic and anoxic environments have existed through geological time, their occurrence in coastal and estuarine areas has increased over the last 40 years. The cause of this increase is related to anthropogenic activities. Synthesis of literature pertaining to benthic hypoxia and anoxia revealed that the oxygen budgets of many major coastal ecosystems have been adversely affected mainly through the process of eutrophication (the production of excess organic matter). The time scales over which hypoxia occurred was a key element in determining the level of benthic effects. Aperiodic hypoxia, the first form to affect a system, occurs at intervals of < 1 year and accounted for 15% of all systems with reported accounts of hypoxia. Annually occurring seasonal hypoxia was the most prevalent and occurred in about 70% of the systems. Persistent hypoxia/anoxia was less common (15%) and a feature of systems with limited circulation or bottom water exchange. Periodic hypoxia with > 1 event per year was uncommon and occurred in < 5% of the systems. Several large systems, with historical data, that never reported hypoxia at the turn of the century (i.e. Kattegat, northern Adriatic Sea) now experience severe seasonal hypoxia. Ecosystems with historic data that are now severely stressed by hypoxia have experienced reduced biodiversity and altered energy flow through the system relative to hypoxic conditions. It is very likely that these types of ecosystem-level alterations have occurred in the other hypoxia-stressed systems that do not have historical data.

Coastal Hypoxia: Consequences for Living Resources and Ecosystems
Coastal and Estuarine Studies, Pages 129-146
Copyright 2001 by the American Geophysical Union

Introduction

Several reviews of literature pertaining to the ecological effects of hypoxia (< 2 mg O_2 l^{-1}) and anoxia (0.0 mg O_2 l^{-1}) in coastal and estuarine ecosystems revealed that the oxygen budgets of many major coastal ecosystems around the world have been adversely affected within the last 40 years [Caddy, 1993; Diaz and Rosenberg, 1995; Johannessen and Dahl, 1996]. Mounting evidence points to the combination of excess nutrients being delivered to coastal ecosystems as causal factors that in turn lead to production of excess organic matter, or eutrophication [Nixon, 1995; Howarth et al., 1996]. Eutrophication produces excess organic matter that fuels the development of hypoxia and anoxia when combined with water column stratification. Many ecosystems have reported some form of long-term decline in annual average dissolved oxygen levels through time with a strong correlation between human activities and declining dissolved oxygen (for example: northern Gulf of Mexico, Texas-Louisiana; northern Adriatic Sea, Italy-Slovenia; Kattegat, Sweden-Denmark). In other ecosystems the linkage of human activity to hypoxia is less obvious (for example: Chesapeake Bay, Maryland-Virginia, USA; Saanich Inlet, British Columbia, Canada; Port Hacking, Australia).

Oxygen is necessary to sustain the life of fishes and invertebrates, but the point at which negative effects from declining dissolved oxygen appear for different species varies with environmental conditions. Generally, behavioral effects start when oxygen concentrations drop below about 3 mg O_2 l^{-1} for mobile fauna and about 2 mg O_2 l^{-1} for sedentary fauna (for summaries, see Diaz and Rosenberg [1995], Craig et al. [this volume], Rabalais et al. [this volume]). In sea water 2 mg O_2 l^{-1} is about 18 to 20% of full saturation. As a point of reference, air contains about 280 mg O_2 l^{-1}. While scientists usually measure dissolved oxygen concentration, from a physiological point of view it is the partial pressure and availability of oxygen, rather than concentration, that determines survival of an organism [Herreid, 1980]. No fish and few invertebrates have extended tolerance of anoxia.

The ecological effects of hypoxia within affected systems vary in magnitude but are surprisingly consistent from system to system in terms of benthic community structure [Diaz and Rosenberg, 1995] and fisheries response [Caddy, 1993]. Initially, both benthos and fishes respond to organic enrichment, which typically can be measured as increased biomass (either standing stock or catch). In addition, there may be a decline in benthic species richness.

In this chapter, we present an overview of the effects of hypoxia on large coastal ecosystems around the world. Unfortunately, few systems have pre/post hypoxia data on both oxygen concentrations and benthic resources that can be used to infer effects. More typically, studies were not initiated until the first reports of hypoxia or anoxia, or obvious signs of an ecological imbalance such as a fish kill. By the time such events occurred, the ecosystems may have already undergone significant changes. Inference to how systems initially respond to hypoxia then must rely on those few systems with long-term data. Most information from hypoxic systems reflects how they currently continue to cope with low dissolved oxygen stress. We will also attempt to draw comparisons to the current northern Gulf of Mexico hypoxia situation.

Linkage Between Nutrients, Eutrophication and Hypoxia

Excess nutrient loading leads to eutrophication of coastal seas, a widespread problem around the globe in general [Nixon, 1995; Howarth et al., 1996]. The primary factor driving marine coastal eutrophication is an imbalance in the nitrogen cycle that can be directly linked to increased urbanization in coastal river drainage basins or expanded agricultural activities [Howarth, 1998]. In many areas hypoxia follows the eutrophication that results from the underlying nutrient problem. The distribution of marine hypoxic zones around the world [Diaz and Rosenberg, 1995] appears to be closely associated with developed watersheds or coastal population centers that deliver large quantities of nutrients, the most important being nitrogen, to coastal seas [Howarth et al., 1996]. Use of industrial fertilizer, combustion of fossil fuels and sewage are regarded as the key generators of nitrogen, but in fact it is the increased population and rising living standards that drive the need for energy and food [Nixon, 1995].

The direct connection between land and sea may be exemplified by the relationship between estuarine and coastal fisheries production and land-derived nutrients. The most productive fisheries zones around the world are associated with significant inputs of either land (runoff) or deep oceanic (upwelling) derived nutrients [FAO, 1997]. While the basic nutrients carried by land runoff and oceanic upwelling are essential elements that eventually support species of economic importance, an excess of nutrients can lead to problems.

The scenario or sequence of events linking nutrient additions to the formation of hypoxia and impacts on benthos and fisheries via eutrophication can be summarized as follows. Excess nutrients lead to increased primary production, which is new organic matter added to the ecosystem. Because shallow estuarine and coastal systems tend to be tightly coupled (benthic-pelagic coupling) much of this organic matter reaches the bottom [Graf, 1992]. This increased primary productivity in many cases leads to increased fisheries production [Caddy, 1993] and benthic standing stock [Rosenberg et al., 1987]. At some point, however, the ecosystem's ability to balance the increased flow of energy through microbial/metazoan metabolic pathways is exceeded and, if physical dynamics permit stratification, hypoxic conditions can develop. Initially, from an economic point of view, the increased fisheries production may offset any obvious detrimental ecological effects of hypoxia. As eutrophication increases and hypoxia expands in duration and space, the demersal fisheries production base is affected and declines. Benthic species respond to hypoxia differently and tend to become pulsed with periods on intense productivity between hypoxic intervals. In the case of both fisheries and benthos, the changes occur through a complex series of ecological interactions initially toward increased organic matter and eventually to declining dissolved oxygen concentration [Diaz and Rosenberg, 1995].

The endpoint of this hypoxia sequence or graded reaction to the complex series of problems related to excess nutrients and organic matter has been documented in many systems around the globe (Table 1). For all these systems we found at least one published account of hypoxia that appeared to be related to some form of anthropogenic disturbance with most authors describing "their" system as affected to varying degrees

TABLE 1. Summary of benthic effects for systems that experience anthropogenic hypoxia. zone within which no macrofauna occur. The absence of fauna from these anoxic zones is not Aperiodic, events that are known to occur at irregular intervals greater than a year; Periodic, round hypoxia. Time trends of hypoxia are: - = improving conditions; + = increasing; 0 = appear similar before and after hypoxic event; Mortality, moderate reductions of populations, Benthic recovery is: No Change, current community dynamics appear unrelated to hypoxia; Multi-year, gradual return of community structure taking more than a year; Annual, recoloni-Stressed, migration out of hypoxia or avoidance; Reduced,

System	Hypoxia Type	Fauna Response	Fauna Recovery
SE North Sea, W. Denmark	Aperiodic	Mortality	Annual
N Gulf of Mexico,Texas Shelf, Deep	Aperiodic	Mortality	Annual
Wismar Bay, Western Baltic	Aperiodic	Mortality	Reduced
Wadden Sea	Aperiodic	Mortality	.
German Bight, North Sea	Aperiodic	Mass Mortality	Annual
New York Bight, New Jersey	Aperiodic	Mass Mortality	Multi-year
Sommone Bay, France	Aperiodic	Mass Mortality	Multi-year
N Gulf of Mecixo, Texas Shelf, Shallow	Aperiodic	Mass Mortality	Multi-year
York River, Virginia	Periodic	None	No Change
Rappahannock River, Virginia	Periodic	Mortality	Annual
Chesapeake Bay Mainstem	Annual	Mortality	Annual
German Bight, North Sea	Annual	Mortality	Annual
Laholm Bay, Sweden	Annual	Mortality	Annual
N Gulf of Mexico, Louisiana Shelf	Annual	Mortality	Annual
Oslofjord, Norway	Annual	Mortality	Annual
Port Hacking, Australia	Annual	Mortality	Annual
Saanich Inlet, British Columbia	Annual	Mortality	Annual
Seto Inland Sea, Japan	Annual	Mortality	Annual
Tome Cove, Japan	Annual	Mortality	Annual
Corpus Christi Bay, Texas	Annual	Mortality	Reduced
Eckernforde Bay, Germany	Annual	Mortality	Reduced
Swedish West Coast Fjords	Annual	Mortality	Reduced
Long Island Sound, New York	Annual	Mortality	?
Marmara Sea	Annual	Mortality	?
Ise Bay, Japan	Annual	Mortality	.
Bilbao Estuary, Spain	Annual	Mortality	.
Sea of Azov	Annual	Mortality	.
Elefsis Bay, Aegean Sea	Annual	Mass Mortality	Annual
Black Sea NW Shelf	Annual	Mass Mortality	Annual
Gullmarsfjord, Sweden	Annual	Mass Mortality	Annual
Hillsborough Bay, Florida	Annual	Mass Mortality	Annual
Kiel Bay, Germany	Annual	Mass Mortality	Annual
Limfjord, Denmark	Annual	Mass Mortality	Annual
Lough Ine, Ireland	Annual	Mass Mortality	Annual
Pamlico River, North Carolina	Annual	Mass Mortality	Annual
Tolo Harbor, Hong Kong	Annual	Mass Mortality	Annual
Århus Bay, Denmark	Annual	Mass Mortality	Multi-year
Baltic Sea, Bornholm Basin	Annual	Mass Mortality	Multi-year
Baltic Sea, Pomeranian Bay	Annual	Mass Mortality	Multi-year
Gulf of Trieste, Adriatic	Annual	Mass Mortality	Multi-year

Several of these systems also experience anoxia. In the case of many fjords there is an anoxic considered a community response but a consequence of stable anoxia. Hypoxia is typed as: events occurring at intervals shorter than a year; Annual, single yearly event; Persistent, year stable; ˙ = no data. Benthic community response is categorized as: None, communities many species survive; Mass Mortality, drastic reduction or elimination of the benthos. Reduced, recolonization occurs but community does not return to prehypoxic structure; zation and return of similar community structure within a year. Fisheries Response is: populations declined; Mortality, loss of populations.

Fisheries Response	Time Trends	Reference
Stressed	.	Westernhangen and Dethlefson [1983]
Stressed	0?	Harper et al. [1981, 1991]
Stressed	.	Prena [1995]
.	.	de Jonge et al. [1994]
.	+	Dethlefsen and Westernhagen [1983]
Mortality, Surf clams	.	Sindermann and Swanson [1980]
Mortality, Cockles	+?	Desprez et al. [1992]
Stressed	+	Harper et al. [1981, 1991]
Stressed	0	Pihl et al. [1991], Diaz et al. [1992]
Stressed	+	Llansó [1992]
Stressed	+	Holland et al. [1987], Dauer et al. [1992]
Stressed	+?	Niermann et al. [1990]
Stressed	+	Rosenberg and Loo [1988]
Stressed	+	Rabalais et al. [this volume]
Reduced	0	Mirza and Gray [1981]
.	.	Rainer and Fitzhardinge [1981]
.	0	Richards, 1965; Tunnicliffe [1981]
.	.	Imabayashi [1986]
.	.	Tsutsumi [1987]
.	.	Ritter and Montagna [1999]
.	.	D'Andrea et al. [1996]
Stressed	+	Josefson and Rosenberg [1988]
Stressed	+	Howell and Simpson [1994], Welsh et al. [1994]
Stressed	+	Orhon and Yüksek, ms
Stressed	.	Nakata et al. [1997]
.	.	Gonzáles-Oreja and Saiz-Salinas [1998]
.	.	Balkas et al. [1991]
.	.	Friligos and Zenetos [1988]
Reduced	+	Zaitsev [1992]
Stressed	+	Nilsson and Rosenberg [2000]
.	.	Santos and Simon [1980]
Stressed	+	Arntz [1981], Weigelt [1990, 1991]
No bottom fishery	+	Jørgensen [1980], Hylleberg [1993]
.	0	Kitching et al. [1976]
Mortality	.	Tenore [1972], Stanley and Nixon [1992]
.	.	Wu [1982]
.	+	Fallesen and Jørgensen [1991]
.	+	Leppäkoski [1969]
.	+	Powilleit and Kube [1999]
Stressed	+	Stachowitsch [1991]

| | Hypoxia | Fauna | TABLE 1.
Fauna |
System	Type	Response	Recovery
Kattegat, Sweden-Denmark	Annual	Mass Mortality	Multi-year
Los Angeles Harbor, Califorinia	Annual	Mass Mortality	Reduced
Mobile Bay, Alabama	Annual	Mass Mortality	?
Hiuchi Sound, Japan	Annual	Mass Mortality	.
Mikawa & Ise Bays, Japan	Annual	Mass Mortality	.
Omura Bay, Japan	Annual	Mass Mortality	.
Neuse River Estuary, North Carolina	Annual	Mass Mortality	.
Elbe Estuary, Germany	Annual	.	.
Loire Estrary, France	Annual	.	.
Baltic Sea, Gotland Basin	Persistent	Mortality	Reduced
Baltic Sea, Northern	Persistent	Mortality	Reduced
Byfjord, Sweden	Persistent	Mortality	Reduced
Idefjord, Sweden-Norway	Persistent	Mortality	Reduced
Caspian Sea	Persistent	Mortality	Reduced
Sullom Voe, Shetland	Persistent	Mass Mortality	No Change
Gulf of Finland, Deep	Persistent	Reduced	Multi-year
Stockholm Inner Archipelago	Persistent	No Benthos	No Change

by eutrophication. Oxygen minimum zones (OMZs) and coastal upwelling systems are not included in this summary because their occurrence is unrelated to anthropogenic activities. In fact, OMZs appear to be stable oceanographic features that have highly specialized faunal associations [Nichols, 1976; Levin and Gage, 1998].

Oxygen Budgets Around the Globe

The increasing input of anthropogenic nutrients to many estuarine and coastal systems over the last 40 to 50 years is likely the main contributor to recent declining trends in bottom water oxygen concentrations both around Europe [Rosenberg, 1990; Johannessen and Dahl, 1996] and North America [Nixon, 1995], including the northern Gulf of Mexico [Turner and Rabalais, 1991; Rabalais et al., 1996]. In systems with long-term oxygen records, like the northern Adriatic or Scandinavian fjords, the declining trends in dissolved oxygen appeared to start in the 1950s or 1960s [Justic' et al., 1987; Rosenberg, 1990; Johannessen and Dahl, 1996]. In the Baltic Sea, because of restricted deep-water exchange, declining dissolved oxygen levels were noted as early as the 1930s [Fonselius, 1969], but hypoxia/anoxia likely occurred in the deepest areas of the Baltic proper as early as the 1870s [Laine et al., 1997, and references therein]. It must be kept in mind, however, that it was not until the late 1880s that a method was developed to measure dissolved oxygen [Winkler, 1888]. Unfortunately, even with the early development of a measurement method, in many systems dissolved oxygen records do not start until after significant environmental changes were observed. The general declining trends in dissolved oxygen lag about 10 to 20 years behind post World War II trends of increased used of chemical fertilizer [Nixon, 1995, and references therein] and emphasize the complex connection between development of hypoxic/anoxic bottoms and the general process of eutrophication.

Continued.

Fisheries Response	Time Trends	Reference
Mortality, Norway lobster	+	Baden et al. [1990a], Josefson and Jensen [1992]
.	.	Reish [1955], Reish et al. [1980]
Stressed	0	May [1973]
.	.	Sanukida et al. [1984]
.	.	Suzuki and Matsukawa [1987]
.	.	Iizuka and Min [1989]
Mortality, oysters	.	Lenihan and Peterson [1998]
Stressed	.	Thiel et al. [1995]
Mortality	+	Thouvenin et al. [1994]
Stressed	-/+	Laine et al. [1997]
Stressed	-/+	Andersin et al. [1978]
Pelagic only	0	Rosenberg [1990]
.	-	Rosenberg [1980]
.	0	Dumont [1998]
.	0	Pearson and Eleftheriou [1981]
.	-	Laine et al. [1997]
.	+	Rosenberg and Diaz [1993]

From an historical perspective, it appears that many coastal systems that are currently hypoxic were not when first studied while many enclosed estuarine, fjordic and embayment systems were. This is possibly related to a combination of geomorphology and closer proximity of enclosed systems to nutrient sources (primarily land-derived) that would tend to lead to earlier development of hypoxia, and the more open nature of coastal systems that would tend to counteract hypoxia development. For example, it is not likely that the hypoxic zone in the northern Gulf of Mexico would exist were it not for close land-sea interaction with the large discharge of the Mississippi River delivering stratifying fresh water and stimulating nutrients directly to the relatively quiescent and shallow continental shelf off Louisiana [Rabalais et al., this volume].

The best examples of systems with long-term data come from Europe where benthic hypoxia was not reported prior to the 1960s in the northern Adriatic [Justic', 1987], 1980 in the Kattegat [Baden et al., 1990a] and the 1980s on the northwestern continental shelf of the Black Sea [Mee, 1992]. The northern Baltic proper, however, represents a special case in that low dissolved oxygen may have existed in deeper areas prior to 1900, but it was not until the 1960s that large areas of bottom were affected by hypoxia [Fonselius, 1969]. Even though the exchange of deep water in the Baltic is episodic, there is convincing evidence that eutrophication had accelerated oxygen consumption in bottom waters [Jonsson and Carman, 1994; Laine et al., 1997].

The Kattegat and the Gulf of Trieste provide good examples of systems with historical data from the 1910s that track the combined affects of eutrophication and hypoxia. The former having long-term benthic data and the latter long-term oxygen data. The Kattegat is the system where the classic descriptions of benthic communities were made in the 1910s and fisheries were well developed [Petersen, 1915]. By the 1970s hypoxia was first documented, and the Kattegat ecosystem was suggested to be in disorder [Rosenberg, 1985]. Seasonal hypoxia began in 1980 and was accompanied by fish and invertebrate mortalities. By 1984, Norway lobster (*Nephrops norvegicus*) populations were affected by hypoxia with high catches, of what were most likely stressed individuals that had migrated out of their burrows with low oxygen [Baden et al., 1990b]. The following year, poor catches were recorded; in 1988 no Norway lobsters were found in the south Kattegat, and mass mortality of benthic and fisheries species were reported [Baden et al., 1990a; Breitberg et al., this volume].

In the northern Adriatic Sea, oxygen measurements from 1911 through 1966 indicated that summer and autumn oxygen concentrations in bottom waters seldom approached hypoxic levels (only three times in the 55-year record). Reports of mass benthic mortality related to hypoxia/anoxia started in 1969 and were irregular up to the 1980s. The sequence of events that led to the northern Adriatic's current state of severe annual hypoxia progressed over the last 40 years and were a direct result of eutrophication leading to increased sedimentation of organic matter from phytoplankton blooms fueled by excess nutrients coming out of the Po River, Italy [Justic', 1987; Justic' et al., 1987; Baramawidjaja et al., 1995].

By the 1970s, ecosystems around the world were becoming over-enriched with organic matter, and many of them manifested hypoxia for the first time. Once it occurred, hypoxia quickly became an annual event and a prominent feature affecting energy flow processes in many of these ecosystems [Baird and Ulanowicz, 1989; Pearson and Rosenberg, 1992; Caddy, 1993]. Over a period of 10 to 15 years, the northern Adriatic Sea went from experiencing aperiodic to annual seasonal hypoxia. From the 1980s to the present, the distribution of systems experiencing hypoxia around the world has not changed in a positive way (Table 1). If anything, it appears that more systems are being affected by hypoxia through time. Only in systems that have experienced intensive regulation of nutrient or carbon inputs have spatial and/or temporal trends in dissolved oxygen concentrations either stabilized or improved. A good example is Chesapeake Bay where about a 20% reduction in controllable nitrogen loads may have stopped the spread of hypoxic waters [Boesch et al., in press]. The first investigations of bottom water quality in Chesapeake Bay in the 1930s reported hypoxia in deep channel areas of the mainstem [Newcombe and Horne, 1938]. Geochronologies from the mainstem Chesapeake Bay pointed to early European settlement of the Bay watershed as a key feature that led to changes in many paleoenvironmental indicators as long as 300 years ago [Cooper and Brush, 1991; Zimmerman and Canuel, 2000]. The first signs of eutrophication appeared around 1850-1880 and the first hypoxic/anoxic event in the early 1890s. Prior to this date, and as far back as 1610, there is no indication of hypoxia occurring in Chesapeake Bay [Zimmerman and Canuel, 2000].

More dynamic systems do not have a natural tendency towards hypoxia. For example, the Kattegat and suprahalocline Baltic Sea (< 70 m) do not have a history of hypoxia with oxygen measurements that go back to the turn-of-the-century [Andersin et al., 1978; Rosenberg and Loo, 1988]. It was not until the 1980s that oxygen was found to be a problem in these areas with significant effects on the benthic fauna [Pearson et al., 1985; Elmgren, 1989]. Similarly, the northern Adriatic, with oxygen data from the 1910s and a geochronology from sediment cores starting in the 1830s, did not exhibit effects attributable to hypoxia until the 1960s [Justic' et al., 1987; Barmawidjaja et al., 1995]. While hypoxia was present when dissolved oxygen was first measured in the early 1970s, an historical picture of oxygen conditions for the northern Gulf of Mexico derived from geochronology of sediment cores indicated that hypoxia was not a prominent feature of the shallow continental shelf prior to the 1920s or possibly even the 1950s [Rabalais et al., 1999; Rabalais et al., 1996; Sen Gupta et al., 1996]. There are, however, examples of small-scale hypoxia reversals associated with improvements in treatment of sewage and pulp mill effluents [Rosenberg, 1972, 1976]. Even the deeper subhalocline regions of the Baltic Sea, where the persistence of hypoxia/anoxia is related to the aperiodic renewal of bottom water from inflows of North Sea water, are not exceptions to the nutrient regulation statement. Eutrophication of surface waters has accelerated the formation of bottom hypoxia in these deep areas [Laine et al., 1997].

Hypoxia and System Response

Of the 57 systems we compiled with published accounts of hypoxia and benthic effects (Table 1), eight experienced aperiodic hypoxia, the first form of anthropogenic hypoxia to affect a system. Annual summer-autumn hypoxia was the most common form of low dissolved oxygen event recorded around the globe and occurred in 39 (68%) systems that ranged from warm temperate to boreal. Persistent hypoxia, a more advanced and disruptive form of hypoxia, had developed in eight boreal systems. Periodic hypoxia, more than one low oxygen event per year, was relatively uncommon and reported from two estuaries in the warm temperate Chesapeake Bay.

Interestingly, the degree of reported ecological effects related to hypoxia varied from system to system and appeared to be a function of the minimum dissolved oxygen concentrations, duration of exposure to hypoxic conditions and areal extent of hypoxia. The York River, Virginia, experiences about five moderate to severe periodic hypoxic events per summer, each lasting form about two to 12 days, yet there were no discernible changes in benthic community structure or secondary production over the affected period [Diaz et al., 1992]. Fishes and crabs were temporarily displaced from the hypoxic areas but returned to forage when oxygen conditions improved [Pihl et al., 1991; 1992; Nestlerode and Diaz, 1998].

The response of the benthos in systems with annual summertime hypoxia was split into two categories, either annual recovery or multi-year recovery. Annual recovery, defined as a return to a similar community structure from one year to the next, was more prevalent in temperate and boreal systems that had already shifted community structure in response to hypoxia (for example, Chesapeake Bay or Kiel Bay). Muti-year recovery from annual hypoxia was more prevalent in boreal coastal systems (for example, Laholm Bay). Aperiodic hypoxic events tend to elicit drastic responses at the ecosystem level and multi-year recovery intervals (for example the 1976 hypoxic event off the coast of New York and New Jersey that caused mass mortality of many commercial and noncommercial species [Azarovitz et al., 1979; Boesch and Rabalais, 1991]). Unfortunately, many systems that now have annual hypoxia started out reporting aperiodic hypoxic events with significant reductions in successional stages of the benthos (for example, northern Adriatic Sea). Subsequently, a reduced benthic community structure responds less to the annual events (for example, the northern Gulf of Mexico). Persistent hypoxia obviously has the greatest effect on benthos and demersal fishes by removing all habitat value of the bottom for extended periods (for example, the Baltic Sea proper).

The general effects of eutrophication, which would tend to favor species with opportunistic life histories [Pearson and Rosenberg, 1978], appeared to have conditioned benthic fauna, both fishes and invertebrates, to lessen their response to hypoxic stress. This would account for the lack of measured response to short-term periodic hypoxia and the annual recovery of systems with severe annual hypoxia. The most serious effects of the combined problems associated with eutrophication and hypoxia are seen in the Black Sea and Baltic Sea, where demersal trawl fisheries have either been eliminated or severely stressed [Mee, 1992; Elmgren, 1984]. Since the 1960s increasing hypoxia and anoxia on the northwestern continental shelf of the Black Sea (which is not part of the deep central basin anoxia) have been blamed for the replacement of the highly valued demersal fish species with less desirable planktonic omnivores. Of the 26 commercial species fished in the 1960s, only six still support a fishery [Mee, 1992]. In the Kattegat, initially in the 1980s, hypoxia caused mass mortality of commercial and non-commercial benthic species. Since then, large-scale migrations and/or mortality among demersal fish and lobster continued, resulting in lower species richness and reduced growth and biomass [Rosenberg and Loo, 1988; Pihl, 1989]. Hypoxia in this area is believed to be partly responsible for the overall decline in stock size, recruitment and landings of commercial demersal fish over the last two decades [Baden et al., 1990b; Pihl, 1989]. Given that many fishery species are mobile, about 70% of hypoxic systems reported avoidance of areas with low dissolved oxygen and 30% mortality or reduction of fishery populations (Table 1).

The northern Gulf of Mexico, a warm temperate system, has been affected by severe annual hypoxia but not to the same degree as boreal systems previously described. Over the last several decades hypoxia has affected benthic invertebrate communities, but there is no clear signal of hypoxic effects in fishery landings statistics [Diaz and Solow, 1999; Chesney et al., 2000]. Fishery yields have remained high for the last 40 y [Chesney and Baltz, this volume] despite occasional low catch years for individual species, such as menhaden (*Brevoortia patronus*) in 1995 [Smith, this volume] and the fact that hypoxia annually displaces mobile species [Craig et al., this volume]. There may be a negative relationship, however, between brown shrimp (*Farfantepenaeus aztecus*, formerly *Penaeus aztecus*), the most offshore of commercial shrimp species in the Gulf of Mexico, yield and hypoxia. From the mid 1980s to the late 1990s brown shrimp catch per unit effort (CPUE) declined about 22% concomitant with an increase in the areal extent of hypoxia [Zimmerman and Nance, this volume]. White shrimp (*Litopenaeus setiferus*, formerly *Penaeus setiferus*), a more inshore species than brown shrimp, CPUE did not decline over this same interval.

To a point, nutrient enrichment may increase fishery yields, but beyond a certain level, it is negative in effect [Caddy, 1993]. At first increased nutrients lead to increased fisheries production. As organic matter production increases, changes occur in the flow of energy that lead to different endpoints such as increased microbial populations and reductions of higher-level benthic consumers. These changes are very predictable and have followed the same path in many marine ecosystems. For example, the relationship between nutrient loads delivered to northern Gulf of Mexico by the Mississippi River, hypoxia and basic ecological responses (i.e., increased primary productivity in the water column, increased flux of organic matter to the bottom, bottom water hypoxia, altered energy flow and stressed fisheries [Rabalais et al., 1999]) are typical of other system responses around the world (see reviews by Brongersma-Sanders [1957], Caddy [1993], Diaz and Rosenberg [1995]). However, hypoxia is not the only stress factor to which fisheries populations respond. Other factors implicated in declining stocks or populations are incidental catch of juvenile individuals [Andrew and Pepperell, 1992; Chesney et al., 2000; Chesney and Baltz, this volume], trawl disturbance [Currie and Parry, 1996], fishing pressure [Turkstra et al., 1991], habitat loss [Chesney et al., 2000], harmful algal blooms [Karup et al., 1993] and altered trophic pathways [Purcell et al., 1999].

Summary

No other environmental variable of such ecological importance to estuarine and coastal marine ecosystems around the world has changed so drastically, in such a short period of time, as dissolved oxygen. While hypoxic and anoxic environments have existed through geological time, their occurrence in estuarine and coastal areas has clearly increased due to anthropogenic activities.

Up to the 1950s, reports of mass mortality of marine animals caused by lack of oxygen were limited to small systems that had histories of oxygen stress, such as Eckernford Bay or fjordic systems, or natural upwelling areas [Brongersma-Sanders, 1957]. Starting in the 1960s, the number of systems with reports of hypoxia-related problems increased with most first-time reports from the late 1970s and 1980s. By the 1990s most estuarine and marine systems in close proximity to population centers had reports of hypoxia or anoxia. Unfortunately, hypoxic conditions in about two-thirds of the systems listed in Table 1 appear to be worsening with time. The main cause for development of hypoxia in these systems can be linked to the delivery of excess nutrients that leads to eutrophication. Except in areas influenced by OMZs or upwelling, coastal hypoxia does not appear to be a natural condition, except in enclosed fjordic systems.

Oxygen deficiency (hypoxia and anoxia) may very well be the most widespread anthropogenically-induced deleterious effect in estuarine and marine environments around the world, but determination of hypoxia effects is complicated by many factors. Effects of hypoxia are masked by inadequate data on historic trends of species populations and dissolved oxygen concentrations, as well as the interaction of multiple stressors, which include factors such as fishing pressure and habitat loss.

Acknowledgments. The authors would like to thank Bob Carney, Nancy Rabalais and a third reviewer for their insightful review of our manuscript. Support for this work was provided by NOAA's Coastal Ocean Program via the Gulf of Mexico Program.

References

Andrew, N. L. and J. G. Pepperell, The by-catch of shrimp trawl fisheries, *Oceanogr. Mar. Biol. Ann. Rev.*, *30*, 527-565, 1992.

Andersin, A. B., J. Lassig, L. Parkkonen, and H. Sandler, The decline of macrofauna in the deeper parts of the Baltic proper and the Gulf of Finland, *Kiel. Meeresforschungen, Sonderheft, 4*, 23-52, 1978.

Arntz, W., Zonation and dynamics of macrobenthos biomass in an area stressed by oxygen deficiency, in Stress Effects on Natural Ecosystems, edited by G. W. Barrett and R. Rosenberg, pp. 215-225, John Wiley & Sons, Chichester, 1981.

Azarovitz, T. R., C. J. Byrne, M. J. Silverman, B. L. Freeman, W. G. Smith, S. C. Turner, B. A. Halgren, and P. J. Festa, Effects on finfish and lobster, in Oxygen Depletion and Associated Benthic Mortalities in New York Bight, 1976, edited by R. L. Swanson and C. J. Sindermann, pp. 295-314, NOAA Professional Paper 11, 1979.

Baden, S. P., L. O. Loo, L. Pihl, and R. Rosenberg, Effects of eutrophication on benthic communities including fish - Swedish west coast, *Ambio, 19*, 113-122, 1990a.

Baden, S. P., L. Pihl, and R. Rosenberg, Effects of oxygen depletion on the ecology, blood physiology and fishery of the Norway lobster *Nephrops norvegicus, Mar. Ecol. Prog. Ser., 67*, 141-155, 1990b.

Baird, D. and R. E. Ulanowicz, The seasonal dynamics of the Chesapeake bay ecosystem, *Ecol. Monogr., 59*, 329-364, 1989.

Barmawidjaja, D. M., G. J. van der Zwaan, F. J. Jorissen, and S. Puskaric, 150 years of eutrophication in the northern Adriatic Sea: evidence from a benthic foraminiferal record, *Mar. Geol., 122*, 367-384, 1995.

Boesch, D. F. and N. N. Rabalais, Effects of hypoxia on continental shelf benthos: Comparisons between the New York Bight and the northern Gulf of Mexico, in *Modern and Ancient Continental Shelf Anoxia,* edited by R. V. Tyson and T.H. Pearson, pp. 27-34, *Geological Society Special Pub., 58,* 1991.

Boesch, D. F., R. B. Brinsfield, and R. E. Magnien, Chesapeake Bay eutrophication: Scientific understanding, ecosystem restoration, and challenges for agriculture, J. Envtl. Qual., in press.

Brongersma-Sanders, M., Mass mortality in the sea, in *Treatise on Marine Ecology and Paleoecology, Vol. 1,* edited by J. W. Hedgpeth, pp. 941-1010, Waverly Press, Baltimore, Maryland, 1957.

Caddy, J., Toward a comparative evaluation of human impacts on fishery ecosystems of enclosed and semi-enclosed seas, *Rev. Fishery Sci., 1*, 57-96, 1993.

Chesney, E. J., D. M. Baltz, and R. G. Thomas, Louisiana estuarine and coastal fisheries and habitats: Perspectives from a fish's eye view, *Ecol. Appl., 10*, 350-366, 2000.

Cooper, S. R. and G. S. Brush, Long-term history of Chesapeake Bay anoxia, *Science, 254*, 992-996, 1991.

Currie, D. R. and G. D. Parry, Effects of scallop dredging on a soft sediment community: A large-scale experimental study, *Mar. Ecol. Prog. Ser., 134*, 131-150, 1996.

D'Andrea, A. F., N. I. Craig, and G. R. Lopez, Benthic macrofauna and depth of bioturbation in Eckernföfde Bay, southwestern Baltic Sea, *Geo-Mar. Letters, 16*, 155-159, 1996.

Dauer, D. M., A. J. Rodi, and J. A. Ranasinghe, Effects of low dissolved oxygen events on the macrobenthos of the lower Chesapeake Bay, *Estuaries 15*, 384-391, 1992.

de Jonge, V. N., W. Boynton, C. F. DiElia, R. Elmgren, and B. L. Welsh, Responses to developments in eutrophication in four different North Atlantic estuarine systems, in *Changes in Fluxes in Estuaries: Implications from Science to Management*, edited by K. R. Dyer and R. J. Orth, pp. 179-196, Olsen and Olsen, Fredensborg, Denmark, 1994.

Desprez, M., H. Rybarczyk, J. G. Wilson, J. P. Ducrotoy, F. Sueur, R. Olivesi, and B. Elkaim,

Biological impact of eutrophication in the Bay of Somme and the induction and impact of anoxia, *Netherlands J. Sea Res., 30*, 149-159, 1992.

Dethlefsen, V. and H. v. Westernhagen, Oxygen deficiency and effects on bottom fauna in the eastern German Bight 1982, *Meeresforschung, 60*, 42-53, 1983.

Diaz, R. J., and R. Rosenberg, Marine benthic hypoxia: A review of its ecological effects and the behavioural responses of benthic macrofauna, *Oceanogr. Mar. Biol. Ann. Rev., 33*, 245-303, 1995.

Diaz, R. J. and A. Solow, Ecological and economic consequences of hypoxia: Topic 2 Report for the Integrated Assessment of Hypoxia in the Gulf of Mexico, NOAA Coastal Ocean Program Decision Analysis Series No. 16, NOAA Coastal Ocean Program, Silver Springs, Maryland, 45 pp, 1999.

Diaz, R. J., R. J. Neubauer, L. C. Schaffner, L. Phil, and S. P. Baden, Continuous monitoring of dissolved oxygen in an estuary experiencing periodic hypoxia and the effect of hypoxia on macrobenthos and fish, *Sci. Total Environ. Supplement 1992*, 1055-1068, 1992.

Dumont, H. J., The Caspian Lake: history, biota, structure, and function, *Limnol. Oceanogr., 43*, 44-52, 1998.

Elmgren, R., Trophic dynamics in the enclosed, brackish Baltic Sea, *Rapports et Proces-Verbaux des Réunions, 183*, 152-169, 1984.

Elmgren, R., Man's impact on the ecosystem of the Baltic Sea: Energy flows today and at the turn of the century, *Ambio, 18*, 326-332, 1989.

Fallesen, G. and H. M. Jørgensen, Distribution of *Nephtys hombergii* and *N. ciliata* (Polychaeta: Nephtyidae) in Århus Bay, Denmark, with emphasis on the effect of sewage oxygen deficiency, *Ophelia Supplement 5*, 443-450, 1991.

FAO, Review of the state of world fishery resources: marine fisheries, FAO Fisheries Circular. No. 920, FAO, Rome, 173 pp., 1997.

Fonselius, S. H., Hydrography of the Baltic deep basins III, Series Hydrography report No. 23, Fishery Board of Sweden, 97 pp., 1969.

Friligos, N. and A. Zenetos, Elefsis Bay anoxia: nutrient conditions and benthic community structure. *Mar. Ecol., 9*, 273-290, 1988.

González-Oreja, J. A. and J. I. Saiz-Salinas, Exploring the relationships between abiotic variables and benthic community structure in a polluted estuarine system, *Wat. Res., 32*, 3799-3807, 1998.

Graf, G., Benthic-pelagic coupling: A benthic view, *Oceanogr. Mar. Biol. Annu. Rev., 30*, 149-190, 1992.

Harper, Jr., D. E., L. D. McKinney, R. B. Salzer, and R. J. Case, The occurrence of hypoxic bottom water off the upper Texas coast and its effects on the benthic biota, *Contrib. Mar. Sci. 24*, 53-79, 1981.

Harper, Jr., D. E., L. D. McKinney, J. M. Nance, and R. B. Salzer, Recovery responses of two benthic assemblages following an acute hypoxic event on the Texas continental shelf, northwestern Gulf of Mexico, in *Modern and Ancient Continental Shelf Anoxia*, edited by R. V. Tyson & T.H. Pearson, pp. 49-64, *Geological Society Special Pub., 58*, 1991.

Holland, A. F., A. T. Shaughnessy, and M. H. Hiegel, Long-term variation in mesohaline Chesapeake Bay macrobenthos: Spatial and temporal patterns, *Estuaries, 10*, 370-378, 1987.

Herreid, C. F., Hypoxia in invertebrates, *Comparative Biochem. Histology, 67A*, 311-320, 1980.

Howarth, R. W., An assessment of human influences on fluxes of nitrogen from the terrestrial landscape to the estuaries and continental shelves of the North Atlantic Ocean, *Nutrient Cycling in Agroecosystems*, *52*, 213-223, 1998.

Howarth, R. W., G. Billen, D. Swaney, A. Townsend, N. Jaworski, K. Lajtha, J. A. Downing, R. Elmgren, N Caraco, T. Jordan, F. Berendse, J. Freney, V. Kudeyarov, P. Murdoch, and Z. Zhao-Liang, Regional nitrogen budgets and riverine N & P fluxes for the drainage to the North Atlantic Ocean: natural and human influences, *Biogeochem.*, *35*, 75-139, 1996.

Howell, P. and D. Simpson, Abundance of marine resources in relation to dissolved oxygen in Long Island Sound, *Estuaries*, *17*, 394-402, 1994.

Iizuka, S. and S. H. Min, Formation of anoxic bottom waters in Omura Bay, *Engan Kaiyo Kenkyu Note*, *26*, 75-145, 1989.

Imabayashi, H., Effect of oxygen-deficient water on the settled abundance and size composition of the bivalve *Theora lubrica*, *Bull. Japanese Soc. Scient. Fish,*. *5*, 391-397, 1986.

Johannessen, T. and E. Dahl, Declines in oxygen concentrations along the Norwegian Skagerrak coast, 1927-1993: A signal of ecosystem changes due to eutrophication?, *Limnol. Oceanogr.*, *41*, 766-778, 1996.

Jonsson, P. and R. Carman, Changes in deposition of organic matter and nutrients in the Baltic Sea during the twentieth century, *Mar. Pollut. Bull.*, *28*, 417-426, 1994.

Jørgensen, B. B., Seasonal oxygen depletion in the bottom waters of a Danish fjord and its effect on the benthic community, *Oikos*, *34*, 68-76, 1980.

Josefson, A. B. and J. N. Jensen, Effects of hypoxia on soft-sediment macrobenthos in southern Kattegat, in *Marine Eutrophication and Population Dynamics*, edited by G. Colombo, I. Ferrari, V. U. Ceccherelli and R. Rossi, pp. 21-28, Olsen & Olsen, Fredensborg, 1992.

Josefson, A. B. and R. Rosenberg, Long-term soft-bottom faunal changes in three shallow fjords, west Sweden, *Netherlands J. Sea Res.*, *22*, 149-159, 1988.

Josefson, A. B. and B. Widbom, Differential response of benthic macrofauna and meiofauna to hypoxia in the Gullmar Fjord basin, *Mar. Biol.*, *100*, 31-40, 1988.

Justic', D., Long-term eutrophication of the northern Adriatic Sea, *Mar. Pollut. Bull.*, *18*, 281-284, 1987.

Justic', D., T. Legovic', and L. Rottini-Sandrini, Trends in the oxygen content 1911-1984 and occurrence of benthic mortality in the northern Adriatic Sea, *Estuar. Coast. Shelf Sci.*, *25*, 435-445, 1987.

Karup, H., S. Evans, E. Dahl, J. Klungsoeyer, L. O. Reiersen, J. S. Gray, P. E. Iverson, T. Bokn, and H. R. Skjoldal, Man's impact on ecosystems, *North Sea Subreg. 8 Assess. Rep.*, pp. 59-62. 1993.

Kitching, J. A., F. J. Ebling, J. C. Gable, R. Hoare, A. A. Q. R. McLeod, and T. A. Norton, The ecology of Lough Ine. XIX. Seasonal changes in the western trough, *J. Animal Ecol.*, *45*, 731-758. 1976.

Laine, A. O., H. Sandler, A. B. Andersin, and J. Stigzelius, Long-term changes of macrozoobenthos in the eastern Gotland Basin and the Gulf of Finland (Baltic Sea) in relation to the hydrographical regime, *J. Sea Res.*, *38*, 135-159, 1997.

Lenihan, H. S. and C. H. Peterson, How habitat degradation through fishery disturbance enhances impacts of hypoxia on oyster reefs, *Ecol. Appl.*, *8*, 128-140, 1998.

Leppäkoski, E., Transitory return of the benthic fauna of the Bornholm Basin, after extermination by oxygen insufficiency, *Cahiers de Biologie Marine*, *10*, 163-172, 1969.

Levin, L. A. and J. D. Gage, Relationships between oxygen, organic matter and the diversity of bathyal macrofauna, *Deep-Sea Res. II*, *45*, 129-163, 1998.

Llansó, R. J., Effects of hypoxia on estuarine benthos: the lower Rappahannock River (Chesapeake Bay), a case study, *Estuar. Coast. Shelf Sci., 35*, 491-515, 1992.

Mee, L. D., The Black Sea in crisis: A need for concerted international action, *Ambio, 21*, 278-286, 1992.

Mirza, F. B. and J. S. Gray, The fauna of benthic sediments from the organically enriched Oslofjord, Norway, *J. Exp. Mar. Biol. Ecol., 54*, 181-207, 1981.

Nakata, K., M. Takei, T. Nakane, G. Maxwell, and D. Torpie, Dissolved oxygen depletion analysis and visualisation in Ise Bay, Japan, using a GIS approach, in The 11th Annual ESRI & ERDAS Users Conference, Canberra, Australia, 1997.

Nestlerode, J. A. and R. J. Diaz, Effects of periodic environmental hypoxia on predation of a tethered polychaete, *Glycera americana*: implication for trophic dynamics, *Mar. Ecol. Prog. Ser., 172*, 185-195, 1998.

Newcombe, C. L. and W. A. Horne, Oxygen poor waters of the Chesapeake Bay, *Science, 88*, 80-81, 1938.

Nichols, J. A., The effect of stable dissolved-oxygen stress on marine benthic invertebrate community diversity, *Int. Rev. ges. Hydrobiol., 61*, 747-760, 1976.

Niermann, U., E. Bauerfeind, W. Hickel, and H. v. Westernhagen, The recovery of benthos following the impact of low oxygen content in the German Bight, *Netherlands J. Sea Res., 25*, 215-226, 1990.

Nilsson, H. C. and R. Rosenberg, Succession in marine benthic habitats and fauna in response to oxygen deficiency analyzed by sediment profile imaging and grab samples, *Mar. Ecol. Prog. Ser., 197*, 139-149, 2000.

Nixon, S. W., Coastal marine eutrophication: A definition, social causes, and future concerns, *Ophelia, 41*, 199-219, 1995.

Orhon, S. and A. Yüksek, The oxygen deficiency and the benthic community in the Marmara Sea, manuscript, 16 pp.

Pearson, T. H. and A. Eleftheriou, The benthic ecology of Sollom Voe, *Proc. Royal Soc. Edinburgh, 80B*, 241-269, 1981.

Pearson, T. H. and R. Rosenberg, Macrobenthic succession in relation to organic enrichment and pollution of the marine environment, *Oceanogr. Mar. Biol. Ann. Rev. 16*, 229-311, 1978.

Pearson, T. and R. Rosenberg, Energy flow through the SE Kattegat: a comparative examination of the eutrophication of a coastal marine ecosystem, *Netherlands J. Sea Res., 28*, 317-334, 1992.

Petersen, C. G. J., On the animal communities of the sea bottom in the Skagerak, the Christiania Fjord and the Danish waters, *Report from the Danish Biological Station, 23*, 1-28, 1915.

Pihl, L., Effects of oxygen depletion on demersal fish in coastal areas of the south-east Kattegat, in *Reproduction, Genetics and Distribution of Marine Organisms*, edited by J. S. Ryland, pp. 431-439, Olsen & Olsen, Fredensborg, 1989.

Phil, L., S. P. Baden, R. J. Diaz, and L. C. Schaffner, Hypoxia-induced structural changes in the diet of bottom feeding fish and crustacea, *Mar. Biol., 112*, 349-361, 1992.

Phil, L., S. P. Baden, and R. J. Diaz. Effects of periodic hypoxia on distribution of demersal fish and crustaceans, *Mar. Biol., 108*, 349-360, 1991.

Powilleit, M. and J. Kube, Effects of severe oxygen depletion on macrobenthos in the Pomeranian Bay (southern Baltic Sea): a case study in a shallow, sublittoral habitat characterised by low species richness, *J. Sea Res., 42*, 221-234, 1999.

Prena, J., Temporal irregularities in the macrobenthic community and deep-water advection in

Wismar Bay (western Baltic Sea), *Estuar. Coast. Shelf Sci.*, *41*, 705-717, 1995.

Purcell, J. E., A. Malej, and A. Benovic', Potential links of jellyfish to eutrophication and fisheries, in *Ecosystems at the Land-Sea Margin: Drainage Basin to Coastal Sea*, edited by T. C. Malone, A. Malej, L. W. Harding, Jr., N. Smodlaka, and R. E. Turner, pp. 241-263, *Coastal and Estuarine Studies, 55*, 1999.

Rabalais, N. N., W. J. Wiseman, Jr., R. E. Turner, D. Justic', B. K. Sen Gupta, and Q. Dortch, Nutrient changes in the Mississippi River and system responses on the adjacent continental shelf, *Estuaries, 19*, 386-407, 1996.

Rabalais, N. N., R. E. Turner, D. Justic', Q. Dortch, and W. J. Wiseman, Jr., Characterization of hypoxia: Topic 1 Report for the Integrated Assessment of Hypoxia in the Gulf of Mexico. NOAA Coastal Ocean Program Decision Analysis Series No. 16. NOAA Coastal Ocean Program, Silver Springs, Maryland, 167 pp, 1999.

Rainer, S. F. and R. C. Fitzhardinge, Benthic communities in an estuary with periodic deoxygenation, *Austr. J. Mar. Freshwater Res., 32*, 227-243, 1981.

Reish, D. J., The relation of polychaetous annelids to harbor pollution, *Public Health Report 70*, 1168-1174, 1955.

Reish, D. J., D. F. Soule, and J. D. Soule, The benthic biological conditions of Los Angeles-Long Beach Harbours: Results of 28 years of investigations and monitoring, *Helgoländer Meeresuntersuchungen 34*, 193-205, 1980.

Richards, F. A., Oxygen in the ocean, in *Treatise on Marine Ecology and Paleoecology, Vol. 1*, edited by J.W. Hedgpeth, pp. 185-238, Waverly Press, Baltimore, Maryland, 1957.

Ritter, C. and P. A. Montagna, Seasonal hypoxia and models of benthic response in a Texas bay, *Estuaries, 22*, 7-20, 1999.

Rosenberg, R., Benthic faunal recovery in a Swedish fjord following the closure of a sulphite pulp mill, *Oikos, 23*, 92-108, 1972.

Rosenberg, R., Benthic faunal dynamics during succession following pollution abatement in a Swedish estuary, *Oikos, 27*, 414-427, 1976.

Rosenberg, R., Effects of oxygen deficiency on benthic macrofauna in fjords, in *Fjord Oceanography*, edited by H. J. Freeland, D. M. Farmer and C. D. Levings, pp. 499-514, Plenum Publishing Corp., New York, 1980.

Rosenberg, R., Eutrophication - the future marine coastal nuisance?, *Mar. Pollut. Bull., 16*, 227-231, 1985.

Rosenberg, R., Negative oxygen trends in Swedish coastal bottom waters, *Mar. Pollut. Bull., 21*, 335-339, 1990.

Rosenberg, R. and R. J. Diaz, Sulphur bacteria (*Beggiatoa* spp.) mats indicate hypoxic conditions in the Inner Stockholm Archipelago, *Ambio, 22*, 32-36, 1993.

Rosenberg, R. and L. O. Loo, Marine eutrophication induced oxygen deficiency: effects on soft bottom fauna, western Sweden, *Ophelia, 29*, 213-225, 1988.

Rosenberg, R., J. S. Gray, A. B. Josefson, and T. H. Pearson, Petersen's benthic stations revisited. II. Is the Oslofjord and eastern Skagerrak enriched? *J. Exper. Mar. Biol. Ecol. 105*, 219-251, 1987.

Santos, S. L. and J. L. Simon, Marine soft-bottom community establishment following annual defaunation: larval or adult recruitment, *Mar. Ecol. Prog. Ser., 2*, 235-241, 1980.

Sanukida, S., H. Okamoto, and M. Hitomi, Alteration of pollution indicator species of macrobenthos during stagnant period in eastern Hiuchi Sound, *Bull. Japanese Soc. Scient, Fish., 50*, 727, 1984.

Sen Gupta, B. K., R. E. Turner, and N. N. Rabalais, Seasonal oxygen depletion in continental-shelf waters of Louisiana: Historical record of benthic foraminifers, *Geology,* *24,* 227-230, 1996.

Stachowitsch, M., Anoxia in the Northern Adriatic Sea: rapid death, slow recovery, in *Modern and Ancient Continental Shelf Anoxia,* edited by R. V. Tyson and T. H. Pearson, pp. 119-129, *Geological Society Special Pub., 58,* 1991.

Stanley, D. W. and S. W. Nixon, Stratification and bottom-water hypoxia in the Pamlico River Estuary, *Estuaries, 15,* 270-281, 1992.

Suzuki, T. and Y. Matsukawa, Hydrography and budget of dissolved total nitrogen and dissolved oxygen in the stratified season in Mikawa Bay, Japan, *J. Oceanogr. Soc. Japan, 43,* 37-48, 1987.

Tenore, K. R., Macrobenthos of the Pamlico River estuary, North Carolina, *Ecol. Monogr., 42,* 51-69, 1972.

Thiel, R., A. Sepúlveda, R. Kafemann, and W. Nellen, Environmental factors as forces structuring the fish community of the Elbe Estuary, *J. Fish Biol., 46,* 47-69, 1995.

Thouvenin, B., P. Le Hir, and L. A. Romana, Dissolved oxygen model in the Loire estuary, in *Changes in Fluxes in Estuaries: Implications from Science to Management,* edited by K. R. Dyer and R. J. Orth, pp. 169-178, Olsen and Olsen, Fredensborg, Denmark, 1994.

Tsutsumi, H., Population dynamics of *Capitella capitata* (Polychaeta; Capitellidae) in an organically polluted cove, *Mar. Ecol. Prog. Ser., 36,* 139-149, 1987.

Tunnicliffe, V., High species diversity and abundance of the epibenthic community in an oxygen-deficient basin, *Nature, 294,* 354-356, 1981.

Turner, R. E. and N. N. Rabalais, Changes in Mississippi River water quality this century and implications for coastal food webs, *BioScience, 41,* 140-147, 1991.

Turkstra, E., M., C. Th. Scholten, C. T. Bowmer, and H. P. M. Schobben, A comparison of the ecological risks from fisheries and pollution to the North Sea biota, *Water Sci. Tech., 24,* 147-153, 1991.

Westernhagen, H. v. and V. Dethlefsen, North sea oxygen deficiency 1982 and its effects on the bottom fauna, *Ambio, 12,* 264-266, 1983.

Weigelt, M., Oxygen conditions in the deep water of Keil Bay and the impact of inflowing salt-rich water from the Kattegat, *Meeresforschung, 33,* 1-22, 1990.

Weigelt, M., Short- and long-term changes in the benthic community of the deeper parts of Kiel Bay (Western Baltic) due to oxygen depletion and eutrophication, *Meeresforschung 33,* 197-2241, 1991.

Welsh, B. L., R. I. Welsh, and M. L. DiGiacomo-Cohen, Quantifying hypoxia and anoxia in Long Island Sound, in *Changes in Fluxes in Estuaries: Implications from Science to Management,* edited by K. R. Dyer and R. J. Orth, pp. 131-137, Olsen and Olsen, Fredensborg, Denmark, 1994.

Winkler, L., The determination of dissolved oxygen in water, *Berichte der Deutschen Chemischer Gesellschaft, 21,* 28-43, 1888.

Wu, R. S. S., Periodic defaunation and recovery in a subtropical epibenthic community, in relation to organic pollution, *J. Exper. Mar. Biol. Ecol., 64,* 253-269, 1982.

Zaitsev, Y. P., Recent changes in the trophic structure of the Black Sea, *Fisheries Oceanogr., 1,* 180-189, 1992.

Zimmerman, A. R. and E. L. Canuel, A geochemical record of eutrophication and anoxia in Chesapeake Bay sediments: anthropogenic influence on organic matter composition, *Mar. Chem., 69,* 117-137, 2000.

9

Benthic Foraminiferal Communities in Oxygen-Depleted Environments of the Louisiana Continental Shelf

Emil Platon and Barun K. Sen Gupta

Abstract

Many species of benthic Foraminifera survive the extreme oxygen depletion of Louisiana shelf waters in spring and summer. The dominant species (e.g., *Buliminella morgani* and *Brizalina lowmani*) are infaunal and have an adaptive tolerance to near anoxia or anoxia. An overall increase in the intensity and duration of seasonal hypoxia is reflected in the *Ammonia-Elphidium* index, but generally in waters shallower than about 30 m. New data on changing species dominances and assemblage density indicate somewhat variable historical trends of progressive hypoxia at different water depths in the Louisiana Bight.

Introduction

Numerous species of benthic Foraminifera have the ability to withstand very low oxygen concentrations or even anoxic conditions. In addition, there is ample evidence that in dysoxic to anoxic marine sediments, benthic Foraminifera may migrate through the sediment to find optimum habitat conditions, although factors other than oxygen also affect this migration. On the Louisiana continental shelf, the effects of seasonal hypoxia and the Mississippi River plume on benthic Foraminifera have been documented by Nelsen et al. [1994], Sen Gupta et al. [1996], Blackwelder et al. [1996], Rabalais et al. [1996] and Platon et al. [1997]. High abundances of certain species in substrates from the

Coastal Hypoxia: Consequences for Living Resources and Ecosystems
Coastal and Estuarine Studies, Pages 147-164

vicinity of the Mississippi River suggest their tolerance to rapid sedimentation. Furthermore, the historical record of species distribution shows a trend of increasing relative abundances of putatively opportunistic species with increasing hypoxia. In particular, the *Ammonia-Elphidium* index (A-E index) is a useful proxy for present and past hypoxia, and the stratigraphic trend of the A-E index provides a historical record of paleohypoxia on the shelf [Sen Gupta et al., 1996]. The aim of this chapter is to (1) examine the summer (hypoxic) and fall (non-hypoxic) distributions (and microhabitats) of living Foraminifera present in sediment cores from the Louisiana continental shelf, and (2) assess the historical trend of common benthic foraminiferal species from the stratigraphic record of the last few decades.

Foraminiferal Response to Low Oxygen Concentrations

Microaerophiles and Facultative Anaerobes in Controlled Experiments

Both field and laboratory observations show that many benthic foraminiferal species can survive extreme oxygen depletion or even anoxia, at least for a short time [reviewed in Sen Gupta and Machain-Castillo, 1993; Bernhard, 1996; Bernhard and Sen Gupta, 1999]. Many laboratory experiments on foraminiferal subsurface activity have been reported in the literature, but concomitant oxygen levels in sediment pore water have been measured in only a few studies. In a study of nearshore species from McMurdo Sound, Antarctica, Bernhard [1989] used an adenosine triphosphate (ATP) assay to recognize living individuals and found that several agglutinated and calcareous species were viable even in anoxic layers. Furthermore, she subjected these Antarctic Foraminifera to anoxic or reducing conditions for 30 days and determined that 30% of the analyzed individuals could withstand such conditions [Bernhard, 1993]. Ultrastructural investigations revealed the presence of chemolithotrophic-type bacteria beneath the organic lining of *Globocassidulina* sp. cf. *G. biora*, leading to the suggestion that these bacteria may help in anaerobic survival of this foraminifer [Bernhard, 1993].

Using foraminiferal species from a Norwegian fjord in which bottom waters are periodically dysoxic, Alve and Bernhard [1995] performed a laboratory experiment in which the oxygen level was reduced to < 0.3 mg l^{-1} three months and maintained at this low level for five weeks. In response, the species migrated upward in the sediment into pore water with higher oxygen concentrations. In another experiment, four species were kept in anoxic seawater (purged with nitrogen) for over three weeks. These species, apparently facultative anaerobes, survived anoxia (based on ATP assays) in different (mostly unknown) ways, including dormancy [Bernhard and Alve, 1996]. In a comparable experiment with Foraminifera collected from the Adriatic Sea shelf, Moodley et al. [1997] showed that many species can withstand anoxia (and the presence of H_2S) for several weeks; the conclusion that the individuals were still alive was based on rose Bengal staining and on life activity checks. Migratory activity of selected common species and mixed foraminiferal assemblages through sediment of varying pore-water oxygen was examined in the laboratory by Moodley et al. [1998]. Upward migration was evident when species were buried in sediments without an oxic zone, indicating that anoxic conditions may stimulate migration. The investigators suggested that the presence of particular species in deeper sediment layers is related to their tolerance to prolonged anoxia and to sediment mixing by physical and biological factors.

Microhabitat Considerations

A foraminiferal "microhabitat" is defined as a micro-environment characterized by a combination of physical, chemical and biological conditions (oxygen, food, toxic substances, biological interactions, etc. [Jorissen, 1999]). A species has a microhabitat preference if it shows greater abundances at the sediment surface or at some depth beneath the surface [Hunt and Corliss, 1993]. Recognition of foraminiferal microhabitats and factors responsible for vertical distribution in sediment is of great importance in assessing foraminiferal responses to environmental changes and in reconstructing past environments of sedimentation.

In the absence of animal burrows, pore-water oxygen concentration usually diminishes sharply with substrate depth, falling below detection limits within a few centimeters [Fenchel and Finlay, 1995]. Thus, species with microhabitats in deeper parts of the sediment column (e.g., *Globocassidulina* sp. cf. *G. biora* living at 7 cm [Bernhard, 1993]) may be microaerophiles or facultative anaerobes. Certain benthic Foraminifera may be able to track seasonal fluctuation of oxygen levels within the sediment, and their temporal distribution pattern should reflect such seasonal variations [Jorissen, 1987; Van der Zwaan and Jorissen, 1991]. Distribution data are available from the northern Adriatic Sea, where ample nutrient availability during the summer results in phytoplankton blooms, which, together with terrestrial material, lead to very high loads of organic matter. Water stratification and decomposition of organic matter cause severe depletion of bottom-water oxygen during late summer and autumn [Jorissen et al., 1992]. In order to assess foraminiferal response to this seasonal lowering of oxygen, Barmawidjaja et al. [1992] analyzed data from a time series of seven samples (December 1988 to November 1989) from one station and showed that bottom-water or pore-water oxygen is a major factor controlling the faunal densities. Three microhabitat-related species groups were recognized: epifaunal, predominantly infaunal and potentially infaunal. Larger variations of epifaunal taxa support the idea that the tolerance of this group to low oxygen levels is less than that of infaunal taxa. The "potentially infaunal" species (with seasonally adjusted microhabitats, from sediment surface to a few centimeters below the surface) apparently follow some critical oxygen level.

Methods

Sediment core samples from six stations located on the Louisiana continental shelf (Louisiana Bight west of the Mississippi River birdfoot delta, Fig. 1) were analyzed for this study. Box cores were collected in (1) October 1995 from 35-, 40- and 80-m water depths (stations F35, G40 and D80, respectively); (2) April 1996 from 20-m water depth (station F20); and (3) July 1996 from 11- and 30-m water depths (stations AA1 and A5, respectively). Each box core was examined for disturbance during retrieval and was considered undisturbed if the overlying water was clear and surface tracks and open burrows were visible. Cylindrical sediment cores were obtained from these box cores

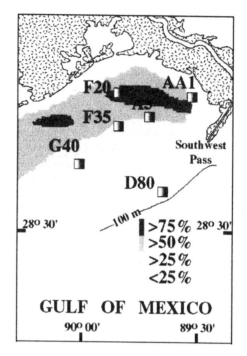

Figure 1. Hypoxia frequency of occurrence mid-summer of 1985 – 1999; Core locations, Louisiana Bight. October 1995: F35, G40, D80; April 1996: F20; July 1996: AA1, A5.

with 7-cm diameter plastic tubes that were sliced at 1-cm intervals. These 1-cm sections became the primary samples for the foraminiferal study. Samples from the top 15 cm of the sediment cores were stained with a solution of rose Bengal in ethanol. All samples were washed over a 63-μm sieve and dried. For the microhabitat study, specimens with bright red coloration of one or more chambers were considered living at the time of collection, and 300 stained specimens (if present) were picked from the sample. If the number of stained specimens was less than 300 (the usual case), all such specimens were picked. For the historical trend study, the residue was split with a microsplitter to provide an aliquot of approximately 300 shells of benthic Foraminifera. Identification and counting of these specimens yielded the species census; in samples with fewer than 300 specimens, all specimens were counted.

Two measures of species diversity were computed from the foraminiferal counts: (1) species richness (S), i.e., the number of species present in the sample or aliquot, and (2) equitability (E), given by the equation

$$E=e^{H(S)}/S \qquad (1)$$

[Buzas and Gibson, 1969, equation (1)], where H(S) is the Shannon Wiener function (= $-\Sigma p_i \ln p_i$, where p_i is the proportion of the ith species). Absolute ages of samples were obtained from ^{210}Pb data (R. E. Turner, unpubl. data).

Hypoxia and Foraminiferal Distribution Patterns on the Louisiana Continental Shelf

Microhabitats of Common Species

Two groups of samples were analyzed to assess the vertical distribution of living Foraminifera in the sediment. The first group consisted of samples collected from stations F35, G40 and D80 in October 1995; the second group (AA1 and A5) consisted of July 1996 samples (Fig. 1). Seventy-three species of benthic Foraminifera (Table 1) were identified, but most of these were rare, and their microhabitat preference could not be ascertained with the present data. The distributions of dominant species is discussed below. Scanning electron micrographs of these are given in Plate 1. The absolute abundance data on living Foraminifera are summarized in Table 2.

Relative abundance data from October 1995 core samples are depicted in Figure 2. In general, *Buliminella morgani* dominated the living assemblages of benthic Foraminifera from the sediment-water interface down to a sediment depth of 5-10 cm, possibly in severe dysoxic, or even anoxic, microhabitats. Based on species richness as well as on living foraminiferal density, three microhabitats were discriminated between the sediment-water interface and the 15-cm depth level. A high dominance of *Buliminella morgani*, a relatively high species richness, and presence of both calcareous and agglutinated forms characterized the first few centimeters (usually 3-5 cm) beneath the sediment-water interface. A severe reduction in species richness and an increase in the proportion of *Brizalina lowmani* marked a second vertical zone, extending to about the 10-cm depth level. Below this level, *Brizalina lowmani* was the dominant or the only living foraminifer.

The number of rose Bengal stained specimens (i.e., living foraminifer assemblage density) was high at the top of each core, but the maximum, in all cases, was recorded 2-4 cm below the sediment-water interface. This number decreased continuously from this level down to a depth of about 15 cm. The mean species richness, computed from values recorded in F35, G40 and D80, decreased steadily from the core top to the depth of 15 cm (Fig. 2b), but several differences regarding the number of species identified at different locations and different sediment depths were noted. A total of 15 living species was identified in core G40. Within these samples, the species richness varied between 10 at the core top and 6 at the deepest level of foraminiferal occurrence, but the decrease was not linear, and the mean value was 6. At D80, the total number of living species was 11, and the maximum (at core top), minimum (at deepest levels), and mean values of species richness were 7, 1 and 5, respectively. Low values of richness (maximum 4, minimum 2, mean 3) were recorded from the F35 samples, and the total number of living species in this core was only 6. These data suggest that water depth as well as the distance from the Mississippi River plume influence microhabitat distribution.

As mentioned earlier, it is generally agreed that factors such as bottom-water dissolved oxygen concentration, food availability, competition and predation, and bioturbation are major determinants of foraminiferal microhabitats. The high nutrient amounts delivered by the Mississippi River in the study area [Blackwelder et al., 1996] should ensure sufficient food supply for both epifaunal and infaunal foraminifers. In general, organic matter within the sediment is less labile, and food quality decreases with

TABLE 1. Identified species.

Species	Species
Ammonia parkinsoniana	*Laterostomella* sp. ?
Ammotium salsum f. *dilatatum*	*Lenticulina calcar*
Angulogerina bella	*Lenticulina* sp.
Bigenerina irregularis	*Lenticulina* sp. cf. *L.* sp. *"F"*
Bolivina albatrossi	*Lenticulina* sp. *L.* sp. *"B"*
Bolivina barbata	*Marginulina obesa*
Bolivina cf. *B. fragilis*	*Marginulina* sp. cf. *M. glabra*
Bolivina goessi	Miliolidae sp.
Bolivina lanceolata	*Miliolinella werreni*
Bolivina lowmani	*Neoeponides antilarum*
Bolivina pusilla	*Nodosaria* sp.
Bolivina sp.	*Pseudoonion grateloupi*
Bolivina sp. *B. fragilis*	*Nonionella basiloba*
Bolivina sp. cf. *B. pusilla*	*Nonionella morgani*
Bolivina subanaerensis	*Prolixoplecta parvula*
Brizalina lowmani	*Pseudononion atlanticus*
Bulimina aculeata	*Pseudononion grateloupe*
Bulimina marginata	*Pullenia buloides*
Bulimina morgani	*Pyrgo* sp.
Bulimina sp.	*Pyrgo carinata* ?
Bulimina tenuis	*Q* sp. *Anderson*
Bulimina translucens	*Q. lamarkiana*
Cancris sagra	*Rectobolivina advena*
Cassidulina subglobosa	*Reophax fusiformis*
Discammina compressa	*Reussella miocenica*
Discorbinella bertholdi	*Saccammina difligiformis*
Dorothia sp.	*Saracenaria* sp.
Elphidium gunteri/excavatum	*Sigmoilina distorta*
Elphidium poeyanum	*Siphotextularia affinis*
Epistominella vitrea	*Spiroloculina* sp. cf. *S. antilarium*
Fursenkoina elegantisima	*Textularia candeiana*
Fursenkoina pontoni	*Textularia porrecta*
Globobulimina sp. ?	*Triloculina trigonula*
Globocassidulina subglobosa	*Uvigerina hispido-costata*
Hanzawaia concentrica	*Uvigerina laevis*
Haplophragmoides sp.	*Uvigerina peregrina*
Lagena sp.	*Uvigerina* sp
Lagena sp. cf. *L. laevis*	*Uvigerina* sp. cf. *U. laevis*
Lagena sp. cf. *L. spicata*	*Vaginulinopsis* sp.

sediment depth. Nevertheless, bioturbation by macrofauna may introduce a fresh food supply within the sediment and create conduits for oxygen penetration. At the top of each core, the species richness maximum and relatively high assemblage density indicate "normal" environmental conditions. More sensitive species such as *Bolivina lanceolata*, *B. albatrossi*, *Fursenkoina pontoni* and *Uvigerina laevis* decline with sediment depth due to possible pore-water oxygen depletion and lower food quality. More tolerant taxa are able, to a certain degree, to cope with both reduced oxygen concentration and low-nutrition food. Thus, the association of maximum assemblage density and a diminished number of species recorded a few centimeters beneath the sediment-water interface is apparently controlled by the combination of at least three factors: very low oxygen concentrations, reduced ratios of fresh/old food supplies and, presumably, reduced predation and competition. Species such as *Buliminella morgani*, *Brizalina lowmani* and *Nonionella basiloba* flourish in this microhabitat.

Foraminiferal data from two July 1996 stations (AA1 and A5) are presented in Figure 3. Bottom-water oxygen data collected during this cruise [N. N. Rabalais, unpubl. data] show that the summer of 1996 was a season of widespread hypoxia in the Gulf of Mexico, with an estimated bottom extent of 17,920 km^2. Dissolved oxygen was measured at 2.52 mg l^{-1} at A5 (water depth 30 m), but fell to a much lower value of 0.56 mg l^{-1} at AA1 (water depth 11 m). We assume that both of these measured values were much lower than the corresponding October 1995 levels. Oxygen data were not collected at the time of the October 1995 cores, but bottom-water oxygen concentrations at a station in 20-m water depth within 50 km of the Louisiana Bight was consistently between 4 and 5 mg l^{-1} for the month of October [N. N. Rabalais, unpubl. data]. The number of species identified in the top 15 cm of the AA1 core ranged from 2-3 in the 8-15 cm interval to 6 in the 4-7 cm interval. The assemblage density maximum was located in the 2-3 cm interval and followed the trend seen in the cores retrieved during the fall of 1995. The species distribution pattern showed a striking dominance of *Brizalina lowmani* over species such as *Nonionella basiloba*, *Buliminella morgani* and *Ammonia parkinsoniana*. This dominance was persistent throughout the core, and no vertical zonation was discerned. In contrast, in A5 samples (species richness 1-5, mean 2.5), the assemblage was dominated by *Buliminella morgani* from the sediment-water interface to a sediment depth of 15 cm. Species such as *Nonionella basiloba*, *Ammonia parkinsoniana* and *Brizalina lowmani* were present, but scarce, in the 0-8 cm interval. Judged by rose Bengal staining, *Buliminella morgani* was the only putative survivor below the depth of 8 cm. The maximum assemblage density was in the 0-1 cm interval.

Reduced species numbers and high relative abundances of *Brizalina lowmani* (maximum 98%) are two conspicuous features of the foraminiferal community at AA1. High relative abundances of *B. lowmani* (up to 56%) have been reported previously from a nearshore (< 20 m) location in this area [Blackwelder et al., 1996], and we infer that the microhabitat distribution of the species at AA1 is strongly affected by both water depth and environmental stress related to the oxygen depletion. A comparison of microhabitat data from F35 (fall 1995, 35 m water depth) with that of nearby A5 station (summer 1996, 30 m water depth) reveals differences that may be related to seasonal hypoxia. (1) *Buliminella morgani* was persistently dominant in A5 (July), but not in F35 (October), in which *Brizalina lowmani* dominated the samples below a depth of 6 cm. (2) The mean density of living individuals was less in A5 (July) that in F35 (October).

Historical Trends of Species Abundances

Historical records of benthic foraminiferal abundance variations in the past 2-14 decades were obtained from three dated sediment cores (F35, G40 and D80) from

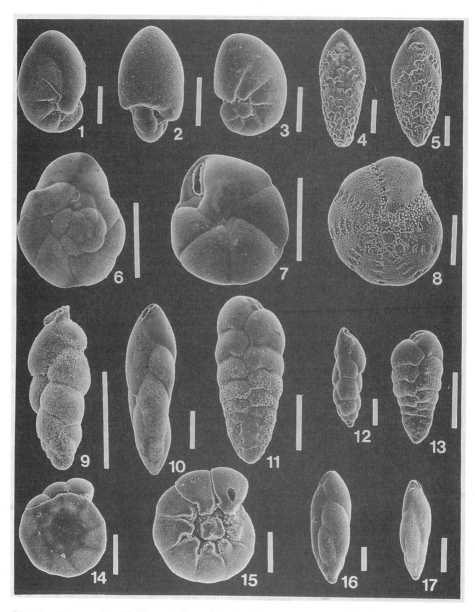

Plate 1. Common benthic Foraminifera of the Louisiana Bight; scale bars = 100 μm, 1, 2, 3, *Nonionella basiloba;* 4, 5, *Bolivina albatrossi;* 6, 7, *Epistominella vitrea;* 8, *Elphidium gunteri/excavatum;* 9, *Uvigerina laevis;* 10, 16, 17, *Fursenkoina pontoni;* 11, 13, *Brizalina lowmani;* 12, *Buliminella morgani;* 14, 15, *Ammonia parkinsoniana.*

TABLE 2. Absolute abundances of common living foraminiferal species in: October 1995 cores (F35, G40, D80); July 1996 cores (A5, AA1); B l = Brizalina lowmani, B m = Buliminella morgani, E v = Epistominella vitrea, N b = Nonionella basiloba, A p = Ammonia parkinsoniana, B ma = Buliminella marginata, F e = Fursenkoina elegantisima.

Depth (cm)	F35				G40				D80					A5					AA1				
	B l	B m	E v	N b	B l	B m	E v	N b	B l	B m	E v	N b	A p	B l	B m	B ma	N b	A p	B l	B m	N b	F e	E v
0	6	24	0	8	28	77	0	145	15	113	3	95	1	2	165	0	63	0	117	9	0	0	0
1	4	31	0	10	5	76	0	80	18	135	6	69	2	1	154	1	50	0	256	2	1	1	0
2	4	10	0	4	4	90	0	107	3	498	8	42	3	2	154	0	16	2	147	9	1	2	0
3	2	8	0	30	7	99	0	99	1	98	2	8	1	1	150	0	11	0	75	4	1	1	0
4	0	2	0	20	5	113	0	105	2	48	3	4	1	0	34	1	4	4	78	4	5	1	1
5	1	0	2	3	25	26	0	32	4	25	1	2	0	0	55	0	33	6	83	2	1	4	1
6	1	0	0	0	26	32	0	21	9	12	0	5	0	1	67	0	50	8	182	1	6	8	1
7	2	0	0	1	25	24	0	10	5	12	1	1	2	2	34	0	23	7	179	2	2	0	2
8	2	0	0	0	22	5	0	7	4	10	1	1	1	1	16	0	11	0	48	0	0	4	0
9	2	0	1	0	26	4	0	3	3	7	1	0	1	0	7	0	2	9	116	3	3	0	1
10	2	0	1	0	16	5	0	0	3	1	0	0	2	0	5	0	1	9	137	3	0	0	0
11	2	0	0	0	15	5	0	0	7	2	0	0	0	0	1	0	1	1	137	0	0	1	1
12	2	0	0	0	3	3	0	1	7	1	0	0	0	0	2	0	1	7	125	0	0	1	3
13	2	0	2	0	4	1	0	0	5	1	0	0	0	0	2	0	1	1	113	0	0	3	0
14	2	0	1	1	4	0	0	0	7	0	0	0	0	0	1	0	2	2	47	0	0	2	0

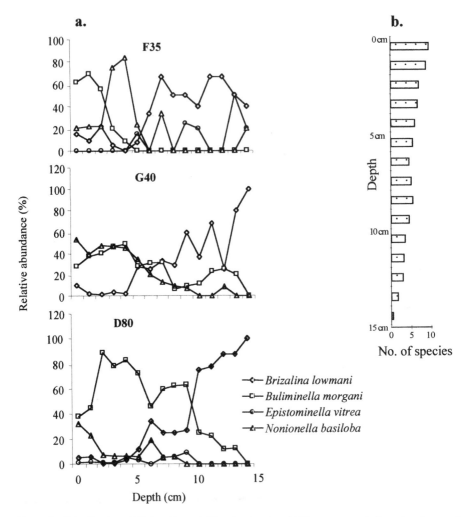

Figure 2. Distribution of living Foraminifera in October 1995 cores: a, relative abundance of dominant species; b, mean species richness.

different water depths, different distances from the Mississippi River plume and differing frequency of bottom-water hypoxia (Fig. 1). In this exercise, the contamination of older (^{210}Pb-dated) foraminiferal assemblages by living populations of infaunal species was not considered a problem, because the number of dead individuals was always much higher (below a substrate depth of 4 cm, orders of magnitude higher) than that of living individuals.

Locations at which foraminiferal assemblages exhibit the lowest species richness and the highest species dominance were considered the most stressful or most recently stressed, whereas those with the highest richness and equitability (a measure of evenness of distribution) were considered to represent the most stable environments [Engen, 1979].

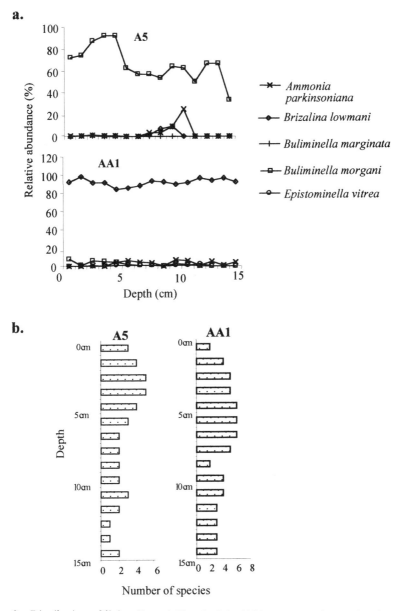

Figure 3. Distribution of living Foraminifera in July 1996 cores: a, relative abundances of dominant and common species; b, species richness.

In the area impacted by the Mississippi River plume, previous workers have found considerably large ranges of species richness (5 to 39 species) and evenness, indicating different degrees of ecological stress [Blackwelder et al., 1996]. Within the core samples from D80, equitability varied between 0.26 and 0.51, and the species richness ranged

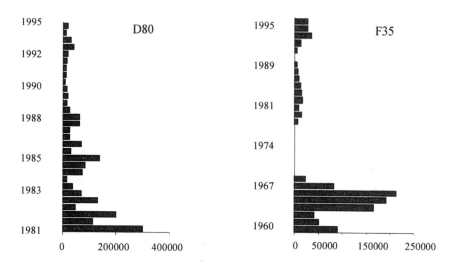

Figure 4. Variations in foraminiferal assemblage density (x axis, shells per 100 cc of sediment).

between 8 and 15 (mean 11.7). In the G40 samples, equitability values were 0.26 to 0.58, and the richness values were 9 to 25 (mean = 15.4). The station at 35 m water depth (F35) was the most frequently affected by seasonal hypoxia, and core samples from this station had relatively low species richness (3 to 11, mean = 8). Corresponding equitability values showed a wide range (0.23 to 0.89), but the very high value of 0.89 (1 being the highest value for this measure) was associated with an extremely low species count of 3 and just a few individuals, and thus may be anomalous.

Total (living + dead) foraminiferal assemblages at these three stations were generally dominated by *Epistominella vitrea, Buliminella morgani* and *Nonionella basiloba*. *Epistominella vitrea* was the dominant species at all levels in the cores, but its relative abundance was considerably less in the living assemblages. Because of the known high abundance of *E. vitrea* in the Mississippi River plume [Blackwelder et al., 1996], we infer that NE-SW and E-W transport processes have enhanced the concentration of shells of this species in our samples. On the basis of ^{210}Pb dating, the estimated sedimentation rate in the core varied from 0.12 to 2.20 cm y^{-1}. Progressively restrictive environmental conditions are suggested by an overall decrease of foraminiferal density from 1981 to 1990 at the deepest (80 m) site, and from 1958 to 1994 at the shallowest (35 m) site (Fig. 4). Furthermore, in the core from 35 m, the foraminiferal density is highest for the mid 1960s and the lowest (zero values) for most of the 1970s.

No conspicuous dominance shifts were observed in the stratigraphic records of the benthic Foraminifera, but several trends were discernible in relative abundances of species. At the 80-m station, the proportion of *Bolivina lanceolata* decreased, whereas that of *Bolivina albatrossi* increased, from 1985 to 1995 (Fig. 5). At the 40-m station, the almost continuous decrease in the relative abundance of *Elphidium gunteri/excavatum* from 1853 to 1995 was interrupted by a positive trend between 1929 and 1945 (Fig. 6). Because of the low proportions of these species in the assemblages (usually < 5%) and small numbers of the individuals counted, the observed trends may or may not have paleoenvironmental significance. More pronounced trends were recognized at the 35 m

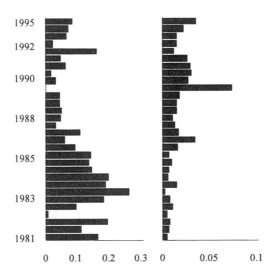

Figure 5. Stratigraphic variations in the proportions of *Bolivina lanceolata* (left) and *Bolivina albatrossi* (right) in the foraminiferal assemblage of D80 in the past two decades.

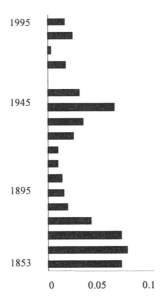

Figure 6. Stratigraphic variations in the proportion of *Elphidium gunteri* in the foraminiferal assemblage of G40 from 1853 to 1995.

station (Fig. 7). After a slight decrease from 1960 to 1971, the density of *Buliminella morgani* increased substantially between 1971 and 1975 and then remained steady until 1995 (core-top date). The 1971-1975 time interval was marked by a drastic reduction in the relative abundance of *Episominella vitrea*, *Nonionella basiloba* and *Brizalina*

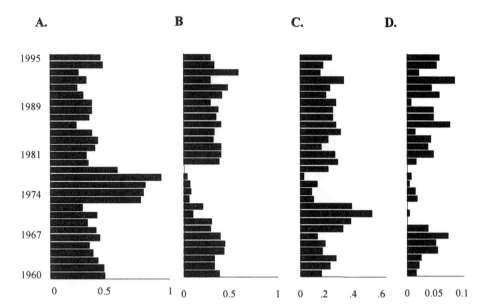

Figure 7. Stratigraphic variations in the proportions of *Buliminella morgani* (A.), *Epistominella vitrea* (B.), *Nonionella basiloba* (C.) and *Elphidium gunteri/excavatum* (D.) in the foraminiferal assemblage of F35 in the past four decades.

lowmani. The reduction in *Epistominella vitrea* may be related to diminishing E-W transport processes and/or environmental changes in the Mississippi River plume area.

The Ammonia-Elphidium Index

Ammonia parkinsoniana and *Elphidium gunteri/excavatum*, common foraminiferal species of continental shelves, may be facultative anaerobes [Kitazato, 1994], but the dominance of *A. parkinsoniana* over *E. gunteri/excavatum* in nearshore surface sediments of the Gulf of Mexico during maximum hypoxia [Sen Gupta et al., 1996] indicates a higher tolerance of the former to oxygen depletion. An *Ammonia-Elphidium* (or A-E) index, given by

$$(N_A \times 100)/(N_A + N_E) \qquad (2)$$

whose distribution in the surface sediments of Louisiana Bight correlates well with the abundance of total organic carbon, was proposed as a proxy measure for paleohypoxia by Sen Gupta et al. [1996]. Because significant proportions of living *Ammonia* and *Elphidium* are typical only of marginal marine and inner shelf habitats, the reliability of this index is reduced beyond a water depth of about 30 m. This is evident in Figure 8. The increase of A-E index correlates well with the overall rise in oxygen stress on Louisiana continental shelf in water depths of 20-30 m, but this index is more variable in greater depths, especially at the margins of the area affected by spring/summer hypoxia (Fig. 1).

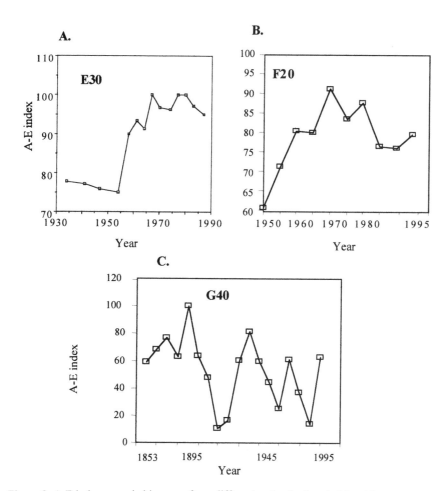

Figure 8. A-E index recorded in cores from different water depths: A, 30 m (Core E 30, Sen Gupta et al., 1996); B, 20 m (this study); and C, 40 m (this study).

Nevertheless, hypoxia is most frequently recorded between bottom-water depths of 5 and 30 m [Rabalais et al., 1996], and the lack of a trend in the A-E index in deeper waters may indicate a much reduced frequency, or an absence, of severe oxygen depletion.

Conclusions

1. Many species of benthic Foraminifera survive under seasonal hypoxia on the Louisiana continental shelf. This finding is consistent with field and laboratory investigations of previous workers on foraminiferal tolerance of severe oxygen depletion. At any given location, the composition of the present foraminiferal community is particularly dependent on water depth, distance from the Mississippi River plume and

dissolved oxygen at the sediment-water interface and in pore water (which was not measured).

2. Judging by their microhabitat preferences and dominances within living foraminiferal assemblages, several species are microaerophilitic or facultative anaerobes. The most notable are *Buliminella morgani* and *Brizalina lowmani*.

3. The *Ammonia-Elphidium* index (based on species proportions in core samples) is a sensitive proxy for paleohypoxia in water depths less than about 30 m. In greater depths, the reduced abundances of both genera diminish the reliability of this proxy, but the high variability of the A-E index here may be related to a decrease in the frequency of hypoxia.

4. On the basis of the present data, progressively restrictive environmental conditions (related to the intensity of seasonal hypoxia) are suggested by a decrease of foraminiferal density from 1981 to 1990 at the deepest (80 m) site and from 1958 to 1994 at the shallowest (35 m) site.

Acknowledgments. Sampling opportunities provided by R. E Turner and N. N. Rabalais made this study possible. We are further indebted to R. E. Turner for ^{210}Pb dates of sediment samples, and to N. N. Rabalais for bottom-water oxygen data. We thank P. Blackwelder, J. M. Bernhard, N. N. Rabalais and an unknown reviewer for their constructive criticisms of the manuscript. The project was supported by Minerals Management Service Grant 14-35-0001-30660.

References

Alve, E. and J. M. Bernhard, Vertical migratory response of benthic Foraminifera to controlled decreasing oxygen concentrations in an experimental mesocosm, *Mar. Ecol. Prog. Ser.,* *116,* 137-151, 1995.

Barmawidjaja D. M., F. J. Jorrisen , S. Puskaric, and G. J. Van der Zwaan, Microhabitat selection by benthic Foraminifera in the northern Adriatic Sea, *J. Foraminiferal Res., 22,* 137-151, 1992.

Bernhard, J. M., The distribution of benthic Foraminifera with respect to oxygen concentration and organic carbon levels in shallow-water Antarctic sediments, *Limnol. Oceanogr., 134,* 1131-1141, 1989.

Bernhard, J. M., Experimental and field evidence of Antarctic foraminiferal tolerance to anoxia and hydrogen sulfide, *Mar. Micropaleontol., 20,* 203-213, 1993.

Bernhard, J. M., Microaerophilic and facultative anaerobic benthic Foraminifera: A review of experimental and ultrastuructural evidence, *Rev. Paléobiol., 15,* 261-275,1996.

Bernhard, J. M. and E. Alve, Survival, ATP pool, and ultrastructural characterization of benthic Foraminifera from Drammmensfjord (Norway): response to anoxia, *Mar. Micropaleontol., 28,* 5-17, 1996.

Bernhard, J. M., and B. K. Sen Gupta, Foraminifera of oxygen-depleted environments, in *Modern Foraminifera,* edited by B. K. Sen Gupta, pp. 201-216, Kluwer Academic Publishers, Dordrecht, 1999.

Blackwelder, P., T. Hood, C. Alvarez-Zarikian, T. A. Nelsen, and B. McKee, Benthic Foraminifera from the NECOP study area impacted by the Mississippi River plume and seasonal hypoxia, *Quaternary International, 31,* 19-36, 1996.

Buzas, M. A. and T. G. Gibson, Species diversity: Benthonic Foraminifera in western North Atlantic, *Science, 163*, 72-76, 1969.

Engen, S., Some basic concepts of ecological equitability, in *Ecological Diversity in Theory and Practice*, edited by J. F. Grassle, G. P. Patil, W. Smith, and C. Taillie, pp. 37-50, International Cooperative Publishing House, Fairland, Maryland, 1979.

Fenchel, T. and B. J. Finlay, *Ecology and Evolution in Anoxic Worlds*, 276 pp., Oxford University Press, Oxford, 1995.

Hunt, A. S. and B. H. Corliss, Distribution and microhabitats of living (stained) benthic Foraminifera from the Canadian Arctic Archipelago, *Mar. Micropaleontol., 20*, 321-345, 1993.

Jorissen, F. J., The distribution of benthic Foraminifera in the Adriatic Sea, *Mar. Micropaleontol., 12*, 21-48, 1987.

Jorissen, F. J., Benthic foraminiferal microhabitats below the sediment-water interface, in *Modern Foraminifera*, edited by B.K. Sen Gupta, pp. 161-179, Kluwer Academic Publishers, Dordrecht, 1999.

Jorissen, F. J., D. M. Barmawidjaja, S. Puskaric, and G. J. Van der Zwaan, Vertical distribution of benthic Foraminifera in the northern Adriatic Sea: The relation with the organic flux, *Mar. Micropaleontol., 19*, 131-146, 1992.

Kitazato, H., Foraminiferal microhabitats in four marine environments around Japan, *Mar. Micropaleontol., 24*, 29-41, 1994.

Moodley, L., G. J. Van der Zwaan, P. M. J. Herman, L. Kempers, and P. van Bruegel, Differential response of benthic meiofauna to anoxia with special reference to Foraminifera (Protista: Sarcodina), *Mar. Ecol. Prog. Ser., 158*, 151-163, 1997.

Moodley, L., G. J. Van der Zwaan, G. M. W. Rutten, R. C. E. Boom, and L. Kempers, Subsurface activity of benthic Foraminifera in relation to porewater oxygen content: laboratory experiments, *Mar. Micropaleontol., 34*, 91-106, 1998.

Nelsen, T. A., P. Blackwelder, T. Hood, B. McKee, N. Romer, C. Alvarez-Zarikian, and S Metz, Time-based correlation of biogenic, lithogenic and authigenic sediment components with anthropogenic inputs in the Gulf of Mexico NECOP study area, *Estuaries 17*, 873-885, 1994.

Platon, E., B. K. Sen Gupta, R. E. Turner, and E. B. Overton, Influence of the Mississippi River plume on the distribution of modern benthic Foraminifera on the Louisiana continental shelf, *Geol. Soc. Amer., 1997 Ann. Mtg., Abstracts with Programs*, p. A95, 1997.

Rabalais, N. N., R. E. Turner, D. Justic, Q. Dortch, W. J. Wiseman, Jr., and B. K. Sen Gupta, Nutrient changes in the Mississippi River and system responses on the adjacent continental shelf, *Estuaries, 19*, 386-407, 1996.

Sen Gupta, B. K. and M. L. Machain-Castillo, Benthic Foraminifera in oxygen-poor habitats, *Mar. Micropaleontol., 20*, 183-201, 1993.

Sen Gupta, B. K., R. E. Turner, and N. N. Rabalais, Seasonal oxygen depletion in continental shelf waters of Louisiana: Historical record of benthic foraminifers, *Geology, 24*, 227-230, 1996.

Van der Zwaan, G. J. and Jorissen, F. J., Biofacial patterns in river-induced shelf hypoxia, in *Modern and Ancient Continental Shelf Anoxia*, edited by R. V. Tyson and T. H. Pearson, pp. 65-82, *Geological Society Special Pub., 58*, 1991.

10

Effects of Hypoxia and Anoxia on Meiofauna: A Review with New Data from the Gulf of Mexico

Markus A. Wetzel, John W. Fleeger and Sean P. Powers

Abstract

Hypoxic/anoxic sediments occur throughout the world's oceans. Hypoxic events in estuarine and coastal waters have generally increased in severity, length and frequency recently as a result of increasing anthropogenic eutrophication. All meiofauna appear to have some sensitivity to extended periods of hypoxia, but a wide range in tolerance (from hours to days, weeks or months of exposure) has been observed. Within a given habitat certain species of foraminiferans and nematodes are typically most tolerant to hypoxia/anoxia while crustacean meiofauna often are least tolerant. Tolerance is probably related to exposure history, phylogenetic constraints and lifestyle. Low oxygen demand, a high surface to volume ratio and anaerobic metabolism enable some species to survive hypoxia/anoxia for extended times. Field studies of hypoxic episodes typically demonstrate a reduction in overall meiofaunal abundance, a reduction in species and higher-taxon diversity and an increase in the relative abundance of nematodes. Similar trends occur over water depth in shallow seas that become hypoxic at greater depths. After summer hypoxia, gradual abundance increases through the next spring often occur. Not all meiobenthic communities respond to hypoxia equally; some assemblages appear to be composed of a high proportion of tolerant species or display unique or unexpected adaptations to survive hypoxia. For example, declines in abundance during hypoxic events have been interpreted to be caused mainly by mortality. Our data from the Gulf of Mexico suggest that meiobenthic nematodes emigrate into the water column in high numbers where they survive hypoxic events until normoxic conditions are re-established.

Introduction

Hypoxic (dissolved oxygen concentration below 2 mg l^{-1}) and anoxic (dissolved oxygen concentration of 0 mg l^{-1}) events have been shown to have a great impact on all elements of marine benthic communities [Diaz and Rosenberg, 1995], including the meio-

Coastal Hypoxia: Consequences for Living Resources and Ecosystems
Coastal and Estuarine Studies, Pages 165-184

fauna. Meiofauna are ubiquitous in the marine benthos, and are defined as protozoans and metazoans small enough to pass through a 0.5 – 1 mm diameter mesh but large enough to be retained on a mesh of 0.042 – 0.063 mm. Metazoan meiofauna belong to 22 phyla, with the Loricifera, Kinorhyncha, Gastrotricha, and Gnathostomulida present only in the meiobenthos [Higgins and Thiel, 1988; Giere, 1993]. In addition to the "permanent" meiofauna, animals which spend their entire life span within the size boundaries of the meiofauna, many macrofauna have juvenile stages in the size range of the meiobenthos. These "temporary" meiofauna include economically important taxa (such as bivalves) as well as other invertebrates.

Meiofauna are also diverse and abundant members of the benthic assemblage in continental shelf and shallow sea sediments. Under normoxic conditions, nematodes (51 – 98 % of all metazoan meiofauna) and harpacticoid copepods (1.4 – 19.1 %) usually dominate the meiofauna in numbers and biomass [cf., Coull et al., 1982; Heip et al., 1990; Montagna, 1991; Guidi-Guilvard and Buscail, 1995]. Nematode densities on continental shelves have been reported to range from 310 to 6295 ind. 10 cm^{-2}, whereas reports of copepod densities range from 5 to 1328 ind. 10 cm^{-2} [Alongi, 1989; Tietjen, 1991; Herman and Dahms, 1992; Montagna and Harper, 1996; Radziejewska et al., 1996; de Bovée et al., 1996]. Species numbers over small spatial scales have been reported to range from 10 to 25 for copepods and from 31 to 56 for nematodes [Radziejewska and Drzycimski, 1988; Murrell and Fleeger, 1989; Tietjen, 1991; Vanreusel et al., 1992]. Vertically, total meiofaunal abundance typically declines with increasing sediment depth through the upper 4 – 6 cm in muddy sediments [Coull, 1988]. On a fine scale, however, harpacticoids are the most abundant meiofaunal taxon in the upper 4 mm, while many nematode species have deeper abundance maxima [Fleeger et al., 1995]. Continental-shelf meiofauna play a significant role in energy flow and nutrient recycling [McIntyre, 1969; Bell and Coull, 1978; Hicks and Coull, 1983; Grant and Schwinghamer, 1987], and they serve as an important food source for juvenile and small fish and invertebrates [Gee, 1989; Coull, 1990; McCall and Fleeger, 1995]. Probably because of their more epibenthic/hyperbenthic life style and their more surface-oriented vertical distribution, benthic copepods are much more frequently found in the gut contents of bottom-feeding fishes [Gee, 1989; McCall, 1992], and are more influenced by erosional events than are nematodes and other meiobenthic taxa [Foy and Thistle, 1991]. Copepods frequently feed on freshly deposited phytodetritus at the sediment-water interface [Decho, 1986; Rudnick, 1989; Montagna et al., 1989], and therefore are likely to disproportionately influence the burial of sedimented phytodetritus [Webb and Montagna, 1993; Fleeger et al., 1995].

The development of hypoxic and anoxic conditions in coastal oceans worldwide appears to have increased in frequency, duration and severity over the last few decades (see Diaz and Rosenberg [1995] for a recent review on the impact of hypoxia on macrofauna), as has our awareness of these events. Hypoxia is often related to the combination of eutrophication and density stratification resulting in oxygen deficiency in sediments and bottom water especially during the summer months. Eutrophication has increased with a higher input of nitrogen and phosphate from agricultural fertilizers [e.g., Rabalais et al., 1991; Rosenberg et al., 1991; Turner and Rabalais, 1991] creating increasing problems for highly productive areas by extending periods of hypoxia or anoxia. Many bodies of water develop low oxygen concentrations on a seasonal basis, mainly associated with thermal stratification. Permanently anoxic water can be found in some fjords and in shallow seas such as the deeper basins of the Baltic and Black Seas. In the open ocean, oxygen concentration in the water column declines with depth reaching a minimum at 500 – 1000 m where oxygen concentrations can be less than 2 mg l^{-1}. Below these depths, oxygen values typically increase, forming an oxygen minimum zone. This zone is often well developed beneath surface waters with high productivity. This oxygen deficient water can drift against the continental slope and even move upwards reaching the continental shelf, thus causing hypoxic or anoxic events there [Rosenberg et al., 1983; Arntz et al., 1991].

In addition, naturally-occurring anaerobic marine habitats are widespread and unrelated to anthropogenic activities. Many marine sediments throughout the world are permanently anoxic a few mm to cm below the surface, forming an "enormous, globally-continuous anoxic environment" [Fenchel and Finlay, 1994]. It has been suggested that the earliest metazoan evolution occurred in this habitat [see Fenchel and Riedl, 1970; Boaden and Platt, 1971; Boaden, 1975, 1989; Mangum, 1991]. A specially-adapted meiofauna, the thiobios, that lives preferentially in this anoxic and sulfidic sediment can be distinguished from the oxybiotic fauna of the surface oxidized sediment [e.g., Fenchel and Riedl, 1970; Boaden and Platt, 1971; Ott, 1972; Boaden, 1977, 1980; Reise and Ax, 1979, 1980; Powell and Bright, 1981; Meyers et al., 1987, 1988; Ott and Novak, 1989; Ott et al., 1991; Wetzel et al., 1995]. Thiobiotic meiofauna live in intertidal and subtidal sandy and muddy sediments, but stable ecological systems at the interface of oxic and anoxic conditions are perhaps most prominent in intertidal and shallow subtidal sandy sediments. Some meiofauna from continental shelves and shallow seas are undoubtedly thiobiotic, but the majority appear likely to be oxybiotic metazoans. Because of the unique nature of thiobiotic meiofauna, several studies examining the relationship between the thiobios and oxygen have been performed (see below).

The purpose of this chapter is to summarize the known influence of hypoxia and anoxia on marine meiobenthos, with special emphasis on studies from continental shelves and shallow seas. We will also give an overview of tolerance studies and the physiological adaptations of meiofauna to low oxygen concentrations, and we will present new data regarding meiofaunal abundance and distribution during and after a hypoxic event off the Louisiana USA coast. Unfortunately, to date, only a few studies have examined meiofauna in hypoxic waters, and most have examined hypoxia only anecdotally; only a small percentage of more recent studies are specifically designed to assess the effects of hypoxia.

Tolerance and Adaptations to Low Oxygen

No metazoan species has been shown to complete its life cycle in the absence of oxygen; obligate anaerobes, which live their entire life cycle under anoxic conditions, have only been found among free-living ciliated protozoans [Fenchel and Finlay, 1991]. In metazoans, oxygen is used as the final electron acceptor in aerobic metabolism and is required for catabolic and anabolic reactions involving fatty acids [e.g., Withers, 1992].

However, tolerance to low oxygen is quite variable among the meiofauna. At least three factors are responsible for these differences: (1) Exposure regime undoubtedly influences tolerance; meiofauna from environments that are permanently (e.g., thiobios) or regularly low in oxygen tend to be tolerant of low oxygen conditions [Vernberg and Coull, 1981]. A good example is found among muddy- and sandy-sediment dwelling meiofauna. Oxygen penetration into muddy sediments is much more shallow than into sandy sediments, and, especially at night, many shallow muddy sediments regularly become hypoxic even at the sediment-water interface [Vopel et al., 1996]. As a result, mud-dwelling meiofauna appear to have developed a greater tolerance to hypoxic and anaerobic conditions compared with sand-dwellers from the same geographic location [Vernberg and Coull, 1975]. (2) Differences among higher taxa of meiofauna strongly suggest that phylogenetic constraints influence tolerance. Compelling evidence comes from two recent, well-designed microcosm studies (Table 1), which show that nematodes, as a group, are highly tolerant of hypoxia/anoxia while crustacean meiofauna (harpacticoid copepods are most commonly studied), are much less resistant [Moodley et al., 1997; Modig and Ólafsson, 1998]. In addition, field studies have shown that nematodes increase and copepods decrease in relative abundance after organic-enrichment (which also leads to an increased prevalence of low oxygen conditions) [Coull and Chandler, 1992; Peterson et al., 1996], and that macrofaunal crustaceans are generally intolerant to low oxygen [Stickle et al., 1989; Diaz and Rosenberg, 1995]. Certainly some crustaceans (especially copepods and ostracods) are very tolerant to hypoxia [Vopel et al., 1996, 1998; Modig and Ólafsson, 1998], and some nematodes are intolerant of hypoxia (see below). Within a habitat, however, nematodes are generally the metazoan taxon most resistant to low oxygen concentrations. (3) Swimming and dispersal ability probably also influence tolerance. Typically, copepods are relatively capable swimmers and rapid dispersers [Sun and Fleeger, 1994] compared to strongly infaunal taxa such as nematodes, and they may use this ability to avoid hypoxia/anoxia by emigration, thereby reducing the likelihood for evolution of resistance. Although nematodes are generally thought to be poor swimmers, new data (below) strongly suggest that some nematode assemblages contain species capable of migrating from sediments under extreme conditions.

High tolerance in selected meiofauna to periods of low oxygen has been recognized for a long time. Moore [1931] maintained unidentified copepods and nematodes from muddy sediments with oxygen concentrations below 0.2 mg l^{-1} in the laboratory. She reported that 25% of all nematodes survived for 9 days, but copepods showed a greatly reduced survivorship; after 14 h, only 10% survived. This high tolerance of meiobenthic nematodes to reduced oxygen has been confirmed by many subsequent studies. Wieser and Kanwisher [1961] reported survivorship for selected nematode species of up to 60 d in anoxic muds, and one freshwater species (*Eudorylaimus andrassyi*) survived up to 6 mo under anoxic laboratory conditions [Por and Masry, 1968]. Jensen [1984] maintained nematodes from different sediment depths in anoxic conditions by flushing the overlying water with argon. He found that non-thiobiotic nematodes living in the topmost centimeter displayed almost 90% mortality within 5 d. Thiobiotic nematodes (i.e., *Sabatieria pulchra* and *Desmolaimus zeelandicus*) from deeper layers survived up to seven weeks in the anoxic treatment and were unable to survive in an oxygen-rich

TABLE 1. Summary of microcosm and field studies examining the effect of hypoxic/anoxic events on meiofaunal communities.

Author	Location	Water depth (m)	Sediment Type	Oxygen (mg/l) (lowest)	Duration of hypoxia	Meiofaunal Taxon	Mean Density (Ind. 10 cm^{-2})	
							Normoxia	Hypoxia/Anoxia
Microcosm experiments								
Moodley et al. [1997]	Northwestern Adriatic Sea (Italy)	19	mud	Normoxia: 7.6 — Anoxia: 0.0	80 d	Nematoda	~500 10 cm^{-3}	~20 10 cm^{-3}
						Copepoda	~5 10 cm^{-3}	0 10 cm^{-3}
						Hard-shelled Foraminifera	~300 10 cm^{-3}	~300 10 cm^{-3}
						Soft-shelled Foraminifera	~300 10 cm^{-3}	~50 10 cm^{-3}
Modig and Ólafsson [1998]	Northwestern Baltic Sea (Sweden)	33	mud	Normoxia: 11 – 12 — Hypoxia: 0.2	2 mo	Nematoda	2,096	1,113
						Copepoda	~600	0
						Amphipoda	~2	0
						Ostracoda	~300	~200
						Kinoryncha	~100	~80
							Prior hypoxia	During hypoxia
Field studies								
Coull [1969]	Devils Hole (Bermuda)	27	mud	0.0 – 0.6	3 mo	Total	1,500 – 2,680	0
Josefson and Widbom [1988]	Gullmar Fjord (Sweden)	115	mud	0.3	~5 mo	Total	1,855 – 10,234	3,156 – 5,727
						Nematoda	1,008 – 5,415	1,318 – 2,536
						Copepoda	13 – 40	8 – 26
						Temporary (Bivalvia)	1 – 22	6 – 11
Murrell and Fleeger [1989]	Gulf of Mexico (USA)	8-13	mud	< 2	5 mo	Total	~800 – 3,800	~500 – 1,100
						Nematoda	~600 – 3,100	~500 – 1,100
						Copepoda	~100 – 410	~0 – 20
						Kinorhyncha	~150 – 220	~10 – 110
Hendelberg and Jensen [1993]	Gullmar Fjord (Sweden)	10	mud	0.3	2 mo	Nematoda	8,900 – 10,200	1,300 – 1,700
							After hypoxia	During hypoxia
Present Study	Gulf of Mexico (USA)	18	mud	0.0	> 3 mo	Nematoda	392	50 – 117
						Copepoda	26	0

treatment. Modig and Ólafsson [1998] exposed sediment from the northwestern Baltic proper with natural populations of meiofauna to a 2 mo period of hypoxia in microcosms and found a tolerant fauna to include two ostracod species, the taxon Kinorhyncha and three nematode species. In a similar microcosm experiment by Moodley et al. [1997], nematodes declined in abundance but survived anoxia for up to two months, a time much longer than other metazoans. Hard-shelled Foraminifera were not influenced in this experiment (see Table 1), leading Moodley et al. [1997] to predict that seasonal anoxia lasting upwards of 2 – 3 mo would not eliminate these foraminiferans.

   Thiobiotic meiofauna are highly tolerant to low oxygen conditions and appear to have developed very complex relationships with oxygen tolerance and unique life history patterns. The nematode, *Theristus anoxybioticus*, lives its relatively long juvenile period (9 mo compared to 13 d to a year in other nematodes studied [cf., Gerlach and Schrage, 1971; Tietjen and Lee, 1977, 1984; Ferris and Ferris, 1979; Alongi and Tietjen, 1980; Smol et al., 1980; Jensen, 1984, 1995; Bouwman et al., 1984; Heip et al., 1985] in deep anoxic and sulfidic sediment layers whereas adults live in surface, oxidized sediments. Oxygen appears to be toxic to the juveniles of *T. anoxybioticus*, and the high survival in anoxic water suggests that juveniles are able to utilize anaerobic metabolism throughout this life stage [Jensen, 1995].

   On the other hand, many oxybiotic meiofauna studied have shown little resistance to hypoxia/anoxia. Lasserre and Renaud-Mornant [1973] studied the tolerance of six meiofauna species to anoxic conditions alone and with hydrogen sulfide. They reported a 50% mortality in two crustaceans after only 2 – 5 h of exposure to anoxia, while the four annelids studied showed survival times of 48 h and more. Vernberg and Coull [1975] reported that mud-dwelling copepods tolerate anoxia for up to 120 h while sand-dwelling copepods tolerate anoxia for as little as 1 h. Moodley et al. [1997] found that copepod densities are reduced in less than 11 d in microcosms exposed to anoxia, and Modig and Ólafsson [1998] identified a group of meiofauna sensitive to long-term (2 mo) and short-term (7 d) bursts of hypoxia that included one species of juvenile amphipod, two species of harpacticoid copepods and a surface-dwelling nematode species.

   Much less is known about multiple-factor effects (and possible synergisms among environmental factors) on tolerance of meiofauna to hypoxia/anoxia. Vernberg and Coull [1975] report that tolerance to anoxia by meiobenthic ciliates and copepods is greatest under near-optimal temperature and salinity conditions; salinity or temperature stress greatly enhanced the toxicity of anoxia. Hypoxia and anoxia are well correlated with the presence of hydrogen sulfide, especially within the sediment. The toxicity of hydrogen sulfide is well known [National Research Council, 1979] and is mainly due to the reversible inhibition of the cytochrome c oxidase, a key enzyme in aerobic respiration [e.g., Vismann, 1991]. Except for Lasserre and Renaud-Mornant [1973], we know of no investigations designed to examine the possible synergistic effects of hypoxia and hydrogen sulfide on meiofauna. Results from this single study were inconclusive; some species showed a pronounced lower tolerance but others slightly higher survival rates when exposed to both hypoxia and H_2S. Results from macrofauna studies show that the presence of sulfide decreases tolerance to hypoxia, sometimes by a large factor [e.g., Hagerman and Vismann, 1995; Vistisen and Vismann, 1997]. Another important synergism that has been overlooked is the combined effect of hypoxia/anoxia and sediment contamination.

Preliminary evidence from tolerance experiments suggests that *Pseudostenhelia wellsi*, a harpacticoid copepod, is much more sensitive to the combination of polycyclic aromatic hydrocarbons and low oxygen than to low oxygen alone [Fleeger, unpublished data]. What adaptations allow meiofauna to survive hypoxic/anoxic conditions? One adaptation is to lower oxygen consumption thus enabling the animal to withstand short periods of anoxia or hypoxia. Oxygen consumption varies greatly among taxa. In general, the respiration rate of benthic copepods is significantly higher than for nematodes after rates are standardized by mass [Coull and Vernberg, 1970], and oxygen uptake is considerably reduced in individuals exposed to anoxia [Ott and Schiemer, 1973; Fox and Powell, 1987; Schiemer et al., 1990]. Many animals with a high tolerance for anoxia can continue oxic metabolism at extremely low oxygen tension [Powell, 1989]. Oxygen consumption is also strongly correlated to activity pattern [Coull and Vernberg, 1970]; very active copepods consume more oxygen than lethargic species. Because respiration rates are influenced by several factors such as stress induced by extraction, presence or absence of sediment during the measurements, temperature, salinity, and culture prior to measurements [e.g., Lasker et al., 1970; Vernberg and Coull, 1974, 1975; Hummon, 1974; Vernberg et al., 1977; Teare and Price, 1979; Schiemer, 1983; Schiemer et al., 1990], caution is advised when comparing respiration rates from different studies.

Special enzyme systems and physiological adaptations could also enhance the ability of an animal to live under low oxygen conditions. For example, hemoglobin with a very high affinity for oxygen is present in a few meiobenthic nematodes [Ellenby and Smith, 1966] and meiobenthic gastrotrichs [Colacino and Kraus, 1984]. Hemoglobin may serve as a reservoir of oxygen and therefore enable animals to continue aerobic metabolism for a short period of time when oxygen is unavailable or as a long-term source of oxygen in animals with reduced metabolic rates [Colacino and Kraus, 1984].

Meiofauna lack respiratory organs, and thus they depend on diffusive oxygen uptake. A higher surface to volume ratio enhances oxygen uptake by diffusion, and body shape may therefore serve as an adaptation to low oxygen environments. For example, animals living under low oxygen concentrations may be more slender to increase surface area relative to volume [e.g., Powell, 1989]. This is especially true for species living within or below the redox potential discontinuity layer (RPD layer), the transition zone between oxidized and anoxic sediment and under hypoxic or anoxic conditions. Thiobiotic nematodes have a significantly greater length to diameter ratio, i.e., they are longer and more slender, than oxybiotic nematodes from oxidized sediment horizons [Jensen, 1986, 1987; Wetzel et al., 1995]. This feature also enhances the uptake of dissolved organic matter.

Finally, if no oxygen is available for extended periods, animals capable of surviving must "switch" to anaerobic metabolism. The high survivorship of some species under anoxic conditions strongly suggests such anaerobic capabilities. A frequent pathway in anaerobic invertebrates, the reverse Krebs cycle, has been suggested for the gastrotrich, *Thiodasys sterreri* [Maguire and Boaden, 1975]. This species can survive anoxic conditions for up to 4 mo. The capability of anaerobic metabolism has also been documented in three thiobiotic turbellarians [Fox and Powell, 1987]. All metazoan meiofauna, including thiobiotic species, appear to undergo aerobic metabolism in the presence of oxygen, despite the potential to survive extended periods of anoxia in their usual anoxic and sulfidic habitats [Powell, 1989].

Most of the research highlighted above was conducted on species from shallow water (< 10 m) and from thiobiotic systems. Much less is known about continental shelf and shallow sea meiofauna. It is possible that meiofauna living in periodically exposed tidal zones with tidal pumping and a migrating RPD layer are much better adapted to periods of environmental stress than meiobenthos from sediments which are continually sub-merged. Tidal exposure, for example, could also lead to conditions of oxygen deficiency in sediment strata normally well oxygenated by the overlying water cover. This is espe-cially true for fauna associated with the oxic halo around macrofaunal worm burrows [e.g., Meyers et al., 1987, 1988; Wetzel et al., 1995]. Macrofauna inhabiting these burrows must pause ventilating their burrows during periods without water cover.

Effect on Meiobenthic Communities

Seasonal Hypoxia and Meiofauna

In shallow coastal areas, hypoxic events are often correlated with water-column strati-fication induced by thermal and/or salinity discontinuities. Nutrient additions can in-crease primary production and therefore contribute to depletion of oxygen in the bottom waters especially in the summer. The effects of seasonal hypoxia on meiobenthic com-munities have been described in studies in the Skagerrak [Rosenberg et al., 1977, Josefson and Widbom, 1988; Austen and Wibdom, 1991; Hendelberg and Jensen, 1993], one study in the Gulf of Mexico [Murrell and Fleeger, 1989] and another in Bermuda [Coull, 1969]. See Table 1 for comparative information on these studies.

Hypoxia in the bottom waters is a common phenomenon in the northern Gulf of Mexico on the Louisiana continental shelf linked to the formation of a well-defined halo-cline [e.g., Rabalais et al., 1991, 1992, 1994]. Hypoxic conditions typically begin in April and continue until October. Oxygen-deficient water masses can usually be found 5 to 60 km offshore in water depths of 5 to 60 m. Hypoxic bottom waters can extend up to 20 m above the bottom. Murrell and Fleeger [1989] (see Table 1) surveyed the meiofaunal assemblage at three shallow stations (8 – 13 m) over an annual cycle and related their results to hypoxia. Total meiofaunal abundances ranged from approximately 800 – 3800 ind. 10 cm^{-2} before a hypoxic event. Meiofaunal abundance increased throughout spring until hypoxia developed. As the hypoxic water mass moved shoreward, they observed dramatic declines in abundance and diversity of major meiofauna taxa. Meiobenthic copepods showed the highest sensitivity to hypoxia, while nematodes and kinorhynchs were less affected. This effect was observed first at the deeper sample sites where hypoxic water masses developed first. Nematode abundance ranged from 600 – 3100 ind. 10 cm^{-2} before hypoxia and from 500 – 1100 ind. 10 cm^{-2} after hypoxia, but declines due to season and hypoxia could not be separated. Copepods, however, showed declines from highs ranging from 100 – 410 to 0 ind. 10 cm^{-2} at all sites when hypoxia developed.

In the Devils Hole, a 27-m deep depression in Harrington Sound, Bermuda, a summer thermocline extends from May to September/October with oxygen values of 0.0 – 0.6 mg l^{-1} and the generation of hydrogen sulfide in bottom sediments. The sediments in the

Devils Hole are coarse silts consisting primarily of poorly-sorted bivalve fecal pellets. Coull [1969] conducted a year-long study of the meiobenthos and found that during the period of hypoxia/anoxia no meiofauna occurred in the anoxic areas of the trough. With the reestablishment of oxic conditions, total meiofauna population density increased quickly to $20 - 150$ ind. 10 cm^{-2}, but did not reach the high abundance found prior to the anoxia (1500 to 2680 ind. 10 cm^{-2}) until the spring [Coull, 1969].

The fjords of the Skagerrak coast of Sweden experience hypoxia and even anoxia below a halocline for most of the year. The Gullmar Fjord, the largest of the Swedish fjords, has a maximum water depth of 120 m. Water exchange with the Skagerrak is limited by a sill at 42 m water depth. Hypoxic conditions in the deep basin are common, because water renewal takes place only once a year, typically in the spring. This leads to a stagnation period with oxygen depletion in the winter with values typically around 2.9 mg l^{-1} O$_2$ or below. Josefson and Widbom [1988] studied the benthic fauna at a depth of 115 m in the Gullmar Fjord over a period of almost five years. One of the low oxygen periods was particularly severe with oxygen depletion in the bottom water to 0.3 mg l^{-1}. The authors noted a strong effect on the macrofauna, but in contrast to studies in the Gulf of Mexico and Bermuda, they found no significant decrease in abundance for any meiofaunal taxa following the severe hypoxic event. Only the temporary meiofauna showed considerable, but not statistically significant, decreases in abundance. By analyzing the nematode species composition from the same data set with both univariate and multivariate statistics, Austen and Wibdom [1991] reported that there was no change in community structure concurrent with the hypoxic event; however, a significant decrease in diversity and a change in the community assemblage occurred during the subsequent year. Even though the observed changes were probably related to the hypoxic event, no clear causal relationship between hypoxia and changes in nematode composition could be drawn from this study, due to limitations in experimental design.

Hendelberg and Jensen [1993] (Table 1) studied the vertical distribution of a nematode community at a 10-m deep subtidal area near the opening of the Gullmar Fjord, Sweden. During a 2-mo hypoxic period, from August to September, they observed a general upward migration of nematodes towards the surface that were normally found in the deeper anoxic and sulfidic sediment layers. At the same time, they observed a pronounced decrease in overall nematode abundance and species diversity. After one month of hypoxia the nematode assemblage consisted only of the bacterial feeder, *Sabatieria pulchra*, a thiobiotic species often associated with the RPD layer [Jensen, 1981a, 1983, 1984; Bouwman et al., 1984]. Laboratory experiments showed that this species is able to survive anoxia for several months [Wieser and Kanwisher, 1961; Jensen, 1984; Modig and Ólafsson, 1998].

Comparing the small data set available from the Skagerrak and the Baltic Sea with the Gulf of Mexico, we hypothesize that meiofaunal assemblages differ geographically in their response to hypoxia. Meiofauna from the Skagerrak appear to be minimally influenced by hypoxia, while the subtropical locations studied so far are strongly influenced. In shallow-water sediments of the Skagerrak and the Baltic Sea, nematode communities appear to be composed of a high proportion of tolerant species. *Sabateria pulchra* is common in these sediments and is clearly tolerant to low dissolved oxygen concentrations [Jensen, 1984; Hendelberg and Jensen, 1993; Modig and Ólafsson, 1998]. Communities at similar depths in other locations appear to have a low frequency of tolerant species. Certainly studies from a broader range of locations are needed to begin to understand why such different community responses occur. Evolutionary history may be important (perhaps the frequency of disturbance/hypoxia over evolutionary time scales has selected for a resistant community in the Skagerrak), or environmental factors perhaps related to the prevalence of a thiobiotic fauna (why do some locations have a high proportion of thiobiotic species while other locations have only normoxic species?) could be responsible.

Changes in Community Structure as a Function of Water Depth

Changes in total meiofaunal abundance and community structure with water depth in shallow seas frequently appear to be associated with hypoxia [Baçesco, 1963; Coull, 1969; Elmgren, 1975; Nichols, 1976; Rosenberg et al., 1977; Arlt et al., 1982; Radziejewska, 1989; Murrell and Fleeger, 1989]. Oxygen concentration tends to decline with water depth in many shallow seas and often reaches hypoxic or anoxic conditions in deeper regions. For example in a survey of the entire Baltic Sea, Arlt et al. [1982] observed that oxygen concentration declined with depth reaching hypoxia between 60 – 80 m with meiofaunal abundance following this decline. This decrease in overall abundance with depth followed a logarithmic pattern and the relationship between the logarithm of the total meiofauna abundance and the logarithm of the oxygen concentration was linear. Several additional studies [Coull, 1969; Nichols, 1976; Radziejewska, 1989; Murrell and Fleeger, 1989] found changes in community structure with decreasing oxygen concentrations resulting in a distinct taxonomic zonation with depth. Typically, nematodes are the only group found in the most hypoxic zone in the deeper areas, whereas meiobenthic copepods were not found at depths below 60 m where dissolved oxygen concentrations tend to be below 2 mg l^{-1} [Coull, 1969, Arlt et al., 1982; Radziejewska, 1989; Murrell and Fleeger, 1989]. Of course other factors, e.g., organic input, vary with increasing water depth, but meiofaunal declines and compositional changes relate well to tolerance studies, suggesting that dissolved oxygen is the causative agent in the studies cited above. On the other hand, meiofauna communities under stable low oxygen concentrations may show relatively high diversity. Nichols [1976] found a highly diverse meiofauna dominated by nematodes ($H' = 1.6 - 2.8$) at 90 to 200 m depth in the Fosa de Cariao, a hypoxic (oxygen values ranged from 0.3 – 1.0 mg l^{-1}) basin off the coast of Venezuela.

Levin et al. [1991] present interesting results that further suggest a strong link between meiofaunal abundance and hypoxia. They surveyed an underwater sea mount in the eastern tropical Pacific with a summit depth of 730 m that extends into the oxygen minimum zone. Linear increases in oxygen concentration with water depth were observed down the sea mount. Correspondingly, the relative contribution of harpacticoid copepods to total meiofauna increased downslope, while the nematode to copepod ratio decreased. On the upper summit (oxygen concentration 0.1 – 0.7 mg l^{-1}), harpacticoid copepods were extremely rare (1 ind. 10 cm^{-2}), while their abundance was highest on the sea mount flank (35 ind. 10 cm^{-2}) at 1000 – 2000 m depth and an oxygen concentration of 4.3 mg l^{-1}

in the water column. These results reinforce the common observation of high sensitivity of copepods to low oxygen values, and strongly suggest that the decreases in meiofaunal abundance and diversity are frequently related to hypoxia.

New Data from the Gulf of Mexico

During the summer of 1995 Powers et al. [this Volume] conducted a study to examine larval supply and settlement of macrobenthic invertebrates in relation to a hypoxic water mass that develops seasonally in the Gulf of Mexico. Because the techniques used to process larval macrofauna samples are identical to those of meiofauna (most macrofauna are classified as meiofauna at the larval stage), the study provided a unique opportunity to examine both sediment densities and near-bottom water column processes related to the response of permanent meiofauna to hypoxic/anoxic conditions.

Methods

Settlement traps were placed at two stations (C-6B and C-SP-2) from June to September 1995; C-6B was also sampled by Murrell and Fleeger [1989], and C-SP-2 is in close proximity to C-6B (see Powers et al. [this Volume] for a complete description of both stations). Traps were deployed for four intervals: 22 June to 13 July (22 days), 13 July to 8 August (23 days), 8 August to 28 August (21 days), and 28 August to 14 September 1995 (17 days). Each settlement trap consisted of three polybuturate tubes (60 cm in length and 5 cm in diameter) attached to separate reinforcement rods (rebar). The three reinforcement rods were set in a 90-kg, 0.7-m^2 concrete block [cf., Yund et al. 1991]. The opening of the tubes was approximately 0.8 m above the sediment surface. The bottom half of each tube was filled with a brine solution (40 ‰) containing rose Bengal and 10% formalin. Settlement traps were retrieved and replaced by SCUBA divers at the end and beginning of each interval. After retrieval of tube samples, divers collected three 5-cm sediment cores from the area around the settlement trap. Sediment was extruded to a depth of 5 cm and preserved in 10% formalin on-board ship. After 24 – 48 h both settlement trap and core samples were washed on a 63-μm sieve, and the contents remaining on the sieve were placed in 70% ethanol. Nematode abundances from trap samples were transformed to a flux measurement (ind. cm^{-2} d^{-1}) to allow comparison between different collection intervals (Fig. 1). The abundance of nematodes and copepods from the sediment cores was standardized to surface area (ind. 10 cm^{-2}). Mean (one standard deviation) dissolved oxygen concentrations in water column above the site were measured [Rabalais et al., unpubl. data] (Fig. 1). We further analyzed nematode generic composition for two intervals (22 June – 13 July; severe hypoxia/anoxia, and August 5 – August 28; normoxic conditions) in trap samples and sediment cores. Samples were examined under a stereo-dissection microscope and 100 – 150 nematodes were removed for preparation of permanent slides according to Riemann [1988] and identified to genus using differential interference contrast microscopy. We cannot give the exact number of species, but at least some genera are represented by more than one species.

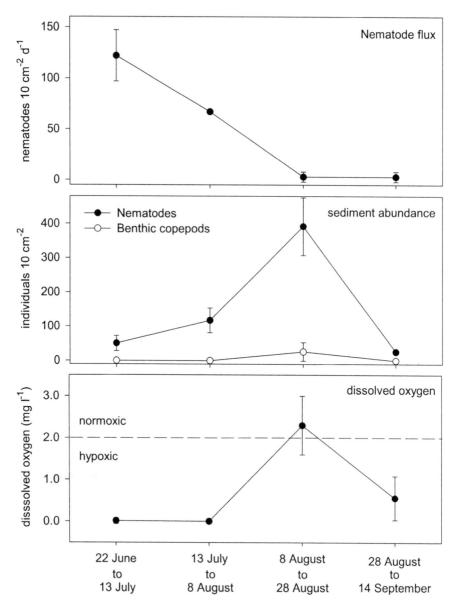

Figure 1. Nematode flux in the water column, nematode and copepod abundance in the sediment, and dissolved oxygen concentration at station C-6B in the Gulf of Mexico from 22 June 1995 to 14 September 1995. Sediment traps were placed at a depth of 18 m. Dissolved oxygen concentration (mg l^{-1}) was measured in the water column 1 m above the sea floor. On the abscissa start and end dates for the sediment traps are given. Sediment cores were taken on the end dates. Error bars represent one standard deviation.

Results and Discussion

A well-defined halocline was present at depths between 2 – 8 m from 22 June to 8 August with hypoxic/anoxic conditions in the water column and sediment below 15 m (see Fig. 4 in Powers et al. [this Volume]). During this time of hypoxia/anoxia, nematodes occurred in the water column in very high abundance (flux ranged from 67 – 122 ind. 10 cm^{-2} d^{-1}), and sediment nematode abundance was remarkably low (50 – 177 ind. 10 cm^{-2}). No copepods were found in sediment samples. From 8 August – 14 September, stratification broke up at the sampling site and oxygen was present at 6 mg l^{-1} levels in the water column at 14 m. Concomitantly, nematode flux was greatly reduced (to 3 ind. 10 cm^{-2} d^{-1}), and nematode sediment abundance increased to 26 – 392 ind. 10 cm^{-2}, a value low for normoxic conditions [see Murrell and Fleeger, 1989], but much higher than values found during hypoxia. Although benthic copepods were not found in settlement traps, they were found in low abundance (26 ind. 10 cm^{-2}) in the sediment after dissolved oxygen concentrations increased. Further analysis of the nematode composition revealed that 23 nematode genera belonging to 16 families (Table 2) were present in the water and sediment. Under severe hypoxia, 19 nematode genera were found in the settlement traps but only one nematode genus was present in the sediment (represented by a single specimen). When normoxic conditions were re-established no nematodes were found in traps and 6 of the 19 nematode genera observed earlier in the settlement traps were present in sediment samples.

The ability of meiofauna to actively and passively enter the water column is now widely recognized [e.g., Walters and Bell, 1986, 1994; Palmer, 1988; Armonies, 1994; Sun and Fleeger, 1994]. Jensen [1981b] demonstrated that some nematode species actively swim up from the benthos. Furthermore, nematodes have also been found associated with marine snow [Shanks and Edmondson, 1990; Shanks and Walters, 1997]. It is difficult to compare water column abundance among studies; however, our results appear to be the highest water column flux (67 – 122 ind. 10 cm^{-2} d^{-1}) yet recorded for nematodes. Shanks and Edmonds [1990] report a nematode flux of 62 – 99 ind. cm^{-2} d^{-1} in a shallow bay in 3-m deep sediment traps). Given the large area of the Gulf of Mexico influenced by hypoxia, these high flux rates suggest that nematodes migrated from the sediment when conditions worsened (under severe hypoxia) into the water column where dissolved oxygen concentrations were probably sufficient to maintain survival. Active emergence rather than erosion and passive transport seem most likely. Bottom currents at the site rarely exceeded 5 cm s^{-1} during the hypoxic event, and underwater visibility ranged from 12 – 30 m during the period of severe hypoxia [Rabalais, personal communication], indicating low resuspension. The similarity in species composition (Table 2) and the rapid increase in abundance (Fig. 1) after the return of normoxic water further suggests settlement or emigration rather than reproduction accounted for the rapid increase in nematode density. In all previous field studies, the absence of species during hypoxia and the recovery after hypoxic conditions has been interpreted in terms of mortality and not migration, especially for meiobenthic nematodes. Our results suggest that the interactions are much more complex and need further investigation.

TABLE 2. Presence (•) and absence (—) of nematode genera from settlement traps and sediment samples during severe hypoxia and following the return of normoxic conditions at station C-6B in the Gulf of Mexico. Sediment traps were place from 22 June 1995 to 13 July 1995 under hypoxic conditions and from 5 August 1995 to 28 August 1995 under normoxic conditions. Sediment cores were collected at the date when settlement traps were emptied, 13 July 1995 (hypoxia), and 28 August 1995 (normoxic), respectively. Bray-Curtis similarity between settlement trap from 22 June 1995 to 13 July 1995 and sediment sample from 28 August 1995 is 55.56%.

Order/Genus	Severe hypoxia		Normoxic	
	Settlement Traps	Sediment	Settlement Traps	Sediment
Enoplida				
Dolicholaimus	•	—	—	—
Halalaimus	•	—	—	•
Nemanema	—	—	—	•
Oncholaimus	•	—	—	—
Viscosia	•	—	—	—
Chromadorida				
Cyartonema	•	—	—	•
Leptolaimus	—	—	—	•
Microlaimus	•	—	—	—
Neochromadora	•	—	—	—
Paracanthonchus	•	—	—	•
Prochromadorella	•	—	—	—
Pselionema	•	—	—	—
Sabatieria	—	—	—	•
Spirinia	•	—	—	—
Synonchiella	•	—	—	—
Monhysterida				
Ascolaimus	—	—	—	•
Cobbia	•	—	—	—
Daptonema	•	•[a]	—	•
Metadesmolaimus	•	—	—	—
Odontophora	•	—	—	•
Sphaerolaimus	•	—	—	—
Terschelingia	•	—	—	•
Parodontophora	•	—	—	—

[a] represented by a single specimen

Conclusions

All metazoan meiofauna appear to show some sensitivity to prolonged periods of low oxygen concentration; however, some meiofauna are sensitive to even short-term (hours) exposures to hypoxia while many meiofauna are quite tolerant to long-term (days-weeks) exposure (as shown by field and laboratory studies). Tolerance is probably related to exposure history (over both ecological and evolutionary time scales), phylogenetic constraints and lifestyle. Studies within a given habitat typically conclude that of all the permanent meiobenthic taxa, foraminiferans and nematodes are best adapted to hypoxia/anoxia and that crustaceans are the least tolerant. Low oxygen demand, hemoglobin with a high affinity to oxygen, a high surface to volume ratio and anaerobic metabolism enable some species to survive hypoxia/anoxia for extended times. Field studies of hypoxic episodes typically demonstrate a reduction in overall meiofaunal abundance, a reduction in species and higher taxon diversity and an increase in the relative abundance of nematodes. Similar trends occur with increasing water depth in shallow seas that become hypoxic at greater depths. Time of recovery after hypoxia has been examined from seasonal collections and in areas that become hypoxic during the summer months, abundance increases by the next spring have typically been shown to occur. Not all areas respond to hypoxia equally; some assemblages appear to be composed of a high proportion of tolerant species. New data from the Gulf of Mexico suggests that meiofauna, including nematodes, are capable of migration from hypoxic sediments into the water column and may settle to the sediments with the return of normoxic conditions. Future research is particularly needed on three major topics: (1) The frequency of migration of meiofauna into the water column as a refuge from periods of hypoxia needs to be studied further to fully understand the impact of hypoxic events. (2) Differences in community response to hypoxia from a broader range of geographic locations should be examined to determine if these differences are due to evolutionary adaptations or ecological factors. (3) Synergistic effects of hypoxia/anoxia with contaminants and hydrogen sulfide seem important, especially because many hypoxic/anoxic events are strongly associated with anthropogenic pollution.

Acknowledgments. We thank N. N. Rabalais, R. E. Turner and W. J. Wiseman, Jr. for permission to use their unpublished oxygen data. W. B. Stickle made valuable remarks on the manuscript. Two anonymous referees are acknowledged for their comments. Support for the 1995 sampling was provided by grants from the LUMCON Foundation, Inc. and the PADI Foundation to Powers.

References

Alongi, D. M., Benthic processes across mixed terrigenous-carbonate sedimentary facies on the central Great Barrier Reef continental shelf, *Cont. Shelf. Res., 9*, 629-663, 1989.
Alongi, D. M. and J. H. Tietjen, Population growth and trophic interactions among free-living marine nematodes, in *Marine Benthic Dynamics.*, edited by K. R. Tenore and B. C. Coull, pp. 151-165, University South Carolina Press, Columbia, 1980.
Arlt, G., B. Müller, and K.-H. Warnack, On the distribution of meiofauna in the Baltic Sea., *Int. Rev. Gesamten Hydrobiol., 67*, 97-111, 1982.
Armonies, W., Drifting meio- and macrobenthic invertebrates on tidal flats in Königshafen: A review, *Helgoländer Meeresunters., 48*, 299-320, 1994.
Arntz, W. E., J. Tarazona, V. A. Gallardo, L. A. Flores, and H. Salzwedel, Benthos communities in oxygen deficient shelf and upper slope areas of the Peruvian and

Chilean Pacific coast, and changes caused by El Nino., in *Modern and Ancient Continental Shelf Anoxia.*, edited by R. V. Tyson and T. H. Pearson, pp. 131-154, Geological Society Special Publication No. 58, 1991.

Austen, M. C. and B. Wibdom, Changes in and slow recovery of a meiobenthic nematode assemblage following a hypoxic period in the Gullmar Fjord Basin, Sweden, *Mar. Biol.*, *111*, 139-145, 1991.

Baçesco, M., Contribution a la Biocoenologie de la Mer Noir L'Étage Périazoique et le Faciès Drissenifère Leurs Caractéristiques, *Rapp. P. -v. Reun. Commn. int. Explor. Scient. Mer Mediter.*, *17*, 107-122, 1963.

Bell, S. S. and B. C. Coull, Field evidence that shrimp predation regulates meiofauna, *Oecologia, 35*, 141-148, 1978.

Boaden, P. and H. Platt, Daily migration patterns in an intertidal meiobenthic community, *Thalassia Jugosl.*, *7*, 1-12, 1971.

Boaden, P. J. S., Anaerobiosis, meiofauna and early metazoan evolution, *Zool. Scr.*, *4*, 21-24, 1975.

Boaden, P. J. S., Thiobiotic facts and fancies (aspects of the distribution and evolution of anaerobic meiofauna), *Mikro. Meeres.*, *61*, 45-63, 1977.

Boaden, P. J. S., Meiofaunal thiobios and "the *Arenicola* negation": case not proven, *Mar. Biol.*, *58*, 25-29, 1980.

Boaden, P. J. S., Meiofauna and the origins of the Metazoa, *Zool. J. Linn. Soc.*, *96*, 217-227, 1989.

Bouwman, L. A., K. Romeijn, and W. Admiraal, On the ecology of meiofauna in an organically polluted estuarine mudflat, *Estuar. Coast. Shelf Sci.*, *19*, 633-653, 1984.

Colacino, J. M. and D. W. Kraus, Hemoglobin-containing cells of *Neodasys* (Gastrotricha: Chaetonotida), *Comp. Biochem. Physiol.*, *79*, 363-370, 1984.

Coull, B. C., Hydrographic control of meiobenthos in Bermuda, *Limnol. Oceanogr.*, *14*, 953-957, 1969.

Coull, B. C., Ecology of the marine meiofauna., in *Introduction to the Study of Meiofauna*, edited by R. P. Higgins and H. Thiel, pp. 18-38, Smithsonian Institution Press, Washington D.C., 1988.

Coull, B. C., Are members of the meiofauna food for higher trophic levels?, *Trans. Am. Microsc. Soc.*, *109*, 233-246, 1990.

Coull, B. C. and G. T. Chandler, Pollution and meiofauna: Field, laboratory and mesocosm studies, *Oceanogr. Mar. Biol. Ann. Rev.*, *30*, 191-271, 1992.

Coull, B. C. and W. B. Vernberg, Harpacticoid copepod respiration: *Enhydrosoma propinquum* and *Longipediaia helgolandica*, *Mar. Biol.*, *5*, 341-344, 1970.

Coull, B. C., Z. Zo, J. H. Tietjen, and B. S. Williams, Meiofauna of the southeastern United States continental shelf, *Bull. Mar. Sci.*, *32*, 139-150, 1982.

de Bovée, F., P. O. J. Hall, S. Hulth, G. Hulthe, A. Landén, and A. Tengberg, Quantitative distribution of metazoan meiofauna in continental margin sediments of the Skagerrak (Northeastern North Sea), *J. Sea Res.*, *35*, 189-197, 1996.

Decho, A. W., Water-cover influence on diatom ingestion rates by meiobenthic copepods, *Mar. Ecol. Prog. Ser.*, *33*, 139-146, 1986.

Diaz, R. J. and R. Rosenberg, Marine benthic hypoxia: A review of its ecological effects and the behavioural responses of benthic macrofauna, *Oceanogr. Mar. Biol. Ann. Rev.*, *33*, 245-303, 1995.

Ellenby, C. and L. Smith, Haemoglobin in *Mermis subnigrescens* (Cobb), *Enoplus brevis* (Bastian), *E. communis* (Bastian), *Comp. Biochem. Physiol.*, *19*, 871-877, 1966.

Elmgren, R., Benthic meiofauna as indicator of oxygen conditions in the northern Baltic proper, *Mereturkimuslait. Julk. /Havsforskningst. Skr.*, *239*, 265-271, 1975.

Fenchel, T. and B.J. Finlay, The biology of freeliving ciliates, *Euro. J. Protistology, 26*, 201-215, 1991.

Fenchel, T. and B. J. Finlay, The evolution of life without oxygen, *Am. Sci.*, *82*, 22-29, 1994.

Fenchel, T. and R. J. Riedl, The sulfide system: a new biotic community underneath the oxidized layer of marine sand bottoms, *Mar. Biol.*, *7*, 255-268, 1970.

Ferris, V. R. and J. M. Ferris, Thread worms (Nematoda), in *Pollution Ecology of*

Estuarine Invertebrates, edited by C. W. Hart and S. L. H. Fuller, pp. 1-33, Academic Press, New York, 1979.

Fleeger, J. W., T. C. Shirley, and J. N. McCall, Fine-scale vertical profiles of meiofauna in muddy sediments, *Can. J. Zool., 73*, 1453-1460, 1995.

Fox, C. A. and E. N. Powell, The effect of oxygen and sulfide on CO_2 production by three acoel turbellarians are thiobiotic meiofauna aerobic?, *Comp. Biochem. Physiol., 86*, 509-514, 1987.

Foy, M. S. and D. Thistle, On the vertical distribution of a benthic harpacticoid copepod - Field, laboratory, and flume results, *J. Exp. Mar. Biol. Ecol., 153*, 153-163, 1991.

Gee, J. M., An ecological and economic review of meiofauna as food for fish, *Zool. J. Linn. Soc., 96*, 243-261, 1989.

Gerlach, S. A. and M. Schrage, Life cycles in marine meiobenthos. Experiments with *Monhystera disjuncta* and *Theristus pertenuis, Mar. Biol., 9*, 274-281, 1971.

Giere, O., Meiobenthology. *The Microscopic Fauna in Aquatic Sediments*, 328 pp., Springer-Verlag, Berlin, 1993.

Grant, J. and P. Schwinghamer, Size partitioning on microbial and meiobenthic biomass and respiration on Brown's Bank, South-west Nova Scotia., *Estuar. Coast. Shelf Sci., 25*, 647-661, 1987.

Guidi-Guilvard, L. D. and R. Buscail, Seasonal survey of metazoan meiofauna and surface sediment organics in a non-tidal turbulent sublittoral prodelta (northwestern Mediterranean), *Cont. Shelf Res, 15* , 633-653, 1995.

Hagerman, L. and B. Vismann, Anaerobic metabolism in the shrimp *Crangon crangon* exposed to hypoxia, anoxia and hydrogen sulfide, *Mar. Biol., 123*, 235-240, 1995.

Heip, C., R. Huys, M. Vincx, A. Vanreusel, N. Smol, R. Herman, and P. M. J. Herman, Composition, distribution, biomass and production of North Sea meiofauna, *Neth. J. Sea Res., 26*, 333-342, 1990.

Heip, C., M. Vincx, and G. Vranken, The ecology of marine nematodes., *Oceanogr. Mar. Biol. Ann. Rev., 23*, 399-489, 1985.

Hendelberg, M. and P. Jensen, Vertical distribution of the nematode fauna in a coastal sediment influenced by seasonal hypoxia in the bottom water, *Ophelia, 37*, 83-94, 1993.

Herman, R. L. and H. U. Dahms, Meiofauna communities along a depth transect off Halley Bay (Weddell Sea-Antarctica), *Polar Biol., 12*, 313-320, 1992.

Hicks, G. R. F. and B. C. Coull, The ecology of marine meiobenthic harpacticoid copepods, *Oceanogr. Mar. Biol. Ann. Rev., 21*, 67-175, 1983.

Higgins, R. P. and H. Thiel, Prospectus, in *Introduction to the Study of Meiofauna*, edited by R. P. Higgins and H. Thiel, pp. 11-13, Smithsonian Institution Press, Washington D.C., 1988.

Hummon, W. D., Respiratory and osmoregulatory physiology of a meiobenthic marine gastrotrich, *Turbanella ocellata* Hummon 1974, *Cah. Biol. Mar., 26*, 255-268, 1974.

Jensen, P., Phyto-chemical sensitivity and swimming behaviour of the free-living marine nematode *Chromadorita tenuis, Mar. Ecol. Prog. Ser., 4*, 203-206, 1981b.

Jensen, P., Species distribution and a microhabitat theory for marine mud dwelling Comesomatidae (Nematoda) in European waters, *Cah. Biol. Mar., 22*, 231-241, 1981a.

Jensen, P., Meiofaunal abundance and vertical zonation in a sublittoral soft bottom, with a test of the Haps corer, *Mar. Biol., 74*, 319-326, 1983.

Jensen, P., Ecology of benthic and epiphytic nematodes in brackish waters, *Hydrobiologia, 108*, 201-217, 1984.

Jensen, P., Nematode fauna in the sulphide-rich brine seep and adjacent bottoms of the East Flower Garden, NW Gulf of Mexico IV. Ecological aspects, *Mar. Biol., 92*, 489-503, 1986.

Jensen, P., Feeding ecology of free-living aquatic nematodes, *Mar. Ecol. Prog. Ser., 35*, 187-196, 1987.

Jensen, P., Life history of the nematode *Theristus anoxybioticus* from sublittoral muddy sediment at methane seepages in the northern Kattegat, Denmark, *Mar. Biol., 123*, 131-136, 1995.

Josefson, A. B. and B. Widbom, Differential response of benthic macrofauna and meiofauna to hypoxia in the Gullmar Fjord basin, *Mar. Biol., 100*, 31-40, 1988.

Lasker, R., J. B. J. Wells, and A. D. McIntyre, Growth, reproduction, respiration and carbon utilization of the sand-dwelling harpacticoid copepod, *Asellopsis intermedia.*, *J. Mar. Biol. Ass. U. K., 50*, 147-160, 1970.

Lasserre, P. and J. Renaud-Mornant, Resistance and respiratory physiology of intertidal meiofauna to oxygen-deficiency, *Neth. J. Sea Res., 7*, 290-302, 1973.

Levin, L. A., C. L. Huggett, and K. F. Wishner, Control of deep-sea benthic community structure by oxygen and organic-matter gradients in the eastern Pacific Ocean, *J. Mar. Res., 49*, 763-800, 1991.

Maguire, C. and P. J. S. Boaden, Energy and evolution in the thiobios: an extrapolation from the marine gastrotrich *Thiodasys sterreri, Cah. Biol. Mar., 26*, 635-646, 1975.

Mangum, C., Precambrian oxygen levels, the sulfide biosystem, and the origin of the Metazoa, *J. Exp. Zool., 260*, 33-42, 1991.

McCall, J. N., Source of harpacticoid copepods in the diet of juvenile starry flounder, *Mar. Ecol. Ser., 86*, 41-50, 1992.

McCall, J. N. and J. W. Fleeger, Predation by juvenile fish on hyperbenthic meiofauna: A review with data on post-larval *Leiostomus xanthurus, Vie Milieu, 45*, 61-73, 1995.

McIntyre, A. D., Ecology of marine meiobenthos, *Biol. Rev., 44*, 245-290, 1969.

Meyers, M. B., H. Fossing, and E. N. Powell, Microdistribution of interstitial meiofauna, oxygen and sulfide gradients, and the tubes of macro-infauna, *Mar. Ecol. Prog. Ser., 35*, 223-241, 1987.

Meyers, M. B., E N. Powell, and H. Fossing, Movement of oxybiotic and thiobiotic meiofauna in response to changes in pore-water oxygen and sulfide gradients around macro-infaunal tubes, *Mar. Biol., 98*, 395-414, 1988.

Modig, M. and E. Ólafsson, Responses of Baltic benthic invertebrates to hypoxic events, *J. Exp. Mar. Biol. Ecol., 229*, 133-148, 1998.

Montagna, P. A., J. E. Bauer, D. Hardin, and R. B. Spies, Vertical distribution of microbial and meiofaunal populations in sediments of a natural coastal hydrocarbon seep, *J. Mar. Res., 47*, 657-680, 1989.

Montagna, P. A., Meiobenthic communities of the Santa-Maria Basin on the California continental shelf, *Cont. Shelf. Res., 11*, 1355-1378, 1991.

Montagna, P. A. and D. E. Harper, Benthic infaunal long term response to offshore production platforms in the Gulf of Mexico, *Can. J. Fisheries Aquat. Sci., 53*, 2567-2588, 1996.

Moodley, L., G. J. van der Zwaan, P. J. Herman, L. Kempers, and P. van Breugel, Differential response of benthic meiofauna to anoxia with special reference to Foraminifera (Protista: Sarcodina), *Mar. Ecol. Progr. Ser., 158:151-163*, 151-163, 1997.

Moore, H. B., The muds of the Clyde Sea Area. III. Chemical and physiological conditions, rate and nature of sedimentation, and fauna, *J. mar. biol. Ass. UK, 17*, 325-358, 1931.

Murrell, M. C. and J. W. Fleeger, Meiofauna abundance on the Gulf of Mexico continental shelf affected by hypoxia, *Cont. Shelf Res., 9*, 1049-1062, 1989.

National Research Council, Division of Medical Science Subcommittee on Hydrogen Sulfide, Hydrogen Sulfide, 183 pp., University Park Press, Baltimore, 1979.

Nichols, J. A., The effect of stable dissolved-oxygen stress on marine benthic invertebrate community diversity, *Int. Rev. Gesamten Hydrobiol., 61*, 747-760, 1976.

Ott, J. and F. Schiemer, Respiration and anaerobiosis of free living nematodes from marine and limnic sediments, *Neth. J. Sea Res., 7*, 233-243, 1973.

Ott, J. A., Determination of fauna boundaries of nematodes in a intertidal sand flat, *Int. Rev. Gesamten Hydrobiol., 57*, 645-663, 1972.

Ott, J. A. and R. Novak, Living at an interface: Meiofauna at the oxygen/sulfide boundary of marine sediments, in *Reproduction, Genetics and Distributions of Marine Organisms*, edited by J. S. Ryland and P. A. Tyler, pp. 415-422, Olsen and Olsen, Fredensborg, Denmark, 1989.

Ott, J. A., R. Novak, F. Schiemer, U. Hentschel, M. Nebelsick, and M. Polz, Tackling the sulfide gradient - A novel strategy involving marine nematodes and chemoautotrophic ectosymbionts, *P.S.Z.N. Mar. Ecol., 12*, 261-279, 1991.

Palmer, M. A., Dispersal of marine meiofauna: A review and conceptual model explaining passive transport and active emergence with implications for recruitment, *Mar. Ecol. Prog. Ser., 48*, 81-91, 1988.

Peterson, C. H., M. C. Kennicutt, R. H. Green, P. Montagna, D. E. Harper, E. N. Powell, and P.F. Roscigno, Ecological consequences of environmental perturbations associated with offshore hydrocarbon production: A perspective on long-term exposures in the Gulf of Mexico, *Can. J. Fisheries Aquat. Sci., 53*, 2637-2654, 1996.

Por, F. D. and D. Masry, Survival of a nematode and an oligochaete species in anaerobic benthal of Lake Tiberias, *Hydrobiol. Bull., 17*, 103-109, 1968.

Powell, E. N., Oxygen, sulfide and diffusion: Why thiobiotic meiofauna must be sulfide-insensitive first-order respirers, *J. Mar. Res., 47*, 887-949, 1989.

Powell, E. N. and T. J. Bright, A thiobios does exist - Gnathostomulid domination of the canyon community at the East Flower Garden brine seep, *Int. Rev. Gesamten. Hydrobiol., 66*, 675-683, 1981.

Rabalais, N. N., R. E. Turner, W. J. Wiseman, Jr. and D.F. Boesch, A brief summary of hypoxia on the northern Gulf of Mexico continental shelf: 1985-1988, in *Modern and Ancient Continental Shelf Anoxia.*, edited by R. V. Tyson and T. H. Pearson, pp. 35-47, Geological Society Special Publication No. 58, The Geological Society, London, 1991.

Rabalais, N.N., R. E. Turner and W. J. Wiseman, Jr., Distribution and characteristics of hypoxia on the Louisiana shelf in 1990 and 1991, in *Nutrient Enhanced Coastal Ocean Productivity*, pp. 62-67 Texas Sea Grant College Program, Galveston, Texas, 1992.

Rabalais, N. N., W. J. Wiseman, Jr. and R. E. Turner, Comparison of continuous records of near-bottom dissolved oxygen from the hypoxia zone along the Louisiana coast, *Estuaries, 17*, 850-861, 1994.

Radziejewska, T., Large-scale spatial variability in the southern Baltic meiobenthos distribution as influenced by environmental factors, *Proceeding of the 21st EMBS Gdansk, 14-19 Sept 1986*, 403-412, 1989.

Radziejewska, T. and I. Drzycimski, Meiobenthic communities of the Szczecin Lagoon, *Kiel. Meeresforsch. Sonderh., 6*, 162-172, 1988.

Radziejewska, T., J. W. Fleeger, N. N. Rabalais, and K. R. Carman, Meiofauna and sediment chloroplastic pigments on the continental shelf of Louisiana, USA, *Cont. Shelf. Res., 16*, 1699-1723, 1996.

Reise, K. and P. Ax, A meiofaunal "thiobios" limited to the anaerobic sulfide system of marine sand does not exist, *Mar. Biol., 54*, 225-237, 1979.

Reise, K. and P. Ax, Statement on the thiobios-hypothesis, *Mar. Biol., 58*, 31-32, 1980.

Riemann, F., Nematoda, in *Introduction to the Study of Meiofauna*, edited by R. P. Higgins and H. Thiel, pp. 293-301, Smithsonian Institution Press, Washington, D.C., 1988.

Rosenberg, R., W. E. Arntz, E. Chuman de Flores, L. A. Flores, G. Carbajal, I. Finger, and J. Tarazona, Benthos biomass and oxygen deficiency in the Peruvian upwelling system, *J. Mar. Res., 41*: 263-279, 1983.

Rosenberg, R., B. Hellman, and B. Johansson, Hypoxic tolerance of marine benthic fauna, *Mar. Ecol. Prog. Ser., 79*, 127-131, 1991.

Rosenberg, R., I. Olsson, and E. Ölundh, Energy flow model of an oxygen-deficient estuary on the Swedish West Coast, *Mar. Biol., 42*, 99-107, 1977.

Rudnick, D.,T., Time lags between the deposition and meiobenthic assimilation of phytodetritus, *Mar. Ecol. Prog. Ser., 50*, 231-240, 1989

Schiemer, F., Comparative aspects of food dependence and energetics of freeliving nematodes, *Oikos, 41*, 32-42, 1983.

Schiemer, F., R. Novak, and J. Ott, Metabolic studies on thiobiotic free-living nematodes and their symbiotic microorganisms, *Mar. Biol., 106*, 1209-138, 1990.

Shanks, A.,L. and E.,W. Edmondson, The vertical flux of metazoans (holoplankton, meiofauna, and larval invertebrates) due to their association with marine snow, *Limnol. Oceanogr., 35*, 455-463, 1990.

Shanks, A.,L. and K. Walters, Holoplankton, meroplankton, and meiofauna associated with marine snow, *Mar. Ecol. Prog. Ser., 156*, 75-86, 1997.

Smol, N., C. Heip, and M. Govaert, The life cycle of *Oncholaimus oxyuris* (Nematoda) in its habitat, *Annls. Soc. r. zool. Belg., 110*, 87-103, 1980.

Stickle, W. B., M. A. Kapper, L. Liu, E. Gnaiger, and S. Y. Wang, Metabolic adaptations of several species of crustaceans and molluscs to hypoxia: Tolerance and microcalorimetric studies, *Biol. Bull. 177*, 303-312, 1989.

Sun, B. and J. W. Fleeger, Field experiments on the colonization of meiofauna into sediment depressions, *Mar. Ecol. Prog. Ser., 110*, 167-175, 1994.

Teare, M. and R. Price, Respiration of the meiobenthic harpacticoid copepod, *Tachidius discipes* Giesbrecht, from an estuarine mudflat, *J. Exp. Mar. Biol. Ecol., 41*, 1-8, 1979.

Tietjen, J. H., Ecology of free-living nematodes from the continental shelf of the central Great Barrier Reef Province, *Estuar. Coast. Shelf Sci., 32*, 421-438, 1991.

Tietjen, J. H. and J. J. Lee, Life histories of marine nematodes. Influence of temperature and salinity on the reproductive potential of *Chromadorina germanica* Butschli, *Mikro. Meeres., 61*, 263-270, 1977.

Tietjen, J. H. and J. J. Lee, The use of free-living nematodes as a bioassay for estuarine sediments, *Mar. Environ. Res., 11*, 233-251, 1984.

Turner, R. E. and N. N. Rabalais, Changes in Mississippi River water quality this century, *BioScience, 41(3)*, 140-147, 1991

Vanreusel, A., M. Vincx, D. Vangansbeke, and W. Gijselinck, Structural analysis of the meiobenthos communities of the shelf break area in 2 stations of the Gulf-of-Biscay (NE Atlantic), *Belg. J. Zool., 122*, 185-202, 1992.

Vernberg, W. B. and B. C. Coull, Respiration of an interstitial ciliate and benthic energy relationships, *Oecologia, 16*, 259-264, 1974.

Vernberg, W. B. and B. C. Coull, Multiple factor effects of environmental parameters on the physiology, ecology and distribution of some marine meiofauna, *Cah. Biol. Mar., 16*, 721-732, 1975.

Vernberg, W. B. and B .C. Coull, Meiofauna., in *Functional Adaptations of Marine Organisms*, edited by F. J. Vernberg and W. B. Vernberg, pp. 147-177, Academic Press, Inc., New York, 1981.

Vernberg, W. B., B. C. Coull, and D. D. Jorgensen, Reliability of laboratory metabolic measurements of meiofauna, *J. Fish. Res. Bd. Can., 34*, 164-167, 1977.

Vismann, B., Physiology of sulfide detoxification in the isopod *Saduria* (*Mesidotea*) *entomon*, *Mar. Ecol. Prog. Ser., 76*, 283-293, 1991.

Vistisen, B. and B. Vismann, Tolerance to low oxygen and sulfide in *Amphiura filiformis* and *Ophiura albida* (Echinodermata: Ophiuridea), *Mar. Biol., 128*, 241-246, 1997.

Vopel, K., J. Dehmlow, and G. Arlt, Vertical distribution of *Cletocamptus confluens* (Copepoda, Harpacticoida) in relation to oxygen and sulphide microprofiles of a brackish water sulphuretum, *Mar. Ecol. Prog. Ser., 141*, 129-137, 1996.

Vopel, K., J. Dehmlow, M. Johansson, and G. Arlt, Effects of anoxia and sulphide on populations of *Cletocamtus confluens* (Copepoda, Harpacticoida), *Mar. Ecol. Prog. Ser., 175*, 121-128, 1998.

Walters, K. and S. S. Bell, Diel patterns of active vertical migration in seagrass meiofauna, *Mar. Ecol. Prog. Ser., 34*, 95-103, 1986.

Walters, K. and S. S. Bell, Significance of copepod emergence of benthic, pelagic, and phytal linkages in a subtidal seagrass bed, *Mar. Ecol. Prog. Ser., 108*, 237-249, 1994.

Webb, D. G. and P. A. Montagna, Initial burial and subsequent degradation of sedimented phytoplankton - Relative impact of macrobenthos and meiobenthos, *J. Exp. Mar. Biol. Ecol., 166*, 151-163, 1993.

Wetzel, M. A., P. Jensen, and O. Giere, Oxygen/sulfide regime and nematode fauna associated with *Arenicola marina* burrows: New insights in the thiobios case, *Mar. Biol., 124*, 301-312, 1995.

Wieser, W. and J. Kanwisher, Ecological and physiological studies on marine nematodes from a small salt marsh near Woods Hole, Massachusetts, *Limnol. Oceanogr., 6*, 262-270, 1961.

Withers, P. C., *Comparative Animal Physiology*, 949 pp., Saunders College Publishing, Fort Worth, Texas, 1992.

Yund, P. O., S. D. Gaines, and M. D. Bertness, Cylindrical tube traps for larval sampling, *Limnol. Oceanogr., 36*, 1167-1177, 1991.

11

Effect of Hypoxia/Anoxia on the Supply and Settlement of Benthic Invertebrate Larvae

Sean P. Powers, Donald E. Harper, Jr., and Nancy N. Rabalais

Abstract

Recovery of benthic animals following large-scale disturbances is primarily a function of larval recruitment. Given the large number of recent studies that have demonstrated the potential importance of larval supply in limiting or regulating populations of benthic animals, understanding how the supply of meroplanktonic larvae is affected by such disturbances is critical in developing a complete understanding of the dynamics of benthic communities. During the summers of 1994 and 1995, we measured the flux of meroplanktonic larvae and holoplankton at three positions in the water column during both stratification and low oxygen events, and during periods when the water column was well mixed. We found that benthic polychaete larvae were distributed throughout the water column and that this pattern did not appear to change in response to low oxygen. We found evidence, however, that at least one polychaete species, *Paraprionospio pinnata*, delayed settlement and remained in the water column until oxygen values returned to a level above 2.0 mg l^{-1}. Barnacle cyprid larvae and many holoplanktonic species were present in reduced densities below the pycnocline when oxygen concentrations were low. We interpreted the differences in response of plankton to low oxygen conditions to be related to differences in the vertical swimming abilities of these organisms or physiological tolerances to hypoxia and anoxia. Overall, species composition and relative abundance of organisms in the sediment reflected patterns of pelagic larval abundance. These results demonstrate that the supply of meroplanktonic larvae appears to determine the recovery population and that the response of plankton to low oxygen waters varies among taxa.

Coastal Hypoxia: Consequences for Living Resources and Ecosystems
Coastal and Estuarine Studies, Pages 185-210
Copyright 2001 by the American Geophysical Union

Introduction

The direct effects of hypoxia and anoxia on marine benthic systems are well documented and include dramatic decreases in the densities of demersal, epibenthic and infaunal animals in areas that experience hypoxia and anoxia [Diaz and Rosenberg, 1995; Rabalais et al., this volume; Wetzel et al., this volume]. Such findings can be placed within our current understanding of the role of disturbance in structuring biological communities [e.g., Sousa, 1985; Menge and Sutherland, 1987]. These models not only describe direct impacts from disturbance, but also encourage researchers to examine potential longer-term indirect impacts. Depending on the duration and severity of hypoxia, the relative importance of processes such as predation [Breitburg et al., 1994], competition and adult-juvenile interaction may be altered. Predators that are typically common, such as penaeid shrimp, crabs and demersal fish, are absent during hypoxia [Renaud, 1986a, 1986b; Pihl et al., 1991]. Because faunal densities are reduced during hypoxia, competition for space and resources should be decreased during and for some time after the occurrence of hypoxia. Adult-juvenile inhibitory behavior such as predicted by the trophic group amensalism hypothesis [Rhoads and Young, 1970; Woodin, 1976] should also be reduced as a result of the decreased faunal densities and longer recovery rates of amphipod populations (see Elmgren et al. [1986] for an example of inhibitory action by amphipods). While reduced faunal densities may be advantageous to potential prey settling in the area, the reduction of prey density both during and following the hypoxic event may result in changes in predator/prey ratios [Breitburg, 1992]. If predators return to the area faster than prey species, an increase in the predator/prey ratio will result. Such changes may result in increased competition between predators and further reduction in prey populations.

Given that reductions in animal densities may result in decreased competition, predation and inhibitory behavior of organisms (i.e., post-settlement factors), such large-scale disturbances may also alter the importance of recruitment in regulating or limiting the community [Menge and Sutherland, 1987]. Whereas infaunal populations may be limited normally by post-settlement factors [Ólafsson et al., 1994], the large decrease in population density as a result of prolonged anoxia may cause the population to be limited by the number of new recruits (i.e., recruitment limitation). The role of disturbance in altering the importance of recruitment is well established for rocky intertidal areas [Menge and Sutherland, 1987] and has been demonstrated for the bay scallop, *Argopecten irradians* [Peterson and Summerson, 1992; Peterson et al., 1996]. Further, the presence of a large layer of low oxygen water above the sediment may adversely affect larval abundance in the water column during a period in which many benthic invertebrates recruit. Decreases in abundance due to increased mortality or avoidance of low oxygen waters may further the possibility of recruitment limitation. Few studies have been performed that examine the response of larvae to hypoxic or anoxic conditions. When conducted, these studies have focused on a limited number of species, primarily bivalves [Widdows et al., 1989; Huntington and Miller, 1989; Wang and Widdows, 1991] and crabs [Palumbi and Johnson, 1982] and have been performed under laboratory conditions. We know of no field studies that have examined the effects of hypoxic/anoxic conditions on the distribution of invertebrate larvae.

Several studies have documented the effect of a large layer of hypoxic/anoxic water on holopanktonic organisms. Longhurst [1967] and Judkins [1980] investigated the vertical profiles of zooplankton in oceanic oxygen minimum regions off the western South American coast. Their results indicated that zooplankton densities were reduced in areas of depleted oxygen. Roman et al. [1993] in a study of copepods in Chesapeake Bay found that during hypoxic conditions copepods and copepod nauplii aggregated near the well-oxygenated surface water. Additionally, they found that diel vertical migrations of copepods to deeper depths did not occur during hypoxia. These studies provide strong evidence that hypoxia may induce changes in the vertical distribution of zooplankton and result in modifications of behavior. The implications of such findings are profound. Because planktonic organisms are the central point in marine food webs, hypoxia-induced changes in plankton dynamics may cascade through the entire food web. Populations that receive their supply of new recruits from the plankton (i.e., most invertebrates and fishes) or animals that feed on the plankton, may be adversely affected by hypoxia/anoxia induced changes in plankton abundance. The consequences of such changes may be manifested in longer recovery times for populations and long-term changes in the dynamics of systems that experience low oxygen events. Further, if larvae are concentrated in the surface waters during hypoxic conditions, an indirect effect of hypoxia may be increased predation on the concentrated larvae and zooplankton. If this is coupled with decreased abundance of larvae below the oxycline, which may be 10-15 m above the bottom, it is possible that the community may become recruitment limited for some time following the end of hypoxia.

Recovery following such large-scale disturbance events is a function of both the resistance of the organisms to the disturbance event and the resilience of the community following the disturbance [Kaufman, 1982]. Resistance of organisms to low oxygen events has been the focus of many studies, several of which have documented the tolerance range for select species (see Diaz and Rosenberg [1995]; Rabalais et al. [this volume]). The resilience of the community is a function of two factors (1) the resistance of these organisms to the disturbance, and (2) the rate of new arrivals following the disturbance. For areas in which anoxic conditions persist for extended periods, resilience of marine organisms would reflect primarily the rate of arrival and growth of new organisms, as opposed to any contributions from the surviving population, which would be very low. For many benthic organisms this arrival is primarily a function of larval recruitment [Santos and Simon, 1980; Harper et al., 1981, Gaston et al., 1985]. Harper et al. [1981] found that species that recruited by planktonic larvae (e.g., polychaetes) recovered more quickly than species without a planktonic larval stage (e.g., amphipods). Santos and Simon [1980] demonstrated that recovery of infaunal populations reflected patterns of larval recruitment, as opposed to juvenile or adult immigration into the area. Their observations fit well into our current understanding of infaunal recovery. When small patches are disturbed, recovery is primarily driven by emigration from adjacent populations; however, for larger areas larval recruitment appears to determine recovery rate [Hall, 1994; Thrush et al., 1998]. Given the importance of larval recruitment in determining recovery rate of benthic communities following these low oxygen events, studies of how such events affect the supply of meroplanktonic larvae are well justified.

Over the last fifteen years, the role of larval supply and the entire recruitment process has been at the center of ecological research and debate [Caley et al., 1996]. The primary focus of most of this research has been to examine the role of recruitment in determining spatial and temporal patterns of abundance in marine populations. The results of this research have been inconsistent across habitats; some studies have shown recruitment limitation (see Caley et al. [1996] for a recent review), and others have failed to detect such limitation (see Ólafsson et al. [1994] for a recent review). Several studies have also demonstrated the potential role of disturbance events in altering the importance of larval recruitment [e.g., Menge and Sutherland, 1987; Peterson et al., 1996]. For the most part, previous studies of recruitment have focused on the settlement of larvae; few studies have examined water column processes that may affect the supply of larvae over an area. Studies that have examined larval supply and settlement have demonstrated the necessity of integrating these components [Gaines and Bertness, 1993]. Studies of zooplankton have demonstrated that water column conditions can significantly affect their abundance and distribution [Roman et al., 1993; Qureshi and Rabalais, this volume]. One such water column condition is the presence of seasonal stratification and low dissolved oxygen events. The purpose of this chapter is to examine how water column conditions affect the distribution and relative abundance of meroplankton and some holoplankton in a coastal system that experiences prolonged hypoxia/anoxia.

Methods and Materials

Study Area

Field studies were performed in the northwestern Gulf of Mexico on the Louisiana continental shelf west of the Mississippi River. Two sites were selected in 1994 (C-6B, 28°52.18'N, 90°28.04'W and C-SP-1, 28°49.93'N, 90°23.53'W) and in 1995 (C-6B and C-SP-2) for field work; all three of which are located on a southeasterly transect off Terrebonne Bay (transect C in Rabalais et al. [this volume]). This area has been the site of several studies on the effect of nutrient inputs from the Mississippi and Atchafalaya Rivers on coastal productivity. The plume from the Mississippi River flows westerly over the study area in the late spring and early summer months following high spring runoff. The increased nutrient load enhances productivity of the surface waters and flux of organic material to the seabed during the spring and summer. If climatic conditions allow stratification of the water column, bottom waters in the area become hypoxic or anoxic for prolonged periods. Over the last 15 years the study area has become hypoxic for some portion of the summer [Rabalais et al. 1998; Rabalais et al., this volume].

Sampling Methods

We used a modification of the passive plankton collector of Yund et al., [1991] to measure the flux of invertebrate larvae. The device, similar in design to a sediment tube trap, has proven effective in giving comparative measurements of the flux of larvae [Yund et al., 1991; Gaines and Bertness, 1993]. Similar collectors have been used in other studies [e.g., Setran, 1992] to determine larval supply. In our study, a set of three tubes, 60-cm long with a diameter of 5 cm, was placed in tube holders and positioned at various depths to measure the flux of zooplankton and invertebrate larvae. The bottom 30-cm of each tube was filled with a 40-psu solution of 10% formalin/rose Bengal. In

1994, tube holders were made of two 1.25-cm thick gray PVC circular sheets (1-m diameter) held 0.3 m apart by PVC rods (Fig. 1). The holders were attached to a line (1-cm diameter) that was held in place by a 60-kg concrete base and a subsurface buoy. One tube holder was placed above the pycnocline (4 to 5-m depth) and one holder below the pycnocline (10- to 12-m depth). The concrete base (at 19- to 20-m depth) had three imbedded pieces of reinforced rod to which three tubes were attached with cable ties.

The design was modified for the 1995 field season (Fig. 1). Instead of a circular design, the tube holders were rectangular. The two layers of PVC were separated by a distance of 0.3 m. To the back of the holder was attached a vane that constantly oriented the apparatus into the current. The apparatus was kept in the water column by buoys attached to each end of the rectangular tube holder. Each holder contained four 7-cm diameter PVC pipes. At the top of each pipe segment was a PVC fitting reducing the 7-cm diameter to 5 cm. A tube identical in design to the tubes used in 1994 was then inserted into each PVC pipe segment. The four tubes were placed equidistant from each other on a 0.65-m bi-layered holder constructed of PVC gray sheet. One PVC sheet was welded to the bottom of each of the PVC pipes and the second sheet was placed 0.3 m above the bottom sheet. The tube extended a distance of 0.6 m from the bottom rack. The array was attached to a 90-kg concrete base with a 0.9-cm diameter nylon line. Swivels between the line and the holder allowed the holder to rotate freely. Each holder was attached to a separate base, allowing the top position to rotate without affecting the middle trap.

To understand how the patterns seen in the larval supply data related to the recovery of the benthic community, we also measured settlement of competent larvae and adult macrobenthic community structure [Powers, 1997]. Settlement of encrusting invertebrate larvae was assayed using settlement of barnacle cyprid larvae onto PVC panels placed on the array. Abundance of juvenile infaunal organisms (between 0.063 mm and 0.5 mm) was measured by taking samples of bottom sediments using 5-cm diameter core tubes taken to a depth of 10 cm. Abundance of adult infauna (> 0.05 mm) were taken following the methods of Harper et al. [1981, 1991] using a 0.0232-m^2 Ekman grab samples taken by divers.

Water column physical/chemical parameters were measured through a combination of moored instruments and discrete vertical profiles. Vertical profiles for dissolved oxygen, salinity, conductivity, temperature, and pH were taken with either a Hydrolab Surveyor 3 or a Seabird CTD. At C-6B an Endeco T1184 dissolved oxygen and temperature logger was deployed throughout both summers and recorded at 15-min intervals.

Sampling Protocols

In 1994 traps were first deployed at two stations (C-6B and C-SP-1) on 1 July. Traps were placed 20 m away from the outer leg of a natural gas platform. The base of the bottom trap at both sites was at a depth of 19-20 m on the sediment surface. The middle and top traps were located at a depth of 11 m and 4 m, respectively. On 21 July, 24 August, and 27 September tubes were removed and replaced at each of the trap positions by SCUBA divers. With the exception of the 27 September collection trip, a second dive was performed to collect three Ekman grabs and three 5-cm diameter core samples. Ekman grab and core samples were collected at both sites on 20 October in lieu of the

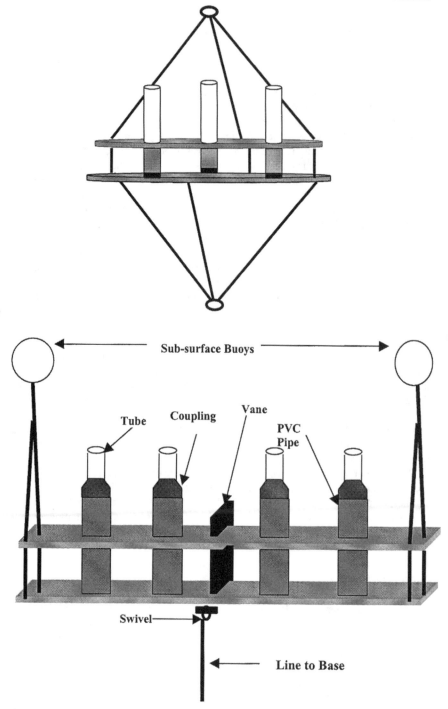

Figure 1. Design of passive plankton collector used in 1994 (upper) and 1995 (lower).

samples on 27 September which could not be collected because of poor weather conditions. Hydrocasts were made at each station prior to sample and tube collection. Because of technical problems with trap deployment, a small number of tube samples were not recovered and are excluded from the data analysis. Samples from the top and middle trap positions at C-SP-1 on 21 July and from C-6B on 24 August were recovered at a depth of 18 m and consequently not used in the analysis. After each trip, samples were washed on a 0.063 mm sieve and placed in 70% ethanol solution. The larvae were then identified to the lowest practical taxa.

In 1995 traps were first deployed on 22 June at two stations (C-6B and C-SP-2). Traps were again placed at least 20 m from the outer leg of a natural gas platform. Bottom traps were again placed on the sediment surface and top (5 m to 6 m) and middle traps (11 m to 12 m) were placed in the water column. On 13 July, 8 August, 28 August and 14 September tubes were removed and replaced by SCUBA divers at each station. Grab and core samples were collected on subsequent dives. Hydrocasts were made at each station prior to tube retrieval. An additional hydrocast was made on 8 September at both stations, but weather conditions prevented diving operations and tube recovery. Again technical problems prevented the use of some tube samples. Samples from the top trap position at C-SP-1 on 13 July and 8 August and from C-6B on 14 September were recovered at a depth of 18 m and were not used in the analysis.

Data Analysis

Data from tube trap samples were transformed to a flux measurement (numbers of ind. $m^{-2} d^{-1}$) to allow comparisons between different collection intervals. Data were then tested using the Kolmogorov-Smirnov (K-S) normality procedure and Bartlett's test for homogeneity of variances. Data were log(x + 1) transformed to meet the assumption of normality and homogeneity of variances. To determine the effect of low oxygen and stratification on the vertical distribution of larvae, two-way ANOVA using vertical position (top, middle and bottom) and date as the effects were performed on $\log_{10}(x + 1)$ transformed flux data. Data from both stations were combined for this analysis because (1) initial inspections of the hydrographic data showed no large differences in the conditions between the two sites, (2) ANOVA on the combined data also showed no significant effect of station, and (3) the exclusion of tube samples due to technical problems resulted in too few replicates at each position for each station. *Post hoc* analyses were conducted on significant effects using the Bonferroni/Dunn test [Day and Quinn, 1989]. If an interaction between the main effects was shown to be significant, then *post hoc* comparisons were made within groups (i.e., whether the dependent variable differed between position within a date).

Results

Physical and Chemical Parameters

1994. Overall, both stations had similar hydrographic patterns in 1994. From 1 July to 21 July 1994 the area around sites C-6B and C-SP-1 remained highly stratified (Fig. 2).

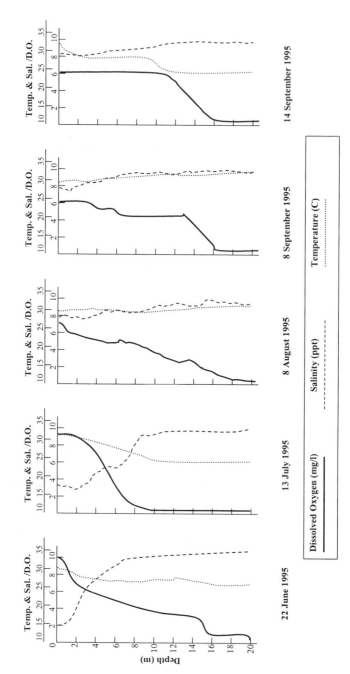

Figure 2. Vertical profiles of dissolved oxygen, salinity and temperature for designated dates during 1994 at C6B. Patterns were similar at C-SP-1 except where noted in the text.

At C-6B a distinct oxycline was present between 2 m and 8 m with hypoxic/anoxic water below 8 m. C-SP-1 showed a similar oxycline, however, dissolved oxygen values on 21 July were higher ranging from 3.5 to 1.0 mg l^{-1} between 8 m and 16 m. Bottom dissolved oxygen values recorded by the Endeco meter at C-6B were at or near 0 mg l^{-1} throughout the period (Fig. 3). For the second period of deployment, 21 July to 24 August, stratification presumably continued until 8 August when the area was influenced by a strong tropical depression. The depression moved from the area after several days of high winds and seas. Bottom dissolved oxygen, which ranged from 0 mg l^{-1} to 0.5 mg l^{-1} before the storm, was above 2 mg l^{-1} for 10 days following the break up of the depression. After 15 August dissolved oxygen values returned to 2 mg l^{-1} or less until early September at C-6B; however, bottom dissolved oxygen values of less than 2 mg l^{-1} were not detected at C-SP-1 after 15 August. Inspection of the temperature and salinity profiles for the same date revealed no distinct stratification or oxycline. Bottom dissolved oxygen values as measured by the Endeco meter remained above 2 mg l^{-1} throughout the third deployment period (24 August to 27 September).

1995. Hydrographic conditions were again similar at both stations during the deployments. The first deployment of traps in 1995 (22 June to 13 July) was completed during a period of water column stratification and hypoxia. Dissolved oxygen showed mark differences between surface (8 mg l^{-1} to 12 mg l^{-1}) and bottom water (0 mg l^{-1}). At C-6B water below 15 m was hypoxic or anoxic on 22 Jun and water below 12 m was anoxic on 13 July (Fig. 4). For both dates, water below 12 m was anoxic at C-SP-2. Bottom dissolved oxygen readings at C-6B recorded anoxic conditions for most of the period (Fig. 3). Stratification persisted for a portion of the second deployment, 13 July to August 8. Vertical profiles for salinity and temperature detected no strong halocline or thermocline at either site on 8 August. Dissolved oxygen values did remain hypoxic/anoxic under 14 m at both stations. Bottom dissolved oxygen values remained at 0 mg l^{-1} from 13 July until 18 July. Dissolved oxygen readings were unavailable from 28 July to 14 August due to failure of the Endeco meter; however, oxygen values of 3.2 mg l^{-1} were recorded on 8 August. From 8 August to 14 September (third and fourth deployments) there was no evidence of stratification at either station. Dissolved oxygen values remained under 1 mg l^{-1} in waters below 15 m at C-6B between 8 August and 8 September with a distinct oxycline between 14 and 16 m. Dissolved oxygen values of less than 2 mg l^{-1} were not detected at C-SP-2 from 8-August until 14 September. Bottom dissolved oxygen values recorded at C-6B were around the 2 mg l^{-1} level for most of the period between 8 August and 28 August and below 1 mg l^{-1} for the period between 28 August and 14 September.

Flux of Larvae and Zooplankton

1994. Larvae of the spionid polychaete, *Paraprionospio pinnata*, accounted for the majority of benthic larvae (60% to 80% of identifiable polychaete larvae; 60 -70% of all larvae) captured by the tube traps. Because the overall patterns were driven by *P. pinnata*, the majority of our analysis focused on this species. Larvae of capitellids (5% to 10%) and spionid (other than *P. pinnata*) polychaetes (10%), nereid polychaetes (10%), *Sigambra tentaculata* (5%), amphinomid and phyllodocid polychaetes (< 2%) accounted for the majority of other polychaete larval forms captured by the tube traps. Barnacle

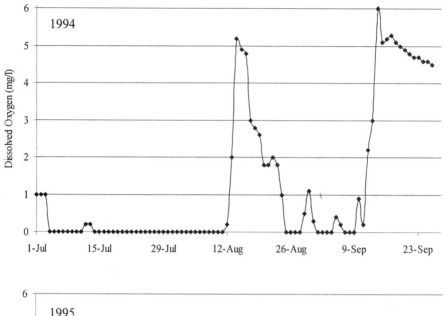

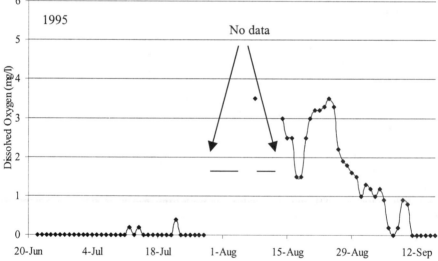

Figure 3. Average daily bottom dissolved oxygen readings at C-6B during 1994 (upper) and 1995 (lower) [Rabalais et al., unpubl. data].

cyprid larvae (15 to 25% of all larvae) accounted for the majority of non-polychaete larvae with decapod larvae and bivalve larvae accounting for less than 5% of the total. ANOVA for the effect of date and position (Table 1) on the flux of *P. pinnata* larvae demonstrated no effect of date or vertical position (Fig. 5); however, a significant interaction effect was demonstrated. Two other groups of animals, cyprid larvae of barnacles and nematodes, were captured in large numbers by the traps. The results of the

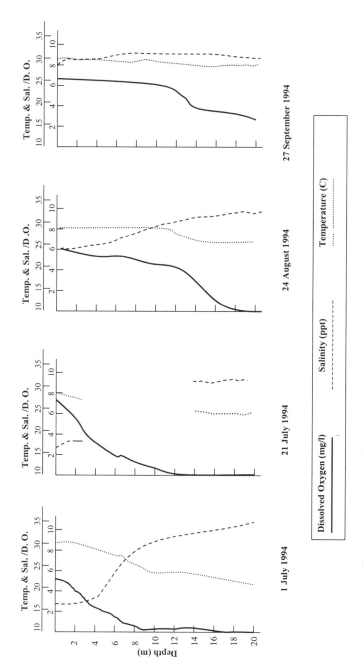

Figure 4. Vertical profiles of dissolved oxygen, salinity and temperature for designated dates during 1995 at C-SP-2 except where noted in the text. Patterns were similar at C-6B.

TABLE 1. Summary of ANOVA results for the effects of vertical position of the particle trap and date on the abundance of select taxa (* indicates statistical significance at p < 0.05).

Year	Taxa	Effect	df	F-Value	p-Value
1994	*Paraprionospio pinnata*	Vertical Position	2	0.99	0.59
		Date	2	0.54	0.38
		Date*Position	4	5.28	< 0.01*
	Copepods	Vertical Position	2	10.72	< 0.01*
		Date	2	2.62	0.09
		Date*Position	4	0.48	0.75
	Chaetognaths	Vertical Position	2	15.11	< 0.01*
		Date	2	0.35	0.71
		Date*Position	4	0.13	0.97
1995	*Paraprionospio pinnata*	Vertical Position	2	10.43	< 0.01*
		Date	3	5.66	< 0.01*
		Date*Position	6	2.58	0.03*
	Nereid larvae	Vertical Position	2	0.48	0.62
		Date	3	1.11	0.36
		Date*Position	6	4.11	< 0.01*
	Copepods	Vertical Position	2	673.33	< 0.01*
		Date	3	184.12	< 0.01*
		Date*Position	6	85.34	< 0.01*
	Chaetognaths	Vertical Position	2	124.99	< 0.01*
		Date	3	31.75	< 0.01*
		Date*Position	6	6.20	< 0.01*

cyprid study are reported elsewhere [Powers, 1997] and a portion of the nematode data appear in the chapter by Wetzel et al. [this volume].

For comparative purposes, a similar two-way ANOVA was performed on holoplankton captured by the traps. A significant effect of vertical position was detected by ANOVA for the flux of copepods and chaetognaths (Table 1). Flux of both copepods and chaetognaths was higher at the middle and top trap positions than the bottom trap position on all three dates. The magnitude of the differences between top and middle compared to bottom was greatest for the 21 July and 27 September samples for copepods and was similar for chaetognaths (Fig. 6). No effect of date or the interaction of date and position was detected for either copepods or chaetognaths.

1995. For the two most common polychaete larvae, *P. pinnata* and Nereidae (the majority of which were probably *Neanthes micromma*), results of ANOVA differed. While the flux of nereid polychaete larvae did not differ by vertical position or date, the flux of *P. pinnata* larvae did differ by vertical position and date (ANOVA, Table 1). The interaction of date and vertical position was also significant for the ANOVA of *P. pinnata* flux. For the period of collection between 22 June and 13 July, flux of *P. pinnata*

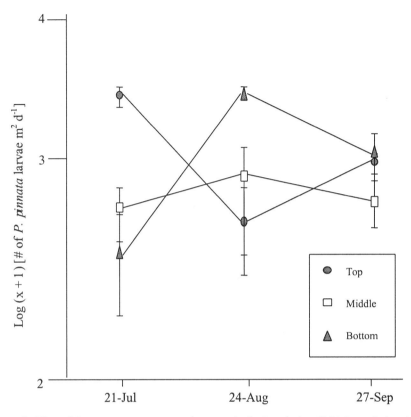

Figure 5. Flux of *Paraprionospio pinnata* larvae at both sites during 1994. Larval abundance is expressed as log(x+1) for # of larvae m^{-2} d^{-1}, bars represent ± 1 standard error.

was highest at the middle trap, followed by the bottom trap, and lowest at the top trap (Fig. 7; $p < 0.02$ for all *post hoc* contrasts). During the second deployment, flux of *P. pinnata* was higher in the top and middle traps compared to the bottom traps ($p > 0.05$ for top versus middle; $p < 0.02$ for top versus bottom and middle versus bottom). For the third deployment (8 August to 28 August), flux did not differ among the three positions ($p > 0.05$ for all *post hoc* contrast). Overall larval flux of *P. pinnata* (i.e., over the three positions) was significantly lower for this period between compared to all other sampling periods (Fig. 7; $p < 0.02$ for August 28 versus 13 July, 8 August, and 14 September). The decrease in larval flux during this period was caused primarily by a decline in large larvae. When separated into two size classes, of larvae retained on a 0.5 mm sieve or larvae passing through the 0.5 mm sieve and retained on a 0.063 mm sieve, only large larvae declined in abundance (Fig. 8). For the final deployment (28 August to 14 September), there was no difference in flux of *P. pinnata* among the three positions.

There were significant effects of date, vertical position and the interaction of date and position for copepods and chaetognaths (ANOVA, Table 1). For both groups, fluxes were higher at all positions from 8 August to 28 August and 28 August to 14 September compared to 22 June to 13 July and 13 July to 8 August (Fig. 9). Both taxa were

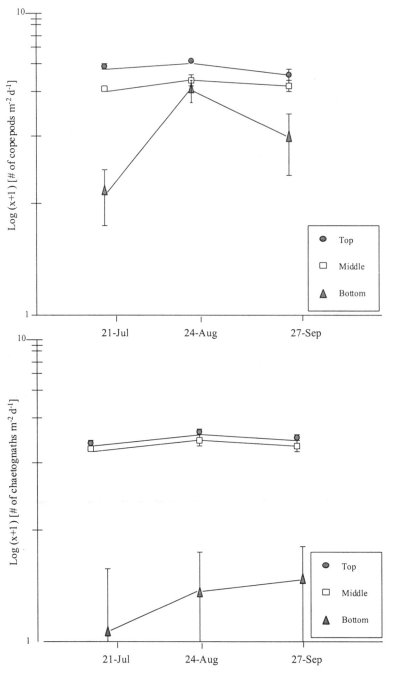

Figure 6.   Flux of copepods (upper) and chaetognaths (lower) at both sites during 1994. Abundance is expressed as log(x+1) for # of organisms m^{-2} d^{-1}.

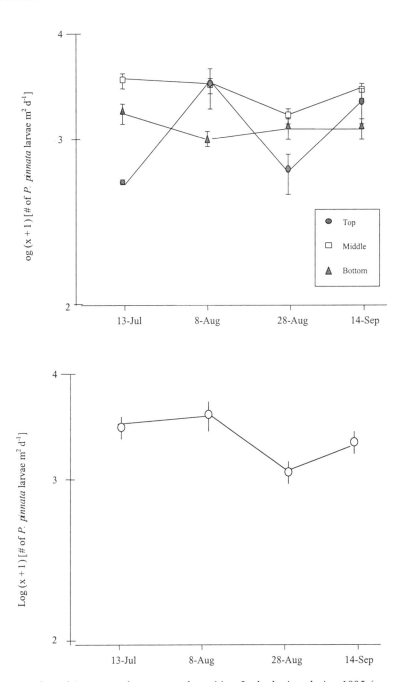

Figure 7. Flux of *P. pinnata* larvae at each position for both sites during 1995 (upper) and across all positions for each date (lower). Larval abundance is expressed as log(x+1) for # of larvae m^{-2} d^{-1}, bars represent ± 1 standard error.

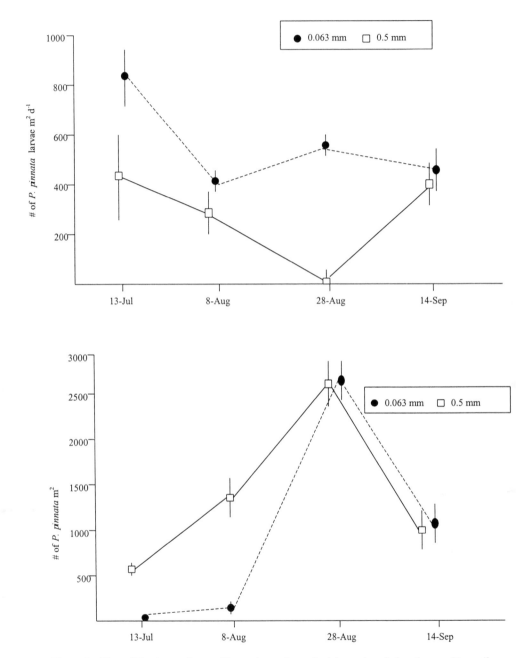

Figure 8. Flux of *P. pinnata* larvae 50 cm above the seabed (upper) and abundance of juvenile *P. pinnata* in the sediment during 1995 for C-SP-2. Larvae are divided by size of sieve they were retained on (0.5 mm and 0.063 mm), bars represent ± 1 standard error.

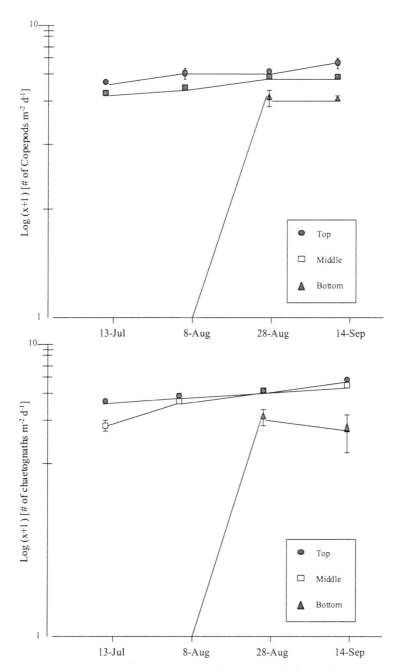

Figure 9. Flux of copepods (upper) and chaetognaths (lower) at both sites during 1995. Larval abundance is expressed as log(x+1) for # of organisms m^{-2} d^{-1}.

common in the top and middle traps with no differences detected between the two positions by *post hoc* analysis, but flux in the upper and middle were significantly higher than the bottom traps. The magnitude of the differences was much greater for the collection from 22 June to 13 July and 13 July to 8 August than the collections from 8 August to 28 August and 28 August to 14 September. An effect of date was seen in analysis of bottom traps for both groups with collection from 22 June to 13 July and 13 July to 8 August differing from 8 August to 28 August and 28 August to 14 September.

Benthic Community Structure

1994. *P. pinnata* dominated the macroinfaunal community at both stations in 1994. With the exception of samples taken at C-SP-1 on 20 October, *P. pinnata* accounted for 40% to 70% of total macrofauna for samples collected at both C-6B and C-SP-1. For samples from C-SP-1 on 20 October the cirratulid polychaete *Chaetezone* sp. B and the flabelligerid polychaete *Piromis roberti* accounted for 60% of the macrofauna with *P. pinnata* accounting for only 10%. Total macrofauna abundance, which primarily reflected the abundance of *P. pinnata*, ranged from 500 ind. m^{-2} to 1500 ind. m^{-2} with the exception of samples from C-6B on 24 August. Densities of *P. pinnata* were 2800 ind. m^{-2} which reflected a large pulse of juvenile *P. pinnata* (1700 ind. m^{-2}) collected in one of the three replicates. Other common taxa found in the samples (i.e., present in multiple samples and representing > 1% of the total animals collected) included the sipunculid, *Aspidosiphon* sp., the hemichordate, *Balanaglossus* sp., the polychaetes, *Sigambra tentaculata, Neanthes micromma, Armandia maculata, Ampharete* sp. A (cf. Vittor [1984]), *Paramphinome* sp. A (cf. Vittor [1984]), *Mellina maculata, Diopatra cuprea* and *Magelona* sp. H (cf. Vittor [1984]).

Core samples at C-6B were dominated by juvenile *P. pinnata* on all three dates sampled; *P. pinnata* comprised 70% of all juveniles identified. The remaining 30% was comprised chiefly of a small cirratulid polychaete, *S. tentaculata,* and *N. micromma.* Juvenile cirratulid polychaetes were the dominant organisms in core samples at C-SP-1 on 21 July accounting for much of the peak in settlement. *P. pinnata* was the dominant species on 24 August and 27 September at C-SP-1. Other common species in the samples included *S. tentaculata, N. micromma, Cirrophorus lyra* and *Mediomastus californiensis.*

1995. Overall macrofaunal abundance ranged from 800 to 1600 ind. m^{-2} at C-6B and from 200 to 1350 ind. m^{-2} at C-SP-2. Macrofaunal density increased from 13 Jul to 8 August, did not change from 8 August to 28 August, and decreased from 8 August to 14 September at C-6B. Macrofaunal density at C-SP-2 showed a similar pattern except a marked decrease in density was seen between 8 August and 28 August. Macrofauna at both stations were dominated by *P. pinnata. P. pinnata* accounted for 50 to 80% at both stations on each of the sampling dates except at C-6B on 28 August and 14 September. The hemichordate *Balanaglossus* sp. and *P. pinnata* each accounted for 40% of the macrofauna on 28 August. *Neanthes micromma* and *Nereis lamellosa* accounted for 60% of the macrofauna with *P. pinnata* accounting for only 20% on 14 September. Other macrofauna species that were common (found on multiple dates) in benthic samples included the sipunculid *Aspidosiphon* sp., the polychaetes *Sigambra tentaculata*, *Capitella capitata*, *Armandia maculata* and *Magelona* sp. H. The most common juvenile/larval form in the core samples at both stations was again *P. pinnata*. Overall abundance patterns for juveniles were driven for the most part by abundance patterns for *P. pinnata*. *P. pinnata* juveniles accounted for 70% - 90% of the larval forms in both size classes (0.063 mm to 0.5 mm and 0.5 mm to 1.0 mm) found in the core samples. Juvenile abundance in sediment cores was low during the first two collections (13 July and 8 August) with very few small larvae found in the cores (Fig. 8). The third collection (28 August) had a large peak in the settlement of polychaetes (primarily *P. pinnata*) in both size classes. Nereid juveniles were the second most common species found in the core samples (5% to 10%) followed by *S. tentaculata, A. maculata, Paramphinome* sp. A and other spinonids (*Dispio uncinata* and *Prionospio* spp.).

Discussion

Effect of Low Oxygen and Stratification on the Supply of Larvae

Based on the physical/chemical data, samples from the summer of 1994 are treated as representative of a hypoxic/anoxic period with stratification present (1 July to 21 July), a period of strong mixing as a result of a tropical depression over the study area (21 July to 24 August) and a normoxic period with infrequent periods of hypoxia and no strong stratification (24 August to 27 September). In 1995 there were two periods of well-defined stratification and hypoxic/anoxic conditions (22 June to 13 July and 13 July to 8 August), one period with ill-defined stratification and bottom oxygen values above 2 mg l^{-1} for most of the period (8 August to 28 August) and one period with both hypoxia and normoxia (28 August to 14 September).

Overall, the study failed to detect any clear patterns of preference in the vertical distribution of polychaete larvae. There was no significant effect of date or position for the flux of *P. pinnata* larvae in 1994. During the first period of deployment (stratification and bottom water hypoxia), the flux of *P. pinnata* larvae was somewhat higher for the top position (Fig. 5), but the pattern was not significant. During the second deployment period, abundances at all three positions were virtually identical as a result of the mixing that occurred from a tropical depression. During 1995, there was a significant effect of vertical position, date and the interaction of date and vertical position. While the overall pattern showed higher flux of larvae at the middle trap position, the flux of *P. pinnata* larvae showed no consistent pattern of preference for any of the positions (Fig. 7) that could be related to the hypoxia. Higher abundance of larvae around well-defined pycnoclines (which would correspond to the middle trap position) have been reported elsewhere [Young and Chia, 1987; Young, 1995]. This same pattern was not evident in the samples from 1994 when a highly stratified water column was present. Nereid polychaete larvae, the second most common polychaete larval form, showed no preference for any vertical position and were common throughout the study area. *P. pinnata* larvae were less abundant in the water column during the one prolonged period of normoxia in 1995 (8 August to 28 August) which correspond to a peak in settlement of

P. pinnata in the sediment cores. The decrease in larval flux near the seabed was primarily caused by the absence of large (retained on a 0.5 mm sieve) *P. pinnata* larvae (Fig. 8). These may represent larvae that have delayed settlement and respond quickly to the return of normoxic conditions. It is also possible that these large larvae represent larvae that have settled, but move back into the water column during periods of anoxia and then settle again when oxygen returns. A similar pattern of emigration from the sediments is suggested for nematodes in the area [Wetzel et al., this volume]. Larvae of the large size class are generally uncommon in the water column, the majority of polychaete larvae settle before reaching a size that would be retained on a 0.5-mm sieve. The results of the flux and settlement data indicate that polychaete larvae were abundant in the water column during the study resulting in a large pool of larvae available for rapid settlement.

There was a significant effect of hypoxia/stratification on the abundance of planktonic copepods and chaetognaths. In 1995 planktonic copepods and chaetognaths were virtually absent from the bottom traps during the two periods of hypoxia and water column stratification; however, during normoxic periods zooplankton were found in much higher densities in the bottom traps. In 1994, copepods and chaetognaths were found at high densities at all trap positions during the second deployment period when a well-mixed water column was present, but were found at reduced densities during the other two periods when hypoxic conditions were present in bottom waters. Similar results showing low densities of copepods in hypoxic water masses have been reported in other areas [Longhurst, 1967; Judkins, 1980; Roman et al., 1993; Qureshi and Rabalais, this volume]. Roman et al. [1993] found that the presence of low oxygen water induced changes in the diel vertical movement of copepods and resulted in reduced densities of copepods in bottom waters and concentration of copepods in well oxygenated surficial waters. Similar results of avoidance of anoxic waters were found in a companion study on supply and settlement of barnacle cyprid larvae [Powers, 1997]. In that study, the larval supply of cyprid larvae, and subsequent settlement of barnacles, was greatly reduced under a well-defined pycnocline in which bottom waters were hypoxic/anoxic. With the return of normoxic conditions, larval supply and settlement increased at depths where no settlement had occurred during anoxic conditions.

Although support for the avoidance of low oxygen by copepods can be found, our data on holoplankton should be viewed with some caution. Given that the tube traps were designed to measure the flux of invertebrate larvae, the efficacy of the tube traps in measuring relative abundance of stronger swimming zooplankton should be conducted. Similar evaluation of the tube traps for the capture of invertebrate larvae has demonstrated the effectiveness of the design for comparative studies of larval abundance [Yund et al., 1991]. Because the results of this study are largely based on the presence or absence of zooplankton, the conclusion reached in this study may be acceptable without such demonstrations. The high correlation between copepods and chaetognaths (r^2 = 0.702) found in this study, prey-predator species which should covary over time, may reflect the usefulness of measuring zooplankton flux using traps.

The differences in response to low oxygen waters by larvae and holoplankton may be explained by differences in their active swimming abilities. Holoplankton generally demonstrate much greater swimming ability than meroplanktonic larvae [Mileikovsky, 1973]. Upward swimming speeds ranging from 25.0 to 110 cm min^{-1} have been reported for copepods and 44.0 cm min^{-1} for chaetognaths. Swimming speeds of 6.6 to 19.8 cm min^{-1} (with most larvae at the lower range) have been reported for polychaetes. Barnacle cyprids exhibit swimming speeds in between these two groups, 10 to 32 cm min^{-1} (see Mileikovsky [1973] for a review). While swimming speeds of these groups are influenced by a host of variables including the size and stage of the organisms, temperature and viscosity of the water, and nutritional condition [Young, 1995; Metaxas and Young, 1998], these reported ranges support the conclusion that greater mobility of the organisms may serve to increase avoidance. A similar trend is seen in the response of adult benthic organisms to low dissolved oxygen; those that are more mobile seem to leave the area as dissolved oxygen decreases [Rabalais et al., this volume]. Meroplankton that cannot swim out of the low oxygen water (e.g., *P. pinnata*) may increase their chances for survival by delaying settlement, thus avoiding the anoxic, H$_2$S rich sediment. Differences in the response of polychaete larvae and holoplankton may also be due to differences in physiological capacity and tolerance to low oxygen between the organisms. Given that polychaetes are primarily sedentary infaunal organisms, they may be better adapted to tolerate low oxygen conditions than holoplanktonic species.

Because hypoxia/anoxia was associated with stratification, most collection periods had either a stratified water column and low oxygen or no stratification and oxygen level above 2.0 mg l^{-1}, a discussion of the synergistic effects of these two on our results seems warranted. In a set of laboratory experiments Metaxas and Young [1998] investigated the effect of haloclines on the vertical position of echinoderm larvae. They found that when large discontinuities were present (21/33 psu), larvae of two species of sea urchins remained in the higher salinity water, a pattern opposite to that found for holoplankton and barnacle cyprids in our studies [Powers, 1997]. Results similar to those of Metaxas and Young [1998] have been reported for ascidian tadpoles, barnacle nauplii, bivalve larvae, blue crab zoea, hermit crab zoea, littorinid larvae and lobster zoea (citations in Metaxas and Young [1998]). Similar responses of larvae to thermoclines, are not well established [Young, 1995]. Studies of larvae of sea scallops [Tremblay and Sinclair, 1990] and blue crab zoea [McConnaughey and Sulkin, 1984] concluded that thermoclines should seldom influence orientation or migration behavior under field conditions. The presence of denser, higher salinity bottom waters may aid in keeping *P. pinnata* larvae suspended in the water column away from anoxic sediments; however, the pycnocline may represent a barrier for the same larvae in escaping the low oxygen water. Species with higher swimming abilities may be able to overcome this barrier and aggregate in well-oxygenated surface waters above the pycnocline.

Implications for the Recovery of the Benthic Community

Relative abundances of benthic polychaetes were predicted by larval supply for many of the polychaete taxa. *P. pinnata* was by far the most common polychaete larval form found in the water column and was also the most common polychaete found in the sediment cores and grab samples. Nereid and *Sigambra tentaculata* larvae were the second most common larval forms and were the second most common polychaetes in the sediment. Cirratulid polychaetes were commonly found in the sediment at C-SP-1 during 1994; however, their water column abundance, which was low, would not have predicted their dominance. Hannan [1981] and Powers [1997] have reported similar correspondence between trap data and sediment abundance. Our results are also in agreement with Santos and Simon [1980] who concluded that recruitment of larvae from the water column was the primary mechanism for benthic recovery following periods of severe hypoxia and/or anoxia. Their results were based on inference from sediment colonization studies, while our study examined both the water column supply and settlement of recruiting larvae.

The dominance of *P. pinnata* in the recovery populations in both years is not surprising given the life cycle and opportunistic nature of the species. *P. pinnata* exhibits high fecundity (~ 6,000 eggs per clutch) and larvae have a 14- to 21-d planktonic period before settlement. Spawning occurs during the summer month with recruitment into the adult population generally occurring between June and October. In many cases multiple generations of worms are produced during the summer months [Mayfield, 1988]. In his study of *P. pinnata* in Galveston Bay, Texas, Mayfield [1988] concluded that the rapid proliferation of this species during the summer and early fall was indicative of an opportunistic life history similar to those seen in other spionid polychaetes. High mortality of large worms occurred in early summer and suggested spawning related mortality. Given these traits, *P. pinnata* seems well suited for colonization following these hypoxic/anoxic events [Holland et al., 1977]. The majority of adults die just prior to or at the onset of hypoxic conditions due to spawning related mortality. The large pool of larvae released by these adults then spend an extended period of time in the plankton and settle with the break-up of anoxic conditions.

The planktonic period may be extended by delayed metamorphosis in response to low oxygen, thus increasing the potential for *P. pinnata* larvae to colonize following the break-up of anoxic/hypoxic conditions. Examination of the flux of *P. pinnata* larvae in bottom water reveals an interesting pattern (Fig. 8). During the first two periods of anoxic conditions and a highly stratified water column, *P. pinnata* larvae were more abundant in the water column than during the third period when oxygen values returned to levels above 2.0 mg l^{-1}. During the final deployment dissolved oxygen values decreased and *P. pinnata* larvae increased in the water column. Settlement of *P. pinnata* retained on a 0.5-mm sieve showed a large peak during the third period that corresponded to the decrease in water column abundance (Fig. 8). Settlement of smaller *P. pinnata* did not show this pattern. It appears that at both stations in 1995, large larvae quickly settled in response to the return of normoxic conditions. The lengthening of the planktonic period has obvious advantages, particularly in a system that can remain hypoxic or anoxic for several months.

Our data on larval supply leads to another pressing problem to be resolved; what is the source of larvae over the area? While this topic was well out of the scope of this paper, some speculations and suggestions for further study seem warranted. Intuitively, only two possibilities exist: the source of larvae is from within the affected area or the source of larvae is from outside the disturbed area. The first possibility seems unlikely given that gravid individuals were not found in the area during hypoxia/anoxia. However, with a planktonic period of 14-21 days (for *P. pinnata*), the possibility exists that the resident population released larvae before the onset of hypoxic/anoxic conditions. Given some kind of delayed settlement, some portion of the recovery populations may come from the resident source. While the most persistent and widespread hypoxia occurs from June through August, intermittent outbreaks of hypoxia may start as early as March or April [Rabalais et al., 1998]. Consequently, delaying metamorphosis may not always be a successful mechanism for colonization. If larvae are advected from adult populations outside the affected areas, questions arise as to their source, the scale of dispersal and the effects of an increasing area affected by hypoxia. While the question of the scale of larval dispersal is one that has taunted investigators for some time, new approaches to the study of larval dispersal may make this question answerable. Methods for tagging larvae [Levin, 1990] offer some promise at addressing this topic, as do, genetic approaches [Palumbi, 1995]. Coupling such techniques with our increasing knowledge of circulation patterns in the Gulf of Mexico may provide the answers to these questions.

Conclusion

During hypoxic or anoxic conditions in the water column, planktonic organisms including some invertebrate larvae (e.g., barnacle cyprid larvae [Powers, 1997]) and most zooplankton (copepods and chaetognaths) avoid the low oxygen bottom waters. Other invertebrate larvae (e.g., polychaetes) are often found in high densities in these hypoxic areas. The differences in response may be due in large part to differences in their vertical swimming abilities and/or physiological capacity to tolerate low oxygen conditions. Barnacle cyprid larvae and copepods are much stronger swimmers than polychaete larvae that lack modified appendages for swimming. For polychaetes, particularly the spionid polychaete, *Paraprionospio pinnata*, large pools of larvae were available for settlement following the return of normoxic conditions. During both years of this study, *P. pinnata* juveniles and adults were the dominant species found during the recovery of the area; *P. pinnata* larvae accounted for approximately 80% to 90% of the polychaete larvae collected by the tube traps. The quick recovery of this species was probably a function of the large pool of larvae found in the water column during and after hypoxic bottom water conditions. Overall, we find that recovery of the benthic community immediately following anoxic/hypoxic events is predicted by larval supply.

Acknowledgments. This research was funded by grants from the Louisiana Universities Marine Consortium Foundation, Inc. (SPP), the NOAA Nutrient Enhanced Coastal Ocean Productivity Program (NNR), the Texas Institute of Oceanography (DEH), and the PADI Foundation (SPP & DEH). We also express our gratitude to the many scientific divers who assisted us in this project, especially Mary Stordal, Monica Dozier, Kathy Hancock, Logan Respess, Randy Robichaux and Lorene Smith. Support during the preparation of this manuscript was provided by a grant to SPP from the National Science Foundation (NSF-97-31718). We thank Linda Schaffner and Daniel Dauer for their review of the paper.

References

Breitberg, D. L., Episodic hypoxia in Chesapeake Bay: interacting effects of recruitment, behavior, and physical disturbance, *Ecol. Monogr., 62*, 525-546, 1992.
Breitberg, D. L., N. Steinberg, S. DuBeau, C. Cooksey, and E. D. Houde, Effects of low

dissolved oxygen on predation on estuarine fish larvae, *Mar. Ecol. Prog. Ser., 104*, 235-246, 1994.

Caley, M. J., M. H. Carr, M. A. Hixon, T. P. Hughes, G. P. Jones, and B. A. Menge, Recruitment and the local dynamics of open marine populations, *Ann. Rev. Ecol. Syst., 27*, 477-500, 1996.

Day, R. W. and G. P. Quinn, Comparisons of treatments after an analysis of variance in ecology, *Ecol. Monogr., 59*, 433-463, 1989.

Diaz, R. J. and R. Rosenberg, Marine benthic hypoxia: a review of its ecological effects and the behavioural responses of benthic macrofauna, *Oceanogr. Mar. Biol. Ann. Rev., 33*, 245-303, 1995.

Elmgren, R., S. Ankar, B. Marteleur, and G. Edjung, Adult interference with postlarvae in soft-sediment communities: the *Pontoporeia – Macoma* example, *Ecology, 67*, 827-836, 1986.

Gaines, S. D. and M. D. Bertness, The dynamics of juvenile dispersal: why field biologists must integrate? *Ecology, 74*, 2430-2435, 1993.

Gaston, G. R., P. A. Rutledge, and M. L. Walther, The effects of hypoxia and brine on recolonization by macrobenthos off Cameron, Louisiana (USA), *Contr. Mar. Sci. 28*: 79-93, 1985.

Hall, S. J., Physical disturbance and marine benthic communities: life in unconsolidated sediments, *Oceanogr. Mar. Biol. Ann. Rev., 32*, 111-178, 1994.

Hannan, C. A. 1981, Polychaete larval settlement: correspondence of patterns in suspended jar collectors and in the adjacent natural habitat in Monterey Bay, California, *Limnol. Oceanogr., 26*, 159-171, 1981.

Harper, Jr., D. E., L. D. McKinney, J. M. Nance, R. R. Salzer and R. J. Case, The occurrence of hypoxia off the upper Texas coast and its effects on the benthic biota, *Contr. Mar. Sci., 24*, 53-79, 1981.

Harper, Jr. D. E., L. D. McKinney, J. M. Nance, and R. R. Salzer, Recovery responses of two benthic assemblages following an acute hypoxic event, in, *Modern and Ancient Continental Shelf Anoxia*, edited by R. V. Tyson and T. H. Pearson, pp. 49-64, Geological Society Special Publication No. 58, The Geological Society, London, 1991.

Holland, A. F., N. K. Mountford, and J. A. Mihursky, Temporal variation in upper bay mesohaline benthic communities: I. The 9-m mud habitat, *Chesapeake Sci., 18*, 370-378, 1977.

Huntington, K. M. and D. C. Miller, Effects of suspended sediment, hypoxia, and hyperoxia on larval *Mercenaria mercenaria* (Linnaeus, 1758), *J. Shellfish Res., 8*, 37-42, 1989.

Judkins, D. C., Vertical distribution of zooplankton in relation to the oxygen minimum off Peru, *Deep-Sea Res., 27 A*, 457-487, 1980.

Kaufman, L. H., Stream aufwuchs accumulation: disturbance frequency and stress resistance and resilience, *Oecologia, 52*, 57-63, 1992.

Levin, L. A., A review of methods for labeling and tracking marine invertebrate larvae, *Ophelia, 32*, 115-144, 1990.

Longhurst, A. R. Vertical distribution of zooplankton in relation to the eastern Pacific oxygen minimum, *Deep-Sea Res., 14*, 51-63. 1967.

Mayfield, S. M., Aspects of the Life History and Reproductive Biology of the Worm *Paraprionospio pinnata*. M. S. thesis, Department of Biology, Texas A&M University, College Station, Texas, 1988.

McConnaughey, R.A. and S. D. Sulkin, Measuring the effects of thermoclines on the vertical migration of larvae of *Callinectes sapidus* (Brachyura: Portunidae) in the laboratory, *Mar. Biol., 81*, 139-145, 1984.

Menge, B. A. and J. P. Sutherland, Community regulation: variation in disturbance, competition, and predation in relation to environmental stress and recruitment, *Am. Nat.*, *130*, 730-757, 1987.

Metaxas, A. and C. M. Young, Behavior of echinoid larvae around sharp haloclines: effects of the salinity gradient and dietary conditioning, *Mar. Biol.*, *131*, 443-459, 1998.

Mileikovsky, S. A., Speed of active movement of pelagic larvae of marine bottom invertebrates and their ability to regulate their vertical position, *Mar. Biol.*, *23*, 11-17, 1973.

Ólafsson, E. B., C. H. Peterson, and W. G. Ambrose, Jr., Does recruitment limitation structure populations and communities of macroinvertebrates in marine soft sediments: the relative significance of pre- and post-settlement processes, *Oceanogr. Mar. Biol. Ann. Rev.*, *32*, 65-109, 1994.

Palumbi, S. R. and B. A. Johnson, A note on the influence of life history stage on metabolic adaptation: the response of *Limulus* eggs and larvae to hypoxia, in, *Physiology and Biology of Horseshoe Crabs: Studies on Normal and Environmentally Stressed Animals*, edited by J. Bonaventura, pp. 115-124, A.R.. Liss, New York, 1982.

Palumbi, S. R., Using genetic and an indirect estimator of larval dispersal, in, *Ecology of Marine Invertebrate Larvae*, edited by L. McEdward, pp. 369-388, CRC Press, New York, 1995.

Peterson, C. H. and H. C. Summerson, Basin-scale coherence of population dynamics of an exploited marine invertebrate, the bay scallop: implications of recruitment limitation, *Mar. Ecol. Prog. Ser.*, *90*, 257-272, 1992.

Peterson, C. H., H. C. Summerson, and R. A. Luettich, Jr., Response of bay scallops to spawner transplants: a test of recruitment limitation, *Mar. Ecol. Prog. Ser.*, *132*, 93-107, 1996.

Pihl, L. S. P. Baden, and R. J. Diaz, Effects of periodic hypoxia on distribution of demersal fish and crustaceans, *Mar. Biol. 108*, 349-360, 1991.

Powers, S. P., Recruitment of Soft-Bottom Benthos. Ph.D. dissertation, Department of Biology, Texas A&M University, College Station, Texas, 1997.

Rabalais, N. N., R. E. Turner, W. J. Wiseman, Jr. and Q. Dortch, Consequences of the 1993 Mississippi River Flood in the Gulf of Mexico, *Regulated Rivers: Research & Management*, *14*, 161-177, 1998.

Renaud, M. L., Hypoxia in the Louisiana coastal waters during 1983: implications for fisheries, *Fish. Bull.*, *84*, 19-26, 1986a.

Renaud, M. L., Detecting and avoiding oxygen deficient seawater by brown shrimp, *Penaeus aztecus* (Ives), and white shrimp, *Penaeus setiferus* (L.), *J. Exp. Mar. Biol. Ecol.*, *98*, 283-292, 1986b.

Rhoads, D. C. and D. K. Young, The influence of deposit feeding organisms on sediment stability and community trophic structure, *J. Mar. Res.*, *28*, 150-178, 1970.

Roman, M. R., A. L. Gauzens, W. K. Rhinehart, and J. R. White, Effects of low oxygen waters on Chesapeake Bay zooplankton, *Limnol. Oceanogr.*, *38*, 1601-1615, 1993.

Santos, S. L. and J. L. Simon, Marine soft-bottom community establishment following annual defaunation: larval or adult recruitment? *Mar. Ecol. Prog. Ser.*, 2, 235-241, 1980.

Setran, A. C., A new plankton trap for use in collection of rocky-intertidal zooplankton, *Limnol. Oceanogr.*, 37, 669-674, 1992.

Sousa, W. P., Disturbance and patchy dynamics on rocky intertidal shores, in, *The Ecology of Natural Disturbance and Patch Dynamics*, edited by P. S. White and S. T. Pickett, pp. 101-124, Academic Press, New York, 1985.

Tremblay, M. J. and M. Sinclair, Diel vertical migration of sea scallop larvae *Placopecten magellanicus* in a shallow embayment, *Mar. Ecol. Prog. Ser.*, *67*, 19-25, 1990.

Thrush, S. F., J. E Hewitt, V. J. Cummings, P. K. Dayton, M. Cryer, S. J. Turner, G. A. Funnell, R. G. Budd, C. J. Milburn, and M. R. Wilkinson, Disturbance of the marine benthic habitat by commercial fishing: impacts at the scale of the fishery, *Ecol. Appl., 8*, 866-879, 1998.

Vittor, B. A., Taxonomic guide to the polychaetes of the northern Gulf of Mexico. Final Report to the Minerals Management Service, contract 14-12-001-29091. Barry Vittor & Associates, Inc. Mobile, Alabama, 7 Vols. 1984.

Wang, W. and J. Widdows, Physiological responses of mussel larvae *Mytilus edulis* to environmental hypoxia and anoxia, *Mar. Ecol. Prog. Ser., 70*, 223-236, 1991.

Widdows, J., R. Newell, and R. Mann, Effects of hypoxia and anoxia on survival, energy metabolism and feeding of oyster larvae (*Crassostrea virginica*, Gmelin), *Biol. Bull., 177*, 154-166, 1989.

Woodin, S. A., Adult-larval interactions in dense infaunal assemblages: patterns of abundance, *J. Mar. Res. 35*, 25-41, 1976.

Young, C. A., Behavior and locomotion during the dispersal phase of larval life, in, *Ecology of Marine Invertebrate Larvae*, edited by L. McEdward, pp. 249-278, CRC Press, New York, 1995.

Young, C. A. and F. S. Chia, Abundance and distribution of pelagic larvae as influenced by predation, behavior and hydrographic factors, in, *Reproduction of Marine Invertebrates, IX, General Aspects: Seeking Unity in Diversity*, edited by A. C. Giese et al., pp. 385-463, Blackwell Scientific, Palo Alto, and the Boxwood Press, Pacific Grove, California, 1987.

Yund, P. O., S. D. Gaines, and M. D. Bertness, Cylindrical tube traps for larval sampling, *Limnol. Oceanogr., 36*, 1167-1177, 1991.

12

Effects of Seasonal Hypoxia on Continental Shelf Benthos

Nancy N. Rabalais, Lorene E. Smith, Donald E. Harper, Jr., and
Dubravko Justic'

Abstract

The benthic communities were characterized for two areas of the southeastern
Louisiana continental shelf—one near the Mississippi River delta with silty sediments
and intermittently affected by hypoxia on time scales of days to weeks and another
farther from the Mississippi River delta in sandier sediments but affected by severe
seasonal hypoxia lasting several months. The composition of the benthic communities
reflected differences in sedimentary regime, seasonal input of organic material and
seasonally severe hypoxia/anoxia. Decreases in species richness, abundance and biomass
of organisms were dramatic at the stations affected by severe hypoxia/anoxia, and lower
than most literature values for similar habitats. Although there were summer/fall declines
in the populations at the intermittently hypoxic site, these were not obviously related to
changes in oxygen. Some macroinfauna, the polychaetes *Ampharete* and *Magelona* and
the sipunculan *Aspidosiphon*, were capable of surviving extremely low dissolved oxygen
concentrations and/or high hydrogen sulfide concentrations. Abundance of
macroinfauna, primarily opportunistic polychaetes (similar to the spring), increased in the
fall following the dissipation of hypoxia, but the numbers of individuals were only
slightly greater than the summer depressed fauna and resulted in no or a negligible
increase in biomass. Fewer taxonomic groups characterized the severely affected stations
throughout the year. Long-lived, higher biomass and direct-developing species were
never members of the severely affected community. Suitable feeding habitat (in terms of
severely reduced populations of macroinfauna that may characterize substantial areas of
the seabed) is thus removed from the foraging base of demersal organisms, including the
commercially important penaeid shrimps.

Coastal Hypoxia: Consequences for Living Resources and Ecosystems
Coastal and Estuarine Studies, Pages 211-240
Copyright 2001 by the American Geophysical Union

Introduction

Bottom-water hypoxia (< 2 mg O_2 l^{-1}) is often a secondary response of an estuarine or coastal system to eutrophication. The dissolved oxygen conditions of many major coastal ecosystems around the world have been affected adversely by the process of eutrophication [Diaz and Rosenberg, 1995, this volume]. Data from many coastal systems where adequate historical data exist indicate a steady decline in dissolved oxygen through time, starting primarily in the 1950s and 1960s. Although there are no consistently collected hydrographic data for the northern Gulf of Mexico prior to the early 1970s and no systematic studies of hypoxia prior to the mid-1980s, geochronologies of biological and chemical remnants in sediments from the Mississippi River delta bight indicate an increase in primary production as well as a worsening of oxygen stress, either in duration or intensity, most dramatically since the mid-1950s [Eadie et al., 1994; Nelsen et al., 1994; Turner and Rabalais, 1994; Rabalais et al., 1996; Sen Gupta et al., 1996].

The zone of bottom-water hypoxia on the northern Gulf of Mexico continental shelf west of the Mississippi River delta is one of the largest zones in the world's coastal ocean, exceeded only by the coastal areas of the Baltic (84,000 km^2; Rosenberg [1985]) and the northwestern shelf of the Black Sea (20,000 km^2; Tolmazin [1985]). From 1993 to 1997, the size of the Gulf of Mexico hypoxic zone was consistently greater than 16,000 km^2 in mid-summer, and reached 20,000 km^2 in mid-summer of 1999 [Rabalais et al., 1999, this volume]. From 1985-1992 the hypoxic bottom-water area covered 8,000-9,000 km^2, with the exception of 1988 when the size was 42 km^2 during mid-summer record low flow of the Mississippi River. Data to define the seasonal sequence of hypoxia are most complete for transect C off Terrebonne Bay on the southeastern Louisiana shelf (Fig. 1), where critically depressed dissolved oxygen concentrations (< 2 mg l^{-1}) occur below the pycnocline from as early as late February through early October and nearly continuously from mid-May through mid-September.

In the course of an analysis of the interactive effects of oil platform discharges and hypoxia [Rabalais et al., 1993, 1995], we documented the variability of benthic communities over two annual cycles of oxygen stress and recovery. Continuous recordings of bottom-water dissolved oxygen concentrations characterized two study areas as one with seasonally severe and persistent hypoxia and the second with aperiodic or moderate hypoxia [Rabalais et al., 1994]. This chapter documents the differences in the seasonal decline of benthos—specifically abundance, species richness, assemblage composition, biomass and vertical distribution—within two differing hypoxia regimes on the southeastern Louisiana shelf. Differences are related to water column and sedimentary environmental variables. We also consider the implications of hypoxia-stressed benthic communities to trophic interactions that might ultimately affect fisheries resources.

Study Areas

Three oil production platforms (two active, WD32E and ST53A, and one inactive, ST53B) in two oil fields (Table 1) were examined initially for the multiple effects of oilfield production effluents and hypoxia on benthic communities [Rabalais et al., 1993].

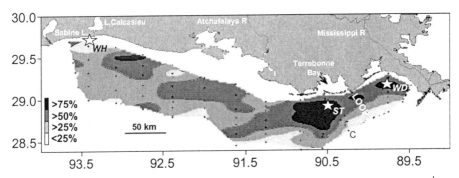

Figure 1. Frequency of occurrence of bottom-water hypoxia (dissolved oxygen < 2 mg l⁻¹) in mid-summer of 1985-1999. Identified are study areas ST53, WD32, West Hackberry (= WH) and LOOP. Bryan Mound, also discussed in the text, is located to the west at 28°45'N and 95°15'W.

Data were obtained in April and June-August 1990 along a gradient of 20 m to 1000 m away from the platforms (including a reference station at the West Delta site). Supplemental data for fall 1990 and spring 1991 focused on the 500- and 1000-m distance stations to avoid platform effects, as well as the reference station at West Delta, and for summer and fall of 1991 focused on a single 500-m distance station at ST53B [Rabalais et al., 1995].

The West Delta site was closer to the Mississippi River delta in a different sedimentary regime than the South Timbalier 53 stations (Table 1). The three stations in the South Timbalier area were within 4 km of each other, and the reference station at West Delta 33 was within 4 km of the West Delta 32E platform. Sediments in the West Delta area were predominantly silts (85 to 90%) with some clay and sand (Fig. 2) and accumulate at a rate of 2 cm y⁻¹ [R. E. Turner, unpublished data]. Sediments at ST53A were predominantly sandy silts with little clay fraction, and those of ST53B were variably sandy silts or silty sands. Although a thin veneer of silt/clay, biogenic particles and fine sand may be deposited in the South Timbalier area [N. N. Rabalais, personal observation], there is little or no long-term sediment accumulation [B. A. McKee,

TABLE 1. Study areas.

Lease Block	Latitude & Longitude	Distance from Mississippi River	Water Depth	Sediment Type
West Delta 32E = WD32E or WD (oxygen meter, 1990) reference station in WD33	29°07.3'N 89°41.7'W	34 km	20 m	silt
South Timbalier 53A = ST53A or STA	28°51.4'N 90°27.7'W	99 km	20 m	sandy silt
South Timbalier 53B = ST53B or STB (oxygen meter, 1990)	28°50.5'N 90°26.0'W	97 km	20 m	variable; sandy silt and silty sand
South Timbalier 53#3 (oxygen meter, 1991)	28°52.2'N 90°28.0'W	101 km	21 m	

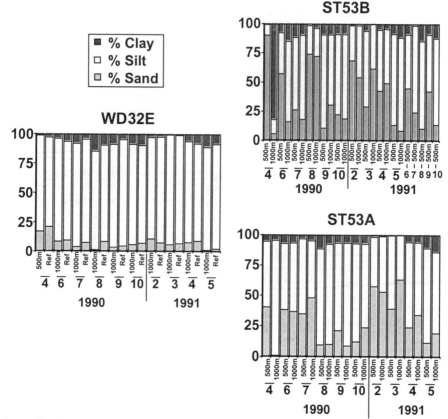

Figure 2. Percent composition of clay, silt and sand for WD32E, ST53A and ST53B for periods indicated (modified from Rabalais et al. [1995]).

personal communication]. A well-consolidated, dense clay is usually located at 10-14 cm below the sediment-water interface, but may become exposed at times (e.g., 1000-m distance station in April, Fig. 2, and following Hurricane Andrew in August 1992, N. N. Rabalais [personal observation]). Sediment TOC values were consistently low across the study area, typically less than 1.0% [Rabalais et al., 1993, 1995]. Sediment chlorophyll a and phaeopigment concentrations were consistent with season and distance from the Mississippi River [Rabalais et al., 1992]. Petroleum related hydrocarbons and trace metals were mostly at background levels for the 500-m and 1000-m distance stations and reference station [Rabalais et al., 1993].

There is a strong seasonal coherence between Mississippi River discharge and nutrient load and surface-water primary production, surface-water net production and bottom-water oxygen depletion on the southeastern Louisiana shelf [Justic' et al., 1993, 1997; Lohrenz et al., 1999]. The longest-term data for the study area are from station C6* (a combination of C6, C6A and C6B within the South Timbalier 53 block) and clearly demonstrate reduced surface water salinity throughout the year, a spring peak in surface water chlorophyll biomass with a smaller peak in the fall, high levels of phytoplankton pigments (chlorophyll a and phaeopigments) in bottom waters, and a seasonal progression of dissolved oxygen decline in the spring and summer (Fig. 3). Similar long-term data do not exist for the West Delta area, but the seasonal cycle is expected to be similar. Carbon flux into a moored particle trap at 15 m in a 21-m water column at station C6B (same as ST53#3) in 1991-1992 was high, approximately 500 to 600 mg C m^{-2} d^{-1}, with flux being greater in the spring and the fall than in the summer [Qureshi, 1995]. The material that sank into the traps was composed of fecal pellets, directly sinking phytoplankton, and other unidentified carbon including molts, dead zooplankton (swimmers and/or dead zooplankton that fell into the traps), marine snow, and particles with adsorbed organic carbon. The high particulate organic carbon flux to the 15-m moored trap was sufficient to fuel hypoxia in the bottom waters below the seasonal pycnocline [Qureshi, 1995; Justic' et al., 1996] and provides the source of carbon for the demersal food web. The flux of organic material in summer, while it sustained hypoxia, was incremental to the majority flux of particulates in the spring [Qureshi, 1995]. The particle traps were not deployed from late fall through early spring when high carbon flux might also occur.

Methods

Field and Laboratory

Benthic samples were collected with an Ekman-type closure 0.1-m^2 box corer with an average penetration of 20 cm and a minimum penetration of 10 cm in sandy sediments. Samples for benthic macroinfaunal analysis were taken with a small, hand-operated Ekman grab (0.023-m^2 surface area) from the larger box corer. Five replicates per station, one from each of five separate box cores, were taken. Organisms retained on a 0.5-mm screen were enumerated to the lowest possible taxon. Biomass was determined on ethanol preserved organisms according to the dry ash-free organic method of Crisp [1984]. Acrylic cores (7.6-cm diameter) were used to subsample the box corer at ST53B, 500-m distance. They were sectioned at 2-cm intervals and enumerated similar to the Ekman grab samples.

Hydrographic profiles were obtained with either a Hydrolab Surveyor II CTD or a SeaBird CTD unit. Oxygen meters (Endeco 1184) were deployed within 1 m of the sediment-water interface at WD32E from mid-June through mid-October 1990 and at ST53B from March to mid-October 1990 (N.B., the first 2-mo record was lost due to instrument malfunction). The removal of the ST53B platform late in 1990 necessitated our moving the oxygen meter to ST53#3 in 1991 (Table 1), where it was deployed from early February through December. Maintenance and calibration of the CTDs and the oxygen meters were described by Rabalais et al. [1994].

Statistics

Ln(x+1) transformed data (species richness and abundance) were analyzed using the General Linear Model procedure for ANOVA [SAS Institute, Inc., 1982]. Significantly different stations/months/sites within a group were identified using Duncan's multiple range test (P < 0.05).

Benthic community data were consolidated into a multi-component database with hydrographic and sediment data. The database was consolidated by season to provide an even number of replicates and months within each category. For the ST53B site for which data were available from April 1990-October 1991, a separate analysis was conducted for monthly data.

For Q-mode analysis (NTSYS-pc, ver. 1.8 [Rohlf, 1993]), biotic data were reduced to a matrix of individual samples described by the mean densities of 51 species (representing 98% of the total number of specimens). These mean densities were computed from logarithmically transformed data (18-30 replicates per sample), and then backtransformed to the original scale. Mean densities were root-root transformed to scale down the scores of abundant species so that they would not overwhelm the other data. We used the Bray-Curtis dissimilarity (=distance) coefficient to produce a similarity matrix and the Unweighted Pair-Group Method, Arithmetic Average to produce a dendrogram from the similarity matrix [Clifford and Stephenson, 1975]. Multi-Dimensional Scaling (MDS) was used for ordination on three dimensions [Rohlf, 1993]. In order to find the environmental factors that were likely to be responsible for patterns found, we superimposed the values of environmental variables on the abundance-based MDS plots.

Results

Conditions of Hypoxia

Bottom waters at ST53B were severely depleted in dissolved oxygen and often anoxic for most of the continuous record from mid-June through mid-August and for much of the month of September in 1990 (Fig. 4). Hydrogen sulfide was detected in bottom-water samples on several occasions in June, July and August at one or more stations along transect C in 1990. There were no strong diurnal patterns in the oxygen time series at ST53B [Rabalais et al., 1994]. In contrast, hypoxia occurred at WD32E for only 50% of the total record, hypoxic events were shorter in duration than at ST53B, and there was a strong diurnal pattern in the oxygen time series [Rabalais et al., 1994]. The record of dissolved oxygen at WD32E was most coherent with the diurnal bottom water pressure signal, which suggested the importance of tidal advection in the variability of that oxygen record [Rabalais et al., 1994]. Wind-induced mixing was insufficient to aerate the water column prior to the outbreak of cold air fronts in late September and early October at which time a relaxation in the stratification also occurred due to thermal cooling. Lack of strong winds and changes in bottom-water temperature suggested that reoxygenation (at ST53B in late August and at WD32E for most of the record) resulted from lateral advection.

The continuous oxygen data for station ST53#3 in 1991 (not continued at WD32E) indicated intermittent hypoxia in March-June (Fig. 5). For most of July and August and into early September, bottom waters were near-anoxic or anoxic, but there was no hydrogen sulfide detected. Beginning in early September, the water column was aerated by wind-induced mixing with the beginning of cold front passages. The early part of the year (March-June) was also characterized by a fairly continuous series of strong weather fronts with strong winds and subsequent wind mixing. The number and duration of hypoxic and anoxic events in the South Timbalier area were not as great in 1991 as in

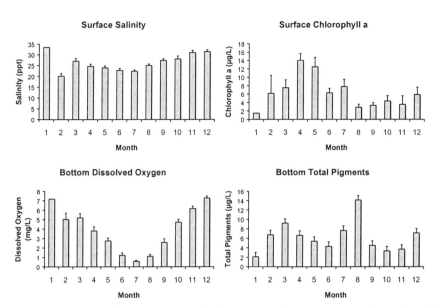

Figure 3. Surface and bottom water quality for station C6* (composite data for stations C6A, C6B and C6, all three are in the ST53 block) for 1985-1997 average conditions (± s.e.). *n* for average condition ranges between 1-10 for winter, 10-20 for spring and fall and 20-40 for summer. (Modified from Rabalais et al. [1998].)

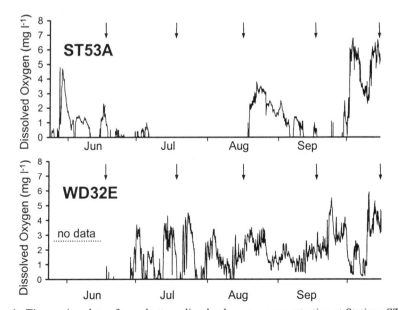

Figure 4. Time series plots of near-bottom dissolved oxygen concentration at Stations ST53A and WD32E in 1990 (modified from Rabalais et al. [1994]). Arrows indicate dates of benthos collections; April not shown.

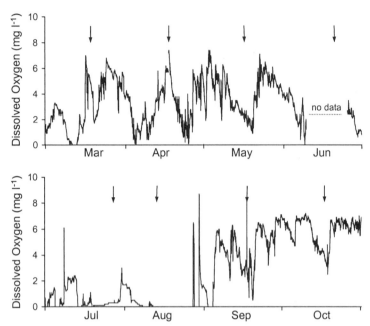

Figure 5. Time series of near-bottom dissolved oxygen for Station ST53#3 in 1991. Arrows indicate dates of benthos collections; February is not shown.

1990. Bottom waters at WD32E were well-oxygenated during the periods of benthic collections in 1991 (February-May) as indicated by CTD casts.

Benthic Communities

Comparison of Three Study Sites

Overall, the two stations at each of the three sites (500- and 1000-m distance stations at ST53A and ST53B; reference and 1000-m distance station at WD32E) were similar with regard to species richness and abundance (ANOVA, $P < 0.05$). Data for the two stations were, therefore, combined at each of the sites for subsequent comparisons. There were statistically significant differences between study sites with regard to both number of species and number of individuals, across all months (April 1990-May 1991) and for most months alone (Appendix 1). There were also significant differences among months at each site (Appendix 2).

There were more species and higher abundances of individuals at all sites in April and June 1990 (Fig. 6). In July, there was a general seasonal decline in populations at all three sites, but the decline was much more precipitous at ST53A and ST53B, the severely hypoxic stations, than at WD32E. The decline in populations at WD32E continued into September and October, but benthic communities at ST53A and ST53B showed slight recovery during that period. Recruitment occurred in the spring of each year but generally lower in 1991 than in 1990.

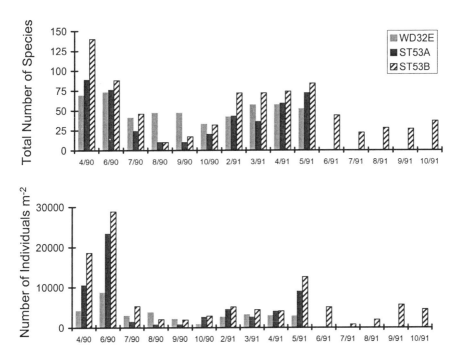

Figure 6. Comparison of species richness (total number for 10 0.02-m^2 cores) and mean number of individuals, n = 10, by study site and month. There are no data for WD32E and ST53A for 6/91-10/91.

During the period of severe hypoxia in 1990 at ST53A&B (August-September), species richness was greater at WD32E. Across all months, number of individuals was greater at ST53B than at either WD32E or ST53A, which were statistically similar to each other. Exceptions in this pattern were seen in August and September 1990, when the infauna was more abundant at WD32E. In both comparisons of species richness and number of individuals across all months, ST53A and ST53B were statistically different from each other (Appendix 1) with variation evident within any month. In any ranking scheme by sample period, ST53B was always greater in number of species and individuals than ST53A, but not always significantly.

Benthic Communities - WD32E

Species richness was similar in April and June 1990, decreased in July and August, then decreased further in September and October (Fig. 6, Appendix 2). The number of species increased in February through May 1991, but not as high as the previous spring. Peak abundance occurred in June 1990 followed by a mid-summer and fall decline and a slight increase the following spring.

Polychaetes were a large component of the benthic community at WD32E, but other major taxonomic groups exceeded the polychaetes in April 1990 and August 1990 and were half in June 1990 (Fig. 7). Polychaetes dominated in July, September and October of 1990 and during February-May 1991. The benthic community at WD32E was diverse,

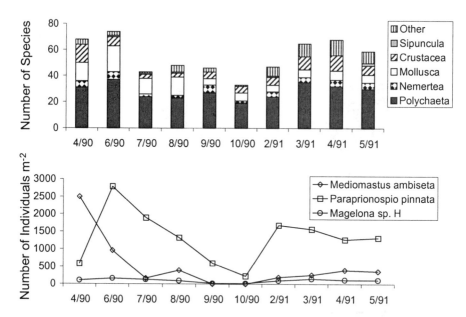

Figure 7. Number of species within taxonomic groups (total for 10 0.02-m² cores) and mean number of individuals m⁻² (n = 10) at WD32E for months indicated in 1990 and 1991.

with a complement of pericaridean crustaceans, bivalves, gastropods and other taxa. Dominant species for most months were *Paraprionospio pinnata* and *Mediomastus ambiseta* (Fig. 7). The abundance of *Armandia maculata* increased in August 1990. Changes in several dominant species through 1990 were evident with *Prionospio cristata*, *Nephtys incisa*, *Magelona* sp. I, *Magelona* sp. H, *Ampharete* sp. A and *Owenia fusiformis*. *Armandia maculata*, *Ampharete* sp. A and *Magelona* sp. I, which were dominants in 1990, were replaced in spring 1991 by *Sigambra tentaculata* and *Cossura soyeri*.

Benthic Communities – ST53A

Species richness was lowest in August and September 1990, and was also low in July and October 1990 (Fig. 6, Appendix 2). Species richness in June and April 1990 was similar and approximately six times greater than in July through October 1990. Species richness increased during the spring of 1991, but not to the level observed in spring 1990 (with the exception of May 1991). Abundance of individuals was high in April 1990, higher in June 1990; then the number of individuals was low from July through September (Fig. 6). There was a slight recovery of individuals in October, and a further increase in spring 1991, but not to the level observed in spring 1990 (with the exception of May 1991).

While polychaetes comprised most of the species at ST53A, composition by other major taxonomic groups was fairly high in April (13 taxa) and June (11 taxa) of 1990, then reduced to four to six major taxa in July through October 1990 (Fig. 8). The polychaetes *Ampharete* sp. A, *Paraprionospio pinnata* and *Mediomastus ambiseta* were

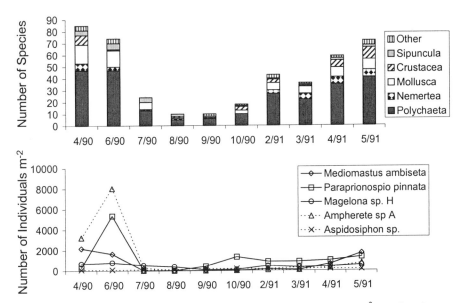

Figure 8. Number of species within taxonomic groups (total for 10 0.02-m^2 cores) and mean number of individuals m^{-2} ($n = 10$) at ST53A for months indicated in 1990 and 1991.

common in spring and early summer of 1990. As hypoxia worsened, the community was reduced to the polychaetes *Ampharete* sp. A and *Magelona* sp. H and the sipunculan *Aspidosiphon* sp. Only *Magelona* sp. H and *Aspidosiphon* sp. maintained any significant population levels in August 1990. During September and October 1990, the overall increase in number of individuals was due primarily to the recruitment of *Paraprionospio pinnata* and *Armandia maculata* and sustained levels of *Magelona* sp. H and *Aspidosiphon* sp. Species richness again increased during the spring of 1991, but polychaetes remained the dominant taxa (Fig. 8). *Owenia fusiformis*, which had been a dominant member of the community in 1990, was replaced by a population of *Sigambra tentaculata* in spring 1991.

Benthic Communities – ST53B

Species richness was lowest in August 1990 with a slight increase in September and October 1990 (Fig. 6). Species richness increased in spring 1991, but not to the level observed in spring 1990. There was again a substantial reduction in number of species in July 1991, but not as great as in mid-summer 1990, and a subsequent increase in species richness in August through October 1991. More species were found in August 1991 compared to August 1990 (Appendix 2). Abundance of individuals was high in April and June 1990, but dropped dramatically in July through September 1990 (Fig. 6, Appendix 2). There was a slight recruitment of individuals in October 1990. Abundance increased somewhat in February-April 1991, then increased substantially in May 1991. A seasonal decrease began in June 1991 with a significant reduction in abundance in July and August. Abundance increased in September and October 1991 to about the same level as early spring 1991.

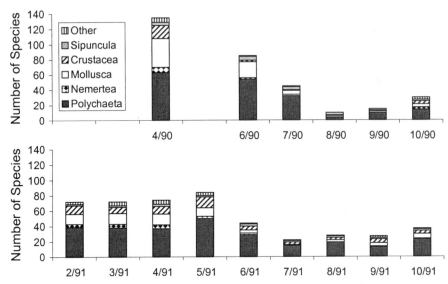

Figure 9. Number of species within taxonomic groups (total for 10 0.02-m^2 cores) at ST53B for months indicated in 1990 and 1991, no data for 2/90, 3/90, 5/90.

Polychaetes were a large component of the benthic community at ST53B, although the number of non-polychaete species was greater in April, August and October 1990. The number of major taxa decreased steadily from April through the period of hypoxia (April, 14 taxa; June, 9 taxa; July, 7 taxa; August, 4 taxa), then increased gradually in September (6 taxa) and October 1990 (10 taxa) (Fig. 9). A diverse fauna was recruited in February-April 1991, as a greater proportion of all taxa was composed of non-polychaetes. Polychaetes dominated the fauna from May through October 1991 (Fig. 9). The polychaetes *Mediomastus ambiseta*, *Paraprionospio pinnata* and *Ampharete* sp. A were common in spring and early summer of 1990 (Fig. 10). The survival of species during hypoxia was similar to ST53A; as hypoxia worsened, the common species were reduced to the polychaetes *Ampharete* sp. A, *Magelona* sp. H and *Clymenella torquata* and the sipunculan *Aspidosiphon* sp. Only *Magelona* sp. H and *Aspidosiphon* sp. maintained any significant population levels in August 1990. The overall fall increase in number of individuals was due primarily to the recruitment of *Paraprionospio pinnata* in September and October and the additional recruitment of *Armandia maculata* in October, as well as maintained population levels of *Magelona* sp. H and *Aspidosiphon* sp. *Owenia fusiformis* and *Clymenella torquata*, which had been dominant members of the community in 1990, were replaced by a population of *Sigambra tentaculata* in spring 1991. Increased abundance in fall 1991 was again due to sustained levels of *Magelona* sp. H and *Aspidosiphon* sp. and recruitment of *Sigambra tentaculata* and *Paraprionospio pinnata*.

Vertical Distribution

Most individuals were within the upper 2 cm of the sediments, especially during peaks in spring of both years (Fig. 11). The smaller recruited individuals at the surface in

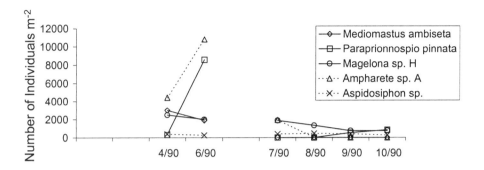

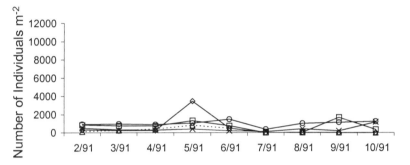

Figure 10. Mean number of individuals m⁻² (n = 10) at ST53B for months indicated in 1990 and 1991, no data for 2/90, 3/90, 5/90.

spring were *Paraprionospio pinnata* and *Ampharete* sp. A, which are surface deposit feeders. Other dominant spring recruits, *Mediomastus ambiseta*, are subsurface deposit feeders/opportunists. Individuals, although low in abundance, were more evenly distributed vertically during mid-summer hypoxia in July-August 1990 and August 1991. As numbers increased in fall (September-October) of both years, they remained more evenly distributed through the sediments (with a few exceptions) as opposed to close to the sediment surface in spring. Species (not illustrated) were more evenly distributed vertically through the sediments across seasons and years, than individuals.

Although it was not measured, the redox potential discontinuity (RPD) layer should move up in the sediment towards the sediment surface as the dissolved oxygen in the overlying waters approached anoxia. The diminishing habitat above the RPD was not, however, reflected by a similar reduction in vertical distribution of the surviving infauna.

Biomass

Biomass generally followed the same pattern as number of individuals (Fig. 12). Biomass at WD32E was linearly related to abundance. The relationships of biomass to abundance for ST53A and ST53B were linear until abundance exceeded 200 replicate⁻¹. These higher abundances were associated with recruitment of smaller *Paraprionospio pinnata* and *Mediomastus ambiseta* where the number of individuals increased, but biomass did not.

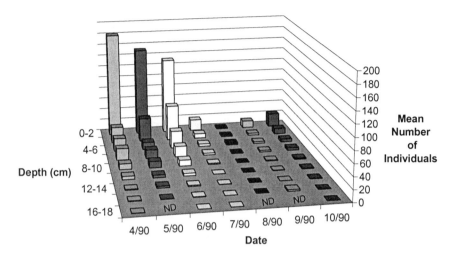

Figure 11. Vertical distribution of mean number of individuals per core section from replicate cores ($n=6$) at ST53B, 500 m, in 1990. ND is no data; zeros are real values.

Relationships with Environmental Variables

Variability in the benthic community parameters of species richness and abundance was correlated with the environmental variables of dissolved oxygen, water temperature and salinity, and sedimentary characteristics (Table 2). Bottom water oxygen concentration was an important environmental variable only when data from all three sites were combined, and then only for species richness. In the case of ST53A and ST53B, there was more of a threshold effect, whereby numbers of species and individuals were reduced linearly when the oxygen concentration fell below 0.5 mg l^{-1},

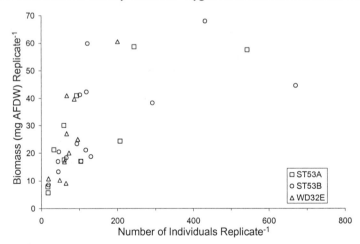

Figure 12. Comparison of benthic biomass (mg ash-free dry weight (AFDW)) replicate^{-1} and number of individuals replicate^{-1} for all three study sites for all individual month/year combinations from April 1990 - October 1991, $n = 10$.

TABLE 2. Results of multiple regression ($P < 0.05$) of number of species and individuals against environmental variables of Mo = month, Sal = bottom-water salinity, °C = bottom-water temperature, %Si = percent silt, Sd:Md = sand:mud ratio, Ox = bottom-water dissolved oxygen, TOC = sediment total organic carbon; sign indicates positive or negative relationship; r^2/P values given below the variables.

WD32E		ST53A		ST53B		All Sites	
#Species	#Indiv	#Species	#Indiv	#Species	#Indiv	#Species	#Indiv
		Mo(-)		Mo(-)	TOC(-)	Mo(-)	%Si(-)
		0.24/0.03		0.36/0.001	0.30/0.008	0.21/0.001	0.12/0.005
		Sal(-)		Sal(-)		Ox(+)	Sd:Md(+)
		0.22/0.04		0.28/0.006		0.06/0.04	0.07/0.03
				°C(-)		°C(-)	TOC(-)
				0.33/0.003		0.17/0.001	0.19/0.001
				%Si(-)		Sal(-)	
				0.32/0.003		0.14/0.002	
				Sd:Md(+)		%Si(-)	
				0.180/0.03		0.12/0.004	
				TOC(-)		Sd:Md(+)	
				0.29/0.009		0.12/0.004	
						TOC(-)	
						0.12/0.008	

rather than a decline through the full range of oxygen values. Grain size was not always paramount in explaining the variation observed in the benthos. Grain size was an important variable for all sites combined, and within ST53B where there was more variability in grain size distribution. Sediment characteristics were fairly uniform at both WD32E and ST53A. The high percentage of silt at WD32E likely contributed to the significant correlations of %Si(-) and Sd:Md(+) for all sites combined (Table 2).

Three groups were distinguished at the hierarchical level of Bray-Curtis dissimilarity > 0.5 (Fig. 13). Four groups, however, were visible at the dissimilarity level of 0.37 when the data were superimposed on the multi-dimensional scaling plots. Oxygen concentration correlated well with cluster groups (Fig. 14). Relationships with other environmental variables, including sediment grain size, were less apparent when overlaid on the same plots (not illustrated). The separation of the WD32E samples from ST53A&B (Fig. 13) likely relates to the higher percentage silt composition at WD32E, but these results were not obvious in the multivariate analyses. Similarly, the separation of ST53A and ST53B within season (Fig. 14) likely relates to the higher sand content and variable sediment composition at ST53B.

Community structure at ST53B was fairly stable during the period February-June (tight clustering of month/year combinations in Fig. 15). Major changes in species composition and species abundance, as evident from the distances between the individual samples in Figure 15, occurred between the winter-spring months and July and August, and then again from mid-summer to the fall (October and September). These shifts were related to periods of recruitment in the spring, reductions of species and individuals with severe hypoxia/anoxia in mid-summer, and a slight recovery of the community in the fall after abatement of hypoxia. The separation of fall and spring samples indicated a difference in the community makeup by season and lower abundances in fall after hypoxia than during spring recruitment.

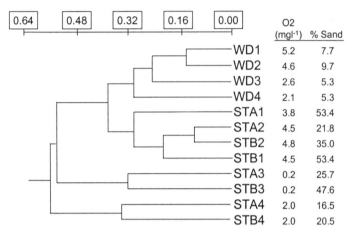

Figure 13. Results of unweighted pair-group method, arithmetic average clustering analysis of species abundance by site and season for Bray-Curtis Dissimilarity Index. WD=WD32E, STA=ST53A, STB=ST53B. 1=winter 1991 (Feb-Mar), 2=spring 1991 (Apr-May), 3=summer 1990 (Jul-Aug), and 4=fall 1990 (Sep-Oct). Mean values for bottom water oxygen ($n = 4$, discrete sample at time of benthos collections) and % sand composition of sediments ($n = 20$) listed on the right.

Discussion

Community Composition

The composition of the benthic communities on the southeastern Louisiana continental shelf reflected differences in sedimentary regime, the seasonal input of organic material and seasonally severe hypoxia/anoxia. There was a precipitous reduction in species, abundance and biomass of macroinfauna at the two stations exposed to severe and continuous hypoxia during mid-summer (Table 3). At the intermittently hypoxic site there was a seasonal decline in both species richness and abundance that was not obviously related to oxygen but could be attributed to a general decrease in organic material supply and/or increased predation. Except during periods of severe hypoxia at the South Timbalier site, WD32E had fewer species and lower abundances, which likely reflected the latter's predominantly silt sediments. Although the benthic communities of both South Timbalier stations demonstrated the effects of severe seasonal hypoxia, the generally higher species richness and abundance at ST53B compared to ST53A was likely a reflection of the higher sand content and sediment variability at ST53B.

The number of major taxonomic groups at WD32E was fairly consistent with time indicating the lack of influence of severe hypoxia (either during the summer or in successive years) on the benthic community. In contrast, at the South Timbalier stations (severe summer hypoxia every year), there was limited diversity of major taxa through most of the year and especially during the period of severe hypoxia despite higher sand content of the sediments. The fauna at WD32E (other than polychaetes) was composed of pericaridean crustaceans, bivalves, gastropods and ophiuroids that were mostly absent

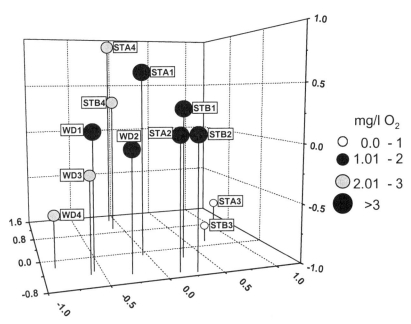

Figure 14. Three-dimensional results of the nonlinear multi-dimensional scaling analysis on matrix with site and season data for species abundances with bottom-water dissolved oxygen concentrations superimposed.

at ST53A and ST53B. The taxonomic diversity at South Timbalier in spring and fall was the result of species with planktonic larvae, not individuals with direct development; e.g., ampeliscid amphipods were essentially nonexistent in the South Timbalier fauna in the two periods studied, 1985-86 and 1990-91 [Rabalais et al., 1989; N. N. Rabalais, unpublished data; this study].

The summer "hypoxia" fauna at the South Timbalier sites was composed mostly of the polychaete *Magelona* sp. H and the sipunculan *Aspidosiphon* sp. Similar population levels of these were maintained throughout the year. *Paraprionospio pinnata* peaked during spring and fall recruitment periods at South Timbalier and dominated the macroinfauna at WD32E similar to other Louisiana-Texas inner shelf areas exposed to intermittent hypoxia (cf. Harper et al. [1981, 1991], Rabalais et al. [1989]). *P. pinnata* is a highly fecund, multiple-spawning, ubiquitous member of the benthic macroinfauna of the northwestern Gulf of Mexico shelf [Mayfield, 1988]. The opportunist capitellid polychaete *Mediomastus ambiseta* and the surface deposit-feeding polychaete *Ampharete* sp. A, which are capable of readily exploiting the freshly deposited organic material, were also dominant spring recruits at the South Timbalier sites. Opportunistic bivalves, such as *Abra aequalis* at the inshore Bryan Mound station [Harper et al., 1981, 1991] and *Mulinia lateralis* at West Hackberry [Gaston, 1985; Gaston and Edds, 1994] and the shallower LOOP stations [Vittor and Associates, 1998], were never common members of the benthic community at WD32E, ST53A or ST53B.

In contrast to other studies [Dauer et al., 1992; Diaz et al., 1992; Hendelberg and Jensen, 1993], the summer "hypoxia" fauna was not restricted to shallow-dwelling organisms. Although most individuals were located within the upper 2 cm of the

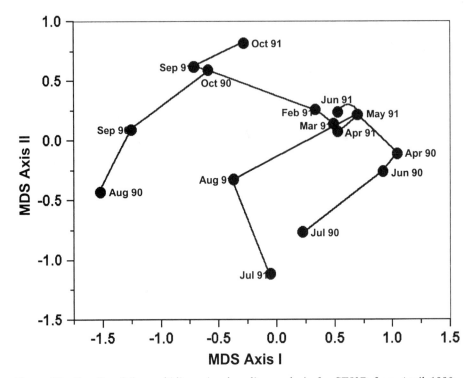

Figure 15. Results of the multidimensional scaling analysis for ST53B from April 1990-October 1991. Months are connected with a minimum spanning tree that was calculated from the same Bray-Curtis dissimilarity matrix as the multidimensional scaling analysis.

sediments during spring recruitment, they were more uniformly distributed vertically through the remainder of the year, even though numbers were drastically reduced during July-August. The vertical distribution of macroinfauna is also not shifted towards the surface at the most hypoxic sites within the Oman margin oxygen minimum zone (OMZ) [Smith et al., 2000]. The more uniform vertical distribution of organisms during severe hypoxia indicates that the survivors (e.g., *Magelona* sp. H or *Aspidosiphon* sp.) are physiologically adapted to severe hypoxia and/or high levels of hydrogen sulfide, and are, therefore, not restricted to the upper 2 cm of the sediments. When fall recruits enter the community, the low-oxygen tolerant species continue to maintain similar levels vertically and are not restricted to the upper sediment layers where most of the smaller recruiting individuals are located.

Differences in abundance of macroinfauna at South Timbalier between 1990 and 1991 for both spring and summer could be attributed to either variable responses to degrees of severity of hypoxia in the two years and/or the supply of organic material to the seabed. Surface-water values of chlorophyll *a*, as a potential indicator of carbon flux [Qureshi, 1995], did not differ between years for either the spring (Apr-Jun) or summer (Jul-Sep) period [N. Rabalais, unpublished data]. Also, the timing and magnitude of river discharge were similar between 1990 and 1991. The number and duration of hypoxic and anoxic events, however, were not as great in 1991 as in 1990, which would indicate that oxygen differences in the two years were most probably responsible for

TABLE 3. Comparison of benthic community data from selected studies, either inner shelf environments and/or hypoxia-affected environments. Details of studies are in the text.

Dauer et al. [1992]	Chesapeake Mainstem		Tributaries	
	Polyhaline Mud	**Hypoxia-affected**	Mesohaline Mud	**Hypoxia-affected**
Density (no. m^{-2})	1,978	**1,723**	3,065	**902**
Biomass (g AFDW m^{-2})	9.9	**1.7**	2.5	**1.1**
Mean taxa (no. 0.02-m^{-2}, $n=4$)	10	**6**	9	**4**

		Intermittent hypoxia		
Bryan Mound	Inshore (15-17 m)		Offshore (21 m)	
[Harper et al., 1981, 1991]	Spring	Summer	Spring	Summer
Density (no. m^{-2})	2,043	457	3,200	1,157
Total taxa (3 0.02-m^2 repls.)	60	36	81	52

		Intermittent hypoxia		
LOOP	Inshore (10 m)		Offshore (27-34 m)	
[Vittor & Assoc., 1998]	Spring	Summer	Spring	Summer
Density (no. m^{-2})	6,052	2,377	1,100	2,088
Total taxa (6 0.1-m^2 repls.)	56	33	72	78

		Intermittent hypoxia	
West Hackberry [Gaston, 1985;	Inshore (10 m)		
Gaston and Edds, 1994]	Spring	Summer	
Density (no. m^{-2})	4,275	1,772	
Total taxa (6 0.1-m^2 repls.)	52	39	

This Study	West Delta 32E		South Timbalier	
(all 20 m)	Intermittent Hypoxia		Prolonged, Severe Hypoxia	
	Spring Recruitment April 1990	Seasonal Low Fall Sep 1990	Spring Recruitment April 1990	**Hypoxia-affected Jul-Aug 1990**
Density (no. m^{-2})	8,637	1,431	18,437	**730**
Biomass (g AFDW m^{-2})	2.59	0.45	2.92	**0.23**
Mean taxa (no. 0.02-m^{-2}, $n=10$)	22	12	51	**4**

	Feb-May 1991		Feb-May 1991	**Jul-Aug 1991**
Density (no. m^{-2})	2,873		6,486	**1,346**
Biomass (g AFDW m^{-2})	0.93		1.55	**0.46**
Mean taxa (no. 0.02-m^{-2}, $n=10$)	16		22	**8**

differences in summer community makeup. Similar differences in taxonomic richness and abundance of both meiofauna and macroinfauna between two summers of variable hypoxia intensity were observed in an area close to ST53A and ST53B (station C5, 15-m water depth) in 1985-1986 [Murrell and Fleeger, 1989; Rabalais et al., 1989; Boesch and Rabalais, 1991; N. N. Rabalais, unpublished data]. The benthic communities were reduced more severely in 1985, when oxygen levels fell lower than in 1986. Of the meiofaunal community, hypoxia virtually eliminated populations of harpacticoid copepods and kinorhynchs, with extensive recovery not occurring until the following winter. Nematodes, however, maintained constant population levels throughout the year ($\sim$ 1000 individuals 10 cm^{-2}), similar to nematode densities found in severely hypoxic sediments in Gullmar Fjord [Josefson and Widbom, 1988] but an order of magnitude higher than meiofaunal densities at the intersection of the OMZ with an eastern tropical Pacific seamount [Levin et al., 1991]. Heavy mortalities occurred in the macroinfauna in both 1985 and 1986, with the persistence of a population of the polychaete *Magelona* sp. H (formerly known as *M. cf. phyllisae*) only in 1986. Also, recruitment and increases in populations at station C5 did not occur after the cessation of hypoxia in fall 1985, and did not begin until the following spring. Differences in the mortality of species depends on the tolerance of the members of the community and the severity of the hypoxia as seen by Llansó [1991], and the recovery of the communities depends on the severity and length of the hypoxia/anoxia exposure as seen by Harper et al. [1981, 1991] and Llansó [1992].

Regions of hypoxic bottom waters have been detected along portions of the Louisiana-Texas coast every summer since 1972 (see reviews of Dennis et al. [1984], Rabalais [1992], Renaud [1985]). The lack of consistently collected hydrographic data in conjunction with benthic studies over longer periods, however, makes it difficult to recreate the development of hypoxia on the Louisiana shelf and subsequent changes in benthic community structure. Analyses of foraminiferal communities in dated sediment cores indicate that while hypoxia may have been present to some degree beginning in the 20th century, there has been a worsening of oxygen stress (duration, intensity) since the 1950s, at least on the southeastern Louisiana shelf [Rabalais et al., 1996; Sen Gupta et al., 1996]. Thus, it is not clear if and when the benthic communities that are now exposed to hypoxia became preconditioned, or whether the pulsing nature of the river-dominated continental shelf has always structured a benthic community composed primarily of smaller, shorter-lived, opportunistic species.

Comparisons with Other Hypoxia-Affected Benthos

A few studies on the Louisiana and Texas shelves have adequate time series to identify benthic community impacts and recovery related to low oxygen events [Harper et al., 1981, 1991; Gaston, 1985; Murrell and Fleeger, 1989; Rabalais et al., 1989; Boesch and Rabalais, 1991]. The benthic studies of the Louisiana Offshore Oil Port (LOOP) [Vittor and Associates, 1998], while long-term in nature (1980-1993), were quarterly and did not capture the annual sequence of events as those listed previously. The studies of Harper and Gaston were designed to examine the effects of brine disposal at Bryan Mound, offshore of Freeport, Texas (15- and 21-m water depths, monthly for 7 years), and West Hackberry, offshore of the Calcasieu estuary on the southwestern Louisiana shelf (10-m water depth, monthly for 4 years with a 4- and 5-y followup), respectively (Fig. 1, Table 3).

Hypoxia occurred infrequently at Bryan Mound, but the event in June-July 1979 was prolonged and severe enough for the production of hydrogen sulfide into the water column above the sediment-water interface at the 15-m station [Harper et al., 1981, 1991]. During this period, benthic infauna declined (Table 3). Ampeliscid amphipods, which at times were a dominant member of the more inshore community, were eliminated and did not become dominant again through the end of the studies in 1984.

Other, less severe, hypoxic events may have occurred at the more inshore stations in 1982-1984 (oxygen data lacking; condition presumed from composition of the benthic community). The populations that maintained numbers during the presumed hypoxic events were *Nereis micromma* and *Lumbrineris verrilli* as opposed to *Magelona* sp. H and *Aspidosiphon* sp. at ST53A&B. The abundances of *N. micromma* and *L. verrilli* increased immediately after the abatement of hypoxia at Bryan Mound, and others that were reduced in the hypoxic event, *Paraprionospio pinnata* and *Magelona* sp. H (*M.* cf. *physillae* in Harper's studies), increased afterwards. The more offshore Bryan Mound benthic community, which was not exposed to the severely low oxygen in 1979 nor hypoxia in any subsequent years, did not suffer the same precipitous decline in species richness or abundance of organisms in 1979 or in subsequent summers. Otherwise the benthic communities at both depth stations were characterized by widely fluctuating seasonal abundances controlled primarily by numbers of *Paraprionospio pinnata*.

The West Hackberry site experienced episodic summertime hypoxia [Gaston, 1985; Pokryfki and Randall, 1987; Gaston and Edds, 1994] which was severely low in some years (e.g., 1981 with discrete values near anoxia in June through August; similar low values in summer 1983, 1984 and 1989) but not as persistent in others (e.g., 1982, 1988). Abundances of benthos reflected summertime oxygen concentrations with individuals dropping below 1000 m^{-2} during severe hypoxia and numbering 2000-3000 m^{-2} when not severe (Table 3). The benthic infauna at West Hackberry was dramatically reduced for most species following severe hypoxia in June-August 1981, with the exception of the polychaete *Magelona* sp. H. (was *M.* cf. *phyllisae* in Gaston's studies) which increased in numbers. As with the Bryan Mound study site, the amphipod crustaceans at the West Hackberry site were dramatically reduced the year following the hypoxic event. Otherwise, a series of opportunistic species, including the polychaetes *Paraprionospio pinnata*, *Magelona phyllisae*, *Sabellides* sp. and *Cirratulus* cf. *filiformis*, the bivalve *Mulinia lateralis* and the phoronid *Phoronis muelleri* were responsible for widely fluctuating seasonal abundances, similar to Bryan Mound.

The two LOOP stations [Vittor and Associates, 1998] experienced intermittent hypoxia at 10 m on the inshore edge of the hypoxic zone (hypoxia in 9 of 14 summers) and in ~30 m on the seaward edge of the hypoxic zone (hypoxia in 8 of 13 summers). The more inshore stations exhibited the effects of summertime hypoxia with reduced abundances of individuals, number of taxa and taxonomic makeup similar to the West Hackberry site, but mean values for the deeper offshore LOOP stations and their taxonomic makeup indicated little effect of hypoxia (Table 3). If stations had been located between the inshore and offshore LOOP sites where bottom-oxygen concentrations are usually lower [N. N. Rabalais, unpublished data], then benthic communities similar to ST53A&B may have been documented.

Within the continuum of benthic communities exposed to a variety of hypoxic/anoxic conditions [Diaz and Rosenberg, 1995, this volume], those of WD32E resemble ones exposed to repeated brief periods (days to weeks) of hypoxia annually and the benthos shows little change during or shortly after hypoxia. These communities are likely stress-conditioned, either from prior hypoxic events, as proposed by Diaz and Rosenberg [1995], or possibly from seasonal pulses of organic matter or high sedimentation rates. The benthos of South Timbalier clearly fall within the category of seasonal hypoxia that lasts for months so that mass mortality occurs despite preconditioning of the communities. Although hypoxia at South Timbalier is not continuous through the year or always anoxic, the severely reduced species richness, abundance and biomass (Table 3) is very reminiscent of benthic communities exposed to persistent hypoxia and eventually anoxia at the extreme end of the Diaz and Rosenberg continuum. The other studies from the Louisiana-Texas shelf (Bryan Mound, West Hackberry, LOOP) fall within the category of aperiodic hypoxia; however, the rate and level of recovery depend on the severity and length of the hypoxia exposure and the community structure which is presumed to be preconditioned to hypoxia stress.

Similarities are apparent between the benthos of both the aperiodic and seasonally severe hypoxic areas on the Louisiana-Texas coast and other areas similarly affected (e.g., lower Chesapeake Bay and several tributaries, Dauer et al. [1992]; Table 3). An altered summertime benthic community composition is characterized by lower species richness, abundance of individuals and biomass. There is a greater dominance of smaller, short-lived, opportunistic species during the period of hypoxia, but especially during spring and fall recruitment on the Louisiana-Texas shelf. In Chesapeake the hypoxia-affected fauna is characterized by a lower proportion of deeper-burrowing equilibrium species such as long-lived bivalves and a greater dominance of short-lived surface-dwelling forms. On the Louisiana-Texas shelf, however, the latter forms the majority of the community year-round, and the survivors of severe hypoxia are evenly distributed vertically through the sediments. Longer-lived, larger, deeper-dwelling, higher biomass organisms were absent from the severely-affected Louisiana shelf. There are many more small individuals, especially during spring recruitment, on the southeastern Louisiana continental shelf than in the Chesapeake Bay mainstem and tributaries under normoxic conditions (Table 3). During hypoxia events, the numbers, species richness and biomass are more drastically reduced on the southeastern Louisiana shelf (South Timbalier area) than in similarly affected environments of Chesapeake Bay.

Long-term trends for the Skagerrak coast of western Sweden in semi-enclosed fjordic areas experiencing increased oxygen stress [Rosenberg, 1990] showed declines in (1) total abundance and biomass, (2) abundance and biomass of mollusks, and (3) abundance of suspension feeders and carnivores. In pre-stressed communities, sensitive faunal groups were already lost from the community before severe hypoxic/anoxic events further depleted the benthic fauna [Josefson and Widbom, 1988]. Holland et al. [1987] examined recurring seasonal hypoxia in the mesohaline Chesapeake and found a reduction in long-lived benthos and dominance by smaller, short-lived species (also Mountford et al. [1977]). Where hypoxia is periodic and intermittent, e.g., Rappahannock River of Chesapeake Bay, Llansó [1992] determined that the benthic community was influenced by the intermittent hypoxia, but in the York River, where hypoxia was intermittent but less severe, there was no effect on the benthic [Llansó, 1991, 1992] where the community may be long-term conditioned to hypoxic stress. Diaz and Rosenberg's [1995] conclusion that, in general, long-term reduction in macrofauna occurs as hypoxic stress increases (e.g., Niermann et al. [1990], Friligos [1976], Friligos and Zenetos [1988], Dauer and Alden [1995]) appears to be true for the South Timbalier hypoxia-impacted benthos.

Besides the anthropogenically-influenced areas of coastal hypoxia worldwide, oxygen-stressed benthos are found in deep-water basins (e.g., Black Sea) and fjords, and where OMZs intersect continental margins and seamounts [Kamykowski and Zentara, 1990]. OMZs are significant mid-water features in the eastern Pacific from California to Chile, the Arabian Sea and off west Africa and result in oxygen and organic matter gradients similar to those found in nutrient enhanced, highly productive, seasonally severe hypoxic estuaries and continental shelve. Where OMZs intersect the seabed, large megafaunal species are sparse, macrofaunal assemblages consist of low density and diversity, low biomass, small-bodied assemblages dominated by nematodes and selective polychaetes, and the bacterial/metazoan biomass ratio is high [Arntz et al., 1991; Levin et al., 1991; Levin and Gage, 1998; Levin et al., 2000]. Many features of OMZ-influenced benthic environments are repeated in the Louisiana continental shelf influenced by the Mississippi River—low oxygen, high organic matter flux and occasionally high sulfide levels—with similar results in benthic assemblages as oxygen stress worsens. For similar levels of oxygen stress on the Oman margin [Levin et al., 2000; Smith et al., 2000] compared to the Louisiana shelf, however, the species richness, abundance and biomass of macroinfauna are much lower here than on the OMZ-affected margin. The hypoxia-affected benthos of the Louisiana shelf is confined to a few species of polychaetes (*Magelona* sp. H and *Ampharete* sp. A) and a sipunculan (*Aspidosiphon* sp.), unlike the spionid, cirratulid and ampharetid polychaetes that dominate in the low oxygen settings on the Oman margin [Levin et al., 2000]. Spionid polychaetes, primarily *Paraprionospio pinnata* and a few other genera, dominated on the Louisiana shelf during oxygenated periods of spring and fall, but did not maintain populations during severe hypoxia. Cirratulids were not common on the severely-affected Louisiana continental shelf, but were members of the intermittently-affected benthic community at the West Hackberry site [Gaston, 1985].

Implications for Fisheries Resources

It is apparent that demersal fish and invertebrates, including the commercially important penaeid shrimps, are not usually found where the oxygen concentration falls below 2 mg l^{-1} [Pavela et al., 1983; Leming and Stuntz, 1984; Renaud, 1986; Craig et al., this volume; Zimmerman and Nance, this volume], although some shrimps and invertebrates such as stomatopods have been seen in submersible video tapes to oxygen values as low as 1.5 mg l^{-1} [Rabalais, Harper and Turner, this volume]. A large area of essential habitat for demersal-feeding organisms (up to 20,000 km^2) is eliminated in summer along the Louisiana shelf. Although these calculations of hypoxic zone size are usually limited to single, 5-d survey estimates, some surveys repeated within two to three weeks indicate a persistence to the distribution and size of the zone, at least in mid-summer [Rabalais et al., 1999; this volume]. Data for the whole shelf are lacking for other times of the summer, but hypoxia can often be widespread and severe along transect C (Fig. 1) on the southeastern shelf for much of May-September.

In other hypoxia-affected estuarine and shelf environments, predators may benefit from a hypoxia-stressed benthos, either during or immediately following hypoxia. Infauna that have moved closer to the sediment-water interface may be more easily preyed upon [Diaz et al., 1992; Pihl et al., 1992; Pihl, 1994; Nestlerode and Diaz, 1998]. This is not likely the case for the severely affected areas of the southeastern Louisiana shelf (e.g., South Timbalier sites) for three reasons: (1) the remaining surviving fauna is not predominantly at the sediment surface, (2) fish predators are excluded from the zone of hypoxia and not seen by either direct observations or video [Rabalais, Harper and Turner, this volume] and (3) the presence of intact moribund and stressed benthic organisms at the sediment surface is evidence for the absence of larger predators [Rabalais, Harper and Turner, this volume]. Following the abatement of hypoxia in the fall, there was either a slight increase in biomass predominantly by small, opportunistic polychaetes (ST53A) or no increase (ST53B) [Rabalais et al., 1995]. Thus, a substantial area of feeding habitat is removed from the foraging base of demersal organisms for months at a time. The proportion of this unsuitable habitat as a whole of the Louisiana shelf is not known. Nematodes, while reduced in abundance at more severely-affected stations (15-m depth) than inshore stations (8-m depth) in the South Timbalier study area averaged about 1200 individuals per 10 cm^2 through the year, but harpacticoid copepods were virtually eliminated by summer hypoxia [Murrell and Fleeger, 1989]. The insensitivity of nematode densities to oxygen deficiency or sometimes increase under severe hypoxia [Josefson and Widbom, 1988; Levin et al., 1991; Cook et al., 2000] may make these remaining meiofaunal organisms potential food for foraging fish. The relative suitability of this potential nematode food to demersal feeders on the shelf compared to harpacticoid copepods and macroinfauna is not known. Fish would not be potential predators during mid-summer severe hypoxia, but nematodes may be suitable prey for some foragers during the fall after hypoxia dissipates.

Periods prior to severe hypoxia during spring recruitment have significantly higher biomass in the form of small, opportunistic surface-dwelling polychaetes that should serve as a readily available food source, but biomass levels vary from spring to spring. Areas on the inshore periphery of severe hypoxia (intermittently or moderately affected) maintain populations of opportunistic species but do experience summer decreases in biomass that may be due either to oxygen stress, reduced food supply or increased predation. Diaz and Solow [1999] pointed out that these types of benthic communities did not store large amounts of energy as biomass to buffer the ecosystem against the pulsing of energy and usually supported boom and bust cycles. On the offshore periphery of Louisiana hypoxia, benthic populations appear to be relatively unaffected, but in general abundances decreased with depth [Gaston et al., 1998], and probably biomass decreased as well if accepted continental shelf depth gradients are applicable to the Louisiana continental shelf. Through an annual cycle, therefore, there are areas potentially without suitable food resources for extended periods and other areas with highly variable populations of opportunistic species that would be suitable prey for demersal feeders.

While biomass in hypoxia-affected habitats on the Louisiana shelf may be periodically high with opportunistic species, the overall productivity of the benthic system, transfer to other trophic levels, and secondary production in general are not known. A high recruitment of larval *Mediomastus, Paraprionospio, Ampharete* and other polychaetes that have high growth rates, utilize the readily available organic matter fluxed to the seabed, and eventually provide suitable food for demersal feeders may contribute to a high, but temporary transfer of carbon to higher trophic levels. These organisms do not persist through severe summer hypoxia, and increase in their biomass in fall is low. Their demise is predicted to be due to low oxygen and not predation (i.e., no transfer of carbon), since the predators vacate the area before the decline in benthos begins. Meroplankton, dominated by larval *Paraprionospio pinnata*, are distributed throughout the water column in the summer and are more abundant when bottom water oxygen is hypoxic than normoxic, but these larvae are not recruiting to the benthos [Powers et al., this volume], or, if they do, die immediately. Larger larvae in the overlying waters may have either delayed metamorphosis, or emigrated from the sediments under extreme oxygen stress (sensu Wetzel et al. [this volume]). A higher secondary production based on high turnover of individuals does not appear to be the case during the period of severe hypoxia.

Despite reduced suitable habitat and apparent reduced food resources at times of the year, demersal fishery production remains high and must be supported by the available benthic production [Chesney and Baltz, this volume]. The overall secondary production, however, may have been affected or shifted within the context of decadal changes in primary production and worsening hypoxia stress. Zimmerman and Nance [this volume] found a correlation between the reduction in total brown shrimp catch in recent years as the mid-summer size of the hypoxic zone increased and a recent decline in the catch per unit effort in the brown shrimp fishery that corresponds with the expansion of hypoxia. Diaz and Solow [1999] provided evidence that annual productivity for some systems with severely stressed habitats as a result of hypoxia was lower but that this trend was not consistent across habitat types. As more estuarine and coastal areas worldwide are exposed to worsening oxygen stress, benthic communities will become more severely stressed. Those of the South Timbalier area are extremely stressed with limited recovery and may be symptomatic of worsening oxygen conditions on the Louisiana shelf. The relative area of such oxygen-stressed habitats on the Louisiana shelf has the potential to affect carbon transfer to higher trophic levels, but at present the relative proportion of such habitats is not known. With the potential for worsening oxygen stress and expanse of the hypoxia zone under scenarios of global climate change [Justic' et al., 1996; this volume], there may be a point at which overall secondary production based on the benthos may be affected if it has not been already.

Appendix 1. Results of general linear model analysis of variance of ln(x+1) transformed data comparison of study sites for number of species and individuals by sample period (April 1990-May 1991) and all months combined (stations 500 m and 1000 m for ST53A=STA and ST53B=STB; 1000 m and Ref for WD32E=WD, 500 m substituted for 1000 m in Apr 1990); * significant; ns = not significant; and Duncan's multiple range test results. Underlined sites are not significantly different from each other, n = number of replicates, ^ indicates fewer replicates than others.

Month	Parameter	P value	n	Duncan's Multiple Range Test		
Apr 1990	Species	P<0.0001*	10	WD	STA	STB
	Individuals	P<0.0001*		WD	STA	STB
Jun 1990	Species	P<0.0041*	10	WD	STA	STB
	Individuals	P<0.0002*		WD	STA	STB
Jul 1990	Species	P<0.0001*	10	STA	WD	STB
	Individuals	P<0.0001*		STA	WD	STB
Aug 1990	Species	P<0.0001*	10,8	STA^	STB^	WD
	Individuals	P<0.0001*		STA^	STB^	WD
Sep 1990	Species	P<0.0001*	10	STA	STB	WD
	Individuals	P<0.0006*		STA	STB	WD
Oct 1990	Species	P<0.1527ns	10	STA	STB	WD
	Individuals	P<0.0001*		WD	STA	STB
Feb 1991	Species	P< 0.0088*	10,9	WD^	STA	STB
	Individuals	P<0.0081*		WD^	STA	STB
Mar 1991	Species	P<0.0009*	10	STA	WD	STB
	Individuals	P<0.0791ns		STA	WD	STB
Apr 1991	Species	P<0.3656ns	10	WD	STA	STB
	Individuals	P<0.4016ns		WD	STA	STB
May 1991	Species	P<0.0020*	10	WD	STA	STB
	Individuals	P<0.0001*		WD	STA	STB
All Months	Species	P<0.0014*	99,98	STA^	WD	STB
Combined	Individuals	P<0.0001*		WD	STA^	STB
				increasing means$\rightarrow$		

Appendix 2. Results of general linear model analysis of variance of ln(x+1) transformed data comparison of months for number of species and individuals and Duncan's multiple range test results. Same station combinations as in Appendix 1. Underlined dates are not significantly different from each other, ^ indicates fewer replicates than others.

West Delta (n = 10,9)

Species P<0.0001*   10/90 2/91^ 7/90 9/90 4/91 3/91 8/90 5/91 6/90 4/90

Individuals P<0.001*   10/90 9/90 2/91^ 5/91 4/91 7/90 3/91 4/90 8/90 6.90

ST53A (n = 10,8)

Species P<0.0001*   8/90^ 9/90 7/90 10/90 3/91 2/91 4/91 5/91 4/90 6/90

Individuals P<0.0001*   9/90 8/90^ 7/90 3/91 10/90 4/91 2/91 5/91 4/90 6/90

ST53B(n = 10,8)

Species P<0.0001*   8/90^ 9/90 10/90 7/90 2/91 4/91 3/91 5/91 6/90 4/90

Individuals P<0.0001*   9/90 8/90^ 10/90 4/91 3/91 2/91 7/90 5/91 4/90 6/90

ST53B(n = 10,5)

Species P<0.0001*   8/90 7/91^ 9/90 10/90 8/91^ 9/91^ 7/90 6/91^ 10/91^ 2/91 4/91 3/91 5/91 6/90 4/90

Individuals P<0.0001*   7/91^ 9/90 8/91^ 8/90 10/90 4/91 6/91^ 3/91 10/91^ 2/91 7/90 9/91^ 5/91 4/90 6/90

Acknowledgments. Funding for this research was provided by the Minerals Management Service/Louisiana Universities Marine Consortium University Research Initiative (Cooperative Agreement 14-35-0001-30470), the National Oceanic and Atmospheric Administration (NOAA) Coastal Ocean Program, Nutrient Enhanced Coastal Ocean Productivity Program (NA90AA-D-SG691, project MAR24 to N. Rabalais and D. Harper for benthic studies and project MAR31 to N. Rabalais, R. Turner and W. Wiseman for the instrument mooring), the Louisiana Sea Grant College Program for supplemental ship funds (NA89AA-D-SG691, project R/HPX-1-PD to N. Rabalais and D. Harper), and the NOAA National Undersea Research Center during July 1990 and July 1991. Department of Energy (DE-FG02-97ER12220) supported the preparation of this manuscript. We thank Ben Cole for help with the figures, and Robert Diaz, Gary Gaston and Lisa Levin for critical reviews of an earlier draft of this manuscript.

References

Arntz, W. E., J. Tarazona, V. Gallardo, L. A. Flores, and H. Salzwedel, Benthos communities in oxygen deficient shelf and upper slope areas of the Peruvian and Chilean Pacific coast and changes caused by El Niño, in *Modern and Ancient Continental Shelf Anoxia*, edited by R. V. Tyson and T. H. Pearson, pp. 131-154, *Geological Society Special Publ., 58*, 1991.

Boesch, D. F. and N. N. Rabalais, Effects of hypoxia on continental shelf benthos: comparisons between the New York Bight and the Northern Gulf of Mexico, in *Modern and Ancient Continental Shelf Anoxia*, edited by R. V. Tyson and T. H. Pearson, pp. 27-34, *Geological Society Special Publ., 58*, 1991.

Clifford, H.T. and W. Stephenson, *An Introduction to Numerical Classification*, Academic Press, New York, 1975.

Cook, A. A., P. J. D. Lambshead, L. E. Hawkins, N. Mitchell, and L. A. Levin, Nematode abundance at the oxygen minimum zone in the Arabian Sea, *Deep-Sea Res., Part II, 47*, 75-85, 2000.

Crisp, D., Energy flow measurements, in *Methods for the Study of Marine Benthos*, edited by N. A. Holme and A. D. McIntyre, pp. 284-372, Blackwell Scientific Publications, London, 1984.

Dauer, D. M. and R. W. Alden III, Long-term trends in macrobenthos and water quality of the lower Chesapeake Bay (1985-1991), *Mar. Pollut. Bull., 30*, 840-850, 1995.

Dauer, D. M., A. J. Rodi, Jr., and J. A. Ranasinghe, Effects of low dissolved oxygen events on the macrobenthos of the lower Chesapeake Bay, *Estuaries, 15*, 384-391, 1992.

Dennis, G. D., T. J. Bright, and C. A. Shalan, Offshore oceanographic and environmental monitoring services for the strategic petroleum reserve: annotated bibliography of hypoxia and other oxygen-depleted literature on the marine environment, Final Report to DOE, Strategic Petroleum Reserve Project Management Office, Contract No. DOE-AC96-83PO10850. Texas A&M University, Texas A&M Research Foundation, College Station, Texas, 1984.

Diaz, R. J. and A. Solow, Ecological and Economic Consequences of Hypoxia: Topic 2 Report for the Integrated Assessment of Hypoxia in the Gulf of Mexico. NOAA Coastal Ocean Program Decision Analysis Series No. 17, NOAA Coastal Ocean Program, Silver Springs, Maryland, 1999.

Diaz, R. J. and R. Rosenberg, Marine benthic hypoxia: A review of its ecological effects and the behavioural responses of benthic macrofauna, *Oceanogr. Mar. Biol. Ann. Rev., 33*, 245-303, 1995.

Diaz, R. J., R. J. Neubauer, L. C. Schaffner, L. Pihl, and S. P. Baden, Continuous monitoring of dissolved oxygen in an estuary experiencing periodic hypoxia and the effect of hypoxia on macrobenthos and fish, in *Marine Coastal Eutrophication*, edited by R. A. Vollenweider, R. Marchetti, and R. Viviani, pp. 1055-1068, *Sci. Total Environ.*, suppl. no. 0048-9697, 1992.

Eadie, B. J., B. A. McKee, M. B. Lansing, J. A. Robbins, S. Metz, and J. H. Trefry, Records of nutrient-enhanced coastal productivity in sediments from the Louisiana continental shelf, *Estuaries, 17*, 754-765, 1994.

Friligos, N., Seasonal variations of nutrients around the sewage outfall in the Saronikos Gulf (1973), *Thalassia Jugoslavica, 12*, 441-453, 1976.

Friligos, N. and A. Zenetos, Elefsis Bay anoxia: nutrient conditions and benthic community structure., *Mar. Ecol., 9*, 273-290, 1988.

Gaston, G. R., Effects of hypoxia on macrobenthos of the inner shelf off Cameron, Louisiana, *Estuar. Coast. Shelf Sci., 20*, 603-613, 1985.

Gaston, G. R. and K. A. Edds, Long-term study of benthic communities on the continental shelf off Cameron, Louisiana: A review of brine effects and hypoxia, *Gulf Res. Repts., 9*, 57-64, 1994.

Gaston, G. R., B. A Vittor, B. Barrett, and P. S. Wolfe, Benthic communities of Louisiana coastal waters, Draft Technical Bulletin No. 45, Louisiana Department of Wildlife and Fisheries, Marine Fisheries Division, Baton Rouge, Louisiana, 1998.

Harper, D. E., Jr., L. D. McKinney, R. R. Salzer, and R. J. Case, The occurrence of hypoxic bottom water off the upper Texas coast and its effects on the benthic biota, *Contr. Mar. Sci., 24*, 53-79, 1981.

Harper, D. E., Jr., L. D. McKinney, J. M. Nance, and R. R. Salzer, Recovery responses of two benthic assemblages following an acute hypoxic event on the Texas continental shelf, northwestern Gulf of Mexico, in *Modern and Ancient Continental Shelf Anoxia*, edited by R. V. Tyson and T. H. Pearson, pp. 49-64, *Geological Society Special Publ., 58*, 1991.

Hendelberg, M. and P. Jensen, Vertical distribution of the nematode fauna in a coastal sediment influenced by seasonal hypoxia in the bottom water, *Ophelia, 37*, 83-94, 1993.

Holland, A. F., A. T. Shaughnessy, and M. H. Hiegel, Long-term variation in the mesohaline Chesapeake Bay macrobenthos: Spatial and temporal patterns, *Estuaries, 10*, 370-278, 1987.

Josefson, A. B. and B. Widbom, Differential response of benthic macrofauna and meiofauna to hypoxia in the Gullmar Fjord basin, *Mar. Biol., 100*, 31-40, 1988.

Justic', D., N. N. Rabalais, R. E. Turner, and W. J. Wiseman, Jr., Seasonal coupling between riverborne nutrients, net productivity and hypoxia, *Mar. Pollut. Bull., 26*, 184-189, 1993.

Justic', D., N. N. Rabalais, and R. E. Turner, Effects of climate change on hypoxia in coastal waters: A doubled CO_2 scenario for the northern Gulf of Mexico, *Limnol. Oceanogr., 41*, 992-1003, 1996.

Justic', D., N. N. Rabalais, and R. E. Turner, Impacts of climate change on net productivity of coastal waters: Implications for carbon budget and hypoxia, *Climate Res., 8*, 225-237, 1997.

Kamykowski, D. and S. J. Zentara, Hypoxia in the world ocean as recorded in the historical data set, *Deep-Sea Res., 37*, 1861-1874, 1990.

Leming, T. D. and W. E. Stuntz, Zones of coastal hypoxia revealed by satellite scanning have implications for strategic fishing, *Nature, 310*, 136-138, 1984.

Levin, L. A., C. L. Huggett, and K. F. Wishner, Control of deep-sea benthic community structure by oxygen and organic-matter gradients in the eastern Pacific Ocean, *J. Mar. Res., 49*, 763-800, 1991.

Levin, L. A., J. D. Gage, C. Martin, and P. A. Lamont, Macrobenthic community structure within and beneath the oxygen minimum zone, northwest Arabian Sea, *Deep-Sea Res., Part II, 47*, 189-226, 2000.

Levin, L. A. and J. D. Gage, Relationships between oxygen, organic matter and the diversity of bathyal macrofauna, *Deep-Sea Res. II, 45*, 129-163, 1998.

Lohrenz, S. E., G. L. Fahnenstiel, D. G. Redalje, G. A. Lang, M. J. Dagg, T. E. Whitledge, and Q. Dortch, The interplay of nutrients, irradiance and mixing as factors regulating primary production in coastal waters impacted by the Mississippi River plume, *Continental Shelf Res., 19*, 1113-1141, 1999.

Llansó, R. J., Tolerance of low dissolved oxygen and hydrogen sulfide by the polychaete *Streblospio benedicti* (Webster), *J. Exper. Mar. Biol. Ecol., 153*, 165-178, 1991.

Llansó, R. J., Effects of hypoxia on estuarine benthos: the lower Rappahannock River (Chesapeake Bay), a case study, *Estuar. Coast. Shelf Sci., 35*, 491-515, 1992.

Mayfield, S. M., Aspects of the Life History and Reproductive Biology of the Worm *Paraprionospio pinnata*. M. S. thesis, Department of Biology, Texas A&M University, College Station, Texas, 1988.

Mountford, N. K., A. F. Holland, and J. A. Mihursky, Identification and description of macrobenthic communities in the Calvert Cliffs region of the Chesapeake Bay, *Chesapeake Sci., 14*, 160-369, 1977.

Murrell, M. C. and J. W. Fleeger, Meiofauna abundance on the Gulf of Mexico continental shelf affected by hypoxia, *Continental Shelf Res., 9*, 1049-1062, 1989.

Nelsen, T. A., P. Blackwelder, T. Hood, B. McKee, N. Romer, C. Alvarez-Zarikian, and S. Metz, Time-based correlation of biogenic, lithogenic and authigenic sediment components with anthropogenic inputs in the Gulf of Mexico NECOP study area, *Estuaries, 17*, 873-885, 1994.

Nestlerode, J. and R. J. Diaz, Effects of periodic environmental hypoxia on predation of a tethered polychaete, *Glycera americana*: implications for trophic dynamics, *Mar. Ecol. Prog. Ser., 172*, 185-195, 1998.

Niermann, U., E. Bauerfeind, W. Hickel, and H. V. Westernhagen, The recovery of benthos following the impact of low oxygen content in the German Bight, *Netherlands J. Sea Res., 25*, 215-226, 1990.

Pavela, J. S., J. L. Ross, and M. E. Chittenden, Sharp reductions in abundance of fishes and benthic macroinvertebrates in the Gulf of Mexico off Texas associated with hypoxia. *Northeast Gulf Sci., 6*, 167-173, 1983.

Pihl, L., Changes in the diet of demersal fish due to eutrophication-induced hypoxia in the Kattegat, Sweden, *Can. J. Fish. Aquat. Sci., 51*, 321-336, 1994.

Pihl, L., S. P. Baden, R. J. Diaz, and L. C. Schaffner, Hypoxia induced structural changes in the diets of bottom-feeding fish and crustacea, *Mar. Biol., 112*, 349–361, 1992.

Pokryfki, L. and R. E. Randall, Nearshore hypoxia in the bottom water of the northwestern Gulf of Mexico from 1981 to 1984, *Mar. Environmental Res., 22*, 75-90, 1987.

Qureshi, N. A., The role of fecal pellets in the flux of carbon to the sea floor on a river-influenced continental shelf subject to hypoxia, Ph.D. Dissertation, Department of Oceanography & Coastal Sciences, Louisiana State University, Baton Rouge, 1995.

Rabalais, N. N., L. E. Smith, and D. F. Boesch, The effects of hypoxic water on the benthic fauna of the continental shelf off southeastern Louisiana, in Proceedings: Ninth Annual Gulf of Mexico Information Transfer Meeting, OCS Study MMS 89-0060, pp. 147-151, U.S. Dept. of the Interior, Minerals Management Service, Gulf of Mexico OCS Region, New Orleans, Louisiana, 1989.

Rabalais, N. N., R. E. Turner, and Q. Dortch, Louisiana continental shelf sediments: Indicators of riverine influence, in *Nutrient Enhanced Coastal Ocean Productivity Workshop Proceedings*, pp. 77-81, Publ. No. TAMU-SG-92-109, Texas A&M University Sea Grant College Program, College Station, Texas, 1992.

Rabalais, N. N., L. E. Smith, E. B. Overton and A. L. Zoeller, Influence of Hypoxia on the Interpretation of Effects of Petroleum Production Activities, OCS Study/MMS 93-0022,

U.S. Dept. of the Interior, Minerals Management Service, Gulf of Mexico OCS Region, New Orleans, Louisiana, 1993.

Rabalais, N. N., W. J. Wiseman, Jr. and R. E. Turner, Comparison of continuous records of near-bottom dissolved oxygen from the hypoxia zone of Louisiana, *Estuaries, 17*, 850-861, 1994.

Rabalais, N. N., L. E. Smith, D. E. Harper, Jr., and D. Justic', The effects of bottom water hypoxia on benthic communities of the southeastern Louisiana continental shelf, MMS/OCS Study 94-0054, U.S. Minerals Management Service, Gulf of Mexico OCS Region, New Orleans, Louisiana, 1995.

Rabalais, N. N., R. E. Turner, D. Justic, Q. Dortch, W. J. Wiseman, Jr., and B. K. Sen Gupta, Nutrient changes in the Mississippi River and system responses on the adjacent continental shelf, *Estuaries, 19*, 386-407, 1996.

Rabalais, N. N., R. E. Turner, D. Justic', Q. Dortch, and W. J. Wiseman, Jr., Characterization of hypoxia: Topic 1 Report for the Integrated Assessment of Hypoxia in the Gulf of Mexico. NOAA Coastal Ocean Program Decision Analysis Series No. 16, NOAA Coastal Ocean Program, Silver Springs, Maryland, 1999.

Renaud, M. L., Annotated Bibliography on Hypoxia and Its Effects on Marine Life, with Emphasis on the Gulf of Mexico, NOAA Technical Report NMFS 21, U.S. Dept. of Commerce, National Oceanic and Atmospheric Administration, National Marine Fisheries Service, Washington, D.C., 1985.

Renaud, M. L., Hypoxia in Louisiana coastal waters during 1983: implications for fisheries, *Fishery Bull., 84*, 19-26, 1986.

Rohlf, F.J., Numerical Taxonomy and Multivariate Analysis System, NTSYS-pc, version 1.80, Applied Biostatistics Inc., New York, 1993.

Rosenberg, R., Eutrophication—The future marine coastal nuisance?, *Mar. Pollut. Bull.,16*, 227-231, 1985.

Rosenberg, R., Negative oxygen trends in Swedish coastal bottom waters, *Mar. Pollut. Bull., 21*, 335-339, 1990.

SAS Institute Inc., *SAS Guide for Statistics*, Version 5 Edition, SAS Institute Inc., Cary, North Carolina, 1985.

Sen Gupta, B. K., R. E. Turner, and N. N. Rabalais, Seasonal oxygen depletion in continental-shelf waters of Louisiana: Historical record of benthic foraminifers, *Geology, 24*, 227-230, 1996.

Smith, C. R., L. A. Levin, D. J. Hoover, G. McMurtry, and J. D. Gage, Variations in bioturbation across the oxygen minimum zone in the northwest Arabian Sea, *Deep-Sea Res, Part II, 47*, 227-257, 2000.

Tolmazin, R., Changing coastal oceanography of the Black Sea. I. Northwestern shelf, *Progress Oceanogr., 15*, 2127-276, 1985.

Turner, R. E. and N. N. Rabalais, Coastal eutrophication near the Mississippi river delta, *Nature, 368*, 619-621, 1994.

Vittor and Associates, LOOP 14-yr Monitoring Program, Synthesis Report to LOOP, Inc. and Louis. Dept. Wildlife & Fisheries, Mobile, Alabama, 1998.

13

Effects of Low Dissolved Oxygen on the Behavior, Ecology and Harvest of Fishes: A Comparison of the Chesapeake Bay and Baltic-Kattegat Systems

Denise L. Breitburg, Leif Pihl, and Sarah E. Kolesar

Abstract

Fish are exposed to low dissolved oxygen in a wide variety of coastal environments ranging from estuaries to fjords. In this chapter, we compare the effects of low dissolved oxygen on fishes of the Chesapeake Bay and Baltic-Kattegat systems. These estuaries are probably the best studied coastal systems for determining effects of low dissolved oxygen on living resources and should provide a basis for predicting effects of low oxygen concentrations on Gulf of Mexico fishes. There are a number of differences between the systems. For example, temperature is lower in the Baltic-Kattegat area, and species that occur there appear to require higher oxygen concentrations for survival. Nevertheless, similarities emerge from the comparison. Because oxygen concentrations are generally not uniform within the water column of these bodies of water, the behavioral responses of mobile animals, as well as physiological tolerances, ultimately determine effects on individuals, populations and multispecies assemblages of coastal fishes. In addition to mortality directly resulting from exposure to low dissolved oxygen concentrations, oxygen depletion likely has important effects on the food web of both systems by altering distributions (and therefore, encounter rates between predators and prey), predator feeding rates, prey vulnerability and growth rates (and thus, size-dependent trophic interactions). In both systems, variation in behavioral responses and physiological tolerances among species are important in determining the effects of hypoxia. In addition, the combination of low oxygen and other natural and anthropogenic stressors can increase mortality and ecological consequences of low oxygen to fish assemblages. Current evidence indicates that effects of

Coastal Hypoxia: Consequences for Living Resources and Ecosystems
Coastal and Estuarine Studies, Pages 241-268
Copyright 2001 by the American Geophysical Union

low oxygen on fishery harvests are more severe in the Baltic-Kattegat than in the Chesapeake.

Introduction

As organisms dependent on aerobic respiration, fish are susceptible to the physiological problems caused by exposure to low dissolved oxygen (hypoxia) or anoxia. As mobile animals with capabilities for complex behavioral responses, however, their vertical movement within the water column as well as lateral migration among habitats, has the potential to modify their exposure and the consequences of otherwise physiologically stressful or lethal environmental conditions. The variation among species and life stages in both physiological tolerances and behavioral capabilities, as well as the variation in tolerances and responses of their predators and prey with which they interact, make predictions of the ultimate effects of low oxygen on fishes in a spatially heterogeneous environment difficult. Nevertheless, the economic and ecological importance of finfish in coastal areas and other aquatic habitats susceptible to seasonal oxygen depletion makes it important that we understand both the responses to low oxygen and the ultimate consequences at both the population and ecosystem levels.

In this review, we examine the behavioral, ecological and fisheries consequences of low dissolved oxygen on fish assemblages. We focus on the North American Chesapeake Bay and European Baltic-Kattegat systems, including their tributaries and neighboring bodies of water. These two regions are probably the coastal systems that have been studied most extensively to determine how low oxygen affects fish as well as other organisms. In both cases, oxygen depletion is a consequence of strong density stratification and increases in anthropogenic loadings of nutrients [Rosenberg et al., 1990]. We believe, however, that the differences between the systems in species composition, the specific focus of historical and current research programs, and the physical nature of the systems make such a comparison useful for increasing our general understanding of low oxygen effects on coastal fishes. We first review the effects and responses of fishes to low oxygen by developmental stage because both tolerances and the capabilities for behavioral response change with ontogeny. We then summarize the consequences of low oxygen to fish at the population level, the ways that low oxygen affects trophic interactions, and the potential for effects on commercial and recreational fisheries. Finally we discuss key similarities and differences in the occurrence and likely consequences of low dissolved oxygen among the Chesapeake Bay, Baltic-Kattegat and Gulf of Mexico systems.

The Physical Systems

Chesapeake Bay

The Chesapeake Bay, located on the mid-Atlantic coast of the United States is the largest semi-enclosed estuary in North America, measuring 320 km in length and 48 km in width at its widest point (Fig. 1A). The Chesapeake Bay and its tributaries are generally

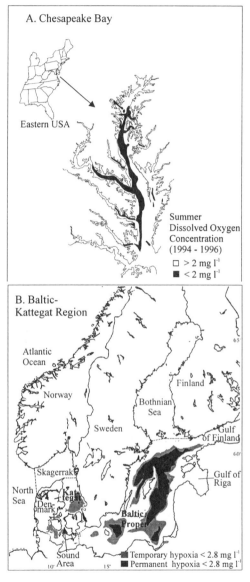

Figure 1. Maps showing the distribution of low dissolved oxygen concentrations in bottom waters of (A) the Chesapeake Bay and (B) the Baltic-Kattegat systems. The Baltic-Kattegat system lies between the Skagerrak and the Bothnian Sea. The Chesapeake Bay map indicates summer dissolved oxygen concentrations in bottom waters of the mainstem Chesapeake Bay and tidal tributaries (redrawn from USEPA 1999). The difference in dissolved oxygen concentrations plotted to represent areas of hypoxia reflects the tendency for US researchers to record field measurements as mg l^{-1} and European scientists to use percent saturation; Baltic-Kattegat data were converted to approximate concentrations in mg l^{-1}.

characterized by their broad shallow flanks and deep central channel. Because of this profile, only about 40% of the system lies below the pycnocline when the water column is strongly stratified during summer. Summer surface temperatures reach 28-30 °C, while bottom waters in mid-summer are a few degrees cooler, generally reaching only 24-26 °C.

The population in the Chesapeake Bay region has grown enormously during the last century, with a doubling of the 1950 population in the watershed expected by the year 2020 [USEPA, 1999]. In 1900, less than 5 million people lived in the Chesapeake Bay basin; by 1990 that number had risen to almost 15 million, and the projected population in 2020 is almost 18 million people. Between 1960 and 1990, some areas surrounding the Bay had increased in population by as much as 500%.

Conversion of forested land to agricultural, residential and other uses has increased nutrient loading to the Bay, leading to eutrophication. Agriculture causes the largest inputs of nitrogen and phosphorus into Chesapeake Bay, contributing 42% of the nitrogen load and 51% of the phosphorus load. Other nonpoint sources also contribute significant nutrient burdens: atmospheric deposition directly to Chesapeake Bay and tributary waters accounts for 7% of the nitrogen and 9% of the phosphorus loadings, urban development and septic systems add about 14% of nitrogen and 10% of phosphorus, and forests contribute 16 and 5% of the nitrogen and phosphorus totals, respectively (atmospheric deposition to land is included in the nonpoint source percentages above). Point sources add 22% of the nitrogen and 25% of the phosphorus loadings to the Chesapeake Bay estuary [data are for 1996: USEPA, 1999].

The average flow into Chesapeake Bay from all its tributaries is just under 2265 m^3 s^{-1} and over 50% of the freshwater flow to Chesapeake Bay comes from the Susquehanna River, which drains portions of Pennsylvania, New York and Maryland [USEPA, 1999]. Streamflow is lowest from July through October and peaks during the spring freshet, which occurs in March or April. The average hydraulic residence time of surface waters originating at the head of the Chesapeake Bay is 50 d [Thomann et al., 1994].

Oxygen depletion is largely restricted to subpycnocline waters and occurs during late spring through early fall. The most severe oxygen depletion occurs in subpycnocline waters of the mesohaline portions of the Chesapeake Bay and its tributaries (surface salinities of 10 to 18 psu and bottom salinities generally ranging from 12 to 24 psu during summer). Bottom waters near the mouth generally have oxygen concentrations above 3.5 to 4.0 mg l^{-1}, which equals approximately 50% saturation, as a result of extensive exchange with the generally well-oxygenated waters of the Atlantic coast. Shallow-water oxygen depletion can occur sporadically in some areas as wind and tidal actions advect bottom water near to shore [Breitburg, 1990; Sanford et al., 1990]. Low oxygen concentrations have also been recorded in some shallow vegetated and soft bottom locations during calm weather and strong stratification, and within the surface layer as large dinoflagellate blooms die off. Extensive oxygen depletion associated with macrophytes or algal mats, however, is not a persistent or widespread feature of the system.

The Baltic-Kattegat system

The Baltic-Kattegat system is a transition area between the almost fresh water Bothnian Sea (2-3 psu) and the marine North Sea (Fig. 1B). Surface salinity varies between 7 and 13 psu in the Baltic and between 15 and 30 psu in the Kattegat. The Baltic-Kattegat area is characterized by a strong stratification of the water with a halocline at 60 to 80 m depth in the Baltic and at 10 to 20 m in the Kattegat. Bottom water salinity is 10 to 20 psu in the Baltic and 32 to 34 psu in the Kattegat. Winter surface water temperature is close to zero and ice normally occurs for some time. During summer, temperatures in the surface water reach 15 to 20 °C, and the water stratification is further strengthened by a thermocline.

The Baltic Sea has a varying bathymetry with deep trenches in the central portion extending to depths of 200 to 400 m, and a sill depth of around 15 m in the Sound area that separates the Baltic and Kattegat. Because of the sill, the water exchange between the Baltic and Kattegat is mainly limited to an outflow of Baltic surface water. Renewal of Baltic bottom water is irregular and on a time scale of decades. The Kattegat is shallow and flat with a mean depth of 23 m. Both surface and bottom water exchange annually with the Skagerrak and the North Sea, especially during winter when the water stratification is less pronounced. The tidal amplitude is low (< 20 cm) in the Baltic-Kattegat area which further limits the mixing of water masses in the region.

The external annual supplies of nitrogen and phosphorus to the Baltic-Kattegat area from rivers, point sources and atmospheric deposition, have been estimated at 800,000 and 40,000 metric tons, respectively [Rosenberg et al., 1990]. Of this, 90% is supplied to the Baltic and 10% directly to the Kattegat. The nutrient inputs to the Baltic were much lower at the turn of the century, and the present supply has been estimated to be higher by a factor of four and eight for nitrogen and phosphorus, respectively [Larsson et al., 1985]. Nitrogen loadings to the Kattegat have increased four-fold from 1930 to 1980 [Aertebjerg, 1985] and have leveled out since then. Phosphorus supply to the Kattegat doubled from 1950 to 1970 [Andersson and Rydberg, 1988], but has declined since then due to the introduction of chemical precipitation in sewage treatment plants in the 1970s. About 50% of the present nitrogen supply to the Baltic is from atmospheric input, including nitrogen fixation by algae, whereas 90% of phosphorus is from land runoff and point sources [Larsson et al., 1985]. In the Kattegat, atmospheric deposition is less important, and external nutrient supply is dominated by river runoff.

Due to its enclosed geographical position with restricted exchange of bottom water, the Baltic has year-round hypoxia (< 2 mg l^{-1}) in bottom waters below approximately 80 m. The bottom-water oxygen concentration fluctuates somewhat over time and space, however, depending primarily on the amount and salinity of incoming deep water from the Kattegat. Bottom water in the deepest parts of the Baltic Sea can remain anoxic for several years at a time. Surface water is generally well oxygenated, and there is a sharp decline in oxygen concentration around the halocline. In the Kattegat, the water mass is well mixed during the winter, with close to 100% saturation of dissolved oxygen in the bottom water. As in the Chesapeake Bay, freshwater inflow and warming of the surface water in the spring create a strong pycnocline in the Kattegat, keeping the water mass stratified until autumn. In the deeper (30 to 60 m) eastern part of the Kattegat, oxygen concentration in the bottom water below the pycnocline declines progressively over the season, and hypoxia usually begins in August or September. Monitoring of oxygen during the autumn from 1984 to 1991 indicated that the duration of hypoxia < 2 mg l^{-1} in the bottom water varied 3-11 weeks [Baden et al., 1990a; Baden and Pihl, 1996]. The zone of hypoxic water in the Kattegat extends from the bottom to 1-10 m above the sediment. Hypoxia also frequently occurs in shallow water in the Baltic-Kattegat area. These events often take place in enclosed coastal areas where water is stagnant, or in connection with the development of green algal mats that reduce water movement and enhance the input of organic matter to the system [Pihl et al., 1999].

Effects of Low Dissolved Oxygen on Chesapeake and Baltic-Kattegat Fishes

Tolerance to Low Oxygen Exposure

Although we do not focus on physiological responses in this paper, it is important to discuss behavioral responses in the context of physiological risk. In the case of low dissolved oxygen, death is the ultimate consequence for fish and other organisms dependent on aerobic respiration if exposure is sufficiently severe and long-lasting. The tolerance of fish to low dissolved oxygen exposure can be expected to vary with life stage, habitat, typical activity level and the presence of other stressors, as well as with fundamental physiological differences among species. Fish are metabolic regulators and, even the early life stages, can respire at rates independent of most ambient oxygen levels except when oxygen concentration is very low. At this low level, called the critical oxygen tension, metabolism varies with the level of oxygen available as well as temperature, activity level, water flow velocity and developmental stage [Rombough, 1988].

Adults and juveniles of most Chesapeake Bay species that have been tested have 24-h LC_{50} values near 1 mg l^{-1} (i.e., approximately 13% saturation at 25 °C and 18 psu). Acute toxicity tests have yielded 50% mortality rates with 24-h exposures at 0.5 - 1.0 mg l^{-1} for species such as hogchoker (*Trinectes maculatus*), northern sea robin (*Prionotus carolinus*), spot (*Leiostomus xanthurus*; but LC_{50} reported as > 1 mg l^{-1} by Pihl et al., 1991), tautog (*Tautoga onitis*), windowpane flounder (*Scopthalmus aquosus*) and fourspine stickleback (*Apeltes quadracus*), and 50% mortality rates between 1.1 and 1.6 mg l^{-1} for Atlantic menhaden (*Brevoortia tyrannus*), scup (*Stenotomus chrysops*), summer flounder (*Paralichthys dentatus*), pipefish (*Syngnathus fuscus*) and striped bass (*Morone saxatilis*) [Pihl et al., 1991; Poucher and Coiro, 1997; Thursby, 1999]. Thus for nearly all species tested, the range of tolerances is quite low; only a 1.0 mg l^{-1} difference separates the most and least sensitive species described above. Some species, however, may be far more sensitive, especially at the high summer temperatures that occur in Chesapeake Bay. For example 10-d tests of juvenile Atlantic sturgeon (*Acipenser oxyrhincus*) yielded 75-88% survival at 3 mg l^{-1} at 19 °C, but 0% survival at the same dissolved oxygen concentration at 26 °C [Secor and Gunderson, 1998].

Although fewer species have been tested during the larval stage, larvae of species that occur in Chesapeake Bay appear to be somewhat more sensitive to low oxygen exposure than are most adults and juveniles. For example, 50% mortality with 24-h exposure occurs between 1.0 and 1.5 mg l^{-1} for skilletfish (*Gobiesox strumosus*), naked goby (*Gobiosoma bosc*) and inland silverside (*Menidia beryllina*) larvae, while 50% mortality occurs at 1.8 to 2.5 mg l^{-1} for larval red drum (*Sciaenops ocellatus*), bay anchovy (*Anchoa mitchilli*), striped blenny (*Chasmodes bosquianus*) and striped bass [Saksena and Joseph, 1972; Breitburg, 1994; Poucher and Coiro, 1997]. Field and laboratory observations indicate that lethal dissolved oxygen concentrations for skilletfish, naked goby and striped blenny adults are ≤ 1.0 mg l^{-1} [Breitburg, unpubl.].

Embryo tolerances vary inconsistently in relation to tolerances of later stages; 50% mortality in 12-96 h occurs at a higher dissolved oxygen concentration than that for larval mortality for bay anchovy (2.8 mg l^{-1}), at a similar oxygen concentration as for larvae for inland silverside (1.25 mg l^{-1}), and at lower concentrations than that leading to larval mortality for winter flounder (*Pleuronectes americanus*; 0.7 mg l^{-1}) and naked goby (approximately 0.6 mg l^{-1}) [Chesney and Houde, 1989; Breitburg, 1992; Poucher and Coiro, 1997].

Baltic-Kattegat species tested tend to require somewhat higher dissolved oxygen concentrations than do Chesapeake Bay species except at very low water temperatures. Lethal dissolved oxygen concentrations for 80-200 g cod (*Gadus morhua*) increased with increasing temperature, ranging from 0.6 mg l^{-1} (5% saturation at 5 °C) to 2.3 mg l^{-1} (29% saturation at 17 °C) in experiments in which dissolved oxygen concentrations were decreased at a rate of 0.5% saturation min^{-1} [Schurmann and Steffensen, 1992]. Thus, fish were less tolerant of low oxygen concentrations at higher temperatures. These experiments also highlight another aspect of the interaction between temperature and dissolved oxygen stress; behavioral responses as well as tolerances can depend on the combined effects of these two physical factors. Cod preferred lower water temperatures at low dissolved oxygen concentrations than at high dissolved oxygen concentrations. The 24-h LC_{50} for the sand goby, *Pomatoschistus minutus* (Pallas), a shallow water benthic species that is often exposed to hypoxia, is also higher than that for the common Chesapeake Bay naked goby; 50% mortality of the sand goby occurred at approximately 1.3 mg l^{-1} or 15.2% saturation at 15 °C and 20 psu [Petersen and Petersen 1990]. Cessation of gill ventilation or LC_{50} values for other Baltic-Kattegat teleost fishes summarized in Magnusson et al. [1998] are (1.0 mg l^{-1}) 11% saturation at 10 °C for saithe *(Pollachius virens)*, 1.8 mg l^{-1} (19% saturation) at 8 °C for plaice (*Pleuronectes platessa*) and 2.0 mg l^{-1} (21% saturation) at 8 °C for dab (*Limanda limanda*).

Experiments with larvae of Baltic-Kattegat species indicate that oxygen tolerance varies with larval development, but that the relationship between development and tolerance varies among species [DeSilva and Tytler, 1973]. The 12-h LC_{50} of herring *(Clupea harengus)* larvae is 2.7 mg l^{-1} at the yolk-sac stage, 5.2 mg l^{-1} after feeding for 5-6 weeks and 3.2 mg l^{-1} for newly metamorphosed herring about 2.5 months posthatch at 13 °C. In contrast, the 12-h LC_{50} of plaice larvae is 3.9 mg l^{-1} at the yolk-sac stage and then steadily decreases to 2.4 mg l^{-1} for 3-4 week postmetamorphosis individuals (2.5 months posthatch) at 13 °C. For both species, the portion of the larval stage most resistant to low oxygen is the one most closely linked to the benthic habitat. Herring lay demersal eggs while plaice eggs are pelagic, but plaice eventually settle and metamorphose into epibenthic juveniles. Experiments by Nissling [1994] indicate that cod larvae will not successfully develop in oxygen conditions less than 3 mg l^{-1}; the 48-h LC_{50} was found to be between 2.5 and 3.0 mg l^{-1} at 7 °C.

Oxygen requirements of Baltic cod eggs are similar to those of cod larvae. Cod eggs do not complete development at oxygen concentrations less than 4.3 mg l^{-1} [Ohldag et al., 1991], and short-term egg survival at 3 mg l^{-1} is only about one-third that at 11 mg l^{-1} [Wieland et al., 1994]. There is also an apparent interaction between oxygen and salinity stress. At oxygen concentrations $\leq$ 2.8 mg l^{-1} at 7 °C, egg survival at 15 psu is roughly twice that at 11 psu; at 100% oxygen saturation there is no difference in survival at 15 and 11 psu [Nissling, 1994]. Kosior and Netzel [1989] predicted a limit for successful spawning of 2.9 mg l^{-1}, based on an evaluation of hydrographic data.

Low Dissolved Oxygen as a Direct Source of Mortality in the Field

Reports of mortality of adult and juvenile fishes directly related to exposure to hypoxia or anoxia in Chesapeake Bay are generally restricted to intrusions of hypoxic water into nearshore habitats [e.g., Breitburg, 1992] and cases where large schools of fish move into small embayments or creeks with restricted circulation. Rapid intrusions of hypoxic or anoxic subpycnocline waters can occur in mesohaline areas of the mainstem Chesapeake Bay (particularly along the western shore of the Maryland portion of the Bay when surface waters are driven by winds toward the eastern shore) and in some tributaries. During these intrusions, oxygen concentrations in nearshore oyster reefs and soft-bottom habitats can decline by 2-3 mg l^{-1} h^{-1} and remain at dissolved oxygen concentrations < 1 mg l^{-1} for up to 14 h [Breitburg, 1990, 1992]. Adults, juveniles and benthic eggs of oyster reef fishes suffer heavy mortality on reefs where dissolved oxygen concentrations reach ≤ 0.6 mg l^{-1}. High mortality of young naked gobies (especially recruits ≤ 2 weeks postsettlement) during intrusions can influence the size distribution of the population in fall samples. Average densities of oyster reef fishes (gobies, blennies, clingfish and toadfish) can decline from as much as 50 ind. m^{-2} to 0 as a result of the combination of mortality and migration in areas that experience dissolved oxygen concentrations ≤ 0.6 mg l^{-1} during severe intrusions.

The extent of mortality caused by episodic hypoxia or anoxia can be severely underestimated in Chesapeake Bay oyster reefs as well as elsewhere. Millions of fish likely die during some episodes, such as those that occurred in the Flag Ponds oyster reef described above. It would have been impossible to detect mortality, however, if one of us (Breitburg) had not coincidentally scheduled a dive and made visual observations towards the end of the intrusion period. During that dive, oyster reef substrate was strewn with the carcasses of dead fish and crabs. Most mortality was of benthic and demersal species, but pelagic species such as bay anchovy and Atlantic menhaden also suffered mortality. Egg masses were abandoned and dead. Within 24 h, however, all signs of the massive mortality were gone except for the reduced densities of organisms in deep areas of the oyster reef. Crabs and highly mobile fishes rapidly recolonized the area and likely scavenged the carcasses of the dead finfish, crustaceans and macrobenthos.

In contrast to the easily overlooked effects of intrusions on oyster reefs, nearshore mortality of pelagic fishes can sometimes be more obvious because the dead fish accumulate at the surface and on beaches. Hypoxia-induced mortality of pelagic schooling fishes is often associated with localized oxygen depletion where large numbers of fish remain in embayments or creeks with limited circulation. In these areas, respiration by fish, algae and other organisms can reduce oxygen concentrations to lethal levels.

As in the Chesapeake Bay, direct low oxygen induced mortality of adults and juveniles in the Baltic-Kattegat system can be difficult to detect. The presence of dead fish in bottom trawls and the lack of a rapid rebound in cod populations when dissolved oxygen concentrations return to higher levels, however, indicate that demersal fishes suffer mortality due to exposure to hypoxia in the southeastern Kattegat, where hypoxia occurs during a 3-11 week period in autumn [Baden et al., 1990a; Baden and Pihl, 1996]. During a period of severe hypoxia in September 1988, demersal fish were almost absent in an approximately 3000 km^2 area of the southeastern Kattegat, and the fish community was not fully recovered in December of the same year. Analysis of stomach contents of five dominant demersal fish species during hypoxic events indicated that death was not due to starvation [Pihl, 1994].

Hypoxia in the southeastern Kattegat may also increase fishery-related mortality. Demersal fishes apparently die from exposure to hypoxia while caught in nets that trap fish as they attempt to emigrate from hypoxic areas [Baden et al., 1990a]. Similar mortality of fish trapped in nets sometimes occurs at other seasons in the Kattegat when oxygen minimum layers form and shift upwards [Kruse and Rasmussen, 1995], as well as in areas of the Baltic proper where winds cause upwelling of hypoxic bottom water [Weigelt and Rumohr, 1986].

Unlike older life stages, fish eggs have no behavioral escape from waters with low dissolved oxygen. Sinking of pelagic eggs into bottom waters where oxygen concentrations are lethal to developing embryos may be a major source of mortality for some species. The extent of such mortality will depend strongly on spawning behavior of adults (i.e., the depth, geographic location and dissolved oxygen concentration at which eggs are released), as well as the sinking rate and stage duration of eggs.

In Chesapeake Bay, the bay anchovy is the most abundant species that spawns pelagic eggs during summer in portions of the Bay with a hypoxic or anoxic bottom layer. In the mesohaline Patuxent River, a subestuary of Chesapeake Bay, sampling within the bottom, pycnocline and surface layers indicated that bay anchovy eggs were randomly distributed throughout the water column regardless of bottom dissolved oxygen concentration (Keister, 1996). Presumably, eggs that remain at lethal oxygen concentrations will die. What is less certain is the fate of eggs that are at or near lethal dissolved oxygen concentrations at the time of hatch but that have not remained there for a sufficient duration to kill the developing embryos. For example, what happens to eggs that sink into a severely hypoxic bottom layer at hour 18 of a typical 20-h development time? Under these conditions, hatch may be delayed (perhaps ultimately leading to mortality) or the newly hatched larva may be unable to swim up to the surface layer. Data collected by E. North and E. Houde [unpubl.] also suggest the potential for variation among sites and species in the proportion of pelagic eggs that occur in hypoxic bottom waters. The vast majority of bay anchovy eggs collected in samples at a mainstem Chesapeake Bay site with a thicker surface layer and stronger density stratification than that found in the Patuxent were in the well-oxygenated surface layer; few eggs were found below the pycnocline. In contrast, the majority of sciaenid eggs at the same site were in the pycnocline and hypoxic bottom layer.

Considerable attention has been focused on the issue of eggs sinking into hypoxic bottom water in the Baltic-Kattegat system because of the particular vulnerability and fisheries importance of cod. In the Baltic Sea there is a limited vertical range of habitat suitable for development of cod eggs, bounded on the top by low salinity and on the bottom by low oxygen. As a result, oxygen depletion, and environmental factors such as salinity and wind-induced turbulence that influence the sinking of eggs into water layers with lethal oxygen concentrations during their approximately 12-14 day development, may be some of the most important factors limiting reproductive success and influencing variation in year-class strength of Baltic cod in the central Baltic [Wieland and Zuzarte, 1991; Wieland et al., 1991, 1994; Plikshs et al., 1993].

The specific gravity of cod eggs is an important determinant of the depth at hatching and the subsequent environmental conditions facing newly hatched cod larvae [Nissling et al., 1994; Nissling and Vallin, 1996] because it makes the eggs neutrally buoyant at salinities that occur only in the deepest basins of the Baltic below the halocline (50-80 m). Cod spawning occurs in three deep basins, the Bornholm, the Gdansk and the Gotland basins [Wieland et al., 1994]. The stratified nature of the Baltic Sea, however, is such that these deep basins experience extensive areas of low oxygen that are unsuitable for egg survival and development. Field surveys of the Baltic from 1987 to 1990 indicated that most of the Baltic cod eggs (50 - 95%) were below the lower oxygen limit required for embryo survival and successful hatching [Ohldag et al., 1991; Wieland et al., 1994].

Salinity distributions indicate that often only the Bornholm can now support viable cod eggs because most or all of the cod eggs in the other basins sink into hypoxic waters [Nissling et al., 1994]. The thickness of the spawning layer for cod (where salinity is > 11 psu and oxygen concentration is > 1.4 mg l^{-1}) decreased significantly between 1977 and 1987 in the three main spawning areas in the Baltic. In the largest area, the Bornholm Basin, the spawning layer was reduced from 36 to 16 m thick between 1977 and 1987, whereas in the Gdansk and Gotland basins the vertical extent of potential spawning area decreased from 150 and 63 m, respectively, to zero in both areas [Nissling et al., 1994]. In addition to variation due to geographic and interannual variation in salinity, the proportion of eggs sinking into hypoxic layers will depend on variation among individuals and clutches in chorion thickness, yolk osmolality, individual female, batch number, egg quality and egg size [Nissling et al., 1994; Nissling and Valin, 1996].

The potential for strong effects of bottom layer oxygen depletion on planktonic eggs of Baltic fishes other than cod varies among species. Sprat eggs in the Bornholm Basin sometimes occur at highest densities within the pycnocline (at 2.2 mg l^{-1}), but more often are primarily found in the surface layer at dissolved oxygen concentrations $\geq$ 5.7 mg l^{-1} [Wieland and Zuzarte, 1991].

In both the Chesapeake and Baltic systems, benthic eggs are susceptible to low dissolved oxygen concentrations that occur prior to hatch. Adults presumably do not deposit eggs in areas that are at lethal oxygen concentrations at the time of spawning. Episodic hypoxia that occurs either as a result of movement of a hypoxic water mass or during calm periods in areas with high macrophyte or macroalgal production, however, are potentially important sources of mortality for benthic fish eggs. In shallow areas of Chesapeake Bay, intrusions of water with oxygen concentrations $\leq$ approximately 0.6 mg l^{-1} lead to mortality of embryos of naked goby and other oyster reef fish both because of direct exposure to low dissolved oxygen concentrations and because males abandon the nests they are guarding, thus exposing eggs to predation as higher oxygen concentrations and predators return [Breitburg, 1992]. In the Baltic, herring reproduce along the coasts and spawn on littoral vegetation. In areas with dense vegetation, low oxygen concentrations may develop for short periods of time, especially during night, and, together with exudates released from the algae, can decrease survival of herring eggs [Aneer, 1987].

Effects of Low Oxygen on the Distribution of Larval, Juvenile and Adult Fishes

Low dissolved oxygen can affect both the vertical distribution of fish within the water column and the lateral distribution within an estuarine system through behavioral habitat selection. Both laboratory and field data suggest that behavioral avoidance of unsuitable oxygen concentrations plays a large role in determining vertical distributions of larvae and older life-history stages, as well as the horizontal distribution of juvenile and adult fishes. Although eggs obviously cannot, by themselves, move in response to low dissolved oxygen, a tendency for adults to avoid spawning in areas with low dissolved oxygen concentrations likely has the same effect.

Field samples collected in the Patuxent River, a tributary of the Chesapeake Bay, indicated that the vertical distributions of larvae of the two most abundant summer spawners (naked goby and bay anchovy) vary with bottom dissolved oxygen concentration [Kesiter et al., in press]. Naked goby larvae were absent or occurred in extremely low densities in samples taken in the bottom layer of the water column at dissolved oxygen concentrations < 2 mg l^{-1}; bay anchovy were absent or scarce in bottom samples taken at ≤ 3 mg l^{-1}. Maximum densities of both naked goby and bay anchovy larvae occurred higher in the water column when bottom oxygen concentrations were low than when they were high.

Behavioral experiments using naked goby and bay anchovy, conducted in aquaria that maintain a stratified water column, indicated that even newly hatched larvae can avoid hypoxic bottom-layer water by moving vertically in the water column [Breitburg, 1994]. In addition, there was a very close correspondence between the behavior observed in these experiments and the vertical distribution of larvae in the field (Fig. 2). Although the possibility of larval mortality in the field due to direct exposure to bottom-layer hypoxia cannot be ruled out, these laboratory experiments and the match between laboratory and field data suggest that behavioral avoidance of bottom-layer hypoxia plays a major role in determining the vertical distribution of larvae in the field. Comparison of data on vertical distributions and mortality indicates that these larvae avoid oxygen concentrations that are somewhat higher than those that cause 50% mortality within 24 h (Fig. 2). Similarly, postlarvae of southern flounder (*Paralichthys lethostigma*) show some avoidance at dissolved oxygen concentrations of 3.7 mg l^{-1} and lower [Deubler and Posner, 1963]. To our knowledge, oxygen tolerance tests have not been done for this species, but the 72-h LC_{50} for newly metamorphosed individuals of its congener, the summer flounder, *Paralichthys dentatus*, is 1.6 mg l^{-1} [Poucher and Coiro, 1997], an oxygen concentration that is considerably lower than that at which avoidance began.

No studies have directly examined behavioral responses of larvae of Baltic species relative to dissolved oxygen concentrations they encounter in the field. It is therefore difficult to determine the extent to which field distributions of larvae are determined by behavioral responses to hypoxia as compared to responses to the distributions of their predators and prey, or the extent to which vertical distributions reflect mortality as well as avoidance. Both cod and sprat larvae have broad depth ranges and can, at times, have highest abundances within the Bornholm Basin in deep hypoxic layers. At other times, larvae of both species can be either distributed throughout the water column, or most abundant above the pycnocline [Wieland and Zuzarte, 1991]. Examination of larval stages from tows indicates that larval cod migrate from the deeper portions of the water column where they hatch to the upper photic zone in order to avoid low oxygen and to feed [Grønkjær and Wieland, 1997]. Sampling within the Bornholm Basin indicates that yolk sac cod larvae are found in the lower water column, but feeding larvae generally congregate above the halocline near zooplankton aggregations [Grønkjær and Wieland, 1997].

Juvenile and adult fish are more capable than larvae of altering their location within an estuarine system in response to unfavorable conditions by both vertical and lateral movement in response to low dissolved oxygen. In nearshore areas of Chesapeake Bay when intrusions bring severely hypoxic bottom water near the surface, many species ranging from striped bass to epibenthic species lacking swim bladders can be seen at the surface employing aquatic surface respiration. In addition, there is a shoreward movement of crabs and demersal fishes, similar to that described for Mobile Bay [Loesch, 1960]. Oyster reef fishes migrate to reefs and other structured habitat shallower and nearer to shore, and then

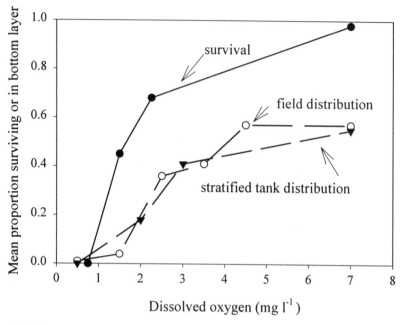

Figure 2. Similarity in laboratory behavior and the vertical distribution of naked goby larvae in the Patuxent River. Laboratory results are of postflexion larvae tested in stratified tanks in aquaria. Data shown are the percent of individuals located below the pycnocline 1 h after larvae were introduced into tanks. Note close correspondence between laboratory and field distributions and that reduced abundances in bottom water layers occurs at higher dissolved oxygen concentrations than those leading to high risk of mortality during 24-h laboratory exposures. [Data are redrawn from Breitburg, 1994 (laboratory distributions and survival) and Keister et al., in press (field distribution)].

generally migrate back to deeper reefs after bottom oxygen concentrations return to more typical levels [Breitburg, 1992]. As described previously, these movements do not totally prevent mortality; large numbers of dead fish are found on the oyster reefs at the end of a severe intrusion event.

Data from bottom trawls conducted on the western shore of the Maryland portion of Chesapeake Bay (near Camp Canoy, Maryland, above the mouth of the Patuxent River) [Breitburg and Kolesar, unpubl.] and in the York River (a western shore tributary in the Virginia portion of Chesapeake Bay) [Pihl et al., 1991] provide evidence for variation among species in bottom dissolved oxygen concentrations that are avoided. In general, decreased abundances and frequency of occurrence, indicating possible avoidance of hypoxic bottom waters, occur at dissolved oxygen concentrations 0.5-2 mg l^{-1} higher than LC$_{50}$s for those species for which tolerance tests have been done. Both the Maryland and Virginia trawl surveys found reduced abundances of spot at bottom dissolved oxygen concentrations less than approximately 1.5 mg l^{-1}, and reduced hogchoker densities at bottom oxygen concentrations < 1 mg l^{-1}. Responses by Atlantic croaker (*Micropogonias undulatus*) varied between the studies; reduced densities at deep stations were found at approximately 3 mg l^{-1} in Virginia and a dissolved oxygen concentrations < 5 mg l^{-1} in Maryland. At the

Maryland site, average number of species collected per trawl sample increased with increasing bottom dissolved oxygen from a low of 1.1 species per trawl at 0.0-0.5 mg l^{-1} to a high of nearly 3.5 species per trawl at 6 mg l^{-1} (Fig. 3).

Changes in movement, distribution and abundance of juvenile and adult fishes in response to low oxygen concentrations within the Baltic-Kattegat system are evident both from changes in catch in commercial nets and from results of research cruises. For example, during the days prior to the peak of the April 1988 upward shift of the pycnocline in the Kattegat, fishermen in the area reported increased capture of young sole, plaice and dab in fishing nets, presumably reflecting movement out of the hypoxic area by these species [Kruse and Rasmussen, 1995]. During periods of hypoxia, however, abundances tend to be reduced. Bottom-trawl surveys from 1984-1991 in the southeastern Kattegat indicated that hypoxic events reduce both the number of fish species (Fig. 3) and overall biomass (from 17-22 species in the spring to 9-20 in the autumn, and from about 200 kg h $^{-1}$ to 50 kg h $^{-1}$) [Baden et al., 1990a; Baden and Pihl, 1996]. In September 1988, bottom trawl surveys conducted when bottom-layer oxygen concentrations were 1.4 mg l^{-1} produced a mean demersal fish biomass that was only 2% of typical levels with many more dead fish in trawls than under higher oxygen conditions [Baden et al., 1990a]. The species composition in these trawls was also quite different from that in trawls done at higher bottom dissolved oxygen concentrations. No live cod, whiting (*Merlangius merlangus*) or plaice were caught in the area during hypoxia, and biomass of dab was reduced to about 10% of that during corresponding periods in other years when bottom oxygen concentrations were higher. In contrast, many more hagfish (100-1000 ind. h $^{-1}$) and benthic invertebrates (200-400 kg h $^{-1}$) were caught when bottom water was hypoxic. Hagfish are rarely caught in trawls when oxygen concentrations are higher, and typical catches of benthic invertebrates are 10-20 kg h $^{-1}$ when dissolved oxygen conditions are less severe.

By the time oxygen concentrations drop to about 3.7 mg l^{-1} (40% saturation) in the Kattegat, the fish fauna below the halocline is reduced [Baden et al., 1990a]. Of the five dominant demersal fish species, cod and whiting are considered the most sensitive, with low numbers in areas where dissolved oxygen falls below 2.8 mg l^{-1} (30% saturation) and absence in waters with 1.4 mg l^{-1} (15% saturation) [Baden and Pihl, 1996]. In general, gadoids (cod and whiting) appeared to be the first to initiate movement out of low oxygen bottom waters and into the more oxygenated waters above the halocline. Several species of flatfish seemed to have higher tolerance for hypoxic waters, although biomass and size distributions were influenced by low oxygen. Plaice were caught in higher numbers than gadoids in areas with reduced oxygen concentrations; dab and long rough dab (*Hippoglossoides platessoides*) appeared to be even more tolerant to hypoxia than plaice and were present at oxygen levels from 0.9 to 1.8 mg l^{-1} (10 to 20% saturation). Mean fish length in the low oxygen bottom waters declined over the years of the study, indicating either reduced growth rate or greater mobility of larger individuals [Petersen and Pihl, 1995].

Some of the changes in the Kattegat fish assemblages may reflect avoidance of algal mats as well as mortality due to exposure to lethal oxygen concentrations within algal mats. After settlement, juvenile flatfish utilize shallow (< 2 m) soft bottoms as nursery grounds during their first year [Pihl, 1990]. On the Swedish west coast the habitat quality of these shallow areas has changed over the last decades as mats of filamentous green algae have developed and now cover the sediment [Pihl et al., 1996]. Inventories of the coastline by means of aerial photography during 1994 to 1996 showed that 30 to 50% of the potential nursery areas for plaice were covered with algal mats [Pihl et al., 1999]. Both field studies

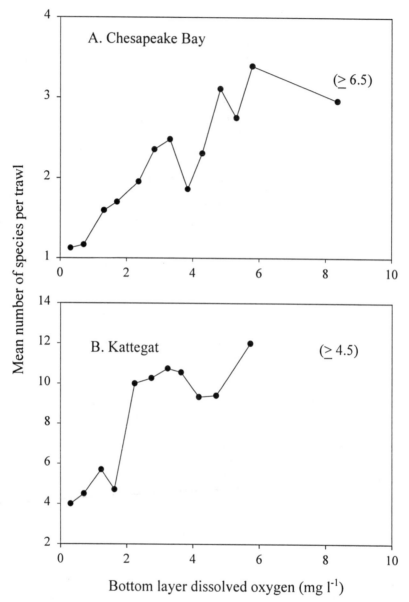

Figure 3. Number of species collected in trawls versus bottom dissolved oxygen for trawls conducted by Academy of Natural Sciences researchers for Baltimore Gas and Electric in the mainstem Chesapeake Bay near Calvert Cliffs, Maryland, and by Baden et al. [1991] and Baden and Pihl [1996] in the Kattegat. Data are mean bottom dissolved oxygen (mg l⁻¹) and mean number of fish species in each 0.5 mg l⁻¹ dissolved oxygen interval. Data are lumped for dissolved oxygen ≥ 6.5 mg l⁻¹ in the Chesapeake and ≥ 5.0 mg l⁻¹ in the Kattegat. Note difference in scale of the vertical axis for the Chesapeake Bay and Kattegat graphs.

and laboratory experiments have shown that juvenile plaice avoid sediment covered with algal mats [Pihl and van der Veer, 1992; Wennhage and Pihl, 1994]. This avoidance behavior could simply be a response to the change in physical structure of the habitat due to the presence of algae, but also may be caused by the reduction in oxygen concentration under the mats due to reduced water exchange and increased loadings of organic matter. Field and laboratory experiments showed a high mortality of larval and juvenile plaice during short term (7 h) exposure to hypoxia (2.2 mg l^{-1}, 25% O_2 saturation) under algal mats, and a significant reduction in growth over 10 days [Larson, 1997].

Trophic Interactions

Low dissolved oxygen has the potential to alter virtually every aspect of predator-prey interactions. Encounter rates between predators and their prey can change due to vertical or horizontal shifts in distributions [Magnuson et al., 1985; Rudstam and Magnuson, 1985; Kolar and Rahel, 1993; Roman et al., 1993; Breitburg, 1994; Howell and Simpson, 1994; Keister, 1996]. Low dissolved oxygen can also affect prey capture rates by modifying the percent of time spent foraging [Bejda et al., 1987], the attack rates of predators [Breitburg et al., 1994], and behaviors of prey that influence their susceptibility to predation [Poulin et al., 1987; Rahel and Kolar, 1990; Pihl et al., 1992; Kolar and Rahel, 1993]. Effects of low oxygen on the escape ability of prey species under hypoxic conditions may influence susceptibility of taxa such as zooplankton and ichthyoplankton [Roman et al., 1993; Breitburg et al., 1994, 1997; Keister, 1996; Stalder and Marcus, 1997] as well as benthic invertebrates [Pihl et al., 1992; Diaz and Rosenberg, 1995] to predation. Reduced growth of prey species caused by exposure to low dissolved oxygen may influence size-dependent mortality. Similarly, changes in population size of either predator or prey populations can influence density-dependent or frequency-dependent predation. Finally, by altering the relative success or importance of various taxa of predators within a system, low dissolved oxygen has the potential to cause changes in the importance of alternate trophic pathways [Breitburg et al., 1997].

Field and laboratory studies of Chesapeake Bay tributaries, as well as individual-based models, indicate that low dissolved oxygen has the potential to alter the relative importance of gelatinous and fish predators within the planktonic food web [see also Purcell et al., this volume], the predation mortality of fish larvae, and the ability of benthic feeding fishes to capture prey that are often otherwise inaccessible or have efficient escape behaviors. Experiments conducted in 1-m^3 mesocosms and smaller containers indicated that under low dissolved oxygen conditions, predation rates on naked goby larvae increased for predation by the scyphomedusa *Chrysaora quinquecirrha*, decreased for predation by fish predators (adult naked goby and juvenile striped bass), and was unaffected for predation by the lobate ctenophore *Mnemiopsis leidyi* [Breitburg et al., 1994; Breitburg et al., 1997; Kolesar, unpubl.]. Spatially-explicit individual-based models that incorporated effects of low dissolved oxygen on vertical distributions of fish larvae and their predators, growth rates of larvae, as well as results of the predation experiments supported the idea that low dissolved oxygen has the potential to alter the relative importance of the dominant types of predators within the planktonic food web, thus altering the relative importance of alternate trophic pathways [Breitburg et al., 1999]. In addition, both experiments and models indicate that low dissolved oxygen in the bottom layer of a stratified water column can strongly affect predation mortality of fish larvae, independent of any direct effect of low oxygen on population densities of either predator or prey. Whether predation mortality increases with a hypoxic bottom layer or the hypoxic layer serves as a partial refuge from predation depends on the behavioral responses and physiological tolerances of the individual species involved in interactions. Predation increases where predators and prey crowd into a smaller volume of water, thus increasing encounter rates, and where the ability of predators to capture prey is less compromised by low oxygen than is the ability of prey to escape. Conversely, predation can decrease in the presence of a moderately hypoxic bottom layer if predators avoid oxygen concentrations in which their prey remain, or if feeding activity of predators decreases.

Bottom-layer hypoxia, especially when episodic, can also allow for opportunistic feeding on benthic species that are normally not accessible as prey. Pihl et al. [1992] found that the diets of spot and hogchoker in the lower York River (Virginia portion of the Chesapeake Bay system) contained significantly larger, deeper-burrowing prey during periods of low oxygen than during alternating periods when bottom oxygen concentrations exceeded 2 mg l^{-1}. They suggested that the slow recovery of lethargic and emerged infauna can potentially be exploited by mobile predators. Spot move into stressful hypoxic bottom waters for short periods of time to feed on infaunal invertebrates not normally available due to their burial depth. Hogchoker, which are less mobile than spot, do not appear to forage in affected areas during the severest periods of hypoxia, but forage on weakened prey during the onset of hypoxia. The increased availability of hypoxia-stressed benthos to predators can have important consequences for energy flow in areas that experience periodic low-oxygen cycles.

These effects also occur in the Kattegat. Short-term hypoxia creates shifts in prey items available to adult fish predators by driving deep-burrowing animals from their refuges in the sediment [Baden et al., 1990; Pihl, 1994]. In addition, the increased abundance of the brittle star, *Amphiura filiformis,* relative to the traditional invertebrate community of molluscs and crustaceans, has reduced the energy available for demersal fish predators during hypoxic episodes. Over longer periods of time, hypoxia in the Kattegat has resulted in shifts in benthic species composition and thus the prey available to fish relying on this resource. This change in benthic infauna community structure may also alter the fish community it supports [Pihl, 1994; Baden and Pihl, 1996]. The shift towards smaller invertebrates with shorter life cycles has been accompanied by a trend towards smaller fish species. In addition, decreases in mean fish length of dab and cod have been observed in the Kattegat and may be linked to changes in food consumption and other stresses attributed to hypoxia [Bagge et al., 1994; Petersen and Pihl, 1995].

Growth and Development

Reduction in growth and feeding occurs at oxygen concentrations higher than those leading to rapid mortality. Changes in size distributions of predators and prey can influence the outcome of size-based trophic interactions. Reduced feeding activity of predators can, in turn, influence population densities and community structure of prey. For Chesapeake Bay species, reduced growth rates occur at 4.3 mg l^{-1} for embryo through larval Atlantic silversides, at 1.9 mg l^{-1} for larval sheepshead minnow (*Cyprinodon variegatus*) and at 3.5 mg l^{-1} for newly metamorphosed summer flounder [Poucher and Coiro, 1997]. Growth of juvenile striped bass is more sensitive to low dissolved oxygen at high temperatures typical of surface waters in the Chesapeake Bay during summer than in cooler water [Gerkin and Brandt, unpubl.], again highlighting the importance of a temperature-dissolved oxygen stress interaction for some species. Laboratory experiments on Kattegat fishes indicate that growth of plaice and dab is reduced at 4.1 to 2.5 mg l^{-1} (50 to 30% oxygen saturation), and feeding of plaice is reduced at 2.5 mg l^{-1} (30% oxygen saturation) at approximately 15 °C and 31-33 psu. There is also some evidence that these flatfish adapt to chronic low oxygen such that growth rates are higher after a period of time than they are during the initial period of exposure [Petersen and Pihl, 1995]. A number of other studies of both Chesapeake and Baltic fishes have shown reduced growth under low dissolved oxygen, but tested too few oxygen concentrations to estimate the maximum dissolved oxygen concentration at which growth reduction begins. Cyclic exposure to low oxygen concentrations near lethal levels can also reduce embryo development rates [Breitburg, 1992].

Adult fish growth may also be affected by hypoxia. Mean length of dab in Århus Bay has decreased during the period of 1953 to 1993. During this same time, the occurrence of low oxygen has increased, indicating a possible link between adult fish growth and low environmental oxygen concentrations [Bagge et al., 1994]. Decreased mean length frequencies of plaice and dab in the Kattegat may be related to low dissolved oxygen during the last two decades, but other factors such as fishing effort and migration patterns could also be important [Mellegaard and Nielsen, 1987; Petersen and Pihl, 1995]. A decrease in the growth of commercially important species can have a large impact on the profit of fisheries because larger individuals are more valuable than those stunted by low oxygen. There are no studies from Chesapeake Bay that examine the relationship between growth rates and oxygen concentrations in the field.

Multiple Stressors

Hypoxia often occurs with other anthropogenic or natural stressors that can modify and often exacerbate effects of exposure to low oxygen waters. Examples of stressors that can co-occur with low oxygen include high temperature, stressful salinity levels, chemical contaminants, disease and algal blooms.

High water temperatures can exacerbate effects of low dissolved oxygen in at least two ways. First, fish that are sensitive to both high temperatures that occur in surface waters and low dissolved oxygen concentrations that occur in bottom waters during summer may find little or no suitable habitat available in areas characterized by summer bottom-layer oxygen depletion. This phenomenon has been termed a 'temperature-oxygen squeeze' and was proposed by Coutant [1985] to explain environmental risks facing striped bass in Chesapeake Bay and elsewhere. Second, a number of species have been shown to be more sensitive to low oxygen at high, but non-lethal temperatures than at lower water temperatures. This issue has been discussed in earlier sections of this paper for striped bass, Atlantic sturgeon and cod. This increased sensitivity to hypoxia at relatively warm temperatures may affect fish abundances and distributions in the Baltic-Kattegat system. Laboratory experiments indicate that cod tolerate low dissolved oxygen better at lower temperatures and may therefore leave the area if no suitable temperature-dissolved oxygen combination exists [Schurmann and Steffensen, 1992].

Salinity can also exacerbate the effects of low dissolved oxygen, particularly in the Baltic where cod reproduction is compromised by both low surface salinities and low bottom oxygen concentrations. In addition to creating a 'squeeze' between two unfavorable habitats, low oxygen acts in concert with salinity. In laboratory experiments, egg survival was much greater at 15 psu than at 11 psu when oxygen concentrations were between 1 and 2.8 mg l^{-1} [Nissling, 1995]. Because of both the salinity-oxygen squeeze and the synergistic effects of salinity and low oxygen, cod egg development in the Baltic is strongly linked to the intrusion of saline, oxygenated water from the North Sea and a link was proposed between inflows and recruitment success of the cod population [Bagge, 1993]. However, virtual population analysis indicated that inflow events alone are insufficient to categorize the early stages of Baltic cod and other factors, such as interactions with other species, may also be important.

Low oxygen may also intensify problems that do not normally have a serious impact on fish. For example, laboratory experiments on hatching of Baltic herring (*Clupea harengus*) eggs in a low oxygen environment in the presence of the filamentous brown algae, *Pilayella littoralis/Ectocarpus siliculosus*, yielded results similar to field observations and indicated that the combination of algal exudates and low oxygen reduces hatching success [Aneer 1987]. Similarly, when plaice larvae are exposed to exudates from the filamentous green algae *Enteromorpha* sp. and hypoxic water (2.2 mg l^{-1}, 25% saturation), their survival is compromised [Larson, 1997]. However, when low density exudate water is present in conjunction with high dissolved oxygen, mortality of plaice larvae is low.

In addition, low dissolved oxygen may lead to increased disease prevalence or intensity. Dab populations in the eastern North Sea and southern Kattegat showed increased instances of diseases such as lymphocystis, epidermal hyperplasia and papilloma during times of low oxygen. Dab lengths also seemed to decrease due to low dissolved oxygen, however, population density was not affected. Three or four years after the hypoxic occurrence, dab disease levels returned to normal [Mellergaard and Nielsen, 1987].

Finally, effects of low dissolved oxygen on geochemical and biogeochemical processes could potentially affect fish populations but has not, to our knowledge, been examined for Chesapeake Bay or Baltic fishes. The example of manganese (Mn) accumulation in Norway lobster (*Nephrops norvegicus*) illustrates this potential. *Nephrops novegicus* is a dominant benthic crustacean and important predator and bioturbator in the Kattegat. Manganese is an abundant metal in most marine soft sediments. Factors controlling sedimentary Mn cycling are oxygen content of bottom water overlying the sediment, and of the pore water, and benthic organic carbon supply [Eriksson and Baden, 1998]. At oxygen concentrations < 1.5 mg l^{-1} (16% oxygen saturation), Mn^{2+} is released from the sediment and manganese concentrations in the bottom water can be two to three orders of magnitude higher than those that are typical under normoxic conditions [Balzer, 1982]. Benthic organisms exposed to these conditions are likely to accumulate manganese in haemolymph and body tissues [Baden and Neil, 1998]. During hypoxic events accumulation of manganese has been recorded in *N. norvegicus* at concentrations 20 times higher than found at normoxia [Eriksson and Baden, 1998]. At high concentrations Mn acts as a nerve toxin, and reduction in neuromuscular performance has been observed after exposure to periodic (weeks) hypoxia [Baden and Neil, 1998].

Synthesis

Despite variations due to the physical systems of the Chesapeake Bay and Baltic-Kattegat systems, and peculiarities of native species, there are common features of the consequences and responses to low dissolved oxygen in estuaries and coastal waters that should have general applications. We believe these features will apply to the area of hypoxia in the northern Gulf of Mexico. Our comparison of the Chesapeake and Baltic-Kattegat systems highlights the ways that characteristics of physical systems, behavioral responses, life history characteristics and physiological tolerances of fish species, their predators and their prey can lead to a wide variety of effects of hypoxia in the bottom layer of stratified coastal systems (Fig. 4). These effects of low dissolved oxygen can likely range from subtle shifts in vertical distributions to dramatic changes in the abundance and distribution of fishes. Both direct effects of low oxygen on growth and mortality, particularly of early life-history stages, and indirect effects such as altered trophic interactions may be common.

The Importance of Behavioral Responses

Behavioral responses of fish to low oxygen, or the lack of appropriate responses, ultimately determine the effects of exposure to low oxygen on survival, growth and ecological interactions. These behavioral responses vary among life history stages and species, and are also affected by constraints imposed by the physical environment. Nonetheless, generalizations emerge. Clearly, the earliest life stages of fishes are the most vulnerable, both because they are often more sensitive to low oxygen than are juvenile and adult fishes, and also because their behavioral responses are more limited, at least spatially. Once eggs are released into the environment, embryo development and survival is entirely dependent on the physical environment in which they find themselves and on the ways in which the physical environment affects their risk of predation mortality. Both benthic eggs deposited in areas where dissolved oxygen declines during development and pelagic eggs that sink into hypoxic or anoxic depths are susceptible to mortality from direct exposure to low oxygen because they have no behavioral response. In addition, predation mortality of eggs can be affected by the influence of low oxygen on the abundance, distribution, and feeding rates of their predators, as well as by the influence of dissolved oxygen on the behavior and presence of parental fish guarding benthic eggs.

Unlike eggs, fish larvae are able to alter their vertical position in response to low oxygen in a fairly stable, stratified water column. This should reduce mortality due to direct exposure to low oxygen, but may result in increased encounter rates with similarly sensitive predators. In addition, nothing is known about how effective this vertical escape response is in a turbulent water column or when oxygen concentrations change rapidly. In addition, the ability of most larvae, especially early larval stages, to change locations by lateral movement that would place them in a different, higher oxygen content water mass, is quite limited.

Juvenile (especially older juvenile stages) and adult fishes have a greater ability to move large distances both vertically and laterally to escape low dissolved oxygen. In doing so, however, they may leave their normal feeding environment and increase their exposure to predators. Remaining in the water column can be energetically expensive for epibenthic

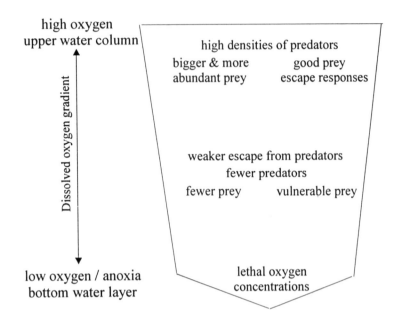

Figure 4. Risks and responses of fish to low dissolved oxygen in a water column with a vertical gradient or stratification. The ultimate effect of low dissolved oxygen in such systems will depend on the presence and form of physiological and behavioral responses by fish, their predators and their prey. A wide variety of consequences of low dissolved oxygen can occur within a single body of water because the vertical gradient or stratification of dissolved oxygen concentrations creates biologically significant physical structure within the water column. In addition, hypoxia, where not too severe, can have both positive and negative effects. For example, water depths with reduced but not severely stressful dissolved oxygen concentrations may have fewer predators of small fish as well as fewer prey on which fish can feed.

species that lack a swimbladder and can expose these fish to predators that are better adapted to the pelagic environment. Flatfish, lacking a swimbladder, are more likely to escape horizontally than vertically and could thereby be forced to migrate long distances to reach better oxygen conditions. In addition, behavioral responses can sometimes increase risk of exposure to low oxygen, as when schools of fish crowd into embayments and contribute substantially to *in situ* oxygen depletion.

Although behavioral responses are critical to the ability of members of individual species to escape from the direct effects of low dissolved oxygen, the variation among species in behavioral responses may be similarly important in determining the ways that low oxygen affects ecological interactions. The relative tolerance, and similarities or differences in the responses of predators and prey, can determine whether low oxygen increases or decreases predation mortality of fishes, and the degree to which mild or episodic low oxygen alters diet and growth rates. Tolerant species may not simply withstand mild or episodic hypoxia, but may be able to respond opportunistically by feeding on normally inaccessible prey or by reaping the advantage of reduced competition or predation. In contrast, sensitive species may become far less important in the food web. Studies in both the Chesapeake and the

Baltic-Kattegat area indicate the potential for low oxygen to cause major changes in the importance of alternate pathways of energy flow within estuarine food webs.

Multiple Effects of Eutrophication

The enhanced production related to high nutrient loadings in eutrophic systems may reduce the otherwise deleterious effects of low oxygen on growth and survival. There is a positive relationship between fisheries landings (including shellfish) and nitrogen loadings in marine systems, including those in which bottom-layer hypoxia occurs [Nixon, 1992]. In addition, individual growth rates of some species may respond positively to anthropogenic nutrient loading [Boddeke and Hagel, 1991]. Recent legislation to reduce nutrient loading into coastal waters has created concern that North Sea production, as well as catches, will decline accordingly [Boddeke and Hagel, 1991]. Increased eutrophication in Laholm Bay in the eastern Kattegat has not negatively impacted the region as an area for young flatfish development; in fact, its significance may be increasing as conditions in the western Kattegat continue to deteriorate [Baden et al., 1990a].

Total fish catches in the Baltic, which are dominated by herring, sprat and cod, increased ten-fold from the 1920s to 1980s [Hansson and Rudstam, 1990], after which a decrease has been observed. The increased landings are mainly due to intensified fishing, but eutrophication as well as decreased predation by seals are also thought to be important factors [Hansson and Rudstam, 1990]. Eutrophication has increased pelagic primary production by 30 to 70% since the turn of the century, and sedimentation of organic carbon has increased by 70 to 190% [Elmgren, 1989]. Increased biomass and productivity above the halocline has resulted in a net doubling of benthic production in the Baltic. At present, the Baltic fishery requires around 10% of the primary production, compared to 1% at the turn of the century [Elmgren, 1989]. For herring and sprat, the effects of continued eutrophication could be positive; however, cod reproductive success is threatened by hypoxia and the stock will probably be negatively affected. The degraded condition of nursery areas in the western Kattegat also points to the limits of the positive relationship between nutrient loadings and fish recruitment. Ultimately, at some level of nutrient loading, hypoxia, the negative effect of eutrophication most directly deleterious to higher trophic levels, exceeds the positive effects of increased production [see Caddy, this volume].

The Chesapeake, Baltic-Kattegat and Gulf of Mexico Systems: Fisheries Harvest and Fish Populations

How different are the effects of low oxygen in the Chesapeake and the Baltic-Kattegat systems? What do these differences suggest about potential effects of low dissolved oxygen on fishes in areas of the Gulf of Mexico in which hypoxia occurs? Available data suggest that both fisheries production and estuarine nursery habitat are more severely affected by oxygen depletion in the Baltic-Kattegat system than in the Chesapeake Bay. This is likely a reflection of real differences between the systems in the biology of important fishery species and in physical characteristics of the systems, as well as in differences in research emphasis and intensity.

In contrast to the Chesapeake Bay, hypoxia in the Baltic-Kattegat directly impacts many commercially important fish species. The nature of the water body as well as characteristics of the target species create an environment in which fish and low oxygen coincide. The five most important commercial species in the Baltic Sea are herring, sprat, cod, flounder and salmon [Bagge, 1993], with whiting and dab being less important. Most of these commercially important finfish are demersal or have a demersal stage at some point during development. The Norway lobster, which supports an important fishery in the Kattegat, is a benthic crustacean [Baden et al., 1990b; Baden and Pihl, 1996; Hagerman et al., 1996]. Much of the focus on hypoxia in the Baltic-Kattegat area was triggered by the near collapse of cod, the most important finfish fishery in the Baltic, as well as by dramatic declines in Norway lobster [Baden et al., 1990b]. Both of these species reproduce within the Baltic-Kattegat and have life histories that make them particularly sensitive to oxygen depletion even when it is limited to a portion of the water column. During hypoxic events, there is often evidence of movement out of the affected area, and sometimes mortality, of many of the important fishery species. These effects of hypoxia reduce stock sizes and alter historic fishing regions [Baden et al., 1990a; Baden and Pihl, 1996; Hagerman et al., 1996]. Although catches in adjacent areas may increase as fish move to higher oxygen habitat, regional catches ultimately decrease as individuals perish and valuable environments are rendered unsuitable for fish.

The most important fishery species in the Chesapeake Bay from an economic standpoint are the eastern oyster (*Crassostrea virginica*), blue crab (*Callinectes sapidus*), Atlantic menhaden and striped bass. Of these, only the oyster spawns in areas of the bay affected by hypoxia. Blue crab spawn near the mouth of the Bay, and Atlantic menhaden, which feed primarily on phytoplankton in the surface mixed layer, spawn along the Atlantic coast. Larvae of both of these species are distributed primarily outside of Chesapeake Bay over the continental shelf. Striped bass are anadramous, and spawn in spring such that times and locations of larvae and hypoxia generally do not coincide, except in some rivers late during larval development when swimming capabilities are well developed. In response to a decline in Chesapeake striped bass populations, Coutant [1985] proposed the temperature-oxygen squeeze hypothesis, which suggested that little suitable habitat remained in Chesapeake Bay during summer that would support striped bass growth because high surface temperatures were physiologically stressful to this species and oxygen concentrations in cooler, bottom waters were too low. Although modeling supports the potential effects of this temperature-oxygen squeeze on juvenile growth potential [Brandt and Demers, unpubl.], a fishing moratorium (but no change in oxygen or temperature) was followed by a rebound of the striped bass population. More recently, Secor and Gunderson [1998] suggested that summertime hypoxia has degraded nursery habitats for Atlantic sturgeon in Chesapeake Bay. However, the relative importance of harvest pressure versus habitat degradation and the potential for sturgeon to behaviorally respond to low oxygen in the field have not yet been fully evaluated. Thus, in contrast to the Baltic, there is no strong evidence that hypoxia is a major factor affecting population abundances of the most important fisheries species, with the possible exception that the lower limit of oyster distributions may have become somewhat shallower than it was in pre-colonial times. The most abundant resident fish that spawn during summer in areas of Chesapeake Bay affected by hypoxia are small species whose densities and population dynamics are poorly understood. In addition, many fish species that use Chesapeake Bay as either a spawning or nursery area spend a portion of the year or a portion of their life cycle outside of the Bay. Linking changes in population size of open, migratory populations is much more difficult than detecting changes in population densities of resident fishes. Nevertheless, there may be important effects of low oxygen on both forage fish and harvested species in Chesapeake Bay that are masked by effects of fishing pressures, other stressors and temporal variation in physical factors (e.g., those affecting transport of larvae into the Bay from the continental shelf). In addition, many of the important ecological changes to Chesapeake Bay apparently accompanied European colonization and agriculture in the region during the 18th century [Cooper and Brush, 1993]. In contrast, hypoxia in the Baltic and Kattegat are relatively recent events that occurred during a time that rigorous environmental monitoring and fisheries research were already in place.

The open nature of the Gulf of Mexico region of hypoxia will likely make distinguishing changes in abundance from changes in distribution more difficult than in the Baltic-Kattegat system. Nevertheless, it is likely that low dissolved oxygen will lead to a wide variety of consequences for Gulf of Mexico fishes, including altered distributions and changes in trophic interactions, as well as mortality at least of early life-history stages. Because of the potential magnitude of the ecological and economic effects of low oxygen on fish populations and harvest potential, an evaluation of behavioral, food web and mortality effects of low dissolved oxygen are critical to consider.

Acknowledgments. The authors would like to thank the editors, N. N. Rabalais and R. E. Turner, for organizing the stimulating workshop that led to this chapter. We would also like to thank Elizabeth North and Edward Houde (University of Maryland Center for Environmental Studies, Chesapeake Biological Laboratory), and Steven Brandt (NOAA Great Lakes Environmental Research Laboratory) for allowing us to cite their unpublished research, and C. C. Coutant for helpful comments on a draft of this manuscript. Support for D. Breitburg and S. Kolesar was provided by NOAA Coastal Ocean Program funding to the COASTES program and the Academy of Natural Sciences Estuarine Research Center. Funding for L. Pihl was provided by the Marine Research Center at Göteborg University.

References

Aertebjerg, G., Causes and effects of eutrophication in the Kattegat and Belt Sea, *22nd Nordic Symp. on Water Research, Nordforsk, 1*, 87-100, Laugarvatn, Iceland, 1986.
Andersson, L. and L. Rydberg, Trends on nutrient and oxygen conditions within the Kattegat: Effects of local nutrient supply, *Estuar. Coast. Shelf Sci., 26*, 559-579, 1988.
Aneer, G., High natural mortality of Baltic herring (*Clupea harengus*) eggs caused by algal exudates?, *Mar. Biol., 94*, 163-169, 1987.
Baden, S. P. and D. M. Neil, Accumulation of manganese in the haemolymph, nerve and muscle tissue of *Nephrops norvegicus* (L.) and its effect on neuromuscular performance, *Comp. Biochem. Physiol., 119A*, 351-359, 1998.
Baden, S. P. and L. Pihl, Effects of autumnal hypoxia on demersal fish and crustaceans in the SE

Kattegat, 1984-1991, *Sci. Symp. on the North Sea, Quality Status Report, 1993*, pp.189-196, Danish Environmental Protection Agency, 1996.

Baden, S. P., L. Pihl, and R. Rosenberg, Effects of oxygen depletion on the ecology, blood physiology and fishery of the Norway lobster (*Nephrops norvegicus* L.), *Mar. Ecol. Prog. Ser., 67*, 141-155, 1990b.

Baden, S. P., L.-O. Loo, L. Pihl, and R. Rosenberg, Effects of eutrophication on benthic communities including fish: Swedish west coast, *Ambio, 19*, 113-122, 1990a.

Bagge, O., Possible effects on fish reproduction due to the changed oceanographic conditions in the Baltic proper, *ICES CM, J:31*, pp. 1-7, 1993.

Bagge, O., E. Steffensen, E. Nielse, and C. Jensen, Growth and abundance of dab and abundance of plaice in Århus Bay in relation to oxygen conditions 1953-1993, *ICES CM, J:15*, pp. 1-11, 1994.

Balzer, W., On the distribution of iron and manganese at the sediment/water interface: thermodynamic versus kinetic control, *Geochim. Cosmochim. Acta, 46*, 1153-1161, 1982.

Bejda, A. J., A. L. Studholme, and B. L. Olla, Behavioral responses of red hake, *Urophycis chuss*, to decreasing concentrations of dissolved oxygen, *Envir. Biol. Fish., 19*, 261-268, 1987.

Boddeke, R. and P. Hagel, Eutrophication of the North Sea continental zone, a blessing in disguise, *ICES CM 1991/E:7*, pp. 1-29, 1991.

Breitburg, D. L., K. A. Rose, and J. H. Cowan, Jr., Linking water quality to larval survival: Predation mortality of fish larvae in an oxygen-stratified water column, *Mar. Ecol. Prog. Ser., 178*, 39-54, 1999.

Breitburg, D. L., Near-shore hypoxia in the Chesapeake Bay: Patterns and relationships among physical factors, *Estuar. Coastal Shelf Sci., 30*, 593-609, 1990.

Breitburg, D. L., Episodic hypoxia in Chesapeake Bay: Interacting effects of recruitment, behavior, and physical disturbance, *Ecol. Monogr., 62*, 525-546, 1992.

Breitburg, D. L., Behavioral response of fish larvae to low dissolved oxygen concentrations in a stratified water column, *Mar. Biol., 120*, 615-625, 1994.

Breitburg, D. L., T. Loher, C. A. Pacey, and A. Gerstein, Varying effects of low dissolved oxygen on trophic interactions in an estuarine food web, *Ecol. Monogr., 67*, 489-507, 1997.

Caddy, J., Toward a comparative evaluation of human impacts on fishery ecosystems of enclosed and semi-enclosed seas, *Rev. Fishery Sci. 1*, 57-96, 1993.

Chesney, E. J. and E. D. Houde, Laboratory studies on the effect of hypoxic waters on the survival of eggs and yolk-sac larvae of the bay anchovy, *Anchoa mitchilli*, in *Population Biology of Bay Anchovy in Mid-Chesapeake Bay*, edited by E. D. Houde, E. J. Chesney, T. A. Newberger, A. V. Vazquez, C. E. Zastrow, L. G. Morin, H. R. Harvey, and J. W. Gooch, Final Report ed., pp. 184-191, Maryland Sea Grant College, Maryland, 1989.

Cooper, S. R. and G. S. Brush, A 2,500-year history of anoxia and eutrophication in Chesapeake Bay, *Estuaries, 16*, 617-626, 1993.

Coutant, C. C., Striped bass, temperature, and dissolved oxygen: A speculative hypothesis for environmental risk, *Trans. Am. Fish. Soc., 114*, 31-61, 1985.

De Silva, C. D. and P. Tytler, The influence of reduced environmental oxygen on the metabolism and survival of herring and plaice larvae, *Neth. J. Sea. Res., 7*, 345-362, 1973.

Deubler, E. E., Jr. and G. S. Posner, Response of postlarval flounders, *Paralicthys lethostigma*, to water of low oxygen concentrations, *Copeia, 2*, 312-317, 1963.

Diaz, R. J. and R. Rosenberg, Marine benthic hypoxia: a review of its ecological effects and the behavioural responses of benthic macrofauna, *Ocean. Mar. Biol. Ann. Rev., 33*, 245-303, 1995.

Elmgren, R., Man's impact on the ecosystem of the Baltic Sea: Energy flows today and at the turn of the century, *Ambio, 18*, 326-332, 1989.

Eriksson, S. P. and S. P. Baden, Manganese in the haemolymph and tissues of the Norway lobster,

Nephrops norvegicus (L.) along the Swedish west coast, *Hydrobiologica, 375/376*, 255-264, 1998.

Grønkjær, P. and K. Wieland, Ontogenetic and environmental effects on vertical distribution of cod larvae in the Bornholm Basin, Baltic Sea, *Mar. Ecol. Prog. Ser., 154*, 91-105, 1997.

Grauman, G. B., Investigations of factors influencing fluctuations in the abundance of Baltic cod, *Rapp. P.-v. Réun. Cons. Int. Explor. Mer., 164*, 73-76, 1973.

Hagerman, L., A. B. Josefson, and J. N. Jensen, Benthic macrofauna and demersal fish, in *Eutrophication in Coastal Marine Systems, Coastal and Estuarine Studies. Vol. 52*, edited by B. B. Jørgensen and K. Richardson, pp. 155-177, American Geophysical Union, Washington, DC, 1996.

Hansson, S. and L. G. Rudstam, Eutrophication and Baltic fish communities, *Ambio, 19*, 123-125, 1990.

Howell, P. and D. Simpson, Abundance of marine resources in relation to dissolved oxygen in Long Island Sound, *Estuaries, 17*, 394-408, 1994.

Keister, J. E., E. D. Houde, and D. L. Breitburg, Effects of bottom-layer hypoxia on abundances and depth distributions of organisms in Patuxent River, Chesapeake Bay, *Mar. Ecol. Prog. Ser.*, in press.

Keister, J. E., Habitat selection and predation risk: Effects of hypoxia on zooplankton and fish larvae in Chesapeake Bay, MS Thesis, University of Maryland, College Park, 1996.

Koisor, M. and J. Netzel, Eastern Baltic cod stock and environmental conditions, *Rapp. P.-v. Réun. Cons. Int. Explor. Mer., 190*, 159-162, 1989.

Kolar, C. S. and F. J. Rahel, Interaction of a biotic factor (predator presence) and an abiotic factor (low oxygen) as an influence on benthic invertebrate communities, *Oecologia, 95*, 210-219, 1993.

Kruse, B. and B. Rasmussen, Occurrence and effects of a spring oxygen minimum layer in a stratified coastal water, *Mar. Ecol. Prog. Ser., 125*, 293-303, 1995.

Larson, F., Survival and growth of plaice (*Pleuronectes platessa* L.) larvae and juveniles in mats of *Enteromorpha* sp. -- the effects of algal exudates and nocturnal hypoxia, MS Thesis, Göteborg University, Sweden, 1997.

Larsson, U., R. Elmgren, and F. Wolff, Eutrophication of the Baltic Sea: Causes and consequences, *Ambio, 14*, 9-14, 1985.

Loesch, H., Sporadic mass shoreward migrations of demersal fish and crustaceans in Mobile Bay, Alabama, *Ecology, 41*, 292-298, 1960.

Magnuson, J. J., A. L. Beckell, K. Mills, and S. B. Brandt, Surviving winter hypoxia: behavioral adaptations of fishes in a northern Wisconsin winterkill lake, *Envir. Biol. Fish., 14*, 241-250, 1985.

Mellergaard, S. and E. Nielsen, The influence of oxygen deficiency on the dab populations in the eastern North Sea and the southern Kattegat, *ICES CM 1987/E:6*, pp. 1-7, 1987.

Nissling, A., Survival of eggs and yolk-sak larvae of Baltic cod (*Gadus morhua*) at low oxygen levels in different salinities, *ICES Mar. Sci. Symp., 198* , 626-631, 1994.

Nissling, A. and L. Vallin, The ability of Baltic cod eggs to maintain neutral buoyancy and the opportunity for survival in fluctuating conditions in the Baltic Sea, *J. Fish. Biol., 48*, 217-227, 1996.

Nissling, A., Salinity and oxygen requirements for successful spawning of Baltic cod, Ph.D. Dissertation, Stockholm University, Sweden, 1995.

Nissling, A., H. Kryvi, and L. Vallin, Variation in egg buoyancy of Baltic cod *Gadus morhua* and its implications for egg survival in prevailing conditions in the Baltic Sea, *Mar. Ecol. Prog. Ser., 110*, 67-74, 1994.

Ohldag, S., D. Schnack, and U. Waller, Development of Baltic cod eggs at reduced oxygen concentration levels, *ICES CM 1991/J:39*, pp. 1-11, 1991.

Petersen, J. K. and G. I. Petersen, Tolerance, behaviour, and oxygen consumption in the sand goby, *Pomatochistus minutus* (Pallas), exposed to hypoxia, *J. Fish. Biol.*, *37*, 921-933, 1990.

Petersen, J. K. and L. Pihl, Responses to hypoxia of plaice, *Pleuronectes platessa*, and dab, *Limanda limanda*, in the south-east Kattegat: distribution and growth, *Envir. Biol. Fish.*, *43*, 311-321, 1995.

Pihl, L., Year-class strength regulation in plaice (*Pleuronectes platessa* L.) on the Swedish west coast, *Hydrobiol.*, *195*, 79-88, 1990.

Pihl, L., S. P. Baden, and R. J. Diaz, Effects of periodic hypoxia on distribution of demersal fish and crustaceans, *Mar. Biol.*, *108*, 349-360, 1991.

Pihl, L., G. Magnusson, I. Isaksson, and I. Wallentinus, Distribution and growth dynamics of ephemeral macroalgae in shallow bays on the Swedish west coast, *J. Sea Res.*, *35*, 169-180, 1996.

Pihl, L., A. Svensson, P.-O. Moksnes, and H. Wennage, Distribution of green algal mats throughout shallow soft bottoms of the Swedish archipelago in relation to nutrient loads and wave exposure, *J. Sea Res.*, *41*, 281-294, 1999.

Pihl, L. and H. van der Veer, Importance of exposure and habitat structure for the population density of 0-group plaice, *Pleuronectes platessa* L., in coastal nursery areas, *Neth. J. Sea. Res.*, *29*, 145-152, 1992.

Pihl, L., Changes in the diet of demersal fish due to eutrophication-induced hypoxia in the Kattegat, Sweden, *Can. J. Fish. Aquat. Sci.*, *51*, 321-336, 1994.

Pihl, L., S. P. Baden, R. J. Diaz, and L. C. Schaffner, Hypoxia-induced structural changes in the diet of bottom-feeding fish and Crustacea, *Mar. Biol.*, *112*, 349-361, 1992.

Poucher, S. and L. Coiro, Test Reports: Effects of low dissolved oxygen on saltwater animals, Memorandum to D. C. Miller. U. S. Environmental Protection Agency, Atlantic Ecology Division, Narragansett, Rhode Island, 02882, 1997.

Poulin, R., N. G. Wolf, and D. L. Kramer, The effect of hypoxia on the vulnerability of guppies (*Poecilia reticulata*, Poeciliidae) to an aquatic predator (*Astronotus ocellatus*, Cichlidae), *Envir. Biol. Fish.*, *20*, 285-292, 1987.

Rahel, F. J. and C. S. Kolar, Trade-offs in the response of mayflies to low oxygen and fish predation, *Oecologia*, *84*, 39-44, 1990.

Roman, M. R., A. L. Gauzens, W. K. Rhinehart, and J. R. White, Effects of low oxygen waters on Chesapeake Bay zooplankton, *Limnol. Oceanogr.*, *38*, 1603-1614, 1993.

Rombough, P. J., Respiratory gas exchange, aerobic metabolism, and effects of hypoxia during early life, in *Fish Physiology. Vol. XI: The Physiology of Developing Fish, Part A: Eggs and Larvae*, edited by W. S. Hoar and D. J. Randall, pp. 59-161, Academic Press, Inc., San Diego, CA, 1988.

Rosenberg, R., R. Elmgren, S. Fleischer, P. Jonsson, G. Persson, and H. Dahlin, Marine eutrophication case studies in Sweden, *Ambio*, *19*, 102-108, 1990.

Rudstam, L. G. and J. J. Magnuson, Predicting the vertical distribution of fish populations: analysis of cisco, *Coregonus artedii*, and yellow perch, *Perca flavescens*, *Can. J. Fish. Aquat. Sci.*, *42*, 1178-1188, 1985.

Saksena, V. P. and E. B. Joseph, Dissolved oxygen requirements of newly-hatched larvae of the striped blenny (*Chasmodes bosquianus*), the naked goby (*Gobiosoma bosci*), and the skilletfish (*Gobiesox strumosus*), *Ches. Sci.*, *13*, 23-28, 1972.

Sanford, L. P., K. R. Sellner, and D. L. Breitburg, Covariability of dissolved oxygen with physical processes in the summertime Chesapeake Bay, *J. Mar. Res.*, *48*, 567-590, 1990.

Schurmann, H. and J. F. Steffensen, Lethal oxygen levels at different temperatures and the preferred temperature during hypoxia of the Atlantic cod, *Gadus morhua* L., *J. Fish. Biol.*, *41*, 927-934, 1992.

Secor, D. H. and T. E. Gunderson, Effects of hypoxia and temperature on survival, growth, and respiration of juvenile Atlantic sturgeon (*Acipenser oxyrinchus*), *Fish. Bull., 96*, 603-613, 1998.

Stadler, L. C. and N. H. Marcus, Zooplankton responses to hypoxia: behavioral patterns and survival of three species of calanoid copepods, *Mar. Biol., 127*, 599-607, 1997.

Thursby, G., D. C. Miller, S. Poucher, L. Coiro, W. Munns, and T. Gleason, Protection of coastal and estuarine animals from low dissolved oxygen: Cape Cod to Cape Hatteras, U. S. Environmental Protection Agency, Office of Water, draft report March 1999.

U. S. Environmental Protection Agency, *The Chesapeake Bay and Its Ecosystem*, http://www.chesapeakebay.net, February 1999.

Weigelt, M. and H. Rumohr, Effects of wide-range oxygen depletion on benthic fauna and demersal fish in Kiel Bay 1981-1983, *Meeresforsch., 31*, 124-136, 1986.

Wennage, H. and L. Pihl, Substratum selection by juvenile plaice (*Pleuronectes platessa* L.): impact of benthic microalgae and filamentous microalgae, *Neth. J. Sea. Res., 32*, 343-351, 1994.

Wieland, K., U. Waller, and D. Schnack, Development of Baltic cod eggs at different levels of temperature and oxygen content, *Dana, 10*, 163-177, 1994.

Wieland, K. and F. Zuzarte, Vertical distribution of cod and sprat eggs and larvae in the Bornholm Basin (Baltic Sea) 1987-1990, *ICES CM 1991/J:37*, pp. 1-12 , 1991.

14

Ecological Effects of Hypoxia on Fish, Sea Turtles, and Marine Mammals in the Northwestern Gulf of Mexico

J. Kevin Craig, Larry B. Crowder, Charlotte D. Gray, Carrie J. Mcdaniel, Tyrrell A. Henwood, and James G. Hanifen

Abstract

Bottom-water hypoxia (< 2.0 mg l^{-1}) is a common effect of nutrient enrichment that impacts the habitat of demersal species. For mobile species, such as fish and many invertebrates, the indirect effects of hypoxia such as energetic costs, habitat loss and altered ecological performance are more important than direct mortality due to lack of oxygen. Air-breathing species such as sea turtles and cetaceans may be indirectly impacted via shifts in the spatial distribution of prey. The northwestern Gulf of Mexico continental shelf is subject to seasonal hypoxia over large areas, but little is known regarding the effects of hypoxia on living resources. We combined information from existing shipboard and aerial surveys to document the spatial distribution of bottom dissolved oxygen and related this to the spatial distribution of demersal fish and invertebrates as well as sea turtles and cetaceans. Hypoxia was widespread during June-July of 1992 and 1993, extending over 8-9% (11,400-12,100 km^2) of the continental shelf. Analysis of trawl data suggests that demersal species are displaced from hypoxic bottom waters to adjacent areas of intermediate dissolved oxygen (2.0-5.0 mg l^{-1}) during June-July but re-occupy these areas by October-November when hypoxia is largely absent. Aerial survey sightings suggest that chronic, large-scale hypoxia should be included among hypotheses to explain the distribution patterns of sea turtles and bottlenose dolphins in the northwestern Gulf of Mexico. Synoptic data on the distribution of bottom-water dissolved oxygen and living resources on appropriate spatial and temporal scales will be necessary to refine and test these hypotheses.

Coastal Hypoxia: Consequences for Living Resources and Ecosystems
Coastal and Estuarine Studies, Pages 269-292
Copyright 2001 by the American Geophysical Union

Introduction

Nutrient enrichment from anthropogenic activities is widely recognized as a major stressor of aquatic ecosystems [Rosenberg, 1985; Nixon, 1990; Vitousek et al., 1997]. Increases in delivery rates of nutrients to aquatic systems have been documented worldwide, with concomitant increases in the intensity of phytoplankton blooms, sediment organic loading and bottom-water hypoxia [Officer et al., 1984; Justic' et al., 1987; Andersson and Rydberg, 1988; Parker and O'Reilly, 1991; Turner and Rabalais, 1991, 1994; Bonsdorff et al., 1997]. While many systems historically experienced some level of nutrient loading and associated water quality problems (e.g., Cooper and Brush [1991], Cooper [1995]), anthropogenic activities have clearly increased the flux of nutrients to aquatic systems this century. Low dissolved oxygen levels in bottom waters (i.e., hypoxia) are one consequence of nutrient enrichment that can impact the habitat of demersal species. Hypoxia (dissolved oxygen < 2.0 mg l^{-1}) may be a permanent feature of some systems (e.g., Black Sea, Mee [1992]) but generally occurs seasonally in association with predictable hydrologic and biological processes [Diaz and Rosenberg, 1995], or aperiodically due to anomalous environmental conditions [Falkowski et al., 1980]. The timing of most severe hypoxia is often coincident with periods of intense annual recruitment and growth of demersal fish and invertebrates [Pihl et al., 1991, 1992; Breitburg, 1992; Lenihan and Peterson, 1998] and potentially impacts fishery production. Some studies have suggested that hypoxia led to the decline of particular fisheries, although definitive evidence is often lacking [Baden et al., 1990; Nissling et al., 1994; Chapman et al., 1995; Lenihan and Peterson, 1998]. Our objective is to evaluate potential impacts of hypoxia on fish, marine mammals and sea turtles in the northwestern Gulf of Mexico. We review ways that hypoxia impacts mobile species and what is known regarding the ecological effects of hypoxia on fish and mobile invertebrates in the northwestern Gulf. We then use information from fishery-independent resource surveys conducted over large portions of the northwestern Gulf continental shelf to document the spatial distribution of bottom dissolved oxygen, as well as coarse-scale patterns in the distribution and abundance of some potentially affected species.

Effects of Low Dissolved Oxygen on Mobile Species

Low dissolved oxygen may impact individuals directly, by affecting their metabolic rate and indirectly, by affecting spatial distribution as well as the energy available for other processes. Direct mortality of highly mobile species has been attributed to low dissolved oxygen concentrations [May, 1973; Garlo et al., 1979; Stachowitsch, 1984; Baden et al., 1990; Paerl et al., 1998], although less often than for species that are sessile or of low motility. Mortality is the extreme result when dissolved oxygen levels are insufficient to maintain the standard metabolic rate. The probability of mortality due to insufficient oxygen is a function of the rate that dissolved oxygen levels decline, the minimum level obtained, the spatial extent and temporal duration, and the rate of return to normoxic levels, as well as individual physiological tolerances, behavioral responses and swimming abilities. Breitburg [1992] used this information for naked gobies (*Gobiosoma bosc*) in Chesapeake Bay to explain the higher mortality of smaller individuals due to periodic intrusions of severely hypoxic bottom water onto oyster reefs. Because smaller individuals are slower swimmers than larger ones, they must either initiate movements away from areas of declining dissolved oxygen before lethal levels are attained or experience a higher probability of mortality than larger, faster swimming individuals.

Low dissolved oxygen levels not sufficient to result in death may impose significant energetic costs on individuals that can affect future survival and reproduction. These energetic costs may be manifested in a variety of ways, such as directed movement [Deubler and Posner, 1963; Petersen and Petersen, 1990; Das and Stickle, 1994], increased swimming speed [Dizon, 1977] or swimming activity [Weber and Kramer, 1983; Bejda et al., 1987; Petersen and Petersen, 1990] presumably to find areas of sufficient oxygen, or physiological and behavioral adjustments such as increased ventilation rates [Randall, 1982; Metcalfe and Butler, 1984] and blood haemoglobin/hemocyanin concentration [Bushnell et al., 1984; Baden et al., 1990; Petersen and Petersen, 1990] to increase oxygen extraction and uptake efficiency. Some species are able to use alternative breathing modes when subjected to low dissolved oxygen that may be ecologically costly [Kramer et al., 1983; Wolf and Kramer, 1987] or less physiologically efficient [Burton and Heath, 1980; Gee, 1980].

Increased energy requirements for the maintenance of metabolism may impact growth, reproduction and ecological performance. In some species increased activity (and presumably energetic expenditures) under low dissolved oxygen stress may increase the probability of finding areas with sufficient oxygen. Other species, however, decrease activity when exposed to low dissolved oxgyen [Metcalfe and Butler, 1984; Fisher et al., 1992; Nilsson et al., 1993; Schurmann and Steffenson, 1994; Crocker and Cech, 1997], possibly to conserve energy until levels increase again. Occupying waters of low dissolved oxygen may result in decreased food intake, food conversion efficiency and possibly nutrient digestibility [Stewart et al., 1967; Pouliot and de la Noue, 1988], leading to decreased somatic growth [Brett and Blackburn, 1981; Cech et al., 1984] and gonadal synthesis [Armstrong et al., 1992]. In addition, auto-immune responses may be impaired leading to increased disease susceptibility [Mellergaard and Nielsen, 1990]. Alternatively, low dissolved oxygen may temporarily increase the availability of some food resources leading to increased foraging rates for species that can tolerate low dissolved oxygen levels or otherwise adjust behaviorally to hypoxic conditions [Rahel and Nutzman, 1994].

Movement away from areas of low dissolved oxygen is the most commonly observed impact of hypoxia on mobile species [Coble, 1982; Magnuson et al., 1985; Suthers and Gee, 1986; Pihl et al., 1991]. These altered spatial distributions may increase [Kramer et al., 1983; Wolf and Kramer, 1987; Breitburg et al., 1997] or decrease [Poulin et al., 1987] encounter rates with predators. Predator capture success and prey escape responses may also be impacted by low dissolved oxygen, for example, via effects on swimming performance [Weber and Kramer, 1983; Bushnell et al., 1984; Breitburg et al., 1997]. How these effects on spatial distribution and ecological performance impact predation rates depend on relative differences in species' responses to low dissolved oxygen. For example, the increased use of aquatic surface respiration by juvenile guppies (*Poecilia reticulata*) as dissolved oxygen levels decline increases vulnerability to green herons (*Butorides striatus*), an aerial predator [Kramer et al., 1983], but decreases vulnerability to predatory cichlids (*Astronotus ocellatus*), an aquatic predator whose feeding activity is significantly impaired by low dissolved oxygen [Poulin et al., 1987]. Similarly, predation rates by juvenile striped bass (*Morone saxatilis*) on larval naked goby decrease at low dissolved oxygen levels while those of medusan jellyfish (*Chrysaora quinquecirrha*) increase because of differences in the relative effects of dissolved oxygen on predator capture success and prey escape responses [Breitburg et al., 1997]. Declines in dissolved oxygen in Lake Victoria have been suggested as contributing factors in the decline of some haplochromine cichlid species, while for other, more tolerant species hypoxic areas may serve as spatial refugia from predation by introduced Nile perch (*Lates niloticus*), a species particularly sensitive to low dissolved oxygen [Chapman et al., 1995]. Model simulations integrating the effects of dissolved oxygen on growth rates, vertical distribution and predator-prey behavioral responses suggest that fish larvae in the Chesapeake Bay suffer the lowest predation mortality when bottom waters are slightly hypoxic compared to anoxic and normoxic conditions [Breitburg et al., 1999]. Similarly, altered spatial distributions may impact reproductive processes such as migration to spawning grounds or finding a mate, particularly if sexes differ in their responses to low dissolved oxygen (e.g., Kramer and Mehegan [1981]) or in the timing of reproductive processes relative to that of low dissolved oxygen events. These examples suggest that low dissolved oxygen can have significant indirect effects on population growth and mortality rates.

The indirect effects of hypoxia on physiological, behavioral and ecological processes are probably more important than direct mortality due to lack of oxygen. Although hypoxia is often defined as a threshold value, effects on mobile species often occur at levels greater than 2.0 mg l⁻¹ and are better characterized as a continuous function of dissolved oxygen. For example, juvenile red hake (*Urophysis chuss*) increase activity levels at less than 4.2 mg l⁻¹ dissolved oxygen, cease antagonistic interactions at less than 3.2 mg l⁻¹, move up into the water column at less than 2.2 mg l⁻¹, remain within 3 cm of the surface at less than 1.5 mg l⁻¹, cease food searching activity at less than 1.2 mg l⁻¹, and die at less than 0.42 mg l⁻¹ [Bejda et al., 1987]. Species vary greatly in their capacity to adjust physiologically to low dissolved oxygen, the level at which such responses are invoked, as well as the type of response. This wide range in response among species to dissolved oxygen can translate into both positive and negative effects on population growth and mortality rates.

Effects of Hypoxia on Mobile Species in the Gulf of Mexico—What is Known

Hypoxia in the Northwestern Gulf of Mexico

Hypoxia is a seasonally dominant feature on the continental shelf of the northwestern Gulf of Mexico and is most intense from June to August but has been detected as early as February and as late as October. The seasonal and spatial dynamics, mechanism of formation and relationship to historical changes in nutrient loading from the Mississippi river have been described [Turner and Rabalais, 1991, 1994; Dortch et al., 1994; Rabalais et al., 1994a, 1996]. Briefly, spring (March-May) freshwater outflow from the Mississippi and Atchafalaya Rivers results in salinity stratification of adjacent continental shelf waters and high nutrient levels (particularly nitrogen) in the upper water column. Spring flow from these systems is predominantly to the west along the Louisiana continental shelf. Weak winds and solar warming maintain stratification through most of the summer months (June-August). Excess carbon is exported to the bottom as phytoplankton cells, cell aggregates and zooplankton fecal pellets. Water column and benthic respiration and bacterial processes result in a net deficit of dissolved oxygen in the lower water column and at the seabed. Wind reversals during late summer maintain hypoxic waters on the Louisiana shelf until strong winds and storms break down stratification in the fall (September-October). The areal extent and shape of the hypoxic zone varies from year to year, but may extend over several thousand square kilometers (up to ~20,000 km^2) off the Louisiana coast, sometimes reaching the eastern Texas coast [Rabalais et al., 1998; Rabalais et al., this volume]). Hypoxic bottom waters generally occur in 5-30 m water depth from 5-30 km from shore but have been documented at depths of up to 60 m and 130 km from shore. The vertical extent of hypoxia in the water column is determined by the depth of the primary and secondary pycnoclines but is generally restricted to the bottom few meters. At times hypoxic water may occupy as much as 50 to 80% of the water column, particularly at shallower depths.

Impacts of Hypoxia on Mobile Fish and Invertebrates

Few studies have addressed the potential impacts of hypoxia on mobile fish and invertebrates in the northwestern Gulf of Mexico. Most inferences regarding ecological effects on mobile species have been drawn from studies with other primary objectives (e.g., resource monitoring surveys, benthic community studies). Using data from a shelfwide resource assessment survey of waters west of Mobile Bay, Alabama to Sabine, Texas (9-110 m) during June of 1982, Leming and Stuntz [1984] found that demersal shrimp and finfish catch per unit effort (CPUE, kg h^{-1}) varied independently of dissolved oxygen for levels above 2.5 mg l^{-1} but declined to near zero at lower oxygen concentrations. In areas of low dissolved oxygen, catches were comprised primarily of pelagic species presumably taken above hypoxic bottom waters as the trawl was deployed or retrieved [Stuntz et al., 1982]. The extent of the hypoxic zone was estimated at 1200 km^2 based on contour plots of bottom dissolved oxygen measurements made from the research vessel. Similar assessment surveys of nearshore waters off Louisiana have reported decreases in the number of individuals and species, as well as total weight of demersal trawl catches in areas of low dissolved oxygen [Hanifen et al., 1997]. Renaud [1986] reported hypoxic (< 2.0 mg l^{-1}) and anoxic (0.0 mg l^{-1}) bottom waters at nearshore (depth < 16 m) and offshore (4-20 m depth) sites off Louisiana during a 2-wk period in June 1983. Simultaneous bottom trawling indicated that demersal fish biomass and the number of penaeid shrimp were generally lower in these areas than at normoxic stations. Pavela et al. [1983] surveyed a transect perpendicular to the Texas coast (5-47 m depth) monthly for four years in an area known to experience hypoxic bottom waters. The abundance of fish and shrimp varied independently of dissolved oxygen concentration when levels were high (> 4-5 mg l⁻¹) but declined sharply when levels dropped below 2.0 mg l⁻¹. Recolonization of previously hypoxic areas by fish and mobile invertebrates was observed within weeks of dissolved oxygen levels returning to normal. Zimmerman et al. [1997] reported increased landings of penaeid shrimp in nearshore waters of the northwestern Gulf during years of extensive hypoxia, presumably due to impeded migration to offshore waters. This conclusion is consistent with results of a mark-recapture study in which shrimp moved greater distances across and along the Louisiana shelf during a year with minimal hypoxia (1979) than during a year when hypoxia was extensive (1978) [Gazey et al., 1982], although different movement patterns could be the result of interannual differences other than hypoxia. Numerous other studies note (anecdotally) the absence of "normal" fish communities in areas subject to hypoxia [Harper et al., 1981; Gaston, 1985; Beardsley, 1997; Malakoff, 1998]. It is clear from the above studies that: (1) the abundance of mobile fish and invertebrates decreases in areas of the Gulf where dissolved oxygen levels are low (~ 2.0 mg l⁻¹), (2) decreases in abundance are most likely due to horizontal displacement, although direct mortality, horizontal movement and vertical movement are often confounded, and (3) in locations where hypoxia is an aperiodic event, mobile species are able to re-occupy previously hypoxic areas fairly rapidly (days-weeks). Little is known, however, regarding the extent and duration of spatial displacement, particularly in areas of chronic hypoxia, variation in response among species, or the population-level consequences of habitat loss. In addition, hypoxia in the northwestern Gulf occurs on a much larger spatial and temporal scale than that of most studies conducted to date, making regional generalizations difficult.

Methods

The scale of hypoxia in the northwestern Gulf of Mexico makes synoptic sampling of water quality parameters, fishery resources and other species of management concern logistically difficult and often prohibitively expensive. Existing fishery-independent resource assessment surveys may contain useful information to address the ecological effects of hypoxia and/or aid in future sampling designs. While these surveys are generally conducted with the goal of developing a time series of abundance for species of management concern, there is significant spatial and temporal overlap with areas subject to seasonal hypoxia in the northwestern Gulf. We used information from shipboard sampling of bottom dissolved oxygen as well as demersal fish and crustaceans, and related these data to results from aerial surveys of cetaceans and sea turtles to investigate potential impacts of hypoxia on demersal species and upper trophic levels. Quasi-synoptic data were available for 1992 and 1993.

Our general hypotheses are that hypoxia impacts the spatial distribution of demersal species and that the distribution of upper trophic levels responds to these changes in the distribution of potential food resources. Specifically, we hypothesize that the biomass of demersal resources during June-July will be low in areas experiencing hypoxia and high in adjacent areas due to spatial displacement. Because hypoxia is generally absent from the continental shelf by October, fall surveys provide a "pseudo-control" for patterns in the distribution of demersal biomass observed during summer. The effects of low dissolved oxygen, however, may persist after hypoxic bottom waters have dissipated. For example, the biomass of benthic communities that are a major food resource for demersal species may remain depressed after dissolved oxygen in overlying bottom water returns to normal levels [Harper et al., 1981; Stachowitsch, 1984]. If habitats subject to hypoxia during June-July remain sufficiently degraded until October-November (see Rabalais et al. [this volume]), fall demersal biomass in areas that were hypoxic the previous summer should remain low relative to other areas.

Several species of seas turtles and marine mammals inhabit the northwestern continental shelf. As air breathers, the potential effects of hypoxia on these species are indirect (e.g., spatial displacement of food resources). We used sightings data from aerial surveys in the region to document the spatial distribution of sea turtles and two species of cetaceans, bottlenose dolphins (*Tursiops truncatus*) and Atlantic spotted dolphins (*Stenella frontalis*), that inhabit the continental shelf. Because aerial surveys (September-October) were not synoptic with the period of most intense hypoxia (June-August) and potential effects are most likely secondary responses, we did not have clear expectations of how the distribution of dolphins and sea turtles might be related to hypoxia. Individuals of these species inhabiting the northwestern shelf feed on a wide range of demersal and pelagic fish and crustaceans [Barros and Odell, 1990; Shaver, 1991; Plotkin et al., 1993; Perrin et al., 1994] and are highly mobile [Dodd, 1988; Renaud and Carpenter, 1994; Mate and Worthy, 1995; Schmid, 1995; Davis et al., 1996]. Their distribution may be unaffected by hypoxia if sufficient prey resources remain in hypoxic areas or alternative prey can be exploited (e.g., pelagic species). Alternatively, movement of prey resources may result in altered spatial distributions such as high concentrations on the edge of hypoxic areas or displacement to adjacent habitats.

Shipboard Surveys

Measurements of bottom dissolved oxygen and demersal catch per unit effort were obtained for the summer (June-July) and fall (October-November) of 1992 and 1993 from the Southeast Area Monitoring and Assessment Program database (Table 1; SEAMAP; Eldridge [1988]). This database is comprised of trawl surveys conducted semi-annually on the northwestern Gulf of Mexico continental shelf by the National Marine Fisheries Service (NMFS) and state agencies (Alabama, Mississippi, Louisiana). The surveys cover the area between Mobile Bay, Alabama and Brownsville, Texas in the 9-110 m depth range (5-60 fathoms). Sampling stations are selected randomly within five longitudinal strata, five depth strata and two diel strata. Bottom dissolved oxygen is measured at each station by sensors attached to a conductivity, temperature and depth (CTD) probe. The CTD is calibrated prior to each survey, and readings are periodically checked during the survey against those obtained from a YSI meter. Bottom trawls (12.8-m shrimp trawl with mud rollers and 2.4-m by 1.0-m wooden chain doors) for demersal fish and invertebrates are conducted at each sampling station. The trawl begins at the site of the oxygen measurement and is towed perpendicular to the depth contour for a minimum of 10 min and a maximum of 60 min or until the appropriate depth interval is covered. If the depth interval is not covered in 60 min, subsequent tows are conducted to cover the interval. The individual states sample nearshore strata (9.1-18.3

TABLE 1. Data from 1992 and 1993 shipboard and aerial surveys in the northwestern Gulf of Mexico. "Observations" refers either to the number of stations occupied during shipboard surveys, or the number of sea turtle or cetacean group sightings (bottlenose dolphins, Atlantic spotted dolphins, unidentified dolphins, and unidentified bottlenose/Atlantic spotted dolphins combined) during aerial surveys. Shipboard surveys extended from Mobile Bay, Alabama to Brownsville, Texas. Aerial surveys extended between the United States-Mexico border and the mid-Louisiana coast in 1992, and east of the mid-Louisiana coast in 1993.

Year	Survey	Dates	Species	Observations
1992	Shipboard	Jun 12-Jul 13	Fish/Invertebrates	255
	Shipboard	Oct 14-Nov 19	Fish/Invertebrates	211
	Aerial	Sept 16-Oct 24	Cetaceans	290
			Sea Turtles	41
1993	Shipboard	Jun 19-Jul 21	Fish/Invertebrates	228
	Shipboard	Oct 15-Nov 18	Fish/Invertebrates	253
			Cetaceans	338
	Aerial	Sept 17-Oct 19	Sea Turtles	32

m) using the same sampling gear and methodology (Alabama, Mississippi, Louisiana). We included state SEAMAP data collected within the time period defined by NMFS summer and fall surveys (Table 1). Additional information regarding survey design and procedures can be found in Donaldson et al. [1998].

Aerial Surveys

Sightings of sea turtles, bottlenose dolphins and Atlantic spotted dolphins were taken from regional aerial surveys conducted in September-October of 1992 and 1993 (Table 1). In 1992 the survey covered continental shelf waters between the United States-Mexico border and the mid-Louisiana coast. In 1993 the survey covered the area east of the mid-Louisiana coast. Details of the survey design, objectives and temporal/spatial coverage can be found in Anonymous [1992, 1993]. Briefly, the survey is divided into three sub-regions based on depth strata: (1) bays and sounds, (2) nearshore strata (coastline-18.3 m isobath), and (3) offshore strata (9.3 km past the 182.9 m isobath). Surveys cover the entire shelf region from shore to the shelf/slope break with transects allocated to sub-regions based on prior density estimates [Blaylock and Hoggard, 1994].

Analysis

We used a geographic information system to integrate information on bottom dissolved oxygen and demersal species abundance from shipboard surveys with cetacean and sea turtle sightings data from aerial surveys. We used an inverse distance weighting procedure to interpolate bottom dissolved oxygen values for the entire northwestern

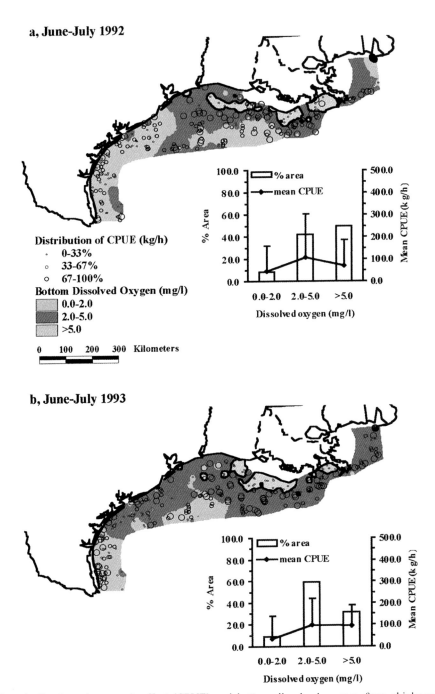

a, June-July 1992

Distribution of CPUE (kg/h)
- · 0-33%
- ○ 33-67%
- ○ 67-100%

Bottom Dissolved Oxygen (mg/l)
- 0.0-2.0
- 2.0-5.0
- >5.0

0   100   200   300   Kilometers

b, June-July 1993

Plate 1. Total catch per unit effort (CPUE) and bottom dissolved oxygen from shipboard surveys during June-July of (a) 1992 and (b) 1993. The legend applies to both maps. Circles refer to the upper, middle and lower third of the distribution of total CPUE (all fish and invertebrates, kg h[-1]) for the entire survey. The graphs depict the percent shelf area in various dissolved oxygen zones (histogram bars) and the mean and standard deviation of total CPUE.

continental shelf (9-110 m). This method assumes spatial autocorrelation in the data and weights locations closer to a given location greater than those further away. We interpolated bottom dissolved oxygen values for 10-km^2 cells using the eight nearest neighbor cells. To examine relationships between demersal biomass and bottom-water dissolved oxygen we overlaid total CPUE (kg h^{-1}) calculated for each bottom trawl with the interpolated bottom dissolved oxygen surface. Similarly, we extracted and overlaid sightings data from the aerial surveys for the area overlapping the shipboard data (U.S.-Mexico border to Mobile Bay, Alabama). To test for a relationship between bottom dissolved oxygen and demersal biomass, we grouped the data into three dissolved oxygen classes (hypoxic: 0.0-2.0, intermediate: 2.0-5.0, normoxic: > 5.0 mg l^{-1}) and tested for effects of dissolved oxygen class and year for summer and fall surveys separately. When significant differences were found, statistical contrasts were used to evaluate differences between particular dissolved oxygen classes (hypoxic versus intermediate, intermediate versus normoxic).

Results and Discussion

Spatial Distribution of Bottom Dissolved Oxygen in the Northwestern Gulf of Mexico

Bottom-water hypoxia was widespread over the northwestern Gulf of Mexico continental shelf during June-July of 1992 and 1993 (Plate 1). Hypoxia was detected at stations extending from 9-31 m depth in an east-west orientation along the majority of the Louisiana coast. The areal extent of hypoxia was estimated as 11,400 km^2 in 1992 and 12,100 km^2 in 1993, corresponding to 8.3 and 8.8% of the continental shelf surveyed (137,100 km^2), respectively (Plate 1). No bottom-water hypoxia was found east of the Mississippi delta region. Isolated areas of hypoxic bottom water were also found further west off the eastern Texas coast. Because hypoxia was detected to the 9.1-m isobath (the landward edge of the survey region), our areal estimates may be minimum estimates of the total shelf area subject to hypoxia. Rabalais et al. [1994a] reported hypoxia in water as shallow as 5 m in this region during these years. Based on transects conducted perpendicular to the shoreline over 5-d periods in July of 1992 and 1993 (7/24-7/28; 3 and 11 d after the end of the 1992 and 1993 surveys), hypoxic bottom waters were documented over an approximate area of 10,804 km^2 and 17,600 km^2, respectively [Rabalais et al., 1994b]. Differences between the surveys in the estimated areal extent of hypoxia may also result from different sampling designs and analytic methods, or from short-term dynamics (days-weeks) in processes resulting in the formation or dissipation of hypoxic bottom waters. Our estimates are based on data over an approximate one month period (32 and 33 days for 1992 and 1993, respectively), while those of Rabalais et al. [1998] are more synoptic over 5 d. In both years the SEAMAP stations were occupied in a west to east direction over the majority of the survey. Sampling off Louisiana where hypoxia was most extensive took place over an approximate 2-wk period in early July (7/2/92-7/13/92, 7/6/93-7/17/93). Given this short time frame, the direction of sampling, and current patterns in the region [Cochrane and

Kelly, 1986], it is unlikely that we overestimated the spatial extent of hypoxia by repeat sampling of moving water parcels.   In contrast, prevailing westerly currents in conjunction with productivity and nutrient regenerative processes are likely to result in underestimates given the sampling design. In addition to comparable estimates of areal extents, both surveys documented a more disjunct hypoxic zone in 1992 than 1993 (Plate 1). These results suggest that bottom-water hypoxia was a relatively widespread and persistent phenomenon on the continental shelf during 1992-93, possibly as early as June. By October-November very little hypoxic bottom water was detected on the shelf (Plate 2). Only one station off Louisiana in 1992 recorded dissolved oxygen levels below 2.0 mg l^{-1} during October-November (Plate 2a). This is consistent with prior studies indicating that hypoxia generally dissipates by October due to the breakdown of stratification by wind and storms [Rabalais et al., 1994a].

     Perhaps more surprising was the extensive shelf area in which bottom dissolved oxygen levels ranged from 2.0-5.0 mg l^{-1} (Plate 1, 58,000 and 81,600 km^2 for June-July 1992 and 1993, respectively). This zone comprised a majority of the survey area during June-July 1993 (59.5%), stretching well onto the Texas continental shelf. In both years this zone was contiguous with hypoxic areas, generally occurring seaward and to the west. During 1993 bottom dissolved oxygen levels in the range 2.0-5.0 mg l^{-1} persisted well into the fall, occurring over 40.2% (55,100 km^2) of the shelf during October-November (Plate 2b). This intermediate region of bottom dissolved oxygen was present along the entire outer continental shelf as well as the inner shelf along the south Texas coast. In contrast, nearly the entire shelf was > 5.0 mg l^{-1} (98.6%) during the same period in 1992 (Plate 2b). The Mississippi river delivered anomalously high freshwater discharge during 1993 resulting in large nutrient fluxes to the adjacent shelf well into mid-summer, and a doubling in the spatial extent of hypoxic bottom waters occurred compared to the 1985-92 average [Rabalais et al., 1994b].   The larger area of intermediate dissolved oxygen (2.0-5.0 mg l^{-1}) and its persistence well into the fall is consistent with a more severe hypoxic event in 1993 than would be concluded from our June-July estimates of the areal extent of bottom-water hypoxia alone.

Relationship between Bottom Dissolved Oxygen and Demersal Biomass

     The relationship between the biomass of demersal species (fish and invertebrates) and bottom dissolved oxygen during summer was similar in 1992 and 1993 (Plate 1, $F_{1,479} = 0.085$, P = 0.77). Although catch rates were highly variable, mean CPUE during June-July was lower in hypoxic areas than adjacent areas of intermediate dissolved oxygen for both years ($F_{1,479} = 13.4$, P < 0.0001). Trawls began at the point where bottom dissolved oxygen was measured and covered considerable distances (~3.3 km), potentially ending in normoxic waters. Therefore, mean catch rates calculated for hypoxic areas may overestimate the amount of demersal biomass in these regions. For example, when two stations in each year that recorded unusually high catches in hypoxic areas were removed, average biomass was 22.8% (1992) and 13.3% (1993) of that in other areas. In addition to exhibiting relatively low mean CPUE, hypoxic regions were the only areas in which catch rates at the majority of stations (63% in 1992 and 87% in 1993) were in the lower third of those for the entire survey region (Plate 1). Further, the

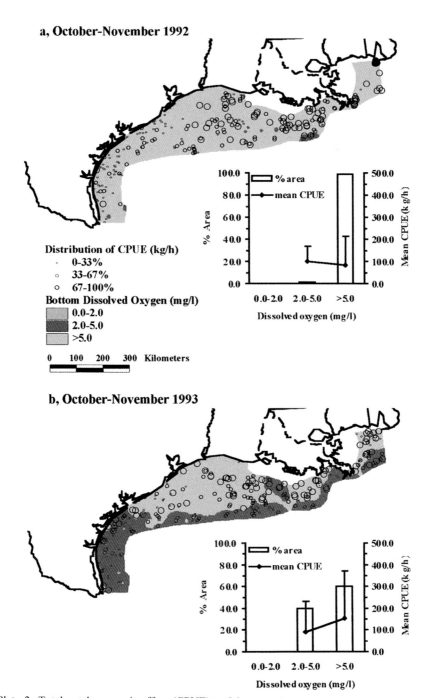

a, October-November 1992

Distribution of CPUE (kg/h)
- · 0-33%
- ○ 33-67%
- ○ 67-100%

Bottom Dissolved Oxygen (mg/l)
- 0.0-2.0
- 2.0-5.0
- >5.0

0   100   200   300   Kilometers

b, October-November 1993

Plate 2. Total catch per unit effort (CPUE) and bottom dissolved oxygen from shipboard surveys during October-November of (a) 1992 and (b) 1993. The legend applies to both maps. Circles refer to the upper, middle and lower third of the distribution of total CPUE (all fish and invertebrates, kg h^{-1}) for the entire survey. The graphs depict the percent shelf area in various dissolved oxygen zones (histogram bars) and the mean and standard deviation of total CPUE.

ten stations in which no catch was obtained occurred where dissolved oxygen concentration was < 2.3 mg l^{-1}. These results support those of Leming and Stuntz [1984] who found consistently low shrimp and finfish biomass in areas where dissolved oxygen levels were below 2.5 mg l^{-1} during 1982 when hypoxia extended over an area approximately 10% that of current estimates, and indicate that extensive areas of the continental shelf are largely unavailable as habitat to most demersal species during a significant portion of June-July.

The highest mean CPUE during June-July of both years occurred in areas that were intermediate in bottom dissolved oxygen concentration with slightly lower catches recorded from adjacent normoxic areas, although differences were marginally insignificant (Plate 1, $F_{1,479}$ = 3.65, P = 0.057). Relatively high mean catch rates in areas of intermediate dissolved oxygen most likely result from spatial displacement of demersal species from adjacent hypoxic waters. Some of the highest catches occurred in areas of intermediate dissolved oxygen on the edge of hypoxic regions (Plate 1). The majority of the continental shelf in which bottom dissolved oxygen was > 5.0 mg l^{-1} occurred off the Texas coast. Sciaenids, such as spot (*Leiostomus xanthurus*), Atlantic croaker (*Micropogonias undulatus*) and seatrout (*Cynoscion* spp.), approach the southern limit of their distribution near south Texas, and this may account for the lower catch rates in normoxic waters [Donaldson et al., 1998].

In contrast to June-July, the relationship between mean CPUE and bottom dissolved oxygen concentration during October-November was not consistent between the two years (Plate 2, $F_{1,461}$ = 3.33, p = 0.037). In 1992 there was no difference in demersal biomass between intermediate and normoxic bottom waters (Plate 2a, $F_{1,209}$ = 0.26, P = 0.61), while CPUE was slightly higher in normoxic waters in 1993 (Plate 2b, $F_{1,251}$ = 6.54, P = 0.01; there was only one hypoxic station during the two fall surveys). The contrasting patterns between summer and fall in conjunction with the relatively low catches in hypoxic areas and high catches in areas > 2.0 mg l^{-1} suggests demersal biomass was displaced from hypoxic waters to adjacent areas of intermediate bottom dissolved oxygen. This displacement was particularly evident off central Louisiana in the Mississippi delta region in both years where biomass seemed to be concentrated on the edge of hypoxic regions (Plate 1). Because hypoxia occurred up to the landward edge of our surveys, displacement to nearshore habitats is difficult to evaluate. Hypoxia is generally less persistent nearshore because of wind and tidal mixing of shallow shelf waters. However, high catches occurred both inshore and offshore of hypoxic bottom waters off western Louisiana in 1992 (Plate 1a), suggesting movement to nearshore areas occurs as well.

Areas experiencing hypoxia during June-July were re-occupied by demersal species at least by October-November (Figure 1). Some of the highest catches observed occurred in areas that were hypoxic the previous summer. In 1992 fall CPUE was unrelated to whether areas were hypoxic, intermediate or normoxic the prior June-July (Figure 1a, $F_{2,208}$ = 1.75, P = 0.18), while in 1993 significant differences in demersal biomass were only found between areas where bottom waters were previously intermediate and normoxic (Figure 1b, $F_{1,251}$ = 7.09, P = 0.008). These results suggest a weak coupling between the distribution of demersal biomass during October-November and the spatial extent of hypoxia during the prior summer. While lag effects of hypoxia on a seasonal time scale may not be important, species presence is a poor surrogate of habitat quality.

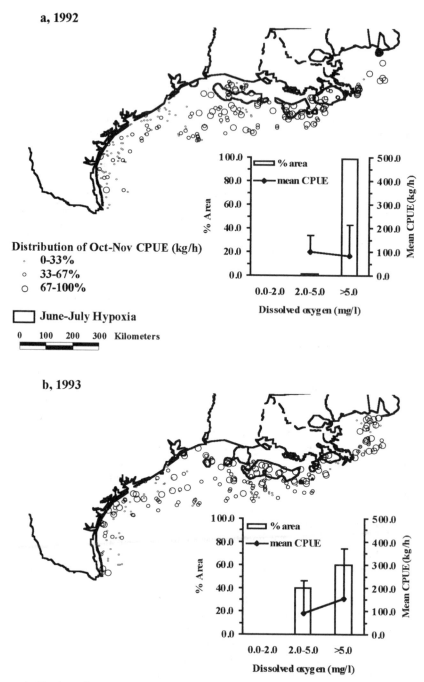

Figure 1. Total catch per unit effort (CPUE) during October-November and distribution of hypoxia (dissolved oxygen < 2.0 mg l^{-1}) during June-July of (a) 1992 and (b) 1993 from shipboard surveys. The legend applies to both maps. The graphs depict the mean and standard deviation of total CPUE (all fish and invertebrates, kg h^{-1}) during fall (October-November) in three dissolved oxygen zones from the previous summer (June-July).

Demersal fish and invertebrates may experience effects on growth and condition factors if benthic food resources remain degraded subsequent to the return of normoxic conditions in bottom waters. In addition, assuming no lag effect of low dissolved oxygen on habitat quality, impacts of hypoxia on growth or condition the prior summer may have implications for mortality or reproduction at later stages, particularly to the extent that these processes are size-dependent.

Spatial Distribution of Sea Turtles and Cetaceans in Relation to Bottom Dissolved Oxygen

A substantial number of sea turtles were sighted during aerial surveys in September-October (Table 1, Figure 2a). Four species of sea turtles were observed, Kemp's ridley (*Lepidochelys kempii*), loggerhead (*Caretta caretta*), leatherback (*Dermochelys coriacea*) and green (*Chelonia mydas*), but none were observed in areas that were hypoxic the prior June-July. Within the survey region in 1992-93, turtle abundance was relatively high off Mississippi/Alabama and off south Texas at depths from 0-90 m, but turtles were most common in the nearshore zones [McDaniel et al., 2000]. Turtle sightings were dominated by Kemp's ridley and loggerhead turtles. Both species feed in the benthos, primarily on crabs [Shaver, 1991; Plotkin et al., 1993] and may be impacted by low oxygen concentrations via effects on the distribution of benthic food resources.

During aerial surveys in September-October both bottlenose dolphins and Atlantic spotted dolphins were sighted on the continental shelf (Figure 2b). Groups of bottlenose dolphins were widely distributed over the shelf and comprised the majority of sightings (92% and 99% of sightings in 1992 and 1993, respectively). The distribution of bottlenose dolphin sightings was more evenly distributed across the shelf off eastern Texas/western Louisiana and Mississippi than in the Mississippi delta region off Louisiana (Figure 2b). In this region herd sightings were disjunct, with sightings in nearshore bays and shallow waters, and offshore of the ~20-m isobath. Few sightings occurred between the 5- and 20-m isobath, an area characterized by significant hypoxia during June-July of 1992 and 1993, although sightings occurred in this depth range in other areas. There is insufficient data to evaluate whether the spatial distribution of cetaceans is impacted by seasonal hypoxia. The disjunct September-October distribution in the vicinity of the Mississippi delta region may be a natural result of the population structure of these species (offshore/coastal stocks; Dowling and Brown [1993]; Waring et al. [1997]), interannual variability in distribution or the result of other oceanographic features such as the freshwater plumes of the Mississippi and Atchafalaya Rivers. In addition, it is unclear if September-October distribution patterns are representative of those during months of intense hypoxia (June-August). Because they are homeothermic, the distribution of marine mammals is generally believed to be driven more by the availability of food resources than physical characteristics of the environment [Davis et al., 1998]. Areas on the shelf in which dissolved oxygen levels were in the 2.0-5.0 mg l^{-1} range had the highest CPUE of demersal fish and

a, Sea Turtles

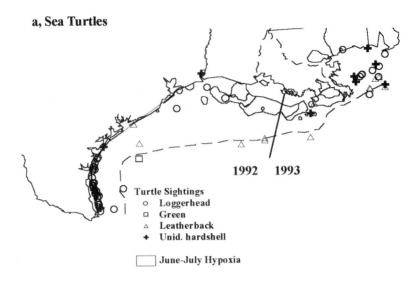

1992   1993

Turtle Sightings
- o   Loggerhead
- □   Green
- △   Leatherback
- +   Unid. hardshell

☐ June-July Hypoxia

0   100   200   300   Kilometers

b, Cetaceans

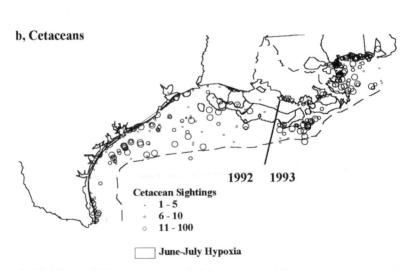

1992   1993

Cetacean Sightings
- ·   1 - 5
- ◦   6 - 10
- o   11 - 100

☐ June-July Hypoxia

Figure 2. Sightings of (a) sea turtles and (b) cetaceans (Atlantic spotted and bottlenose dolphins) during October-November 1992 and 1993 from aerial surveys and distribution of hypoxia (dissolved oxygen < 2.0 mg l^{-1}) during June-July 1992-93 from shipboard surveys. 92% of sea turtle sightings were single animals and 96% of cetacean sightings were confirmed bottlenose dolphins for the two years. Dashed lines delineate the boundary for the 1992 and 1993 aerial surveys.

crustaceans (Plate 1). These catches were dominated by sciaenids, Atlantic croaker (*Micropogonias undulatus*), spot (*Leiostomus xanthurus*) and seatrout (*Cynoscion* spp.), which are a major component of the diet of bottlenose dolphins in this region [Barros and Odell, 1990]. Thus, the distribution of bottlenose dolphins may be indirectly impacted to the extent that hypoxia modifies the spatial distribution of demersal prey resources.

Altered spatial distributions of demersal fish, cetaceans and sea turtles may have implications for regional fisheries. The shrimp fishery in the northwestern Gulf of Mexico is the highest valued fishery in the United States [Condrey and Fuller, 1992]. The height of its execution takes place during the summer when bottom-water hypoxia is most extensive. Bycatch of demersal fish and sea turtles are well documented for this fishery and have been the subject of recent management efforts [Condrey and Fuller, 1992; TEWG, 1998]. In addition, bottlenose and Atlantic spotted dolphins are known to interact with shrimp boats in the northwestern gulf [Waring et al., 1997]. For example, relative to other regions in the Gulf of Mexico and South Atlantic, bottlenose dolphin diets in the northwestern Gulf are more diverse, contain a higher proportion of shrimp, and contain fish that are more similar in size and species composition to shrimp fishery discards [Barros and Odell, 1990]. Based on these observations and numerous sightings of bottlenose dolphins in association with shrimp boats (e.g., Leatherwood [1975]), Barros and Odell [1990] hypothesized that at least some dolphins feed on discards from the shrimp fishery. Zimmerman et al. [1997] demonstrated that shrimping effort is displaced to nearshore waters (< 18 m water depth) during years of extensive hypoxia. Thus, large-scale hypoxia may modify interactions between regional fisheries and nontarget or protected species due to shifts in spatial distribution of prey. These interactions may be intensified if the fishery and these species are concentrated in nearshore (or offshore) areas.

Conclusions

Dissolved oxygen can impact mobile species in a variety of ways. At a minimum, low dissolved oxygen levels impose subtle physiological costs, while at a maximum they can result in direct mortality. Although hypoxia is a widespread, persistent feature of the northwestern Gulf of Mexico continental shelf, little is known regarding its impact on demersal species. Direct mortality has been observed, but its magnitude is difficult to quantify in such a large, open system. It is clear from laboratory studies that low dissolved oxygen can impact individual energy budgets, but the importance of sublethal effects on growth, mortality and reproductive processes in the field is largely unknown.

Our analysis supports previous studies indicating that several thousand square kilometers of the northwestern continental shelf experience hypoxic bottom waters during summer. Average catch rates of demersal fish and invertebrates in these areas were much lower than in other areas, indicating that suitable habitat is greatly reduced during a significant portion of the growing season for many demersal species. In addition, biomass in adjacent areas of intermediate dissolved oxygen was increased relative to other areas, suggesting species move out of hypoxic regions, possibly concentrating on the edge of the hypoxic zone. Thus, the impacts of low dissolved oxygen on mobile species extend well beyond the boundaries of hypoxic areas. A possible consequence of such extensive seasonal habitat loss and corresponding increased density of organisms in adjacent habitats is a decline in overall productivity of demersal species. Although demersal species re-occupy hypoxic areas by fall, it is not known whether sublethal effects of hypoxia persist beyond June-July. In the northwestern Gulf of Mexico upper trophic levels such as sea turtles and cetaceans use demersal fish and invertebrates as food resources. It is unclear how closely the distribution of these species is linked to that of demersal prey. The absence of turtles in the hypoxic zone and the disjunct distribution of bottlenose dolphins in the vicinity of the Mississippi delta region suggest that the effects of large-scale, chronic hypoxia should be included among hypotheses to explain the spatial distributions of sea turtles and cetaceans in the northwestern Gulf. Synoptic data on the distribution of bottom dissolved oxygen, demersal fish and invertebrates, and nontarget species will be necessary to further refine and test this hypothesis.

Acknowledgements. We thank Bruce Comyns, Randall Davis, James Nance, Nancy Rabalais, Andy Read, Eugene Turner and an anonymous reviewer for comments on the manuscript. The data used in the preparation of this paper were collected, in part, under a grant/cooperative agreement from the National Oceanic and Atmospheric Administration to the state(s) of Louisiana, Alabama and Mississippi. The views expressed herein are those of the author(s) and do not necessarily reflect the views of NOAA or any of its sub-agencies.

References

Andersson, L. and L. Rydberg, Trends in nutrient and oxygen conditions within the Kattegat: effects of local nutrient supply, *Estuar. Coast. Shelf Sci.*, *26*, 559-579, 1988.

Anonymous, GOMEX92, Gulf of Mexico regional aerial survey design, September-October, 1992, NMFS, SEFSC Contr. MIA-91/92-72, 1992.

Anonymous, GOMEX93 aerial survey design, SEFSC, September, 1993.

Armstrong, J. D., I. G. Priede, and M. C. Lucas, The link between respiratory capacity and changing metabolic demands during growth of northern pike, *Esox lucius* L., *J. Fish Biol.*, *41(Suppl. B)*, 65-75, 1992.

Baden, S. P., L. Pihl, and R. Rosenberg, Effects of oxygen depletion on the ecology, blood physiology and fishery of the Norway lobster (*Nephrops norvegicus*), *Mar. Ecol. Prog. Ser.*, *67*, 141-155, 1990.

Barros, N. B. and D. K. Odell, Food habits of bottlenose dolphins in the southeastern United States, in *The Bottlenose Dolphin*, edited by S. Leatherwood, and R. R. Reeves, pp. 309-328, Academic Press, San Diego, California, 1990.

Beardsley, T., Death in the deep, *Sci. Amer.*, *277*, 17-20, 1997.

Bejda, A. J., A. L. Studholme, and B. L. Olla, Behavioral responses of red hake, *Urophycis chuss*, to decreasing concentrations of dissolved oxygen, *Env. Biol. Fishes*, *19*, 261-268, 1987.

Blaylock, R. A. and W. Hoggard, Preliminary estimates of bottlenose dolphin abundance in southern U.S. Atlantic and Gulf of Mexico continental shelf waters, NOAA Tech. Memo., NMFS-SEFSC-356, 1994.

Bonsdorff, E., E. M. Blomqvist, J. Mattila, and A. Norkko, Coastal eutrophication: causes and perspectives in the archipelago areas of the northern Baltic Sea, *Estuar. Coast. Shelf Sci.*, *44(Suppl. A)*, 63-72, 1997.

Breitburg, D. L., Episodic hypoxia in Chesapeake Bay: interacting effects of recruitment, behavior, and physical disturbance, *Ecol. Monog.*, *62*, 525-546, 1992.

Breitburg, D. L., T. Loher, C. A. Pacey, and A. Gerstein, Varying effects of low dissolved oxygen on trophic interactions in an estuarine food web, *Ecol.Monog.*, *67*, 489-507, 1997.

Breitburg, D. L., K. A. Rose, and J. H. Cowan, Jr., Linking water quality to fish recruitment: predation mortality of fish larvae in an oxygen-stratified water column, *Mar. Ecol. Prog. Ser.*, *178*, 39-54, 1999.

Brett, J. R. and J. M. Blackburn, Oxygen requirements for growth of young coho (*Oncorhynchus kisutch*) and sockeye (*O. nerka*) salmon at 15° C, *Can. J. Fish. Aquat. Sci.*, *38*, 399-404, 1981.

Burton, D. T. and A. G. Heath, Ambient oxygen tension (P_O) and transition to anaerobic metabolism in three species of freshwater fish, *Can. J. Fish. Aquat. Sci.*, *37*, 1216-1224, 1980.

Bushnell, P. G., J. F. Steffensen, and K. Johansen, Oxygen consumption and swimming performance in hypoxia-acclimated rainbow trout (*Salmo gairdneri*), *J. Exp. Biol.*, *113*, 225-235, 1984.

Cech, J. J. Jr., S. J. Mitchell, and T. E. Wragg, Comparative growth of juvenile white sturgeon and striped bass: effects of temperature and hypoxia, *Estuaries*, *7*, 12-18, 1984.

Chapman, L. J., L. S. Kaufman, C. A. Chapman, and F. E. McKenzie, Hypoxia tolerance in twelve species of east African cichlids: potential for low oxygen refugia in Lake Victoria, *Cons. Biol.*, *9*, 1274-1288, 1995.

Coble, D. W., Fish populations in relation to dissolved oxygen in the Wisconsin river, *Trans. Amer. Fish. Soc.*, *111*, 612-623, 1982.

Cochrane, J. D., and F. J. Kelly, Low-frequency circulation on the Texas-Louisiana shelf, *J. Geophys. Res.*, *91*, 10645-10659, 1986.

Condrey, R. and D. Fuller, The U.S. Gulf shrimp fishery, in *Climate Variability, Climate Change, and Fisheries*, edited by M. H. Glanz, pp. 89-119, Cambridge University Press, London, 1992.

Cooper, S. R. and G. S. Brush, Long-term history of Chesapeake Bay anoxia, *Science*, *254*, 992-996, 1991.

Cooper, S. R., Chesapeake Bay watershed historical land use: impact on water quality and diatom communities, *Ecol. Appl.*, *5*, 703-723, 1995.

Crocker, C. E. and J. J. Cech, Jr., Effects of environmental hypoxia on oxygen consumption rate and swimming activity in juvenile white sturgeon, *Acipenser transmontanus*, in relation to temperature and life intervals, *Env. Biol. Fishes*, *50*, 383-389, 1997.

Das, T. and W. B. Stickle, Detection and avoidance of hypoxic water by juvenile *Callinectes sapidus* and *C. similis*, *Mar. Biol.*, *120*, 593-600, 1994.

Davis, R. W., G. A. J. Worthy, B. Wursig, and S. K. Lynn, Diving behavior and at-sea movements of an Atlantic spotted dolphin in the Gulf of Mexico, *Mar. Mam. Sci.*, *12*, 569-581, 1996.

Davis, R. W., G. S. Fargion, N. May, T. D. Leming, M. Baumgartner, W. E. Evans, L. J. Hansen, and K. Mullin, Physical habitat of cetaceans along the continental slope in the north-central and western Gulf of Mexico, *Mar. Mam. Sci.*, *14*, 490-507, 1998.

Deubler, E. E., Jr. and G. S. Posner, Response of postlarval flounders, *Paralichthys lethostigma*, to water of low oxygen concentrations, *Copeia*, *1963*, 312-317, 1963.

Diaz, R. J. and R. Rosenberg, Marine benthic hypoxia: a review of its ecological effects and the behavioural responses of benthic macrofauna, *Ocean. Mar. Biol. Ann. Rev.*, *33*, 245-303, 1995.

Dizon, A. E., Effect of dissolved oxygen concentration and salinity on the swimming speed of two species of tunas, *Fish. Bull.*, *75*, 649-653, 1977.

Dodd, C. K., Jr., Synopsis of the biological data on the loggerhead sea turtle (*Caretta caretta*, Lineaeus, 1758), USFWS Biol. Rep., 88, 1-110, 1988.

Donaldson, D. M., N. Sanders, Jr., D. Hanisko, and P. A. Thompson, SEAMAP environmental and biologial atlas of the Gulf of Mexico, 1996, Gulf States Mar. Fish. Comm., 52, 1-263, 1998.

Dortch, Q., N. N. Rabalais, R. E. Turner, and G. T. Rowe, Respiration rates and hypoxia on the Louisiana shelf, *Estuaries*, *17*, 862-872, 1994.

Dowling, T. E. and W. M. Brown, Population structure of the bottlenose dolphin (*Tursiops truncatus*) as determined by restriction endonuclease analysis of mitochondrial DNA, *Mar. Mam. Sci.*, *9*, 138-155, 1993.

Eldridge, P. J., The southeast area monitoring and assessment program (SEAMAP): a state-federal-university program for collection, management, and dissemination of fishery-independent data and information in the southeastern United States, *Mar. Fish. Rev.*, *50*, 29-39, 1988.

Falkowski, P. G., T. S. Hopkins, and J. J. Walsh, An analysis of factors affecting oxygen depletion in the New York Bight, *J. Mar. Res.*, *38*, 479-506, 1980.

Fisher, P., K. Rademacher, and K. Kils, *In situ* investigations on the respiration and behaviour of the eelpout (*Zoarces viviparus*) under short-term hypoxia, *Mar. Ecol. Prog. Ser.*, *88*, 181-184, 1992.

Garlo, E. V., C. B. Milstein, and A. E. Jahn, Impact of hypoxic conditions in the vicinity of Little Egg Inlet, New Jersey in summer 1976, *Estuar. Coast. Shelf Sci.*, *8*, 421-432, 1979.

Gaston, G. R., Effects of hypoxia on macrobenthos of the inner shelf off Cameron, Louisiana, *Estuar. Coast. Shelf Sci.*, *20*, 603-613, 1985.

Gazey, W. J., B. J. Ballaway, R. C. Fechhelm, L. R. Martin, and L. A. Reitsema, Shrimp mark release and port interview sampling survey of shrimp catch and effort with recovery of captured tagged shrimp, in Shrimp Population Studies: West Hackberry and Big Hill Brine Disposal Sites off Southwest Louisiana and Upper Texas coasts, 1980-1982, Vol. II, edited by W. B. Jackson, NOAA/NMFS Final Rep. DOE., 1982.

Gee, J. H., Respiratory patterns and antipredator responses in the central mudminnow, *Umbra limi*, a continuous, facultative, air-breathing fish, *Can. J. Zool.*, *58*, 819-827, 1980.

Hanifen, J. G., W. S. Perret, R. P. Allemand, and T. L. Romaire, Potential impacts of hypoxia on fisheries: Louisiana's fishery-independent data, in Proc., First Gulf of Mexico Hypoxia Management Conference, December 1995, New Orleans, Louisiana, pp. 87-100, Publ. No. EPA-55-R-97-001, Gulf of Mexico Program Office, Stennis Space Center, Mississippi, 1997.

Harper, D. E., Jr., L. D. McKinney, R. R. Salzer, and R. J. Case, The occurrence of hypoxic bottom water off the upper Texas coast and its effects on the benthic biota, *Contr. Mar. Sci.*, *24*, 53-79, 1981.

Justic', D., T. Legovic', and L. Rottini-Sandrini, Trends in oxygen content 1911-1984 and occurrence of benthic mortality in the northern Adriatic Sea, *Estuar. Coast. Shelf Sci.*, *25*, 435-445, 1987.

Kramer, D. L. and J. P. Mehegan, Aquatic surface respiration, an adaptive response to hypoxia in the guppy, *Poecilia reticulata* (Pisces, Poeciliidae), *Env. Biol. Fishes*, *6*, 299-313, 1981.

Kramer, D. L., D. Manley, and R. Bourgeois, The effect of respiratory mode and oxygen concentration on the risk of aerial predation in fishes, *Can. J. Zool.*, *61*, 653-665, 1983.

Leatherwood, S., Some observations of feeding behavior of bottlenose dolphins (*Tursiops truncatus*) in the northern Gulf of Mexico and (*Tursiops* cf. *T. gilli*) off southern California, Baja California, and Nayarit, Mexico, *Mar. Fish. Rev.*, *37*, 10-16, 1975.

Leming, T. D. and W. E. Stuntz, Zones of coastal hypoxia revealed by satellite scanning have implications for strategic fishing, *Nature*, *310*, 136-138, 1984.

Lenihan, H. S. and C. H. Peterson, How habitat degradation through fishery disturbance enhances impacts of hypoxia on oyster reefs, *Ecol. Appl.*, *8*, 128-140, 1998.

Magnuson, J. J., A. L. Beckel, K. Mills, and S. B. Brandt, Surviving winter hypoxia: behavioral adaptations of fishes in a northern Wisconsin winterkill lake, *Env. Biol. Fishes*, *14*, 241-250, 1985.

Malakoff, D., Death by suffocation in the Gulf of Mexico, *Science*, *281*, 190-192, 1998.

Mate, B. R. and G. Worthy, Tracking the fate of rehabilitated dolphins, in Proceedings of the Eleventh Biennial Conference on the Biology of Marine Mammals, Dec. 14-18, 1995, abstract, Society for Marine Mammology, 1995.

May, E. B., Extensive oxygen depletion in Mobile Bay, Alabama, *Limnol. Oceanogr.*, *18*, 353-366, 1973.

McDaniel, C. J., L. B. Crowder, and J. A. Priddy, Spatial dynamics of sea turtle abundance and shrimping intensity in the Gulf of Mexico, *Cons. Ecol.*, *4(1)*, 15 [online], <http://www.consecol.org/vol4/iss1/art15>, 2000.

Mee, L. D., The Black Sea in crisis: a need for concerted international action, *Ambio*, *21*, 278-286, 1992.

Mellergaard, S. and E. Nielsen, Fish disease investigations in Danish coastal waters with special reference to the impact of oxygen deficiency, ICES, C.M. 1990/E, 6, Mar. Env. Qual. Comm., 1990.

Metcalfe, J. D. and P. J. Butler, Changes in activity and ventilation in response to hypoxia in unrestrained, unoperated dogfish (*Scyliorhinus canicula* L.), *J. Exp. Biol.*, *108*, 411-418, 1984.

Nilsson, G. E., P. Rosen, and D. Johannson, Anoxic depression of spontaneous locomotor activity in crucian carp quantified by a computereized imaging technique, *J. Exp. Biol.*, *180*, 153-162, 1993.

Nissling, A., H. Kryvi, and L. Vallin, Variation in egg buoyancy of Baltic cod (*Gadus morhua*) and its implications for egg survival in prevailing conditions in the Baltic Sea, *Mar. Ecol. Prog. Ser.*, *11*, 67-74, 1994.

Nixon, S. W., Marine eutrophication: a growing international problem, *Ambio*, *19*, 101, 1990.

Officer, C. B., R. B. Biggs, J. L. Taft, L. E. Cronin, M. A. Tyler, and W. R. Boynton, Chesapeake Bay anoxia: origin, development, and significance, *Science*, *223*, 22-27, 1984.

Paerl, H. W., J. L. Pinckney, J. M. Fear, and B. L. Peierls, Ecosystem responses to internal and watershed organic matter loading: consequences for hypoxia in the eutrophying Neuse river estuary, North Carolina, *Mar. Ecol. Prog. Ser.*, *166*, 17-25, 1998.

Parker, C. A., and J. E. O'Reilly, Oxygen depletion in Long Island sound: a historical perspective, *Estuaries*, *14*, 248-264, 1991.

Pavela, J. S., J. L. Ross, and M. E. Chittenden, Jr., Sharp reductions in abundance of fishes and benthic macroinvertebrates in the Gulf of Mexico off Texas associated with hypoxia, *Northeast Gulf Sci.*, *6*, 167-173, 1983.

Perrin, W. F., D. K. Caldwell, and M. C. Caldwell, Atlantic spotted dolphin *Stenella frontalis*

(G. Cuvier, 1829), in *Handbook of Marine Mammals, Vol. 5, The First Book of Dolphins*, edited by S. H. Ridgway, and R. Harrison, pp. 173-190, Academic Press, San Diego, California, 1994.

Petersen, J. K. and G. I. Petersen, Tolerance, behaviour and oxygen consumption in the sand goby, *Pomatoschistus minutus* (Pallas), exposed to hypoxia, *J. Fish Biol., 37*, 921-933, 1990.

Pihl, L., S. P. Baden, and R. J. Diaz, Effects of periodic hypoxia on distribution of demersal fish and crustaceans, *Mar. Biol., 108*, 349-360, 1991.

Pihl, L., S. P. Baden, R. J. Diaz, and L. C. Schaffner, Hypoxia-induced structural changes in the diet of bottom-feeding fish and crustacea, *Mar. Biol., 112*, 349-361, 1992.

Plotkin, P. T., M. K. Wickstein, and A. F. Amos, Feeding ecology of the loggerhead sea turtle, *Caretta caretta*, in the northwestern Gulf of Mexico, *Mar. Biol., 115*, 1-5, 1993.

Pouliot, T. and J. de la Noue, Apparent digestibility in rainbow trout (*Salmo gairdneri*): influence of hypoxia, *Can. J. Fish. Aquat. Sci., 45*, 2003-2009, 1988.

Poulin, R., N. G. Wolf, and D. L. Kramer, The effect of hypoxia on the vulnerability of guppies (*Poecilia reticulata*, Poeciliidae) to an aquatic predator (*Astronotus ocellatus*, Cichlidae), *Env. Biol. Fishes, 20*, 285-292, 1987.

Randall, D., The control of respiration and circulation in fish during exercise and hypoxia, *J. Exp. Biol., 100*, 275-288, 1982.

Rabalais, N. N., W. J. Wiseman, Jr., and R. E. Turner, Comparison of continuous records of near-bottom dissolved oxygen from the hypoxia zone along the Louisiana coast, *Estuaries, 17*, 850-861, 1994a.

Rabalais, N. N., R. E. Turner, and W. J. Wiseman, Jr., Hypoxic conditions in bottom waters on the Louisiana-Texas shelf, in Coastal Oceanographic Effects of 1993 Mississippi River Flooding, edited by M. J. Dowgiallo, pp. 50-54, Special NOAA Report, NOAA Coastal Ocean Office/National Weather Service, Silver Spring, Maryland, 1994b.

Rabalais, N. N., R. E. Turner, D. Justic', Q. Dortch, W. J. Wiseman, Jr., and B. K. Sen Gupta, Nutrient changes in the Mississippi River and system responses on the adjacent continental shelf, *Estuaries, 19*, 386-407, 1996.

Rabalais, N. N., R. E. Turner, W. J. Wiseman, Jr., and Q. Dortch, Consequences of the 1993 Mississippi River flood in the Gulf of Mexico, *Regulated Rivers: Research & Management, 14*, 161-177, 1998.

Rahel, F. J. and J. W. Nutzman, Foraging in a lethal environment: fish predation in hypoxic waters of a stratified lake, *Ecology, 75*, 1246-1253, 1994.

Renaud, M. L., Hypoxia in Louisiana coastal wates during 1983: implications for fisheries, *Fish. Bull., 84*, 19-26, 1986.

Renaud, M. L. and S. A. Carpenter, Movements and submergence patterns of loggerhead sea turtles (*Caretta caretta*) in the Gulf of Mexico determined through sattelite telemetry, *Bull. Mar. Sci., 55*, 1-15, 1994.

Rosenberg, R., Eutrophication-the future marine coastal nuisance, *Mar. Pollut. Bull., 16*, 227-231, 1985.

Schmid, J., Marine turtle populations on the east-central coast of Florida: results of tagging studies at Cape Canaveral, Florida, 1986-91, *Fish. Bull., 93*, 139-151, 1995.

Schurmann, H. and J. F. Steffensen, Spontaneous swimming activity of Atlantic cod (*Gadus morhua*) exposed to graded hypoxia at three temperatures, *J. Exp. Biol., 197*, 129-142, 1994.

Shaver, D. J., Feeding ecology of wild and headstarted Kemp's ridley sea turtles in south Texas waters, *J. Herp., 25*, 327, 1991.

Stachowitsch, M., Mass mortality in the Gulf of Trieste: the course of community destruction, *Mar. Ecol. (I.C.Z.N.)*, *5*, 243-264, 1984.

Stewart, N. E., D. L. Shumway, and P. Doudoroff, Influence of oxygen concentration on the growth of juvenile largemouth bass, *J. Fish. Res. Bd. Can.*, *24*, 475-494, 1967.

Stuntz, W. E., N. Sanders, T. D. Leming, K. N. Baxter, and R. M. Barazotto, Area of hypoxic bottom water found in northern Gulf of Mexico, *Coast. Ocean. Climat. News*, *4*, 37-38, 1982.

Suthers, I. M. and J. H. Gee, Role of hypoxia in limiting diel spring and summer distribution of juvenile yellow perch (*Perca flavescens*) in a prairie marsh, *Can. J. Fish. Aquat. Sci.*, *43*, 1562-1570, 1986.

Turner, R. E. and N. N. Rabalais, Changes in Mississippi river water quality this century, *BioScience.*, *41*, 140-147, 1991.

Turner, R. E. and N. N. Rabalais, Coastal eutrophication near the Mississippi river delta, *Nature*, *368*, 619-621, 1994.

TEWG (Turtle Expert Working Group), An assessment of the Kemp's ridley (*Lepidochelys kempi*) and loggerhead (*Caretta caretta*) sea turtle populations in the western north Atlantic, NOAA Tech. Memo., NMFS-SEFSC, 1998.

Vitousek, P. M., H. A. Mooney, J. Lubchenco, and J. M. Melillo, Human domination of Earth's ecosystems, *Science*, *277*, 494-499, 1997.

Waring, G. T., D. L. Palka, K. D. Mullin, J. H. W. Hain, L. J. Hansen, and K. D. Bisack, U.S. Atlantic and Gulf of Mexico Marine Stock Assessments 1996, NOAA Tech. Memo., NMFS-NE-114, 1997.

Weber, J. M. and D. L. Kramer, Effects of hypoxia and surface access on growth, mortality, and behavior of juvenile guppies, *Poecilia reticulata*, *Can. J. Fish. Aquat. Sci.*, *40*, 1583-1588, 1983.

Wolf, N. G. and D. L. Kramer, Use of cover and the need to breathe: the effect of hypoxia on vulnerability of dwarf gouramis to predatory snakeheads, *Oecologia*, *73*, 127-132, 1987.

Zimmerman, R., J. Nance, and J. Williams, Trends in shrimp catch in the hypoxic area of the northern Gulf of Mexico, in Proc., First Gulf of Mexico Hypoxia Management Conference, December 1995, New Orleans, Louisiana, pp. 64-74, Publ. No. EPA-55-R-97-001, Gulf of Mexico Program Office, Stennis Space Center, Mississippi, 1997.

15

Effects of Hypoxia on the Shrimp Fishery of Louisiana and Texas

Roger J. Zimmerman and James M. Nance

Abstract

Large-scale hypoxia, recently approaching 20,000 km^2, overlaps with habitat and fishing grounds of commercial shrimp species in Louisiana and Texas shelf waters. It is expected that an environmental impact of this magnitude would have an effect on the shrimp population that is reflected in catch statistics. In this paper, we examine the geographic distribution and amount of shrimp catch in relation to location and size of the hypoxic zone. The results are interpreted in context with what is known about the life cycles and habits of the two shrimp species involved and the behavior of shrimp fisheries. A significant negative relationship was evident between catch of brown shrimp from Texas and Louisiana waters versus the relative size of the mid-summer hypoxic zone. In addition, Texas catch was significantly dependent upon Louisiana catch, a relationship that has become stronger since 1980. Catch per unit effort of brown shrimp also has declined significantly during a recent interval in which hypoxia was known to expand. Importantly, the same relationships were not significant between hypoxia and catches of white shrimp. As hypothesized, owing to their more offshore habitat requirements, brown shrimp were impacted to a greater degree than white shrimp. The annual success of shrimp fisheries, like most commercial fisheries, is highly related to environmental factors. The combined evidence indicates that hypoxia in Louisiana, due to its large area of coverage, has increased as an environmental factor for shrimp in recent decades.

Coastal Hypoxia: Consequences for Living Resources and Ecosystems
Coastal and Estuarine Studies, Pages 293-310
This paper is not subject to U.S. copyright
Published in 2000 by the American Geophysical Union

Introduction

Historical Background

Louisiana and Texas shrimp fisheries depend primarily upon two species of shrimp, the brown shrimp [*Farfantepenaeus* (formerly *Penaeus*) *aztecus*] and the white shrimp [*Litopenaeus* (formerly *Penaeus*) *setiferus*], as revised [Perez-Farfante and Kensley, 1997]. The life cycles of these species are geographically wide ranging, encompassing inshore estuarine habitats as well as offshore shelf waters. The primary white shrimp fishing grounds are inshore (coastal bays) and nearshore (from the beach out to 10 fm (18 m)), extending from the Mississippi River to the upper Texas coast. The main U.S. brown shrimp fishing grounds are nearshore and offshore (deeper than 10 fm (18 m)), extending from the Mississippi River to the Texas/Mexico border. In Louisiana, the offshore life stages of white shrimp and brown shrimp inhabit areas that overlap with the hypoxic zone [Downing et al., 1999].

Commercial fishermen are characteristically opportunistic and are continually searching for new ways to increase their catch. When the shallow-water shrimp fishery of the first half of the 20th century evolved from using beach seines and sailing ships to motorized vessels, the harvest capability extended into deeper shelf waters. The discovery of abundant brown shrimp grounds in deep waters during the 1950s further stimulated expansion of the offshore fishery through use of large trawling vessels with multiple nets. When the nearshore catch of white shrimp declined in the late 1940s, brown shrimp became targeted both nearshore and offshore, and the overall harvest was maintained [Moffett, 1967].

Louisiana shrimp fishery. Louisiana's early shrimp fishing industry, prior to the 1950s, was based upon catches of white shrimp. With its inshore and nearshore life cycle, this species was a favorite of commercial fishermen from the time of early settlements in the region. When white shrimp declined during the late 1940s, the Louisiana shrimp fishery shifted and catches became increasingly dominated by juvenile brown shrimp [Condrey and Fuller, 1992].

Louisiana has a complex system of open and closed fishing seasons that focus on shrimp in inshore and nearshore territorial waters (within 3 nmi (5.5 km) of shore). The Louisiana Wildlife and Fisheries Commission sets the dates of closures annually within each of three coastal zones. Accordingly, Louisiana inshore waters are closed from December until late May, open from late May into July, closed again from a date in July until late August, and re-opened from late August into December [Louisiana Department of Wildlife and Fisheries, 1992]. The annual harvest of juvenile brown shrimp in Louisiana begins during the latter part of May. This allows an unlimited number of fishing craft access to the small brown shrimp as they exit estuarine systems. Importantly, federal waters of the Exclusive Economic Zone (EEZ) offshore of Louisiana are open throughout the year.

Texas shrimp fishery. The Texas shrimp fishing industry was developed primarily during the 1950s with the discovery of adult brown shrimp grounds in offshore waters. The industry is dominated by large trawlers that pull as many as four nets apiece (quad-rigged) and target high valued large shrimp. The substantial growth of the U.S. shrimp fishery after 1960 is based upon brown shrimp [Christmas and Etzold, 1977].

Texas, like Louisiana, opens its inshore bays to commercial shrimp fishing in May, but closes nearshore and offshore waters from mid-May until mid-July. In Texas, the number of fishing craft are limited inshore, thus allowing escapement of juvenile brown shrimp into offshore waters where they continue to grow. When the Texas closure ends in mid-July, an unlimited number of fishing vessels have access to the offshore shrimp population of larger size individuals. The closure in Texas has been in existence in territorial waters (within 9 nmi (16.5 km) of shore) since 1959, and was extended to federal waters (from 9 to 200 nmi (16.5 to 1000 km)) beginning in 1981 [Klima et al., 1982].

Value and characteristics of the fishery. The Louisiana and Texas shrimp fishery is consistently one of the most valuable in the U.S. During 1998, Louisiana and Texas together landed 168.6 million pounds of shrimp tails (60.7% of the nation's total) with a dockside value of $312.9 million [Holiday and O'Bannon, 1999].

Shrimp are an annually renewable resource that is considered fully exploited [Nance and Harper, 1999]. Shrimp fishermen are highly flexible and are ready to expand or reduce fishing effort each year depending upon the magnitude of the shrimp crop. Due to this practice, annual shrimp fishing effort and shrimp landings are closely correlated and are reflective of the annual abundance of shrimp. Shrimp landings/abundance vary widely from year to year, and this variation is generally attributed to environmental factors.

Shrimp Life Cycle

Generalized reproductive cycle. The life cycle of commercial shrimp involves offshore (Gulf shelf) and inshore (estuarine) phases. Adults spawn on the Louisiana and Texas shelf. Planktonic larvae immigrate via currents into coastal estuaries. Within the estuaries, postlarvae metamorphose into small juvenile shrimp that are benthic in habit. After about two months, intermediate size juveniles emigrate from the nursery and return to the offshore shelf to complete their growth into adults. The life cycle from egg to adult takes about six months. Larval, postlarval, subadult and adult shrimp utilize habitats overlapping with the hypoxic zone, and, depending on which stage within the life cycle, their spawning grounds, feeding grounds or migratory pathways may be impacted.

Brown shrimp. Spawning by adult brown shrimp takes place in relatively distant shelf waters, from 8 to 60 fm (15 to 110 m) [Cook and Lindner, 1970; Christmas and Etzold, 1977]. Spawning occurs throughout the year and planktonic larvae move from offshore to inshore estuaries via currents. Peaks in the immigration of postlarvae occur during early spring and early fall. Juveniles use the tidal wetlands as nurseries to enhance growth and survival from March through November. After approximately two months inhabiting estuarine nurseries, the juveniles return as subadults to offshore Gulf waters. Large adults are abundant on the middle shelf where they spawn and renew the reproductive cycle. During migration on the shelf, brown shrimp have been observed to move as far as 335 nmi (620 km) [Sheridan et al., 1987].

White shrimp. Spawning by adult white shrimp occurs mostly near the coast between 4.5 and 17 fm (8 and 31 m) [Lindner and Anderson, 1956; Lindner and Cook, 1970]. Due to their affinity for nearshore habitats, adult white shrimp are commonly found in proximity to larvae, postlarvae and juveniles. Unlike brown shrimp, white shrimp spawning is restricted mostly between April and August [Bryan and Cody, 1975]. Immigration of postlarvae from the Gulf to the estuaries begins in May, and two or more peaks occur from June through September [Baxter and Renfro, 1967]. Juveniles are abundant in estuarine nurseries from May through November. White shrimp have been observed to move as far as 150 nautical miles (278 km) during migration on the shelf [Lyon and Boudreaux, 1983].

Prior Evidence Suggesting Effects of Hypoxia

Past investigations suggest that shrimp and fish species may avoid hypoxic waters. Fishery-independent surveys, using bottom-trawls, reveal reduction or complete absence of shrimp and fishes in waters with very low oxygen content. Both the abundance and biomass of finfishes and shrimps are significantly less where oxygen concentrations in bottom water fall below 2 mg l^{-1} [Leming and Stuntz, 1984; Renaud, 1986a]. Several scenarios are possible – the shrimp or fish may die, they may move away either horizontally along the bottom or vertically upward in the water column, or they simply may not be attracted to the area for reasons other than hypoxia.

For shrimp, laboratory experiments substantiate that commercially dominant species have the ability to detect and avoid water with low oxygen concentration [Renaud, 1986b]. Under experimental laboratory conditions, white shrimp avoided water with dissolved oxygen lower than 1.5 mg l^{-1} and brown shrimp were even more sensitive and avoided water of less than 2.0 mg l^{-1} dissolved oxygen. It has been suggested that this ability to detect and avoid water with low oxygen leads to a blocking effect on shrimp emigrating from inshore nurseries to offshore feeding and spawning grounds. In a mark-recapture study of shrimp migration, juvenile brown shrimp leaving marsh nurseries were sometimes blocked from normal movement offshore by an environmental barrier [Gazey et al., 1982]. Also, offshore migration by brown shrimp was reported to be greater and catches larger during periods when hypoxia was not evident as compared to periods when hypoxia was observed [Renaud, 1986a]. Avoidance, or crowding, of shrimp and fish away from hypoxic waters has been repeatedly observed in phenomena called "jubilees" [Loesch, 1960].

Recent analyses further suggest a localized negative relationship between shrimp catch and hypoxia [Zimmerman et al., 1997]. Where hypoxia is widespread and persistent on the Louisiana shelf, the shrimp catch is always low. If blocking by hypoxia of shrimp migration offshore does occur, shrimp distributions and densities may be modified. Indeed, in Louisiana, the nearshore concentration of shrimp is orders of magnitude higher than in the hypoxic zone. Also, it is hypothesized that due to hypoxia shrimp movement is diverted laterally and parallel to the coast along with the current. The predominant movement appears to be westward. Such large-scale disruptions to shrimp distributions could be expected, because as much as 50% of the Louisiana shelf is affected by hypoxia during the summer months. This is the period when juvenile brown shrimp are migrating offshore and when adult white shrimp are spawning on the inner shelf.

In some instances, shrimp could move up in the water column as a short-term means of escaping bottom hypoxia. Such a move might substantially increase their exposure to predation and increase mortality. Also, it is not known to what degree shrimp are killed directly by oxygen levels too low to sustain life.

Shrimp Fishery Statistical Methods

Statistical Subareas

Shrimp statistics for commercial fisheries are collected by port agents located in coastal ports around the Gulf of Mexico. Currently, there are about 20 port agents employed by state or federal agencies participating in the Gulf Shrimp Statistics Program.

To facilitate the geographic assignment of commercial shrimp catch and catch per unit effort (CPUE), the continental shelf of the Gulf of Mexico has been divided into 21 statistical subareas (Fig. 1). Each of these subareas has been further divided between "inshore" (bays and sounds) and "offshore" (seaward from the shoreline). The inshore area is labeled depth zone zero. The offshore area comprises depth zones in 5-fm (9-m) increments extending to 45 fm (82 m) (identified as depth zone 1, depth zone 2, ... depth zone 9). The first two offshore depth zones are termed "nearshore", encompassing 0 to 10 fm (18 m). All shrimp fishery data collected in depths greater than 45 fm are included in the 45-fm depth zone for analysis. Thus, each of the 21 statistical subareas in the Gulf of Mexico has ten depth zones (one inshore and nine offshore). Each subarea/depth zone combination is a unique location within the Gulf of Mexico, and is termed a "location cell."

Determination of Catch and Effort

Port agents collect shrimp statistics from seafood dealers and fishermen. Data are obtained from dealer records on the species, size, amount and value of the shrimp that are unloaded or landed at the dealer's place of business. These data are referred to as "dealer data" in the landings file. Currently there are about 460 active seafood dealers in the Gulf of Mexico. A monthly canvas of dealers by port agents provides estimates of total weight in pounds of shrimp landed. The second type of data includes detailed information on fishing effort (CPUE) and location for each fishing trip and is collected by interviewing either the captain or a member of the crew. These data are referred to as "interview data" in the landings file.

Because the fishing trip is the basic sampling unit, the ideal situation is to collect both the landings and interview data on a trip-by-trip basis. The total weight caught in the fishery by trip would be considered a census; however, because of the large number of fishing trips that occur in the Gulf shrimp fishery (i.e., currently over 300,000 total trips), it is impossible for the interview record to include every fishing trip. Consequently, data collection procedures include two modifications that allow sampling of the total population. The modifications attempt to facilitate non-bias selection of vessels that have fished in the offshore areas [Nance, 1992, 1993].

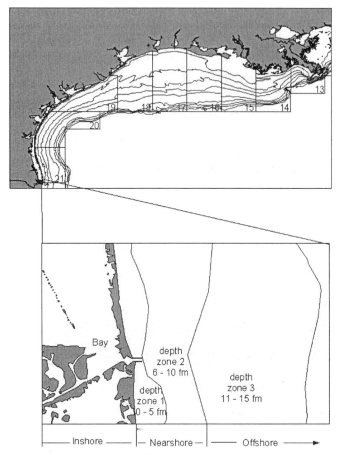

Figure 1. Subareas and depth zones used for collecting fishery data from Texas and Louisiana marine waters. The boundary between Louisiana and Texas is approximately the nearshore border of areas 17 and 18. (U. S. Dept. Commerce, National Oceanic and Atmospheric Administration, National Marine Fisheries, Service)

Estimation of shrimp fishery effort is dependent upon data summarized by location cell. To estimate fishing effort for each location cell on a monthly basis, there must be two elements of data for each cell: 1) the total weight of shrimp caught by species, and 2) the average catch per unit of effort (CPUE; pounds per 24-h fished). As mentioned above, the total pounds caught by species is acquired from commercial seafood dealers located along the Gulf coast, while CPUE is obtained from sample interviews with captains of shrimp vessels at the termination of their trip. Although the interview level has no effect on the collection of total pounds data, it does affect the estimation of average CPUE. Obviously, the more interviews that port agents can gather during a particular month, the more precise the estimate of average CPUE for that month. During peak shrimp production months, about 70-80% of the pounds of shrimp caught have an interview obtained average CPUE associated with them.

Monthly effort (days fished) for each location cell is estimated by dividing the

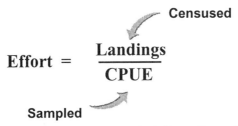

Figure 2. The relationship between shrimp catch, fishing effort and catch per unit effort (CPUE).

monthly shrimp landings from a location cell by the average CPUE obtained from interviews at the same time and location (Fig. 2).

For a few location cells, shrimp landings are reported, but there are no interviews from which to estimate CPUE. To account for the missing data, a statistical model was devised to estimate CPUE for those cells. Both the yearly variation in number of shrimp available and regional differences in shrimp abundance within the Gulf of Mexico play important roles in determining the CPUE for a given location cell. Therefore, a general linear model was developed to predict current CPUE with year and geographic location as the independent variables. Monthly differences in shrimp abundance were accounted for by using a different model for each month. Each of the 12 linear models is in the general form of:

$\log \text{CPUE(ij)} = \mu(ij) + \text{year}(i) + \text{location}(j) + \varepsilon(ij)$, where

CPUE(ij) is the observed CPUE in year i at location cell j;
$\mu(ij)$ is the overall mean;
year(i) is the effect on CPUE due to year i;
location(j) is the effect on CPUE due to location j; and
$\varepsilon(ij)$ is a random error term with expected value 0 and equal
variance for all i and j.

Total effort for any month is estimated by summing the effort estimates for each of the individual location cells. Total annual effort is calculated for descriptive purposes as the sum of the monthly efforts. Total effort is also used to estimate monthly and annual CPUE values.

Catch and Effort in Texas and Louisiana

Annual Trends and Distribution of Catch

The annual catches of brown shrimp and white shrimp have generally increased since 1960 (Fig. 3). Brown shrimp catch steadily increased, peaked at 103.4 million pounds (47 thousand metric tons) in 1990 and declined thereafter through 1998. White shrimp

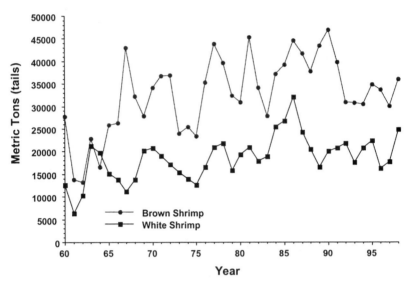

Figure 3. Trends in annual catch of brown shrimp and white shrimp in the western Gulf of Mexico, 1960-1998 (total weight of shrimp tails caught in Texas and Louisiana waters).

catch increased to a lesser degree than brown shrimp and peaked at 70.7 million pounds (32 thousand metric tons) in 1986.

The distribution of catches of white shrimp and brown shrimp from 1992 to 1998, a period of expanded hypoxia, are insightfully different. During the interval, most white shrimp were caught inside 10 fm (18 m) along the Louisiana and upper Texas coast (Fig. 4a). In both states, white shrimp juveniles were predominantly caught in bays and estuaries and the adults were caught nearshore along the coast as has been the case historically. Distribution of brown shrimp catch during this period was quite different from white shrimp and also differed between the two states (Fig. 4b). Louisiana brown shrimp catch was almost entirely of juveniles taken from bays, estuaries and shallow nearshore waters. In Texas, the brown shrimp catch was mainly comprised of subadults and adults taken in offshore waters. Most importantly, the catch of brown shrimp offshore of Louisiana was negligible, coincident with the area of the hypoxic zone (Fig. 4b).

Nearshore versus Offshore Catch in Texas and Louisiana

In Louisiana, the nearshore catch and effort within 10 fm (18 m) is always higher than beyond 10 fm. The large difference between nearshore and offshore catch in Louisiana is remarkable in comparison to nearby Texas (Fig. 5). Since shrimp in Louisiana are caught as they exit the inshore nurseries, they are small in size and, at least in theory, their productivity through growth to a larger size is unrealized. Production models conservatively estimate that several million of pounds are lost each year because of curtailed growth due to early harvest of shrimp [Nance et al., 1994]. In Texas, juvenile brown shrimp are allowed to grow to a larger size as they migrate offshore during a six-week closed season, which normally occurs between May 15 and July 15 of each year.

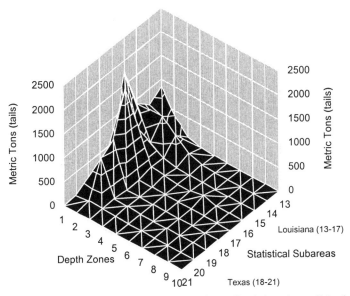

Figure 4a. Mean annual catch of white shrimp by subarea/depth location cell in the western Gulf of Mexico, during the 1992-1998 interval of expanded hypoxia on the Louisiana shelf. Subarea designations extend from the mouth of the Mississippi (Subarea 13) to the mouth of the Rio Grande (Subarea 21). Depth zones are in 5-fm (9-m) increments.

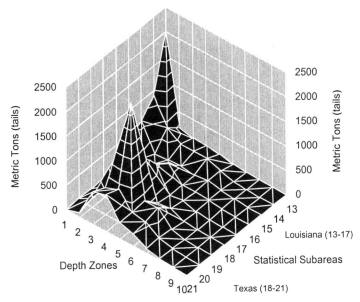

Figure 4b. Mean annual catch of brown shrimp by subarea/depth location cell in the western Gulf of Mexico, during the 1992-1998 interval of expanded hypoxia on the Louisiana shelf. Subarea designations extend from the mouth of the Mississippi (subarea 13) to the mouth of the Rio Grande (subarea 21). Depth zones are in 5-fm (9-m) increments.

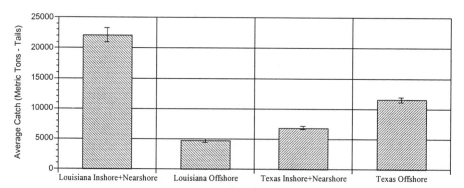

Figure 5. Shrimp catch nearshore within 10 fm (18 m) versus offshore beyond 10 fm from Louisiana and Texas shelf waters, 1960-1998.

It is informative to note that earliest fishery-independent surveys aboard the R/V *Oregon* located concentrations of adult brown shrimp offshore of Texas and to the east of the Mississippi delta, but not offshore of Louisiana [Springer, 1951]. The discontinuity and absence of large brown shrimp offshore of Louisiana is striking and suggests the presence of environmental phenomena unfavorable to adult brown shrimp. The observation also suggests that unfavorable conditions (hypoxia or something else) pre-date recent times.

Relationship of Catch and Effort

Shrimp fisheries exhibit a strong positive relationship between catch and fishing effort. Typically, the geographical (spatial) and monthly or annual (temporal) distributions of catch are proportionally related to hours of fishing effort applied. Trends in increasing catch since the 1960s are mirrored by greater fishing effort. The highest levels of catch (Fig. 3) and correspondingly highest effort occurred in the middle to late 1980s. However, CPUE has not necessarily remained the same over the years but in fact has declined since the late 1970s [Nance, 1989]. The most significant decline in CPUE has occurred recently, in the latter part of the 1980s and early 1990s [Downing et al., 1999].

Relationship of Catch to Hypoxia

Catch versus Percent Hypoxia in a Location

A negative relationship between the magnitude of catch and the degree of hypoxia at locations on the Louisiana shelf has been demonstrated previously [Zimmerman et al., 1997]. Annual maps depicting location and configuration of the hypoxic zone [Rabalais et al., 1991, 1992, 1998, 1999, unpublished data] were transferred to a geographic information system (GIS) format. These hypoxia maps, derived from single mid-summer cruises and despite potential error in representation, are the best spatial data available. The GIS hypoxia data were layered with annual shrimp catch data from location cells identified as geographically defined subareas and depth zones (Fig. 1). The annual percent of hypoxic area within each location cell was calculated for each year from 1985 through 1994. The average of July and August shrimp catch (all species combined) from each location cell was determined for corresponding years. It was assumed that July and August were months in which shrimp catch was most immediately impacted by hypoxia. A step-wise regression was performed using catch in location cells as the dependent variable and depth, subarea, east-to-west location, years and percent area of hypoxia as independent variables. Similar regressions were performed using CPUE and catch per unit area as dependent variables.

These analyses revealed a strong negative relationship between shrimp catch and the amount of area covered by hypoxia within location cells [Zimmerman et al., 1997]. A negative relationship also existed between catch and depth. However, no relationship existed between CPUE and hypoxia or depth. As noted, all species of shrimp were combined in these analyses.

The results were interpreted as evidence supporting hypotheses (1) that localized shrimp catch is negatively related to the amount of local coverage by hypoxia, and (2) that shrimp migration to offshore habitat is blocked by hypoxia. Although declining catch in relation to depth is a confounding factor, in adjacent Texas waters the catch increases with depth (Fig. 5). Also, the absence of shrimp in deeper shelf waters in Louisiana, beyond the hypoxic zone, could be interpreted as evidence of blocked migration. The lack of a relationship between CPUE and hypoxia is easily explained since shrimp fishermen do not put down their large main nets unless shrimp are present (i.e., hypoxic areas are rarely if ever trawled). Small trawls called "try nets" are used first to sample the bottom for shrimp in advance of deployment of full-size commercial trawls.

Overall Catch versus Relative Area of Hypoxia

If hypoxia causes localized catch to diminish, then the cumulative effect of an expanding hypoxic zone should be reflected in a reduction of the overall catch. To test this hypothesis, we examined the relationship between the annual shrimp catch and the area of mid-summer hypoxia between 1985 and 1997. The hypoxic area estimates were provided by N. N. Rabalais (personal communication) and were used as a relative measure of differences in the scale of hypoxia among years. Data on catches of brown shrimp and white shrimp from Louisiana and Texas from the months of July and August combined, as well as the annual total, were used in separate analyses. Regressions of July and August catch by species versus area of hypoxia were performed. The present analyses differ from the previous study [Zimmerman et al., 1997] by using overall catch, rather than catch at specific locations, and total area of hypoxia, rather than local area of hypoxia.

The results disclosed a significantly negative relationship between catch of brown shrimp and the area of hypoxia ($P = 0.02$, $n = 13$) from combined catches of Texas and Louisiana during the months of July and August (Fig. 6). The catch from Louisiana alone during July and August was weakly related ($P = 0.1$, $n = 13$) to the mid-summer

Penaeid Shrimp Fishery

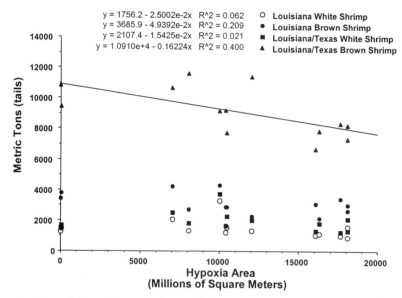

Figure 6. July and August brown shrimp catch from Texas and Louisiana versus relative size of the hypoxic zone on the Louisiana shelf during the years from 1985 through 1997.

area of hypoxia. Texas brown shrimp catch alone during July and August was not significantly related to area of hypoxia. White shrimp catch in either state and in both states combined, during July and August and for the annual total, was not significantly related to the area of hypoxia.

The negative relationship between area of hypoxia and catch of brown shrimp in Louisiana and Texas is relatively strong and remarkable. We suggest that the relationship involves negative environmental effects imposed upon the offshore habitat used by a brown shrimp population common to both states. In this case, the effect of annual size, timing and configuration of hypoxia in Louisiana becomes important to the catch of both states.

In a straightforward cause-and-effect circumstance, when hypoxia expands, the overall population of brown shrimp, as well as catch and CPUE in the fishery, would be reduced. A decadal trend in decline of CPUE in the brown shrimp fishery coincides with expansion of hypoxia. Since 1980, CPUE has decreased significantly even though effort declined through the 1990s. Changes in brown shrimp mean CPUE during each decade since 1960 demonstrate the downward trend of the 1980s and 1990s, i.e., 1960s = 12.9 kg h^{-1}, 1970s = 12.6 kg h^{-1}, 1980s = 11.2 kg h^{-1}, 1990s = 9.1 kg h^{-1} [Downing et al., 1999]. To further explore the possible relationship of expanded hypoxia and reduced CPUE, we examined the annual data of Texas and Louisiana combined. We found that since 1960, brown shrimp CPUE has declined significantly and the greatest decline has occurred after the mid-1980s (Fig. 7). By comparison, white shrimp CPUEs also declined, but not significantly since 1960. The difference in potential effects of hypoxia on CPUE of brown shrimp and white shrimp are consistent with differences in the nearshore versus offshore life cycles of the two species.

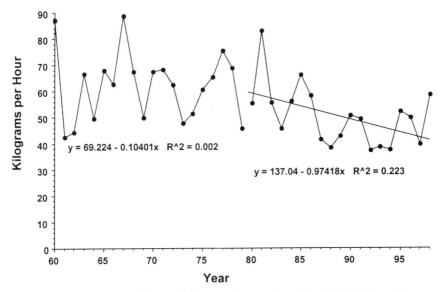

Figure 7. Trend in annual brown shrimp catch per unit effort (CPUE) from Texas and Louisiana from 1960 through 1998.

Jurisdictional Fisheries and Shrimp Populations

Louisiana versus Texas Shrimp Populations

The close correlation between the Texas brown shrimp catch and the Louisiana catch implies a surprisingly strong connection between the populations of each state. We propose that the magnitude of this relationship may, at least in part, result from the effects of hypoxia. Thus, if juvenile brown shrimp are blocked from migrating offshore by hypoxia, they may move laterally down-current into Texas waters. Brown shrimp that are not captured by the Louisiana nearshore fishermen contribute to the catch in Texas.

If blocking of brown shrimp migration into offshore Louisiana waters is a factor, the relationship between Louisiana and Texas could be stronger during years of expanded hypoxia. The larger the number of shrimp that cannot move offshore, the more to go elsewhere including escapement to Texas. Our regression of Texas catch, as the dependent variable, versus Louisiana catch revealed a significant positive relationship during the years from 1980 to 1998 (Fig. 8). During the earlier period, from 1960 to 1979, the relationship was not significant. We interpret this comparison as evidence supporting a greater blocking effect and expansion of hypoxia since 1980.

Unlike brown shrimp, regression analyses of white shrimp catch revealed no significant relationship between Texas and Louisiana. Two factors are likely involved that contribute to the difference between these two species. Firstly, white shrimp do not migrate as far as brown shrimp during their life cycle. Secondly, white shrimp do not

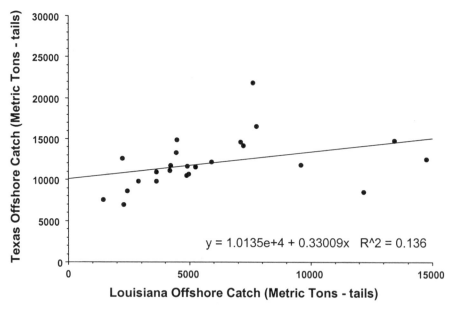

Figure 8a. Louisiana brown shrimp catch versus Texas brown shrimp catch from 1960 through 1979.

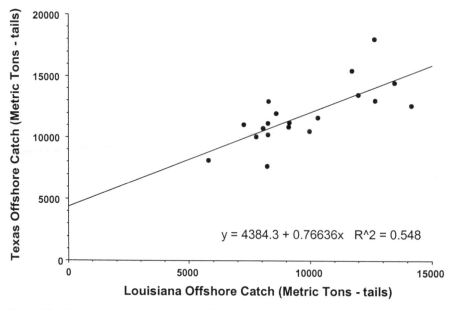

Figure 8b. Louisiana brown shrimp catch versus Texas brown shrimp catch from 1980 through 1998.

depend upon offshore habitats to the same degree as brown shrimp and thus are not as affected by the hypoxic zone.

Consequences of Hypoxic Zone Expansion

Reduction in Shrimp Catch

Recent evidence indicates that the catch in brown shrimp fisheries of Texas and Louisiana may decrease significantly when the hypoxic zone expands. During the years between 1985 and 1998, the area of hypoxia on the Louisiana shelf nearly doubled to 18,000 km² [Rabalais et al., 1998; Downing et al., 1999]. Within the interval, the brown shrimp catch declined from moderately high levels in the late 1980s (approximately 42 thousand metric tons per year) to low levels in the 1990s (approximately 32.5 thousand metric tons per year), coincident with hypoxia expansion. This reduction amounts to 22.6% of the previous catch, representing approximately 9.5 thousand metric tons less brown shrimp per year for Texas and Louisiana fisheries. The recent decline in CPUE in the brown shrimp fishery also corresponds with expansion in hypoxia. Average catch rates of brown shrimp changed from 11.2 kg h⁻¹ to 9.1 kg h⁻¹ from the 1980s to the 1990s [Downing et al., 1999]. Notably, white shrimp did not demonstrate the same recent declining trends in catch and CPUE as brown shrimp, but remained within the expected range of annual variability.

These changes in catch indicate an environmental impact to brown shrimp that did not affect white shrimp. The cumulative evidence suggests that the impact was caused by, or at least associated with, expansion of hypoxia. The negative effect of hypoxia on brown shrimp may co-vary with the negative effect of other factors, such as excessive freshwater in nursery areas or altered nursery habitat. Regardless of co-varying relationships, negative effects on catch appear to have been strengthened by greatly expanded hypoxia since 1990. Also, the details of how hypoxia affects brown shrimp in offshore waters are still unknown. It is not known whether cumulative environmental factors are involved nor whether negative effects on shrimp catch during one year translate to subsequent years.

Influence on Distribution of Catch and Effort

Spatial effect. During the late spring and throughout the summer months, when hypoxia occurs, the Louisiana nearshore and inshore shrimp fisheries are at their peak. These fisheries target juvenile shrimp as they emigrate from estuarine nurseries and when individuals are still relatively small in size and are not yet adults. If, as is hypothesized, expansive hypoxia blocks migration into deeper offshore waters, the effect would increase the concentration of shrimp occupying the nearshore shelf. Since shrimp fishermen target concentrations of shrimp, most of the trawling effort in Louisiana would remain nearshore. This appears to be the case, as exemplified by catch distributions (Figs. 4a and 4b). Thus, the relatively dense concentration of shrimp in the area between the hypoxic zone and the shoreline favors and perhaps helps to maintain the nearshore fishery in Louisiana.

Such is not the case in Texas where juvenile brown shrimp can migrate offshore without impediment. There, an offshore fishery is favored since the population can easily disperse across the shelf where individuals grow to adult size in deeper waters. Juveniles are fished by a restricted number of vessels within Texas bays, and subadult shrimp that emigrate out of the estuaries are protected by a temporary offshore fishery closure. Juveniles that escape the Louisiana fishery also can move down-current into Texas waters. This movement may cause a dependency of the Texas catch on Louisiana production. The apparent dependency has strengthened over the years, and since 1980 the success of brown shrimp in Texas has become significantly related to Louisiana catch (Fig. 8b). At least in theory, Louisiana may be reciprocally dependent upon Texas for offshore spawning by brown shrimp to replenish postlarvae in its estuaries each year.

Temporal effect. Under most circumstances, hypoxia is not a year-long phenomena [Rabalais et al., 1991]. Although the occurrence arrives and disappears, its size, timing and duration may change the consequences for shrimp. Early spring hypoxia could negatively impact recruitment of brown shrimp postlarvae into nurseries. Large-scale, nearshore hypoxia during the spring could impinge on white shrimp spawning areas. Large-scale summer hypoxia can impede brown shrimp juveniles from moving offshore. Extension of summer hypoxia shoreward may adversely crowd shrimp and intensify fishery trawling effort in localized areas.

Influence of Spawning and Feeding Habitat

Shrimp spawning grounds in Louisiana are likely impacted by hypoxia. The spawning area for white shrimp can be reduced during the spring and summer when hypoxia extends close to shore and the timing and location of both events coincide. Spawning grounds of brown shrimp may be eliminated entirely in offshore Louisiana during the months in which hypoxia occurs. The re-routing of juveniles into Texas waters during summer months may also lead to lower numbers of adult brown shrimp offshore of Louisiana after hypoxia disappears. Fall and winter months are the peak spawning period for brown shrimp.

Summer hypoxia in a band along the coast blocks access of juvenile shrimp migrating to offshore feeding grounds. In areas where severe hypoxia has killed infaunal annelid worms [Harper and Rabalais, 1997] which are important prey for brown shrimp [McTigue and Zimmerman, 1998] the forage value of habitat may be diminished (see Rabalais et al. [this volume]). Losses in production due to lost feeding and impairment of growth are difficult if not impossible to determine, but theoretically they may be significant and especially so for brown shrimp. Such losses in productivity only exacerbate when the hypoxic zone expands in space and time.

References

Baxter, K. N. and W. C. Renfro, Seasonal occurrence and size distribution of postlarval brown and white shrimp near Galveston, Texas with notes on species identification, *U.S. Fish Wildl. Serv., Fish. Bull., 66*, 149-158, 1967.

Bryan, C. E. and T. J. Cody, A study of commercial shrimp, rock shrimp and potentially commercial finfish 1973-1975, in *Part I. White Shrimp, Penaeus setiferus (Linn.), Spawning in the Gulf of Mexico off Texas Coast*, Fish. Branch, Texas Parks and Wildl. Dept., P.L. 88-309 Proj. 2-202-R:1-29, 1975.

Christmas, J. Y. and D. J. Etzold, The shrimp fishery of the Gulf of Mexico United States: a regional management plan. *Tech. Rep. No. 2*, Gulf Coast Research Laboratory, Ocean Springs, Mississippi, 1977.

Condrey, R. E. and D. Fuller, The U.S. Gulf shrimp fishery, in *Climate Variability, Climate Change, and Fisheries*, edited by M. H. Glantz, pp. 89-119, Cambridge University Press, New York, 1992.

Cook, H. L. and M. J. Lindner, Synopsis of biological data on the brown shrimp *Penaeus aztecus* Ives, 1891, *FAO Fish. Rep. (57) 4*, 1471-1497, 1970.

Downing, J. A., J. L. Baker, R. J. Diaz, T. Prato, N. N. Rabalais, and R. J. Zimmerman, Gulf of Mexico Hypoxia: Land and Sea Interactions, *Council for Agricultural Science and Technology, Task Force Report No. 134*, 44 pp., 1999.

Gazey, W. J., B. J. Galloway. R. C. Fechhelm, L. R. Martin, and L. A. Reitsema, Shrimp mark and release and port interview sampling survey of shrimp catch and effort with recovery of captured tagged shrimp, in *Shrimp Population Studies: West Hackberry and Big Hill Brine Disposal Sites Off Southwest Louisiana and Upper Texas Coasts, 1980-1982*, Vol. III, edited by W. B. Jackson, 306 pp., Department of Commerce NOAA/NMFS Final Report to U. S. Department of Energy, 1982.

Harper, Jr., D. E. and N. N. Rabalais, Responses of benthonic and nektonic organisms, and communities, to severe hypoxia on the inner continental shelf of Louisiana and Texas, in Proc., First Gulf of Mexico Hypoxia Management Conference, December 1995, New Orleans, Louisiana, pp. 41-65, Publ. No. EPA-55-R-97-001, Gulf of Mexico Program Office, Stennis Space Center, Mississippi, 1997.

Holiday, M. C. and B. K. O'Bannon, *Fisheries of the United States, 1998*, Current Fisheries Statistics No. 9800, National Oceanic and Atmospheric Administration/National Marine Fisheries Service, Washington, D.C., 1999.

Klima, E. F., K. N. Baxter, and F. J. Patella, A review of the offshore shrimp fishery and the 1981 Texas closure, *Mar. Fish. Rev., 44*, 16-30, 1982.

Leming, T. D. and W. E. Stuntz, Zones of coastal hypoxia revealed by satellite scanning have implications for strategic fishing, *Nature, 310*, 136-138, 1984.

Lindner, M. J. and W.W. Anderson, Growth, migrations, spawning and size distribution of shrimp, *Penaeus setiferus, U.S. Fish Wildl. Serv., Fish Bull., 106*, 554-645, 1956.

Lindner, M. J. and H. L. Cook, Synopsis of biological data on the white shrimp *Penaeus setiferus* (Linn.), 1767, *FAO Fish. Rep., 57*, 1439-1469, 1970.

Loesch, H., Sporadic mass shoreward migrations of demersal fish and crustaceans in Mobile Bay, Alabama, *Ecology, 41*, 292-298, 1960.

Louisiana Department of Wildlife and Fisheries, A Fisheries Management Plan for Louisiana's Penaeid Shrimp Fishery, 231 pp., 1992.

Lyon, J. M., and C. J. Boudreaux, Movement of tagged white shrimp, Penaeus setiferus, in the northwestern Gulf of Mexico, *Louisiana Dept. Wildl. Fish. Tech. Bull., 39*, 1-32, 1983.

McTigue, T. A. and R. J. Zimmerman, The use of infauna by juvenile *Penaeus aztecus* Ives and *Penaeus setiferus* (Linnaeus), *Estuaries, 21*, 160-175, 1998.

Moffett, A. W., The shrimp fishery in Texas, *Texas Parks and Wildl. Dept. Bull. 50*, 36 pp., 1967.

Nance, J. M., Stock assessment for brown, white and pink shrimp in the U. S. Gulf of Mexico, 1960-1987, *NOAA Tech. Memo., NMFS-SEFSC-221*, 65 pp., 1989.

Nance, J. M., Estimation of Effort for the Gulf of Mexico Shrimp Fishery, *NOAA Tech. Memo., NMFS-SEFSC-300*, 12 pp., 1992.

Nance, J. M., Effort Trends for the Gulf of Mexico Shrimp Fishery, *NOAA Tech. Memo., NMFS-SEFSC-337*, 37 pp., 1993.

Nance, J. M., and D. Harper, Southeast and Caribbean invertebrate fisheries, in *Our Living Oceans, Report on the status of U.S.Lliving Marine Resources, 1999*, U.S. Dept. Commerce, *NOAA Tech. Memo. NMFS-F/SPO-41*, 143-147, 1999.

Nance, J. M., E. X. Martinez, and E. F. Klima, Feasibility of improving the economic return from the Gulf of Mexico brown shrimp fishery, *N. Amer. J. Fish. Mgt., 14*, 522-536, 1994.

Perez-Farfante, I. and B. F. Kensley, Penaeoid and sergestoid shrimps and prawns of the world, keys and diagnoses for families and genera, *Mem. Mus. natn. Hist. nat., 175*, 1-233, 1997.

Rabalais, N. N., R. E. Turner, W. J. Wiseman, Jr., and D. F. Boesch, A brief summary of hypoxia on the northern Gulf of Mexico continental shelf: 1985-1988, in *Modern and Ancient Continental Shelf Anoxia*, edited by R. V. Tyson and T. H. Pearson, pp. 35-47, *Geological Society Special Publ., 58*, 1991.

Rabalais, N. N., R. E. Turner, and W. J. Wiseman, Jr., Distribution and characteristics of hypoxia on the Louisiana shelf in 1990 and 1991, in Nutrient Enhanced Coastal Ocean Productivity Workshop Proceedings, pp. 15-20, Publ. No. TAMU-SG-92-109, Texas A&M University Sea Grant College Program, College Station, Texas, 1992.

Rabalais, N. N., R. E. Turner, W. J. Wiseman, Jr. and Q. Dortch, Consequences of the 1993 Mississippi River Flood in the Gulf of Mexico, *Regulated Rivers: Research & Management, 14*, 161-177, 1998.

Rabalais, N. N., R. E. Turner, D. Justic', Q. Dortch, and W. J. Wiseman, Jr., Characterization of hypoxia: Topic 1 Report for the Integrated Assessment of Hypoxia in the Gulf of Mexico. NOAA Coastal Ocean Program Decision Analysis Series No. 16, NOAA Coastal Ocean Program, Silver Springs, Maryland, 167 pp., 1999.

Renaud, M. L., Hypoxia in Louisiana coastal waters during 1983: implications for fisheries, *U.S. Fish Wildl. Serv., Fish Bull., 84*, 19-26, 1986a.

Renaud, M. L., Detecting and avoiding oxygen deficient sea water by brown shrimp, *Penaeus aztecus* (Ives), and white shrimp *Penaeus setiferus* (Linnaeus), *J. Exp. Mar. Biol. Ecol., 98*, 283-292, 1986b.

Sheridan, P. F., R. G. Castro M., F. J. Patella, Jr., and G. Zamora, Jr., Factors influencing recapture patterns of tagged penaeid shrimp in the western Gulf of Mexico, *U.S. Fish Wildl. Serv., Fish Bull., 87*, 295-311, 1989.

Springer, S., The *Oregon*'s fishery explorations in the Gulf of Mexico, 1950, *Com. Fish. Rev., 13*, 1-8, 1951.

Zimmerman, R., J. Nance, and J. Williams, Trends in shrimp catch in the hypoxic area of the northern Gulf of Mexico, in Proc., First Gulf of Mexico Hypoxia Management Conference, December 1995, New Orleans, Louisiana, pp. 64-74, Publ. No. EPA-55-R-97-001, Gulf of Mexico Program Office, Stennis Space Center, Mississippi, 1997.

16

Distribution of Catch in the Gulf Menhaden, *Brevoortia patronus*, Purse Seine Fishery in the Northern Gulf of Mexico from Logbook Information: Are There Relationships to the Hypoxic Zone?

Joseph W. Smith

Abstract

The gulf menhaden reduction fishery is conducted by approximately 50 purse seine vessels that range from five ports in the northern Gulf of Mexico from Mississippi to Louisiana. During 1993-1997, total landings of gulf menhaden averaged 571,100 metric tons (1.2 billion lb) annually. The chief products are fish meal, fish oil and fish solubles. The fishery ranks among the top fisheries in the U.S. in terms of volume. A majority of the catch occurs off the Louisiana coast. Since the late 1970s, menhaden captains have completed logbooks of daily fishing activity called Captains Daily Fishing Reports (CDFRs). On CDFR forms, captains enumerate for each set: set time and location, estimated catch, distance to shore and weather conditions. The menhaden fleet averaged 23,425 purse seine sets per year for the 1994-1996 fishing seasons. CDFR information represents an untapped data source to use to examine the effects of hypoxic waters on a major coastal stock of finfish. Herein are selective summaries of 1994-1996 CDFR catch information by 10 x 10 minute cells of latitude and longitude from Alabama to Texas. Preliminary analyses of the CDFR data indicate that catches during summer 1995 were exceptionally low off the Louisiana coast from Southwest Pass west to Marsh Island. Poor catches may have been a result of hypoxic waters impinging upon Louisiana nearshore waters during late July and August 1995.

Coastal Hypoxia: Consequences for Living Resources and Ecosystems
Coastal and Estuarine Studies, Pages 311-320
This paper is not subject to U.S. copyright
Published in 2000 by the American Geophysical Union

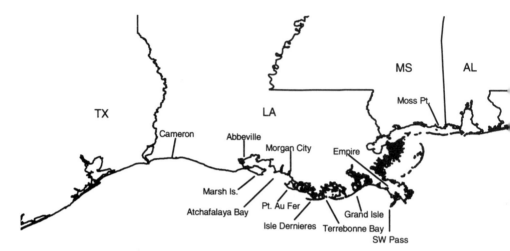

Figure 1. Range of the purse seine fishery for gulf menhaden, *Brevoortia patronus*, in the northern Gulf of Mexico. Labeled locales within the boundaries of Louisiana and Mississippi are sites of extant menhaden reduction factories. Other geographic sites are noted in the text.

Introduction

Gulf menhaden, *Brevoortia patronus*, are small clupeid fishes, generally less than 22 cm in fork length, that form large, dense, near-surface schools in the inshore waters of the northern Gulf of Mexico from spring through fall. Schools of *B. patronus* are harvested by large (up to 61-m) purse seine vessels for an industrial reduction fishery [Smith, 1991]. Gulf menhaden are short-lived, and approximately 95 percent of the commercial catch is comprised of age 1 and 2 fish [Vaughan et al., 1996]. Gulf menhaden perform spring inshore and fall offshore migrations, but are not known to undergo extensive coastal migrations [Ahrenholz, 1991].

The gulf menhaden fishery is conducted from Alabama to eastern Texas, although a majority of the catches are made off the Louisiana coast [GSMFC, 1995]. Fishing occurs in the Gulf of Mexico proper and its contiguous sounds in southeastern Louisiana and Mississippi. The fishery is truly coastal in nature with about 60 percent of the catch occurring within 3 mi of the coastal shoreline and about 90 percent occurring within 10 mi of shore [GSMFC, 1995]. The current 28-wk fishing season extends from mid-April through October [GSMFC, 1995]. Peak monthly landings usually occur in June through August. Since 1994, five factories located at Moss Point, Mississippi, and Empire, Morgan City, Abbeville and Cameron, Louisiana (Fig. 1) have processed *B. patronus* from approximately 50 purse seine vessels. Through 1993 a sixth factory operated at Dulac (between Morgan City and Empire), Louisiana; however, it closed permanently after the 1993 fishing season. Company personnel maintain that its closure was based strictly on economic issues and not on the availability of fish.

In recent years, 1993-1997, annual purse seine landings of gulf menhaden averaged 571,100 metric tons (mt) with an ex-vessel value of $65.2 million. Record harvests occurred in the mid-1980s when annual landings exceeded 800,000 mt for six consecutive years (1982-1987). The chief products of the menhaden industry are fish meal, fish oil and fish solubles. Fishing operations for gulf menhaden occur in daylight, and concentrations of fish schools are located by spotter pilots in small, overhead, fixed-wing aircraft. Spotter pilots direct the purse boat crews via radio in setting the seine around fish schools. Because the carrier vessels are equipped with large fish holds and refrigerated seawater systems, gulf menhaden purse seiners are capable of long-range, multiple-day fishing trips, Monday through Friday. Generally, vessels operate in the vicinity of their home port. Vessels from ports in western Louisiana (Morgan City, Abbeville and Cameron) rarely fish east of the Mississippi River delta. Likewise, vessels from Moss Point, Mississippi, rarely fish west of the delta. Vessels from Empire, Louisiana, fish on both sides of the delta, but rarely farther west than Atchafalaya Bay.

The Beaufort Laboratory of the National Marine Fisheries Service (NMFS) has monitored landings, fishing effort, and size and age composition of the catch in the gulf menhaden fishery since 1964 [Smith, 1991]. Additionally, between 1964-1969 gulf menhaden captains were asked to complete logbooks [Nicholson, 1978] designed to assess daily fishing activities and patterns. Although fleet compliance was incomplete and some vessels kept only partial records, Nicholson [1978] summarized information on over 48,000 purse seine sets for the 6-y period.

During the late 1970s, the menhaden industry agreed to participate in another logbook project called Captains Daily Fishing Reports (CDFRs). The project evolved as a joint industry, state and federal effort with many of the original formats and guidelines developed by Standard Products of Virginia, Inc. CDFRs are deck logs of daily menhaden fishing activities. For each fishing (and non-fishing) day, captains are asked to enumerate time and location of each purse seine set, and for each set the estimated catch, distance and direction from shore, and weather conditions. The gulf menhaden fleet has continuously participated in the program since its inception, and compliance has been near 100%. Through 1991, CDFRs existed primarily as paper files, although limited attempts were made to digitize the data. Beginning in 1992, menhaden program staff at the NMFS Beaufort Laboratory began entering CDFR data into data base files on personal computers. To date, only limited analyses of the CDFR data sets have been performed.

Contributors to a recent conference on hypoxic waters (≤ 2.0 mg l^{-1} of dissolved oxygen) in the northern Gulf of Mexico clearly demonstrated that the inner to mid-continental shelf waters from the western Mississippi River delta to eastern Texas may seasonally harbor the largest area of coastal low oxygenated marine water in the western Atlantic Ocean (e.g., Rabalais et al. [1999]). This hypoxic zone is annual in nature, seasonal in severity (May through September), distributed from nearshore depths of 4 to 5 m out to depths of 60 m, and may encompass 10 to 80 percent of the water column [Rabalais et al., 1999]. Following the Great Mississippi River Flood of 1993, the areal extent of the bottom-water hypoxic zone in the northern Gulf of Mexico during summers 1993-1995 doubled to 16,000 to 18,000 km^2 [Rabalais et al., 1997]. The consequences of hypoxia can be devastating to infaunal and sessile organisms [Harper and Rabalais, 1997], however, hypoxic effects on nektonic communities may be less pronounced. Although their landings data were admittedly limited for the purpose of rigorous statistical analyses, Zimmerman et al. [1997] found evidence of a negative relationship between hypoxia and shrimp catch off the Louisiana coast. Overall, offshore areas of extensive hypoxia yielded lower shrimp catches in summer than nearshore areas with less hypoxia. Hanifen et al. [1997] suspected minimal impact to pelagic finfish, although severe hypoxic events, extending high into the water column, may tend to alter finfish distributions, and subsequently concentrate finfish and fishing effort.

In light of the suspected tendency for inner-shelf hypoxic waters to exclude finfish from nearshore waters or to concentrate finfish in narrow, nearshore "corridors," I examined the gulf menhaden CDFR data bases. Specifically, I summarized information on spatial and temporal distribution of purse seine catches of gulf menhaden in the northern Gulf of Mexico during the summers 1994-1996. This is not a rigorous statistical analysis of the data sets, but is, instead, a preliminary scan of the distribution of gulf menhaden catches relative to the areal extent of the hypoxic zone during the given summers [Rabalais et al., 1997].

Methods

CDFR forms from the 1994-1996 gulf menhaden fishing seasons were key-entered into data base files, edited, then converted into SAS [SAS Institute, Inc., 1995] data sets. At-sea estimates of catch in thousands of "standard fish" (1000 standard fish = 670 lb; [Smith, 1991]) are near-approximations of actual catch ($\pm$ 10% [Smith, 1999]) and were converted into metric tons. Catch locations are generally traditional fishing sites, most of which are adjacent to well-known geographic points. Catch locations were recorded by crew members on CDFR forms using 5-digit codes drawn from a CDFR guidebook. For example, Oyster Bayou on the Louisiana coast is coded as "55279."

Each fishing location in the CDFR guidebook (n = 264) was identified on a nautical chart by its inclusive 10 x 10 minute cell of latitude and longitude. Catch data were summarized monthly by these 10 x 10 minute cells. Catch information was then entered into a desktop geographic information system (GIS). Monthly catches by cell were represented by five markers that graded from light gray (0-2000 mt) to black (> 8000 mt).

Results and Discussion

Fishing for gulf menhaden during the analysis years (1994-1996) was highlighted by landings in 1994 of 761,600 mt, which represented the greatest annual landings since 1987, followed by two years of mediocre landings in 1995 and 1996, 463,900 mt and 479,400 mt, respectively. Landings during each analysis month in 1994 exceeded 120,000 mt and amounted to 123,560 mt, 132,419 mt and 157,042 mt during June through August, respectively. By contrast, monthly landings during June through August, 1995-1996 never exceeded 100,000 mt. Rather, monthly landings in 1995 amounted to 93,734 mt, 85,703 mt and 85,392 mt, and landings in 1996 amounted to 98,618 mt, 50,169 mt and 76,445 mt, June through August, respectively.

Fishing effort, in terms of number of purse seine sets, was greatest during 1994 when 26,234 sets occurred, 3903 in June, 4207 in July and 5618 in August. During 1995, 21,264 sets were made: 4134, 3456 and 3809 during June through August, respectively. Similarly during 1996, 22,776 sets occurred: 4406, 3124 and 3855 June through August, respectively.

For the 1994 gulf menhaden fishing season, catches in June (Fig. 2) were well-distributed throughout central and western coastal Louisiana from Southwest Pass to Cameron and eastern Texas. The greatest catches were adjacent to Atchafalaya Bay. Catch distributions in July and August 1994 (Fig. 2) were similar to the previous month, with catches over 8000 mt clustering near the ports of Morgan City, Abbeville and Cameron.

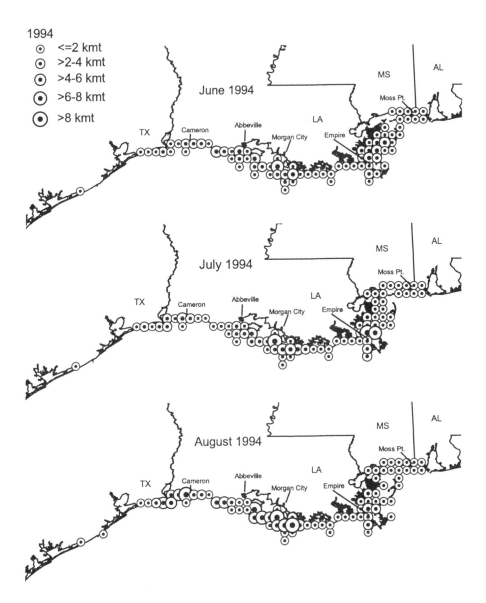

Figure 2. Catch of gulf menhaden, *Brevoortia patronus*, by the purse seine fleet in the northern Gulf of Mexico during June, July and August 1994, by 10 x 10 minute cells of longitude and latitude. Legend units (kmt) are in thousands of metric tons.

Catches in June and July 1995 (Fig. 3) were widespread from Southwest Pass, Louisiana, to eastern Texas, similar to catch distributions in 1994. However, the magnitude of catches by area (10-min cells) was lower, and the greatest catches occurred from Abbeville, Louisiana, to eastern Texas. Catches in the vicinity of Atchafalaya Bay did not improve until

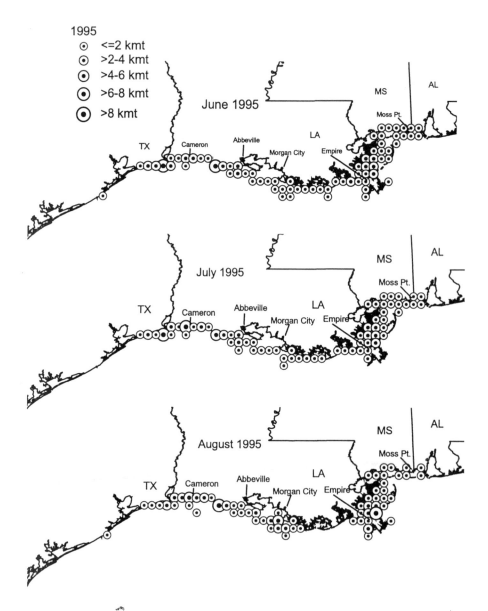

Figure 3. Catch of gulf menhaden, *Brevoortia patronus*, by the purse seine fleet in the northern Gulf of Mexico during June, July and August 1995, by 10 x 10 minute cells of longitude and latitude. Legend units (kmt) are in thousands of metric tons.

August. Moreover, August 1995 (Fig. 3) was notable for its conspicuous lack of any catch from Grand Isle to Isle Dernieres, approximately 110 km of central Louisiana coastline. Good catches of gulf menhaden in June 1996 (Fig. 4) were concentrated in an area west of Abbeville, Louisiana, to eastern Texas. Catches off Atchafalaya Bay were poor, with few fish caught east of Terrebonne Bay to Grand Isle. Catches in July and August 1996 (Fig. 4) were poor to mediocre throughout the fishing grounds west of Southwest Pass, Louisiana, with few fish caught in the area from Terrebonne Bay to Grand Isle.

From the catch distributions of gulf menhaden during summers 1994-1996, a few generalizations can be made relative to fishing patterns of the purse seine fleet west of the Mississippi River delta. (1) Although catches may be distributed during summer throughout the central and western Louisiana coast from Southwest Pass to eastern Texas (as in 1994), by mid-summer catches tend to be greatest off Atchafalaya Bay and west to eastern Texas. Nicholson [1978] noted that during the late 1960s most purse seine activity (number of sets) west of the Mississippi River occurred west of 91°W, or roughly the Atchafalaya Bay and west. (2) Catches tend to be greatest adjacent to ports where menhaden plants are located. Nicholson [1978] also mentioned the tendency of gulf menhaden vessels to fish most often in areas close to their homeport, and Smith [1999] quantified this tendency for Atlantic menhaden vessels fishing from the port of Reedville, Virginia. (3) Catches tend to be poor or non-existent along portions of the central Louisiana coast from Grand Isle to Isle Dernieres during mid- to late summer. Similarly, Nicholson [1978] during the late 1960s showed that the fewest number of purse seine sets along the Louisiana coast occurred between 90° to 91°W, approximately Grand Isle to Isle Dernieres.

Concerning hypoxic waters and menhaden catch distributions, the question arises: Could the increased areal extent of the hypoxic zone in the northern Gulf of Mexico in summer during recent years affect trends in gulf menhaden abundance and distribution? Following the Great Mississippi River Flood of 1993, Rabalais et al. [1997] graphically showed the hypoxic zone impinging upon the Louisiana coastline from near Southwest Pass to Grand Isle to Isle Dernieres during July 1993-1995. To the west, the zone tended farther offshore, but enveloped nearshore waters of the western Louisiana coast. It is tempting to suggest two hypotheses. First, when hypoxic waters impinge upon the shoreline, gulf menhaden catches decline, as dramatically suggested by zero catch of menhaden near Grand Isle in July 1995. Second, if the hypoxic zone forms a continuous band across the inner shelf of western Louisiana, gulf menhaden may be forced into a narrower "corridor" of normoxic waters, as suggested by greater catches off Atchafalaya Bay and western Louisiana in mid-summer.

Admittedly, the current evidence for these hypotheses is circumstantial, and from this preliminary scan of the gulf menhaden CDFR data sets, it is unclear whether there are definite relationships between menhaden catches and the extent of the hypoxic zone of the Louisiana coast in mid-summer. Many factors, such as Mississippi River discharge, turbidity of nearshore waters, and searching habits of vessel captains and spotter pilots, to name a few, affect the fishing patterns of the menhaden fleet and offer alternative explanations for the observed catch distributions.

CDFR data dating back to the late 1970s are archived at the NMFS Beaufort facility, although the pre-1994 data have not been digitized. As over a decade of hypoxia survey data exist for monthly transects south of Terrebonne Bay, Louisiana [Rabalais et al., this volume], future cooperation between federal biologists and scientists in the northern Gulf of Mexico could further explore menhaden catch/hypoxia relationships on a more refined time scale.

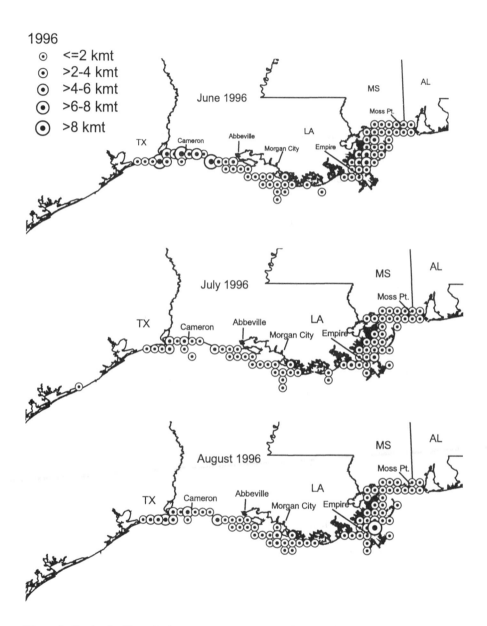

Figure 4. Catch of gulf menhaden, *Brevoortia patronus*, by the purse seine fleet in the northern Gulf of Mexico during June, July and August 1996, by 10 x 10 minute cells of longitude and latitude. Legend units (kmt) are in thousands of metric tons.

Acknowledgments. Several staff members of the Menhaden Program at the NMFS Beaufort Laboratory assisted in computerizing CDFR data: E. Hall, N. Mcneil, B. O'Bier and S. Sechler compiled, key-entered and edited CDFRs; N. Wolfe and C. Krouse provided programming expertise; D. Dudley and R. Clayton identified specific CDFR fishing locations; D. Ahrenholz and D. Vaughan reviewed initial drafts of the manuscript. The Environmental Services Data and Information Management Program of the National Oceanic and Atmospheric Administration provided start-up funding for CDFR key-entry. I also express my appreciation to personnel of the three gulf menhaden companies, Daybrook Fisheries, Inc., Gulf Protein, Inc., and Omega Protein, Inc. (formerly Zapata Protein), for making CDFR data bases available to the NMFS.

References

Ahrenholz, D. W., Population biology and life history of the North American menhadens, *Brevoortia* spp., *Mar. Fish. Rev., 53(4)*, 3-19, 1991.

Gulf States Marine Fisheries Commission [GSMFC]., The menhaden fishery of the Gulf of Mexico, United States: A regional management plan, *Gulf States Mar. Fish. Comm., Fishery Manag. Rep. No. 32*, 1995.

Hanifen, J. G., W. S. Perret, R. P. Allemand, and T. L. Romaire, Potential impacts of hypoxia on fisheries: Louisiana's fishery-independent data, in Proc., First Gulf of Mexico Hypoxia Management Conference, December 1995, New Orleans, Louisiana, pp. 87-100, Publ. No. EPA-55-R-97-001, Gulf of Mexico Program Office, Stennis Space Center, Mississippi, 1997.

Harper, D. E., and N. N. Rabalais, Responses of benthonic and nektonic organisms, and communities, to severe hypoxia on the inner continental shelf of Louisiana, in Proc., First Gulf of Mexico Hypoxia Management Conference, December 1995, New Orleans, Louisiana, pp. 41-56, Publ. No. EPA-55-R-97-001, Gulf of Mexico Program Office, Stennis Space Center, Mississippi, 1997.

Nicholson, W. R., Gulf menhaden, *Brevoortia patronus*, purse seine fishery: Catch, fishing activity, and age and size composition, 1964-73, *NOAA Tech. Rep. NMFS SSRF-722*, 1978.

Rabalais, N. N., R. E. Turner, and W. J. Wiseman, Jr., Hypoxia in the northern Gulf of Mexico: Past, present and future, in Proc., First Gulf of Mexico Hypoxia Management Conference, December 1995, New Orleans, Louisiana, pp. 25-36, Publ. No. EPA-55-R-97-001, Gulf of Mexico Program Office, Stennis Space Center, Mississippi, 1997.

Rabalais, N. N., R. E. Turner, D. Justic', Q. Dortch, and W. J. Wiseman, Jr., Characterization of hypoxia: Topic 1 Report for the Integrated Assessment of Hypoxia in the Gulf of Mexico. NOAA Coastal Ocean Program Decision Analysis Series No. 16. NOAA Coastal Ocean Program, Silver Springs, Maryland, 167 pp., 1999.

SAS Institute, Inc. [SAS], SAS Fundamentals: A Programming Approach, SAS Institute, Inc., Cary, North Carolina, 1995.

Smith, J. W., The Atlantic and gulf menhaden purse seine fisheries: Origins, harvesting technologies, biostatistical monitoring, recent trends in fisheries statistics, and forecasting, *Mar. Fish. Rev., 53(4)*, 28-41, 1991.

Smith, J. W., The distribution of Atlantic menhaden purse seine sets and catches from southern New England to North Carolina, 1985-96, *NOAA Tech. Rep. NMFS, 144*, 1999.

Vaughan, D. S., E. J. Levi, and J. W. Smith, Population characteristics of gulf menhaden, *Brevoortia patronus, NOAA Tech. Rep. NMFS, 125*, 1996.

Zimmerman, R., J. Nance, and J. Williams, Trends in shrimp catch in the hypoxic area of the northern Gulf of Mexico, in Proc., First Gulf of Mexico Hypoxia Management Conference, December 1995, New Orleans, Louisiana, pp. 64-74, Publ. No. EPA-55-R-97-001, Gulf of Mexico Program Office, Stennis Space Center, Mississippi, 1997.

17

The Effects of Hypoxia on the Northern Gulf of Mexico Coastal Ecosystem: A Fisheries Perspective

Edward J. Chesney and Donald M. Baltz

Abstract

    The northern Gulf of Mexico is an economically important coastal zone that produces large yields of fish and shellfish. Because of the documented impacts low oxygen can have on living resources, hypoxic bottom waters (< 2.0 mg l^{-1}) that form along the coast are viewed as a threat to sustained fisheries production in the region. We reviewed factors related to fish habitat and production and evaluated potential effects of the hypoxia on nekton within this system. A complex set of environmental and anthropogenic factors impact nekton in the northern Gulf of Mexico. Fishery yields have remained strong for the northern Gulf over the last 40 years. Effects of hypoxia on distributions of nekton have been documented in the Gulf of Mexico. Also some changes in community structure of nekton are evident, although it is impossible to attribute any of the changes solely or specifically to hypoxia. Given the intensity and extent of hypoxia in the northern Gulf of Mexico other effects of hypoxia on nekton are probable. Nevertheless, we speculate that it is likely that other quantifiable impacts of greater magnitude may currently have more significant effects than hypoxia on the community structure and secondary production of nekton populations in the northern Gulf of Mexico. We hypothesize that the effects of hypoxia on the nekton in the northern Gulf may be buffered by characteristics of the basin, the fauna and the ecosystem. These characteristics may partially offset some of the negative impacts of hypoxia seen in other systems by providing spatial and temporal refuges for demersal nekton.

Coastal Hypoxia: Consequences for Living Resources and Ecosystems
Coastal and Estuarine Studies, Pages 321-354
Copyright 2001 by the American Geophysical Union

Introduction

The northern Gulf of Mexico adjacent to the Mississippi River outflow is an economically important coastal zone that yields large catches of fish and shellfish to both commercial and recreational fishers. Annual landings in 1997 for coastal states adjacent to the Mississippi River outflow (Louisiana, Mississippi, Texas) accounted for 93% of all the U.S. fishery landings in Gulf of Mexico waters. Commercial landings totaled 769 million kg with an ex-vessel value of $575 million [personal communication from the National Marine Fisheries Service, Fisheries Statistics and Economics Division]. The seasonal development of a large zone of hypoxic bottom water on the Louisiana-Texas shelf is a consequence of the substantial primary production of the area coupled with strong stratification along the coast [Rabalais et al., 1996; Wiseman et al., 1997]. Nutrients discharged from the Mississippi River are the primary stimulus for the prolific primary production [Turner and Rabalais, 1991; Rabalais et al., 1996]. Low-oxygen bottom water (hypoxia) has been a persistent seasonal feature of this highly productive area for many years [Pokryfki and Randall, 1987; Turner et al., 1987; Rabalais et al., 1991], but sediment cores and monitoring in recent years suggests the problem may be worsening [Rabalais et al., 1999]. Because of the documented impacts that low oxygen can have on living resources and their habitats [Caddy, 1993; Diaz and Rosenberg, 1995], hypoxia in the Gulf of Mexico is viewed as a potential threat to the sustainability of high fisheries production in the region [Renaud, 1986; Hanifen et al., 1997; Zimmerman et al., 1997; Diaz and Solow, 1999; Rabalais et al., 1999].

The goals of this paper are (1) to review the complex of interacting factors that, together with hypoxia, influence fish habitat and fish production in the area of the Gulf of Mexico that is subjected to hypoxia, and (2) to specifically evaluate potential effects on the dominant nekton (fishes and mobile macroinvertebrates) of this system. If we look strictly at the impacts of hypoxia from the perspective of the nekton (i.e., our fisheries perspective), we can evaluate the likely effects of hypoxia on the prosperity and overall health of nekton populations in the Gulf of Mexico. We focus on two major issues that are relevant to the management of the hypoxia issue. First, what have been the effects of hypoxia on the long-term sustainability of fishery production in the region? Second, have there been any effects on nekton community structure that are clearly attributable to hypoxia? Since we cannot measure nekton production directly and stock assessments are not available for many of the key species, our indicators of production are yields in the fisheries and catch rates in fishery surveys. Because hypoxia in the Gulf of Mexico occurs primarily off the coast of Louisiana, we will focus on Louisiana and bordering coastal regions of Texas and Mississippi. This area constitutes the 'fertile crescent', and any effects on fisheries production and community structure from hypoxia are most likely to be evident for landings data from this area.

Unraveling cause and effect relationships in large complex ecosystems is difficult, and this is especially true for the complex environment of the northern Gulf. Two established linkages complicate the analysis of the effects of hypoxia on fisheries production. First, fisheries production and nutrient enrichment are positively correlated in many marine systems [Nixon et al., 1986; Nixon, 1988; Iveson, 1990; Caddy, 1993; Houde and Rutherford, 1993; Steckis et al., 1995] and nutrient enrichment and hypoxia are also coupled and strongly correlated [Rabalais et al., 1996]. As nutrient enrichment intensifies, increased primary production generally leads to greater secondary production and subsequently greater fish production and landings prior to declines associated with excessive eutrophication or associated factors [Caddy, 1993]. Historically, coastal zones with substantial nutrient loads have experienced eutrophication, followed by bottom hypoxia and anoxia, and these changes have been blamed for shifts in community structure and/or declines in abundance of phytoplankton, benthos and fishes [Caddy, 1993; Hagerman et al., 1996].

Coastal zones are heavily used for many social and industrial activities that can have effects on nekton populations (such as shipping, fishing, industrial pollution, mineral extraction, coastal development and tourism), and these activities can complicate the interpretation of the effects of hypoxia on nekton community structure or fisheries production. The process of harvesting abundant fishes as well as other environmental impacts affecting fishes and their habitats (i.e., wetland loss, harmful algal blooms) are also likely to confound an analysis of hypoxia effects [Kerr and Ryder, 1992; Caddy, 1993; Chesney et al., 2000].

A third complication is that hypoxia does not affect all species or life stages of nekton equally [Breitburg et al., 1997]. Pelagic forms are less likely to be impacted by hypoxia while demersal forms and early life history stages are more likely to be vulnerable to ecosystem level stresses associated with low oxygen, but are unlikely to be affected to the same degree or in the same way.

With these caveats in mind we review factors related to hypoxia that are likely to affect the nekton populations that reside seasonally or continuously within the hypoxic area of the northern Gulf of Mexico. Our objective is a qualitative assessment of whether hypoxia has had any obvious effects on either the production or the community structure of coastal nekton populations in the Gulf of Mexico.

Current Status of the Fisheries in the Northern Gulf of Mexico

Production

The highly productive coastal zone bordering the mouths of the Mississippi and Atchafalaya rivers and the zone where hypoxia typically occurs along the coast during the summer months overlap extensively. Given the large area exposed to hypoxia and the documented impacts it can have on living resources, concerns have been raised about environmental effects and degradation of fisheries and associated habitats. Several recent efforts evaluated the trends in fisheries landings as well as analyzed fishery-independent data for the Gulf of Mexico in an effort to look for environmental impacts on important fish stocks. Some of these were specific attempts to determine whether fish populations showed signs of population declines from environmental impacts such as hypoxia, eutrophication and other factors [Govoni, 1997; Diaz and Solow, 1999; Chesney et al., 2000].

Fishery landings data from 1950-1997 show that the combined commercial catches for Texas, Louisiana and Mississippi increased steadily to more than 1.5 billion pounds (769,000 Mt) and have remained above that level since 1969 [personal communication from the National Marine Fisheries Service, Fisheries Statistics and Economics Division]. The dominant species contributing most to the landings has been Gulf menhaden,

Brevoortia patronus, a pelagic planktivore that might be expected to benefit from eutrophication. If menhaden landings are removed from the total, catches of other species have increased over the same time frame [Chesney et al., 2000]. Louisiana landings, as a percentage of the total commercial catch, have also steadily increased and have generally exceeded 70% of the total catch within U.S. waters of the Gulf of Mexico since 1978. In spite of documented effects of hypoxia on the distribution of important commercial species such as shrimp [Leming and Stuntz, 1984; Renaud, 1986; Zimmerman et al., 1997; Hanifen et al., 1997; Zimmerman and Nance, this volume], trends in the fishery-dependent data suggest that fishery production has remained strong within the 'fertile crescent' (i.e., north-central Gulf of Mexico). This is also in spite of numerous impacts on the fisheries and fish habitat, including those from hypoxia [Chesney et al., 2000]. Several commercially exploited fishes were in decline or severely overexploited less than a decade ago within the Gulf of Mexico. Some of these species are abundant, year-round residents and/or spawn in the coastal areas affected by hypoxia, such as red snapper, *Lutjanus campechanus*, king mackerel, *Scomberomorus cavalla*, Spanish mackerel, *Scomberomorus maculatus*, and cobia, *Rachycentron canadum*. Although strictly managed and still harvested, stock assessments of these important species suggest they have maintained or increased their populations in the Gulf of Mexico over the past decade, even with significant pressure from commercial and recreational fishing and in spite of any influences from hypoxia [Anonymous, 1996; Thompson, 1996; Schirripa and Legault, 1997]. Another species that lives offshore as an adult and is now largely protected from commercial harvest, red drum, *Sciaenops ocellatus*, has also maintained its population size in the northern Gulf of Mexico. Red snapper have been severely overfished through directed fishing for many years and snapper populations also suffer tremendous bycatch mortality as juveniles [Schirripa and Legault, 1997]. Nevertheless, 93% of the allowable commercial catch of red snapper in the Gulf for 1998 was landed in Louisiana, Texas and Mississippi with most of those landed in Louisiana [personal communication from the National Marine Fisheries Service, Fisheries Statistics and Economics Division]. In conclusion, the current status of fish production in the Gulf suggests that fisheries production remains strong in the 'fertile crescent'. It also seems likely that if fisheries production has been affected by the hypoxia, any effects on production are either secondary to the impacts from fishing activities, or that the effects from hypoxia are obscured by fishing effects and/or other impacts to nekton populations.

Community Structure

The vast coastal marshes in the Gulf of Mexico serve as nursery habitat for many nekton species including many of those that spawn off the coast in the zone affected by hypoxia. Thus the patterns observed for inshore populations are relevant to coastal processes because they reflect recruitment patterns for both coastal and inshore species. Chesney et al. [2000] evaluated fishery-independent data from the Louisiana Department of Wildlife and Fisheries' (LDWF) shrimp sampling program and examined trends in nekton community structure for inshore populations.

Population trends in 21 years (1972–1992) of catch per unit effort (CPUE) data were analyzed for the ten most abundant species captured, plus commercial species of special interest. Statistically significant long-term trends emerged for a few of the species [Chesney et al., 2000]. Two pelagic species, bay anchovy, *Anchoa mitchilli*, and Gulf menhaden, along with the least puffer, *Sphoeroides parvus*, showed significant increases in CPUE over the 21-year period [Chesney et al., 2000]. The CPUEs of three groundfish species, Atlantic cutlassfish, *Trichiurus lepturus*, star drum, *Stellifer lanceolatus*, and southern flounder, *Paralichthys lethostigma*, significantly decreased over the same 21-year period. Perhaps most noteworthy was the number of species that showed no increasing or decreasing trend in CPUE over the period. It should be emphasized that although the analysis focused on inshore and estuarine trawl samples and did not address changes occurring prior to 1972, many of the species are important inhabitants of the coastal zone affected by hypoxia. It is also noteworthy that three of the commercially important species, brown shrimp, *Farfantepenaeus aztecus*, white shrimp, *Litopenaeus setiferus*, [both formerly in *Penaeus*] and blue crab, *Callinectes sapidus*, showed no significant increasing or decreasing trends between 1972 and 1992 in the inshore environment, while two other heavily exploited species, Gulf menhaden (a commercial, bycatch and forage species) and bay anchovy (a bycatch and forage species), showed increasing trends over the same period.

The general patterns of long-term change observed in the inshore populations are supported by other studies. Govoni [1997] analyzed Gulf menhaden recruitment in association with the Mississippi River for the years 1964-1989 and found a possible decadal-scale positive correlation between river flow and the number of recruits. He concluded that recruitment became elevated after 1975 and corresponded with increased river flow to the coastal zone and concluded it was possibly a response to increased primary production stimulated by enhanced nutrient flux into the area [Govoni, 1997]. If menhaden recruitment was enhanced during that time frame, then survival of the early life history stages of other species may have been enhanced as well.

Several different trawl studies support the hypothesis that significant structural changes in nekton communities have taken place over time with a general pattern of pelagic species becoming more abundant and some of the dominant demersal species declining in prominence within trawl bycatch. Most compelling is a comparison of the composition of trawl bycatch between the 1930s and 1989 from the shrimp fishery of coastal Louisiana [Gunter, 1936; Anonymous, 1992; Adkins, 1993].

The rank order of 37 fish species in the demersal assemblage reported by Gunter [1936] differed substantially from that found in 1989 [Anonymous, 1992; Adkins, 1993]. By 1989, the second most abundant species, star drum, had dropped to the 25th rank, the 9th ranked, Atlantic moonfish, *Vomer setapinnis*, had dropped to the 20th rank, and the 13th ranked Gulf butterfish, *Peprilus burti*, had dropped to the 31st rank in the bycatch assessment. The Atlantic croaker, *Micropogonias undulatus*, only moved down from first to fourth rank, but the catch rate declined dramatically in the bycatch studies (207.4 fish h^{-1} in 1932–33 verses 16.0 fish h^{-1} in 1989). Two planktivores, bay anchovy and Gulf menhaden, moved from third and sixth rank to first and second. Other formerly low-ranked planktivorous species also made substantial upward climbs in the bycatch distributions: Atlantic bumper, *Chloroscombrus chrysurus*, moved up from 22nd to seventh. The overall rank order of 37 species in the assemblage structures remained correlated (Spearman's $r = 0.674$, n = 37, P < 0.0001) probably due to the inclusion of numerous uncommon species [Chesney et al., 2000]. When only the 15 most common species were analyzed, the correlation was marginal (Spearman's $r = 0.515$, n = 15, P <

0.0496) and became non-significant when only the ten (Spearman's $r = 0.321$, n = 10, P < 0.3655) or the five (Spearman's $r = -0.100$, n = 5, P < 0.8729) most common species were considered [Chesney et al., 2000].

Although there were some differences between years and locations in Gunter's bycatch study, the dominant species were generally similar and, we believe, representative of the Louisiana coastal fish community at that time as shown by the following analysis. Gunter's [1936] surveys characterized fish community structure in coastal Louisiana in inshore and nearshore locations for the years 1932 and 1933. We used Kendall's W [Sokal and Rohlf, 1981] to test for concordance among years and locations (i.e., inshore-1932, nearshore-1932, inshore-1933, and nearshore-1933). An overall ranking of abundance was used to order the 39 fish species. We then examined concordance among the four separate rankings of all 39 species and progressively smaller assemblages of 30, 20, 10 and five species. All 39 species were highly concordant (W = 0.85, $\chi^2 = 128.9$, df = 38, P < 0.005), as were smaller assemblages of the dominant 30, 20, 10 and five species (W ≥ 0.61, $\chi^2 \geq 10.4$, P ≤ 0.05).

We also compared Gunter's [1936] inshore bycatch data (1932-33) with more recent (1972-92) fishery-independent data from the inshore waters of central Louisiana to determine their concordance [Baltz and Chesney, 1995]. Spearman's rank correlations indicated that the assemblages were significantly concordant for the most common 25, 20, 15 and 10 species (r ≥ 0.63, df = 9, P ≤ 0.0289). As you might expect given the suspected differences in community structure between the 1930s and the 1990s, the correlation broke down when the rank abundance of Gunter's five most common species were compared (r = 0.80, df = 4, P > 0.10). In 1932-33, the five dominant species for inshore waters (in rank order) were Atlantic croaker, bay anchovy, white trout, Gulf menhaden and hardhead catfish, but in the 1970s through the early 1990s, bay anchovy and Atlantic croaker swapped positions, white trout were reduced to sixth rank, hardhead catfish to eighth rank, and spot and Atlantic bumper moved into the top five. Thus the dominant inshore bycatch species became bay anchovy, Atlantic croaker, spot, Gulf menhaden and Atlantic bumper in the area we compared [Baltz and Chesney, 1995].

Fishery-independent trawl surveys and fishery-dependent trawl bycatch studies are not ideal sources of data to show trends in abundance or changes in composition of nekton because of the inherent variability in net performance, differences in fishing methods and survey techniques [Hayes et al., 1996]. Nevertheless, the consistency in the trends observed in the various studies corroborate and collectively strengthen the argument that there has probably been a significant change in nekton community structure of coastal Louisiana over the past 60 years. For some of the species, such as menhaden, bay anchovy and a few others, these changes are likely to have come in the last few decades, while for others these changes may have come earlier. The critical question is what anthropogenic and/or environmental factors have caused these changes?

Eutrophication, Hypoxia and Fisheries Production: A Dynamic Interplay

In the coastal ecosystem of the Gulf of Mexico affected by hypoxia, it is impossible to assign changes in abundance or community structure solely or specifically to a single causative factor for two fundamental reasons. First, the complexity and magnitude of environmental changes in the Louisiana coastal ecosystem have been substantial, and second, a variety of factors other than hypoxia affect fish populations of the Gulf of Mexico in both positive and negative ways [Chesney et al., 2000]. Kerr and Ryder [1992] point out that "...the patterns of change often accompanying eutrophication and its effects on fish production systems are recognizable, but not always amenable to differentiation from the effect of associated factors, or to quantification." This is especially true for the northern Gulf of Mexico. Although we focus on the effects of hypoxia on nekton populations, the hypoxia in the Gulf is symptomatic of the eutrophication brought about by nutrients from the Mississippi River discharge.

Nutrient inputs, primary production and fisheries yields are generally positively correlated in estuaries and coastal systems [Nixon et al., 1986; Iveson, 1990; Houde and Rutherford, 1993; Caddy, 1993]. With annual landings by Louisiana, Mississippi and Texas that are in excess of 769 million kg and 93% of Gulf of Mexico landings, the influence of the Mississippi River system on fisheries production of the Gulf of Mexico is apparent. Because the discharge of the river is located within the coastal zone of Louisiana, the 'fertile crescent' is centered around the Louisiana coast. While the nutrients carried by the river enrich the productivity of the region, the river also has the potential to bring pulses of excessive nutrient loads to the Gulf of Mexico, especially during periods of river flow [Atwood et al., 1994; Rabalais et al., 1996]. Large areas of hypoxia develop on the Louisiana and Texas shelf during April and typically continue through October. Hypoxia typically develops in water depths of 5-30 m and has affected as much as 20,000 km^2 or more of coastal bottom waters in the Gulf of Mexico during midsummer [Rabalais et al., 1991; Rabalais et al., 1996; Rabalais et al., 1999; Rabalais et al., this volume].

We explored potential mechanisms for the changes that have occurred in nekton populations of coastal Louisiana (vulnerable to trawls) by looking at their life history characteristics. We examined life history information for some of the dominant species captured historically and currently as bycatch in the shrimp trawl fisheries [Gunter, 1936; Adkins, 1993]. We also included information on three commercially important invertebrates and four commercially or recreationally important fishes that inhabit the coastal zone (Table 1). We examined six variables identified as likely to be critical to determining if hypoxia could potentially have a strong influence on the population of a given species or its habitat. We used a Principal Component Analysis (PCA) to look for patterns in susceptibility to hypoxia off the Louisiana coast (SAS Institute, 1989; Proc Factor). Life history characteristics were verified from the published literature and ranked on ordinal scales[1]. The PCA analysis was based on the correlation matrix with rotation using the Varimax option (SAS, 1989). We then plotted the species' PCA scores in three-dimensional space (Fig. 1) to group similar species and look for patterns in the ordination that might relate to hypoxia[1].

---

[1]Characterization of species principal habitats and life histories in the PCA were scored as follows: (1) Pelagic-Inhabiting upper water column, (2) Nektonic-Inhabiting entire water column, (3) Epi-demersal-Associated with lower water column, (4) Demersal- Strongly associated or living on bottom; (1) Estuary-Landward of beaches in estuary and marsh habitat, (2) Inshore- < 5 m depth near beaches and barrier islands, (3) Nearshore-5-25 m depth, (4) Offshore-25-100 m depth; (1) Planktivore, (2) Piscivore, (3) Small demersal nekton, (4) Small benthos; (1) Winter spawner, (2) Spring, (3) Fall, (4) Summer.

TABLE 1. Listed by relative abundance in bycatch studies from 1932-33 [Gunter, 1936] Trawl fishery plus species of special interest. Life history and habitat characteristics relevant

Species	Relative CPUE 1933 vs 1989	Water Column Distribution	Principal Adult Habitat
Bay anchovy, *Anchoa mitchilli*	29.9 : 55.7	Pelagic	Inshore-Estuary
Gulf menhaden, *Brevoortia patronus*	14.1 : 26.9	Pelagic	Nearshore-Inshore
Sand seatrout, *Cynoscion arenarius*	25.1 : 17.7	Demersal	Offshore-Nearshore
Atlantic croaker, *Micropogonias undulatus*	207.4 : 16.0	Demersal	Offshore-Nearshore
Sea catfish, *Arius felis*	15.1 : 10.6	Demersal	Offshore-Inshore
Spot, *Leiostomus xanthurus*	8.3 : 4.4	Demersal	Offshore-Nearshore
Atlantic bumper, *Chloroscombrus chrysurus*	1.1 : 3.5	Pelagic	Nearshore
Atlantic threadfin, *Polydactylus octonemus*	8.7 : 1.8	Nektonic	Offshore-Nearshore
Fringed flounder, *Etropus crossotus*	3.8 : 1.7	Demersal	Nearshore-Estuary
Silver perch, *Bairdiella chysura*	4.1 : 1.6	Demersal	Nearshore-Estuary
Cutlassfish, *Trichiurus lepturus*	11.6 : 1.5	Epi-demersal	Offshore-Nearshore
Least puffer, *Sphoeroides parvus*	0.8 : 1.3	Epi-demersal	Offshore
Hogchoker, *Trinectes maculatus*	7.2 : 0.7	Demersal	Nearshore-Estuary
Atlantic moonfish, *Selene setapinnis*	8.3 : 0.6	Epi-demersal	Offshore-Nearshore
Southern kingfish, *Menticirrhus americanus*	4.1 : 0.6	Demersal	Offshore-Nearshore
Lined sole, *Achirus lineatus*	1.3 : 0.4	Demersal	Nearshore-Inshore
Southern flounder, *Paralichthys lethostigma*	0.7 : 0.4	Demersal	Nearshore-Estuary
Blackcheek tonguefish, *Symphurus plagiusa*	1.2 : 0.3	Demersal	Nearshore-Estuary
Spotted seatrout, *Cynoscion nebulosus*	2.6 : 0.3	Nektonic	Nearshore-Estuary
Star drum, *Stellifer lanceolatus*	30.6 : 0.3	Demersal	Offshore-Nearshore
Gulf butterfish, *Peprilus burti*	4.3 : 0.1	Nektonic	Offshore
Spanish mackerel, *Scomberomorus maculatus*	0.07 : 0.23	Pelagic	Offshore-Nearshore
Silver seatrout, *Cynoscion nothus*	2.7 : 0.04	Demersal	Nearshore-Estuary
White shrimp, *Litopenaeus setiferus*	—	Demersal	Offshore-Nearshore
Brown shrimp, *Farfantepenaeus aztecus*	—	Demersal	Offshore-Nearshore
Blue crab, *Callinectes sapidus*	—	Demersal	Nearshore-Estuary
Red snapper, *Lutjanus campechanus*	—	Epi-demersal	Offshore
Cobia, *Rachycentron canadum*	—	Nektonic	Offshore
King mackerel, *Scomberomorus cavallas*	—	Pelagic	Offshore
Red drum, *Sciaenops ocellatus*	—	Nektonic	Offshore-Nearshore

In the PCA, six variables were resolved into three principal components with eigenvalues greater than one, and together they explained 83% of the variation among 30 species (Table 2). Principal component 1 (PC1) loaded heavily for spawning location and adult habitat and moderately for nursery habitat. Principal component 2 (PC2) loaded heavily for water column distribution and foraging habits. Principal component 3 (PC3) loaded heavily for spawning season and moderately for nursery habitat. Although most species were found at intermediate levels of PC1 and PC2, several species groups were evident in three-dimensional principal component space (Fig. 1). One group of three offshore species, king mackerel, cobia and red snapper, had high scores on PC1 and PC3. Most notably, two other groups had low and high scores on PC2. The three species and 1989 [Anonymous, 1992] are nekton commonly captured in the Louisiana shrimp to their potential to be affected by hypoxia are listed for each species.

Foraging Habits	Principal Nursery Habitat	Spawning Location	Spawning Season
Zooplanktivorous	Estuary	Inshore-Estuary	April-Sept
Planktivorous	Estuary	Offshore-Nearshore	Dec-April
Piscivorous	Inshore-Estuary	Offshore	March-Sept
Benthos	Inshore-Estuary	Offshore	Oct-March
Omnivorous	Inshore-Estuary	Nearshore-Inshore	May-Aug
Benthos	Estuary	Offshore	Winter
Zooplankton	Estuary	Nearshore	June-Oct
Shrimp-Crustacea	Nearshore-Estuary	Offshore	Dec-April
Benthic copepods	Estuary	Nearshore	May-Aug
Benthos	Inshore-Estuary	Nearshore-Estuary	May-Sept
Piscivorous	Nearshore-Estuary	Offshore-Nearshore	April-Sept
Small benthos	Nearshore-Estuary	Offshore-Nearshore	March-Oct
Small benthos	Nearshore-Estuary	Estuary	May-Sept
Small nekton	Nearshore-Estuary	Offshore-Nearshore	Summer
Small benthos	Nearshore-Inshore	Offshore	Summer
Small benthos	Inshore-Estuary	Nearshore	Summer
Piscivorous-Nekton	Nearshore-Estuary	Offshore	Winter-Spring
Small benthos	Nearshore-Estuary	Nearshore-Estuary	Summer
Piscivorous	Estuary	Inshore-Estuary	May-Aug
Small benthos	Nearshore-Estuary	Offshore-Nearshore	April-Oct
Omnivorous	Nearshore	Offshore-Nearshore	Oct-April
Piscivorous	Nearshore-Inshore	Offshore-Nearshore	May-Sept
Piscivorous-Nekton	Offshore-Nearshore	Offshore-Nearshore	April-Aug
Small benthos	Estuary	Offshore-Nearshore	April-Sept
Small benthos	Estuary	Offshore	Sept-May
Omnivorous	Nearshore-Estuary	Nearshore-Inshore	Spring-Fall
Nekton	Offshore-Nearshore	Offshore	May-Sept
Nekton	Offshore-Nearshore	Offshore	May-Sept
Piscivorous	Offshore-Nearshore	Offshore	May-Sept
Omnivorous	Estuary	Inshore	Fall

on the low end of PC2, are pelagic planktivores or zooplanktivores. They include three of the four species (i.e., bay anchovy, Gulf menhaden and Atlantic bumper) that increased substantially in relative abundance between 1933 and 1989; however, the fourth species, least puffer, increased slightly in abundance. It is a small benthivore and did not cluster with the group that increased in relative abundance.

At the other end of PC2, a group of nine species, including Atlantic croaker, hogchoker, blackcheek tonguefish, silver perch, star drum, lined sole, southern kingfish, brown shrimp, and white shrimp, can be characterized as demersals that forage primarily on small benthic organisms. All, with the exception of brown and white shrimp, have declined in relative abundance since the 1933 surveys. Most notable was the 100-fold

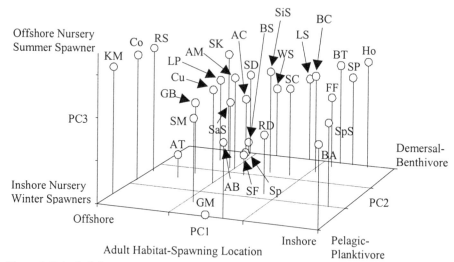

Figure 1. Principal Component Analysis (PCA) for life history characteristics of some selected fishes and macroinvertebrates of coastal Louisiana.   Species are listed in Table 1. Abbreviations designate individual species and are based on common names as follows: BA=bay anchovy, GM=gulf menhaden, SaS=sand seatrout, AC=Atlantic croaker, Sp=spot, AB=Atlantic bumper, AT=Atlantic threadfin, FF=fringed flounder, SP=silver perch, Cu=cutlassfish, LP=least puffer, Ho=hogchoker, AM=Atlantic moonfish, SK=southern kingfish, LS=lined sole, SC=Sea catfish, SF=southern flounder, BT=blackcheek tonguefish, SpS=spotted seatrout, SD=star drum, GB=gulf butterfish, SM=spanish mackerel, SiS=silver seatrout, WS=white shrimp, BS=brown shrimp, BC=blue crab, RS=red snapper, Co=cobia, KM=king mackerel, RD=red drum.

decline of star drum and the 13-fold decline of Atlantic croaker (Table 1). Many of those same species also had high positive scores for PC3 (i.e., they spawn in months/seasons when hypoxia is common and use shallow nursery habitats in or near estuaries). The lack of a documented decline in brown shrimp, white shrimp and blue crab populations may be due to their relatively shorter life histories and high fecundities that may allow them to rebound quickly in response to trawling or other impacts when compared to fishes in the same grouping.

When we compared CPUE data in the bycatch studies from the 1930s to 1989 [Gunter, 1936; Adkins, 1993], it appears that some groundfishes have not been affected to the same degree as others. Some of the reported differences may be due to differences in sampling techniques between the two studies or other artifacts. Nevertheless, virtually all the trawled groundfishes were relatively less abundant while all the small pelagic fishes were relatively more abundant as bycatch (Table 1), and that pattern agrees with the pattern documented by the inshore fishery-independent data analyses [Govoni, 1997; Chesney et al., 2000]. If we look specifically at the life history characteristics of individual species, many of the most heavily impacted species have attributes that would make them potentially vulnerable to impacts of hypoxia. For example, the analyses of both the bycatch data and the fishery-independent data show that star drum populations have declined significantly; they are demersal, spawn offshore in summer and feed on small benthos (Table 1). Hypoxia should have the greatest potential to be detrimental to

TABLE 2. Rotated principal component (PC) analysis of life history and habitat characteristics for selected fisheries species in the northern Gulf of Mexico.

Variable	PC1	PC2	PC3
Water Column Distribution	0.009	0.932	-0.20
Adult Habitat	0.866	-0.103	0.078
Foraging Habits	0.179	0.907	-0.013
Nursery Habitat	0.561	-0.056	0.670
Spawning Location	0.907	-0.030	-0.203
Spawning Season	-0.290	0.078	0.884
Eigenvalues	2.00	1.71	1.28
Variance Explained	0.33	0.29	0.21
Cumulative Variance Explained	0.33	0.62	0.83

a species with these life history characteristics. White shrimp, brown shrimp, and possibly blue crabs however, also seem to have many life history characteristics that would make them vulnerable to hypoxia, yet they have not shown any overwhelming signs of impacts to their populations or total fishery landings (Figs. 2, 3).

In addition to hypoxia, other factors significantly affect nekton populations within the Louisiana coastal zone. Directed fisheries, bycatch, other fishing effects, wetland loss and degradation, water management and eutrophication all interact with hypoxia to affect the habitat and life cycles of coastal nekton [Chesney et al., 2000]. Bycatch can be estimated and has been shown to be a large and significant impact on coastal nekton populations and it has impacted their populations for many years [Gunter, 1936; Lindner, 1936]. Recent annual bycatch estimates from the Louisiana shrimp fishery in 1996 were conservatively estimated at 83 million kg or equivalent to about 10% of the annual directed fishery landings in the U.S. Gulf of Mexico waters [Chesney et al., 2000]. This is a major and persistent impact to groundfish populations because shrimp trawl bycatch consists largely of juvenile groundfishes. In addition, other fisheries directed at non-shrimp species probably have had a significant impact on the present status of groundfish populations. These include undocumented effects, such as ghost fishing by lost fishing gear (National Research Council, 1999).

Another example is the trawl fishery that developed during the 1950s and 1960s in the northern Gulf of Mexico that targeted groundfish such as large croaker (Fig. 4). This fishery also included small Atlantic croaker, banded croaker, star drum, cutlassfish and many other species used as food fish and pet food (Fig. 5) [Gutherz et al., 1975]. This fishery grew quickly and netted landings from the 'fertile crescent' of 30,000-50,000 metric tons annually as pet food and several thousand tons of Atlantic croaker as food fish at its peak, but abruptly died out in the early 1980s (Figs. 4, 5). It is possible that directed fisheries combined with annual bycatch of juvenile groundfishes in the shrimp fishery have contributed significantly to the changes in the groundfish populations that are still apparent[2]. This hypothesis is supported by the fact that there was a Gulf-wide decline in croaker landings rather than just a decline in the region affected by hypoxia

---

[2]Reportedly, fishing pressure by the Atlantic butterfish trawling fleet was redirected to the Gulf of Mexico and added pressure to the existing trawl fishery for groundfishes. This, coupled with economic pressures in the pet food industry, brought about the eventual demise of the fishery [G. Adkins, personal communication].

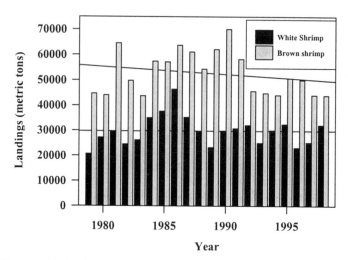

Figure 2. Comparative landings of white shrimp, *Litopenaeus setiferus,* and brown shrimp, *Farfantepenaeus aztecus,* for Louisiana, Mississippi and Texas in the years 1979-1997 [Source: Personal communication from the National Marine Fisheries Service, Fisheries Statistics and Economics Division]. The horizontal lines are regressions for catches of each species.

(Fig. 5) and a continued absence of large croaker in catches off coastal Louisiana since the decline of the fishery in the early 1980s. While a role for hypoxia in these changes cannot be ruled out, it seems most likely that fishing and not hypoxia was the major structuring force for at least some of these changes in the nekton community structure.

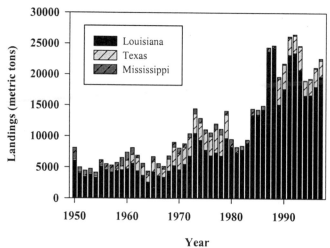

Figure 3. Landings of blue crab, *Callinectes sapidus,* for Louisiana, Mississippi and Texas in the years 1950-1997, listed by state [Source: Personal communication from the National Marine Fisheries Service, Fisheries Statistics and Economics Division]. Note that the data were unavailable for Texas landings between 1979-1988.

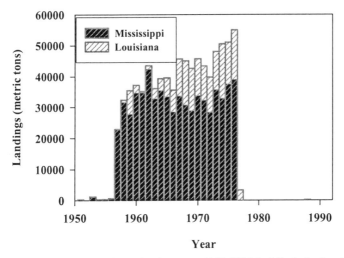

Figure 4. Unclassified finfish landings for the years 1950-1992 in Mississippi and Louisiana [Source: Personal communication from the National Marine Fisheries Service, Fisheries Statistics and Economics Division]. The bulk of the unclassified landings are made up of industrial groundfish, which are composed primarily of small Atlantic croaker, spot, cutlassfish, sea trout and other bottomfishes [Gutherz et al., 1975].

How is Hypoxia Likely to Affect Coastal Nekton of the Gulf of Mexico?

Effects of hypoxia on some living resources such as benthos are well documented [Diaz and Rosenberg, 1995], while direct impacts on nekton are either poorly established, undocumented, or hypothetical and sometimes anecdotal. Unlike the impacts to benthos, not all the influences of hypoxia are necessarily detrimental to nekton populations. The effects of hypoxia on larval, juvenile and adult nekton are likely to depend on the frequency, duration, intensity and pattern of hypoxia as well as the life history characteristics and physiology of each species. The expression and interpretation of individual effects at the population level are even more complex. The potential direct and indirect effects of several environmental impacts (including hypoxia) that can affect coastal nekton, their prey or their habitat along the Louisiana coast were recently reviewed [Diaz and Solow, 1999; Chesney et al., 2000; Craig et al., this volume].

Direct Effects on Nekton

Studies of the effect of hypoxia on the distribution of nekton have documented that low-oxygen waters can limit the distribution of fishes and invertebrates and alter their migratory paths. These effects have been observed in the Gulf of Mexico [Pavela et al., 1983; Leming and Stuntz, 1984; Gaston et al., 1985; Renaud, 1986; Harper and Rabalais, 1997; Diaz, 1997; other chapters, this volume]. Demersal species that reside in the nearshore/offshore environment off Louisiana in water depths typically ranging from 5-

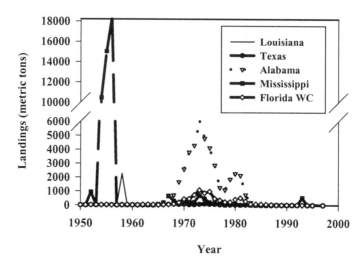

Year

Figure 5. Fishery landings of Atlantic croaker, *Micropogonias undulatus*, in the Gulf of Mexico, listed by State for the years 1950-1997. [Source: Personal communication from the National Marine Fisheries Service, Fisheries Statistics and Economics Division].

30 m are most substantially affected and can be displaced from a large portion of suitable habitat by hypoxia during summer months [Leming and Stuntz, 1984; Renaud, 1986; Hanifen et al., 1997]. Hypoxia is most likely to restrict the vertical distribution of nekton first [D. Stanley and C. Wilson, personal communication; E. Chesney, personal observation] and eventually can displace them from large areas of the bottom as evidenced by diminished trawl catches [Leming and Stuntz, 1984; Renaud, 1986; Hanifen et al., 1997; Craig et al., this volume]. Many schooling pelagic forage fishes, such as menhaden and anchovies, disperse from surface schools at night and distribute themselves throughout the entire water column. Hypoxia may limit their vertical distribution and increase their risk of predation.

Migratory pathways can be affected for estuarine-dependent and other species migrating from offshore spawning areas to inshore or alongshore, especially for those that utilize the lower water column or make vertical migrations [Zimmerman and Nance, this volume]. Significant numbers of estuarine-dependent species spawn offshore during summer, including white shrimp, brown shrimp, blue crab, mangrove snapper, *Lutjanus griseus,* cobia, *Rachycentron canadum,* sand seatrout, *Cynoscion arenarius,* silver seatrout, *C. nothus* and others (Table 1). Their early life history stages migrate inshore to coastal and estuarine nursery habitats. Hypoxia may be an impediment to these migrations, but the effect that hypoxia has on this process for species in the Gulf of Mexico has not been extensively studied.

Because adult nekton in open systems can generally migrate away from areas affected by hypoxia, direct mortality of nekton from hypoxia is likely to be rare in the northern Gulf. Fish kills associated with hypoxia in large basins or coastal embayments are sometimes related to unusual meteorological conditions that trap nekton [Schroeder and Wiseman, 1988; Waller and Fischer, 1989], and these occasionally occur in the Gulf of Mexico. Although the number of nekton killed can sometimes be relatively large and

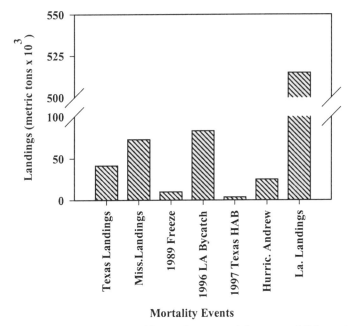

Mortality Events

Figure 6. A comparison of the estimated losses from several documented fish mortality events in the northern Gulf of Mexico. They include estimated losses of nekton in coastal Louisiana from the 1989 freeze (Source: Louisiana Department of Wildlife and Fisheries), nekton losses from Hurricane Andrew (Source: Louisiana Department of Wildlife and Fisheries) and the 1997 harmful algal bloom fish kill in Texas coastal waters [Source: http://www.tpwd.state.tx.us/news/news/971013d.htm]. (Numbers of fish killed were converted to biomass for the Texas fish kill by assuming an average fish weight of 250 grams). For comparison, the total 1997 fishery landing for Louisiana, Texas, and Mississippi are included [Source: Personal communication from the National Marine Fisheries Service, Fisheries Statistics and Economics Division] and the estimated mortality of bycatch in the 1996 Louisiana shrimp season [Chesney et al., 2000].

unsightly, at present the impacts to populations from hypoxia are unlikely to be significant, especially compared to other catastrophic mortality events that have been documented for nekton in the Gulf of Mexico (Fig. 6). For example, the estimated bycatch rates in the Louisiana shrimp fishery alone are in the range of 83-109 million kg annually [Adkins, 1993; Chesney et al., 2000]. If we assume a mean annual bycatch impact to Louisiana's coastal fish populations of 100 million kg, we can contrast the potential for an impact on fish production. If there were a series of fish kills from hypoxia, 548,000 individual fish (each weighing one-half kg) would need to be killed every day of the year to equal the documented bycatch mortality to groundfish from the Louisiana shrimp fishery alone. Clearly, the nekton populations in a highly productive system such as the Gulf of Mexico are capable of compensating for significant mortality from hypoxia or other factors. Especially when we consider that this mortality analogy only accounts for a fraction of the total annual fish landings and other bycatch affecting fish populations in the area. Fish kills due to asphyxiation from low oxygen may have a more important impact on eco-tourism than on the productivity of fish populations themselves, especially in resort areas.

Sublethal exposures of nekton to low oxygen can affect individual health, behavior, physiology or habitat choices [Fischer et al., 1992; Schurmann and Steffensen, 1992; Dalla et al., 1994; Secor and Gunderson, 1998]; however, these effects are species-dependent and difficult to study and document in large open systems. For example, nekton exposed to low oxygen in the laboratory respond in a variety of ways including lowered metabolism, increased ventilation, physiological adaptation and sometimes death [deFur and Mangum, 1989; Fischer et al., 1992; Schurmann and Steffensen, 1992; Dalla et al., 1994; Secor and Gunderson, 1998]. Nevertheless, it is likely that under most circumstances in an open-shelf system, highly-mobile nekton can and do migrate to avoid low oxygen conditions along the coast of the Gulf of Mexico.

Hypoxia has a greater potential to cause mortality to the early life history stages of nekton (i.e., eggs and larvae). Laboratory studies have demonstrated that fish eggs and larvae are very sensitive to low oxygen concentrations [Rombough, 1988]. Because of their small size and poorly developed swimming ability, larvae are poorly equipped to migrate long distances to escape or avoid hypoxic waters. Hatching of fish eggs and behavior of larvae can be affected below 3.5 mg l^{-1} and mortality can result from extended exposure to oxygen concentrations less than 2.0 l^{-1} [Rombough, 1988; Breitburg et al., 1994, 1997; E. Chesney, personal observation]. There are two major questions related to the issue of larval mortality that are relevant to the hypoxia issue in the Gulf of Mexico. Are sufficient numbers of eggs or larvae exposed long enough to low oxygen to cause mass mortality, and, if so, does hypoxia ultimately have an impact on recruitment success that is realized at the population level?

Only a few field studies on impacts of hypoxia on early life history stages of nekton have been made, and none of these were in the Gulf of Mexico. In the Chesapeake Bay, bay anchovy eggs and larvae collected during the summer hypoxic period were significantly higher in concentration at deeper sampling depths and at the lower dissolved oxygen concentrations that occur at those depths during summer [Dalton, 1987]. Many fish eggs are close to neutral buoyancy and can easily mix throughout the water column, especially when surface waters are freshened by river flow. In the Baltic Sea, reduced salinity caused smaller, less buoyant cod eggs, *Gadus morhua,* to sink into hypoxic bottom waters more readily than larger cod eggs, reducing their survival [Wieland, 1988; Nissling et al., 1994; Nissling, 1994; Nissling and Vallin, 1996]. Hypoxia may also change trophic interactions between larvae and their predators, although the outcome may vary dramatically with the types of predators and their behaviors [Breitburg et al., 1994, 1997].

The effects of hypoxia on mortality of early life history stages of nekton in the Gulf of Mexico have not been studied. It may also be difficult to draw extensively on work done in other systems because of the distinctive physical and biological differences between the Gulf of Mexico and other systems. There is little doubt that some mortality of the early life history stages of fish and other nekton due to low oxygen must occur on the Louisiana-Texas shelf. Although there may be significant mortality of early life history stages of nekton related to hypoxia, it is difficult to demonstrate a significant impact on recruitment of most of the major species of nekton in the Gulf of Mexico at present. We view the lack of any clearly definable impacts on nekton production as indicating that there are likely to be some mitigating factors that buffer the effects of hypoxia on early life history stages and their recruitment. The possibilities include the following:

- The enhanced productivity from the eutrophication effect may bolster survival in the early life history stages of nekton.
- Because the water column is highly stratified, the entire water column is rarely hypoxic and hypoxia is never shelfwide.
- Adults of most of the nekton spawning in the offshore areas subjected to low oxygen are widely distributed, highly fecund, and have extended spawning seasons.
- Characteristics of the early life stages of the nekton may be a mitigating factor. Fish eggs of many species are buoyant and may tend to stay in the upper water column. Summer spawned eggs hatch quickly (typically < 20 h), larvae develop quickly, become mobile and may be attracted to the well-oxygenated surface layer by phototaxis.
- The early life history stages of fish and other nekton suffer tremendous natural mortality rates in the absence of hypoxia and other environmental impacts [McGurk, 1986; Houde, 1994]. This provides significant scope for compensation in the mortality process.

The life history characteristics described above are typical of many of the fish and invertebrate larvae inhabiting the Gulf of Mexico and are likely to allow them to sustain some level of mortality from hypoxia (and other mortality factors) without significant consequences for their recruitment at present. The principal mitigating factor for mortality during the early life history of most nekton is that their mortality patterns can be highly compensatory [Goodyear, 1980; Jensen, 1993; Cowan et al., 1999]. Early life history stages of many species of marine nekton favor compensatory responses because they are typically subjected to extremely high mortality rates throughout their early life [Houde, 1994]. Commercially important marine species are typically very abundant and their large populations are sustained by the enormous fecundity and the reproductive potential of their adults. This is especially true in tropical and subtropical regions where fishes and other nekton have protracted spawning seasons and serial spawning is common [Johannes, 1978]. Compensatory reserve should be greatest for short-lived, highly fecund species [Cowan et al., 1999].

It should not be surprising that the meso- and eutrophic stages of nutrient enrichment have the potential to enhance survival and/or recruitment of planktonic fish and invertebrate larvae because there is a demonstrated relationship between primary production and the secondary production of fishes [Nixon et al., 1986; Iveson, 1990; Caddy, 1993; Houde and Rutherford, 1993]. The elevated productivity adjacent to the Mississippi River has been shown to provide a fertile environment for enhanced growth and possibly enhanced recruitment of nekton [Grimes and Finucane, 1991; Govoni and Grimes, 1992; Grimes and Kingsford, 1996; Govoni, 1997].

Given the lack of clear-cut evidence for catastrophic effects of hypoxia at the population level, it seems likely that at present, any mortality from hypoxia in the Gulf of Mexico, even death of a large proportion of eggs or larvae, may be compensated by a concomitant decrease in mortality due to predation or other mortality factors [Jensen 1993]. Because hypoxia in the Gulf of Mexico does not appear to constitute a constant and pervasive source of mortality for larvae under present environmental conditions, it is possible that mortality of early life history stages caused by hypoxia is not a major factor affecting recruitment of most nekton at this time. Nevertheless, this conclusion is speculative and needs to be studied! Life histories that might be especially vulnerable to hypoxia would include nekton that are demersal spawners, have low fecundity, a discreet spawning season, and reproduce offshore during summer with nursery habitats that include the hypoxic zone (see Table 1).

Indirect Effects on Nekton

Hypoxia can be viewed as a physical disturbance that can significantly alter system structure and function [Breitburg, 1997]. One of the most pervasive indirect effects that hypoxia can have on nekton is that it seasonally kills the prey of benthic foragers [Harper et al., 1981; Diaz and Rosenberg, 1995; Diaz, 1997; Harper and Rabalais, 1997; Rabalais et al., this volume]. Although hypoxia can limit the distribution of nekton directly by limiting the availability of normoxic habitats [Coutant, 1985], it can indirectly affect their distribution and growth by limiting the distribution, abundance and availability of their prey. The ultimate effects of these indirect impacts on nekton populations are equivocal and likely to depend on a number of factors.

There is little ambiguity with regard to the effects hypoxia and anoxia can have on benthic community structure and diversity [Diaz and Rosenberg, 1995]. Shallow marine benthic communities that are affected by hypoxia are characterized by lower species diversity and community structure that shifts from equilibrium to opportunistic species [Murrell and Fleeger, 1989; Dauer et al., 1992; Diaz and Rosenberg, 1995; Ritter and Montagna, 1999; Rabalais et al., this volume]. In the Gulf of Mexico, benthic recolonization can be rapid following a return to normoxia and recolonization is dominated by opportunistic species [Gaston et al., 1985; Boesch and Rabalais, 1991; Harper et al., 1981, 1991; Rabalais et al., this volume]. In areas or in years with persistent seasonal hypoxia, community structure of the benthos may not return to pre-hypoxia conditions by the onset of hypoxia in the subsequent year [Gaston et al., 1985; Harper et al., 1981, 1991; Rabalais et al., this volume]. Recolonization of benthic environments may also be affected by the seasonal degradation of the bottom sediments as suitable habitat for recruiting benthos. Low oxygen may disrupt larval settlement patterns, especially for benthic organisms that tend to reproduce synchronously and have discreet spawning seasons such as epitokous polychaetes.

Two important issues relevant to the impacts of hypoxia on nekton is how it affects the overall productivity of the benthic ecosystem and how foraging opportunities upon benthos are modified by the stress associated with low oxygen. The principal dilemma for the northern Gulf of Mexico is that we know that a significant biomass of demersal nekton that feed on benthos is being produced in the coastal zone associated with the hypoxia of the Gulf (see Table 1). We also know that significant amounts of benthic biomass are stressed and eventually killed by the coastal hypoxia. Effects of hypoxia on the behavior of benthos, their secondary production of benthos and transfer rates to higher trophic levels are poorly understood for the northern Gulf. The ultimate outcome should depend on the intensity, distribution and persistence of the hypoxia, as well as the physical characteristics of the basin (circulation, temperature, basin morphology), the life histories of the dominant fauna and the behavior of the fauna in response to oxygen stress [Rhoads et al., 1978; Robertson, 1979; Boesch and Rabalais, 1991; Stachowitsch, 1992; Diaz and Rosenberg, 1995; Chesney et al., 2000].

The onset of the conditions that induce hypoxia or anoxia within the sediments are likely to affect some of the benthic infauna, well before it completely displaces mobile nekton. Exposure of benthos to reduced oxygen and rising Eh in the sediments will eventually cause lethargy and should force benthos that require normoxic conditions or that are sensitive to sulfides, toward the surface of the sediments. We speculate that this should result in unusual foraging opportunities, especially for benthic feeding nekton [Schroeder and Wiseman, 1988; Pihl et al., 1991; Diaz et al., 1992]. These foraging opportunities should be most common in spring and early summer when hypoxia may be patchy and in the early stages of development [Pihl et al., 1992; Pihl, 1994; Diaz and Rosenberg, 1995]. They would also be most likely near the periphery of the hypoxia or in areas where small low oxygen patches develop.

In the Chesapeake Bay two effects were observed for predators of benthos impacted by low oxygen. Initially, demersal-feeding fishes changed their feeding habits to take advantage of stressed macrobenthos that came to the surface of the sediments [Diaz et al., 1992; Pihl et al., 1992]. Eventually the nekton relocated to shallower, better oxygenated areas of Chesapeake Bay [Pihl et al., 1991; Diaz et al., 1992]. Similar diet shifts were observed for demersal-feeding fishes in hypoxic areas of the Kattegat, Sweden [Pihl, 1994]. Shifts in benthic community structure (to opportunists) and reduced faunal diversity have been observed in response to hypoxia on the Texas-Louisiana shelf [Harper et al., 1981, 1991; Gaston et al., 1985; Boesch and Rabalais, 1991; Rabalais et al., this volume]. However, rates of benthic secondary production and shifts in trophic pathways of benthic-feeding nekton associated with eutrophication and hypoxia have not been studied or documented for the Gulf of Mexico hypoxic zone.

Several alternative scenarios for the response of nekton to the onset of hypoxia are possible, but evidence is inconclusive, primarily because causal mechanisms are impossible to identify from presence-absence benthic survey data. At the onset of hypoxia nekton may be forced to move by low oxygen or they may temporarily move higher in the water column to avoid low oxygen on the bottom in the early stages of development. In some instances, after the available benthos are exploited and opportunistic foraging opportunities are exhausted, they may move into alternate areas in search of prey rather than necessarily being driven from an area by hypoxia. Thus, all nekton species may not always be displaced directly by low oxygen conditions but some may be induced to move by the elimination of forage species killed by hypoxia, exploited by opportunistic predators or displaced by unsuitable conditions. With re-oxygenation of the sediments, benthic habitats will be recolonized by opportunistic species first [Rhoads et al., 1978; Dauer et al., 1992; Lake, 1997].

In Long Island Sound, the seasonal effects of hypoxia on the movement and diet of scup, *Stenotomus chrysops* were studied [Lake, 1997]. Lake concluded that seasonal and hypoxic effects were evident in the diet of scup. He found that the prolonged hypoxic event in Long Island Sound severely reduced the diversity and abundance of benthic prey and kept the community at an early successional stage [Lake, 1997]. Lake noted that early in the summer, scup preyed heavily on capitellid and spionid polychaetes. He surmised that this was because their abundance and proximity to the sediment surface made them the most available food item. As the summer hypoxia intensified in Long

Island Sound there was an increase in large polychaetes, such as *Nephtys* sp. and *Pherusa affinis,* in the diet of scup. Lake [1997] speculated that this diet shift resulted from diminishing oxygen and increasing sulfides driving deep burrowing benthos closer to the surface.

Upon cessation of the hypoxia in Long Island Sound, another diet shift was observed as scup quickly re-invaded the hypoxic sites [Lake, 1997]. Returning scup foraged on pelagic copepods, the crab, *Cancer irroratus,* and the shrimp, *Crangon septemspinosa.* Lake [1997] speculated that the scup reinvaded to forage upon the crabs and shrimp, which he believed returned in high numbers to exploit infaunal organisms that were killed or distressed by hypoxia. These results suggest that environmental conditions associated with eutrophication and hypoxia, combined with life history characteristics of opportunistic infauna, can favor high exploitation rates by certain types and sizes of nektonic predators [Pihl et al., 1992; Lake, 1997].

The presence of large numbers of opportunistic infauna may be integral to sustaining the production of demersal feeding nekton in the Gulf of Mexico. Opportunistic benthos dominate stressed environments because they typically have asynchronous reproduction and short generation times that lead to rapid reproductive rates and high production:biomass (P:B) ratios. Secondary production by opportunistic species is often four to five times higher than equilibrium communities that are typical of well-oxygenated, undisturbed sediments [Rhoads et al., 1978; Chesney, 1985; Pihl et al., 1992]. Two aspects of the recolonization of a disturbed environment favor higher P:B ratios, (1) abundant food resources for opportunistic deposit-feeders or scavengers, and (2) little competition for resources.   As planktonic larvae recolonize defaunated sediments, they can grow and mature quickly. Juvenile growth and production rates of colonizing benthos can be high, because little competition exists for abundant organically-rich resources. Also, very little energy is expended for maintaining biomass, which is low, especially during the early phases of recolonization. It is possible that in spite of low biomass and diversity, opportunistic fauna are turned over by predators at a high rate in stressed systems such as the Gulf of Mexico hypoxic zone. Consequently a depauperate fauna and low biomass is not necessarily a good measure of trophic transfer between benthos and nekton in a disturbed or oxygen-stressed system. Most important for foraging fishes and invertebrates, the opportunistic species most suitable as forage for fishes and invertebrates are typically small, surface-oriented, soft-bodied forms that can quickly colonize open sediment habitats [Rhoads et al., 1978; Gaston et al., 1985; Murrell and Fleeger, 1989; Gaston et al., 1995]. This pattern of benthic recolonization has been observed in many marine systems impacted by hypoxia, including the Gulf of Mexico. In the Gulf opportunistic colonizers, such as the polychaetes, *Mediomastus* sp., *Paraprionospio pinnata* or *Streblospio benedicti,* often dominate the macrobenthos [Gaston, 1985; Harper et al., 1991; Diaz and Rosenberg, 1995].

Besides life history characteristics that favor high turnover rates, many opportunistic species have larvae that are moderately or extremely tolerant to low oxygen [Yokoyama, 1995]. This can allow them to quickly colonize organically enriched sediments affected by hypoxia. Thus, after reoxygenation of the bottom, nekton that can prey upon small infauna may experience an offsetting benefit in the winter from recolonization of highly productive benthos that can be rapidly turned over by predation. Some species of nekton may also be favored by a changing benthic community structure in response to the seasonal disturbance caused by the hypoxia. For example, brown shrimp appear to be trophically linked to surface dwelling infauna such as polychaetes and amphipods [McTigue and Zimmerman, 1998]. Winter-spring production, combined with enhanced benthic production in normoxic habitats (stimulated by the high primary productivity of the system), may contribute substantially to the overall productivity of the Louisiana-Texas shelf demersal nekton. Thus at present hypoxia may have a partially beneficial effect for some demersal nekton. This process could be stimulated further in the Texas-Louisiana coastal ecosystem because of man-made disturbances in the form of trawling that continually disturb much of the benthic ecosystem. Like hypoxia, trawling may broadly promote the development of opportunistic communities, even in normoxic benthic sediments.

As in other coastal systems, most of the nekton species that reside inshore along the Gulf coast during the summer typically move farther out on the shelf to spawn or to avoid the cooler inshore water temperatures during the winter months. At least one shelfwide trawl survey shows a summer-winter pattern of distribution of nekton on the Louisiana shelf that supports the possibility of enhanced benthic production [Moore et al., 1970]. In areas with intense productivity and probably hypoxia in summer, nekton densities were found to be much lower within that zone during summer, but shelfwide they were remarkably high in the same area during the winter [Moore et al., 1970; Darnell et al., 1983]. Although Darnell et al. [1983] believed some hypoxic water was present during their summer surveys, no oxygen measurements were made to confirm the extent or intensity of hypoxia.

It is evident that somehow the substantial biomass and production of demersal fish and invertebrates associated with the Gulf of Mexico hypoxic zone are being supported by the available benthic secondary production [Chesney et al., 2000]. We hypothesize that the mid-shelf losses to benthic production associated with hypoxia during summer [Rabalais et al., this volume], may at present, be compensated for by enhanced shelfwide benthic production on the Texas-Louisiana shelf throughout the rest of the year and by downstream effects of eutrophication [Chesney, 1985; Dauer et al., 1992; Chesney et al., 2000]. One of the main goals of fisheries management is to maximize the level of sustainable production and harvest, thus it is imperative to understand how efforts to manage nutrient enrichment and alleviate hypoxia might impact secondary production and particularly fisheries production within this system.

Ecosystem Function and Effects of Hypoxia in the Gulf of Mexico

Low oxygen bottom waters in the Gulf of Mexico are symptomatic of a pattern seen in many marine systems where nutrients from both natural and anthropogenic sources stimulate high rates of primary production resulting in eutrophication [Vollenweider, 1992; Caddy, 1993]. Obviously, most aquatic organisms are not adapted to live in hypoxic environments, but just as we see clear differences in the expression of the impact of eutrophication [Caddy, 1993], we see differences in how different ecosystems respond to hypoxia. Some of these differences are attributable to the stage of the eutrophication process [Caddy, 1993], while other differences are likely to be due to the characteristics of the ecosystems themselves.

A key question is: How does the portion of the Gulf of Mexico ecosystem impacted by hypoxia compare to other ecosystems impacted by hypoxia and how might that be reflected in the structure and secondary production of nekton populations? If we compare the Gulf of Mexico to other systems affected by low oxygen, we can begin to see whether there are distinct attributes of the Gulf of Mexico system that may explain some of the apparent resilience displayed by its nekton populations [Chesney et al., 2000]. Caddy's [1993] comparative evaluation of human impacts on fishery ecosystems provides a basis for contrasting the effects of eutrophication in large ecosystems. We expand on Caddy's basic framework to include several additional system characteristics and use these to compare the Gulf of Mexico to other systems impacted by hypoxia (Table 3).

If we review specific attributes of the Gulf of Mexico coastal ecosystem, we can identify several key characteristics that probably help buffer the nekton of the Gulf of Mexico from some of the deleterious effects of eutrophication and the subsequent low oxygen conditions observed in other ecosystems (Table 3). The most significant characteristics are:

-   The hypoxia in the Gulf of Mexico affects only a part of a shallow open-shelf ecosystem with adjacent offshore and inshore habitats that can serve as suitable refuge for nekton during periods of hypoxia.
-   Conditions favoring the development and persistence of hypoxia in the Gulf of Mexico are strongly seasonal.
-   The Gulf of Mexico basin is large, most of the basin is oligotrophic and the ratio of the total basin area to the hypoxic area is large.
-   The hypoxic area in the Gulf of Mexico is virtually unbounded. Circulation within the Gulf of Mexico system is not restricted by a sill or narrow inlet as it is in many hypoxic basins around the world. Circulation and exchange of shelf water in the northern Gulf can be facilitated by winds, tides, coastal currents and other mechanisms, thus the shelf residence time is relatively short compared to other systems that are heavily impacted by hypoxia.
-   The nekton species of the Gulf of Mexico shelf are diverse, the trawl fishery is directed at shrimp, not groundfishes, and the most productive fisheries (shrimp and menhaden) and many of the groundfishes, are short-lived, highly productive species.

Although the hypoxia in the Gulf of Mexico affects a large area of the continental shelf adjacent to Louisiana and Texas, in a sense it is localized because it does not affect the entire shelf and only affects a small fraction of the entire basin (Table 3). The effects are most pervasive downstream from the entrance of the Mississippi River, and the effects decline as nutrients are gradually diluted with oligotrophic waters (from the open Gulf) and incorporated into the biomass of primary producers [Lohrenz et al., 1990]. The Gulf of Mexico hypoxia is also not a permanent feature, and the intensity varies temporally and spatially from year to year as river flow and nutrient loads vary [Justic' et al., 1993; Rabalais et al., 1991, 1996].

Nutrient reductions in the Mississippi River drainage basin have been proposed as a strategy for reducing coastal eutrophication and the temporal and spatial extent of hypoxia. A dilemma posed by management calling for nutrient reduction strategies is that nutrient enrichment has a significant positive effect on secondary production of coastal marine systems [Nixon et al., 1986; Nixon, 1988; Iverson, 1990; Houde and Rutherford, 1993; Caddy, 1993; Pennock et al., 1994; Steckis, 1995]. Furthermore, the

TABLE 3. A comparison of physical and biological characteristics of large marine ecosystems impacted by eutrophication and hypoxia.

Basin	Basin Morphology	Latitude	Sea-surface Temperature °C	Surface Salinity Range	Stratification Pattern	Annual Fishery Yields (10³ mt)	Major Fisheries
Mediterranean	Mostly enclosed sea[1]	--	--	--	--	--	--
--Adriatic	Semi-enclosed gulf[4]	44°N	6-27[7]	26-38[7,8]	Moderate, seasonal	100[14]	Sardine
Chesapeake Bay	Semi-enclosed bay	38°N	2-28[11]	11-20[11]	Strong, seasonal	275[14]	Menhaden
Seto Inland Sea	Semi-open sea[1]	34°N	10-27	28-32	Moderate, seasonal	830[13,+]	Anchovy, oyster
Baltic Sea	Mostly enclosed sea[1]	57°N	0-17	2-13[12]	Strong, semi-permanent	622[++]	Herring, sprat, cod
Black Sea	Mostly enclosed sea[1]	43°N	4-20	12-18	Strong, permanent	800[16]	Sprat, whiting
Gulf of Mexico	Semi-open gulf[1]	--	--	--	--	--	--
--La, Ms, Tx shelf	Open shelf	29°N	12-30	15-35	Strong, seasonal	769[**]	Menhaden, shrimp

Basin	Principal Watershed (10³ km²)	Average Depth (m)	Suboxic* Area (10³ km²)	Basin to Suboxic Area	Surface Area (10³ km²)	Watershed to Basin Area	Basin Volume (10³ km³)	System Residence Time
Mediterranean	--	1502	--	--	2510	0.6:1[1]	3771	Long
--Adriatic	70[9]	--	1.0	139:1	139[7]	0.5:1	35[7]	Moderate
Chesapeake Bay	166	6.5	1-2[6]	6:1	12	14.4:1	0.075	0.5 y[4]
Seto Inland Sea	25	37[10]	2.2[3]	10:1	22[13]	1.2:1	1.06	1.3 y[18]
Baltic Sea	1604	101	84[15]	5:1	382	4.2:1[1]	38	Long
Black Sea	1400	1191	20	25:1	508	6.2:1[1]	605	Long
Gulf of Mexico	2901	2164[2]	--	290:1	1813	1.6:1	3923	Long
--La, Ms, Tx shelf	--	35[19]	10-20[5]	6:1	126[19]	--	--	Weeks-Months

[1]from Caddy [1993]
[2]from Kennish [1989]
[3]estimated from Nagai and Ogawa [1997]
[4]from Nixon [1988]
[5]from Rabalais et al. [1991, this volume]
[6]from Eaton [1979], Officer et al. [1984]
[7]from Cattani and Corni [1992]
[8]from Vollenweider et al. [1992]
[9]from Franco and Michelato [1992]
[10]from Nakanishi et al. [1992]
[11]at Solomons, Maryland

[12]from Nehring [1992], Rosenberg et al. [1993]
[13]from Nakanishi et al. [1992]
[14]from Houde et al. [1999]
[15]from Rosenberg [1980]
[16]from Gomoiu [1992]
[17]from Takeoka [1984]
[18]estimated from Dimmel and Wiseman [1986]
[19]estimated for the central and western Gulf of Mexico continental shelf out to 200 m

* either hypoxic (<2.0 mg l⁻¹) or anoxic conditions except for Seto Sea (<2.0 ml l⁻¹; see Nagai and Ogawa [1997])
** commercial landings only
+ includes aquaculture
++ for Baltic proper

outcome of nutrient enrichment is not a deterministic process. The effects on a particular ecosystem and its biota are likely to vary with the size and morphology of the basin, the basin exchange rate, the species makeup of key organisms, latitude, the degree of eutrophication and other factors [Nixon, 1988; Iverson, 1990; Caddy, 1993; Houde and Rutherford, 1993; Pennock et al., 1994; Steckis et al., 1995; Pennock, 1997]. For example, immediately downstream from the mouth of the Mississippi River, nutrient enrichment may have the most immediate, persistent and serious ecological impact resulting in intense and persistent hypoxia during summer. Away from the river mouth, the effects gradually subside to the point where the eutrophication effect enriches the productivity of the area (Fig. 7a). Downstream the most significant ecological effect may be the potential change in stoichiometry of nutrients. These changes may cause species composition of phytoplankton assemblages to shift from predominantly diatoms to other less desirable assemblages [Aneer, 1985; Justic' et al., 1995; Rabalais et al., 1996; Burkholder et al., 1999], but this phenomenon is largely unrelated to hypoxia. In fact, fish kills from toxic algae may sometime be mistaken for effects of hypoxia [Burkholder et al., 1999].

Another key question is: In the face of hypoxia and other major factors affecting nekton, how have the nekton populations of the Gulf of Mexico sustained their present level of production without collapse? Hypoxia in the northern Gulf is seasonal and the intensity varies with river flow [Rabalais et al., 1991]. This pattern of impact to the coastal zone promotes both temporal and spatial heterogeneity in the degree of nutrient enrichment which nekton, with their mobility, may use to their advantage to sustain their productivity as well find suitable habitat within the constantly changing conditions of the coastal environment. Figure 7 shows a hypothetical seasonal progression of ecosystem responses to nutrient enrichment and the development of hypoxia in an open-shelf system. As the Mississippi and Atchafalaya rivers spill sediments and nutrients onto the shelf, primary production is stimulated downstream along the coast. In the spring, hypoxia develops as bottom oxygen demand increases (Fig. 7b). Subsequently benthic biomass and benthic production decline in the areas affected by hypoxia (Fig. 7c). Downstream, nutrient concentrations and primary production gradually decline locally to the point where nutrients stimulate primary production, but not to the point that it fuels intense hypoxia (Fig. 7a). In the fall as water temperatures drop and primary production declines, hypoxia gradually subsides shelfwide (Fig. 7b). Benthos recolonize areas affected by hypoxia at rates dictated by the re-oxygenation of the bottom layer of the water column and the sediments (Figs. 7b, 7d).

The spatial and temporal heterogeneity that results from the abrupt across-shelf and along-shelf transition from a eutrophic to an oligotrophic system are likely to provide a mosaic of habitat quality that is highly dynamic and seasonally variable. It seems likely that these characteristics would allow nekton to utilize the habitat in a way that minimizes the effects of temporarily degraded benthic habitats and allow them to sustain their productivity without significant population level effects. We suggest that the open-ended nature of the system helps to buffer the effects of the hypoxia in the Gulf of Mexico. Key elements of this open system are a relatively narrow shelf straddled by shallow abundant estuaries and marsh habitat inshore and well-oxygenated oligotrophic waters offshore that provide a suitable refuge for demersal nekton during the summer hypoxic period when hypoxia restricts the suitability of the mid-shelf demersal environment.

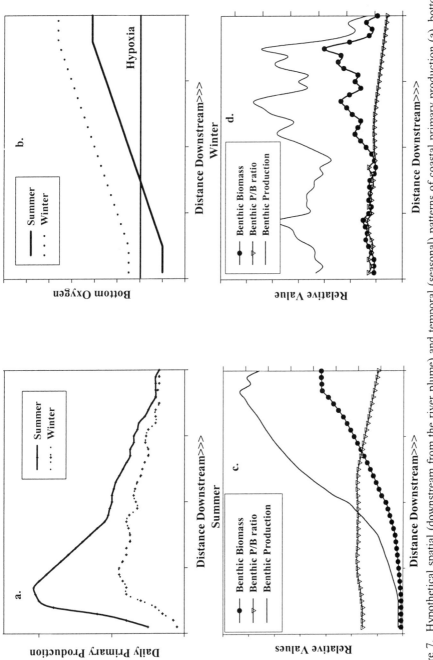

Figure 7. Hypothetical spatial (downstream from the river plume) and temporal (seasonal) patterns of coastal primary production (a), bottom oxygen (b), summer patterns of benthic production, benthos biomass and production:biomass ratio (c), and winter benthic production, benthos biomass and production:biomass ratio (d) in the area influenced by the Mississippi River outflow.

At present the data are not available to quantitatively predict all the effects of the hypoxia on the Gulf of Mexico nekton. Managers need to know at what point the Gulf of Mexico ecosystem has reached along the continuum from mesotrophic to dystrophic conditions. They also need to know how productivity of the entire shelf ecosystem is sustained, before they can decide how to optimize system resources, including nutrient enrichment [Caddy, 1993]. In mostly enclosed systems, such as the Baltic or Black Seas, the transition from a highly-productive eutrophic system to a faltering dystrophic system may be small. In larger more open systems where the eutrophication effects are localized, some degree of local system overload may occur and manifest itself as severe environmental conditions (i.e., hypoxia) without system collapse when adjacent habitats (either temporally or spatially) are mesotrophic and oligotrophic and can serve as a refuge. For example, hypoxia of bottom waters is often observed in upwelling areas and this can impact benthos [Arntz et al., 1985, 1991; Diaz and Rosenberg, 1995]. These effects, however, are not permanent and as primary productivity patterns shift in response to El Niño, benthos and demersal fishes respond as the environment oscillates between upwelling conditions favoring pelagic fisheries and hypoxia and El Niño favoring demersal nekton [Arntz et al., 1991]. In contrast, some closed or semi-enclosed systems, such as the Black and the Baltic Seas, are either permanently or mostly hypoxic at depth and have very little normoxic habitat to act as refuge, either seasonally or spatially. Thus, the impacts to the productivity and the habitat are predictably severe at all trophic levels and life stages in these seas [Nehring, 1992; Hagerman et al., 1996].

Conclusions and Recommendations

We conclude, based on the available information that, at present, the exploited nekton populations in the Gulf of Mexico are able to tolerate the effects of hypoxia without obvious major consequences for their recruitment, production or population health. It is also likely that currently other anthropogenic impacts, such as the direct and indirect effects of fishing, have more significant consequences for the production of nekton populations in the Gulf of Mexico. Nevertheless, these conclusions are speculative because of limitations of available data and the lack of directed research to adequately address the issue of hypoxia impacts on nekton. Important questions remain to be definitively answered. For example, are there signs of changes in the trophic pathways of demersal fishes that could explain some of the nekton community structure changes? Will expansion or intensification of the hypoxic zone reduce or threaten fishery production in Louisiana or throughout the 'fertile crescent'? How important are the other threats, such as toxic algal blooms, to fisheries in the Gulf of Mexico from eutrophication or associated phenomena? Are there species that are not commercially harvested that have been significantly impacted by low oxygen? Are there impacts on long-lived species that are taking place now, that will be realized at the population level in a decade or so? If nutrient reduction strategies are implemented, how will they affect fishery production? This is an especially critical question downstream where the eutrophication effect apparently enhances primary productivity and secondary production, without effects from hypoxia?

We advocate development of an ecosystem model designed to study the effects of hypoxia on fishery resources. This would be a major step toward understanding the complex interaction of this coastal ecosystem. Given the lack of research on hypoxia and living resources for the Gulf of Mexico, the accuracy of the model might be limited because so little is known about the influences of hypoxia on coastal nekton populations. Research on indirect impacts and sublethal stresses of low dissolved oxygen are needed to increase our poor understanding of how hypoxia affects fish populations, fish health and fisheries production in the Gulf of Mexico. While it is possible to hypothesize about potential impacts, virtually nothing is known of how hypoxia affects fisheries productivity or the health of individual fishes. Very little research has been directed at influences of hypoxia on pelagic fishes or the early life history stages of any species within the Gulf of Mexico hypoxic zone. The effect on prey of demersal species is a critical issue. A shelfwide assessment of benthic secondary production coupled with studies of trophic pathways for benthic foragers would be a major step toward understanding the present status and the trends of groundfish populations in the Gulf of Mexico and would provide a baseline from which to predict future changes. Given the value of the living resources in the Gulf of Mexico and all the uncertainty involved in our understanding of ecosystem function, hypoxia and its effects on living resources, the precautionary principle should be followed in the future management of this system.

Acknowledgments. We thank Dr. Glenn Thomas of the Louisiana Department of Wildlife and Fisheries for his assistance with reports and data on coastal Louisiana. We thank Tara Montgomery and Ken Deslarzes of the U.S. Minerals Management Service for their assistance in estimating the area of the coastal shelf in the Gulf of Mexico. We also acknowledge Dr. Kyoichi Tamai for his assistance with providing information on the Seto Inland Sea.

References

Adkins, G., A comprehensive assessment of bycatch in the Louisiana shrimp fishery, Louisiana Department of Wildlife and Fisheries Tech. Bull. No. 42, Bourg, Louisiana, 1993.

Aneer, G., Some speculations about the Baltic herring (*Clupea harengus membras*) in connection with eutrophication of the Baltic, *Can. J. Fish. Aquat. Sci., 42*, 83-90, 1985.

Anonymous, A fisheries management plan for Louisiana's penaeid shrimp fishery, Louisiana Department of Wildlife and Fisheries, Baton Rouge, Louisiana, 1992.

Anonymous, Supplement to the 1996 report of the mackerel stock assessment panel, Gulf of Mexico Fisheries Management Council and South Atlantic Fisheries Management Council, 1996.

Arntz, W. E., J. Tarazona, V. A. Gallardo, L. A. Flores, and H. Salzwedel, Benthos communities in oxygen deficient shelf and upper slope areas of the Peruvian and Chilean coast and changes caused by El Niño, in *Modern and Ancient Continental Shelf Anoxia,* edited by R. V. Tyson and T. H. Pearson, pp. 131-154, *Geological Society Special Publ., 58,* 1991.

Arntz, W. E., L. A. Flores, M. Maldonado, and G. Carbajal, Cambios de factores ambientes, macrobentos y bacteria filamentosas en la zona de minimo de oxigeno frente al Peru, in *El Niño: Su Impacto en la Fauna Marina,* edited by W. Arntz, A. Landa, and J. Tarazona, pp. 65-77, Boletin Instituto del Mar, Calloa, Peru, 1985.

Atwood, D. K., A. Bratkovich, M. Gallagher, and G. L. Hitchcock, Introduction to the dedicated issue, *Estuaries, 17,* 729-731, 1994.

Baltz, D. M. and E. J. Chesney, Spatial and temporal patterns in the distribution and abundance of fishes and macroinvertebrates in coastal Louisiana: Shrimp bycatch species in fishery-independent trawl samples, Final Report Submitted to the U. S. Dept. of Commerce, NOAA, National Marine Fisheries Service, Southeast Regional Center, MARFIN Project NA37FF044-01, 1995.

Boesch, D. F. and N. N. Rabalais, Effects of hypoxia on continental shelf benthos: comparisons between New York bight and the northern Gulf of Mexico, in *Modern and Ancient Continental Shelf Anoxia*, edited by R. V. Tyson and T. H. Pearson, pp. 27-34, *Geological Society Special Publ., 58*, 1991.

Breitburg, D. L., Episodic hypoxia in Chesapeake Bay: Interacting effects of recruitment, behavior and physical disturbance, *Ecol. Monogr., 62*, 525-546, 1992.

Breitburg, D. L., T. Loher, C. A. Pacey, and A. Gerstein, Varying effects of low dissolved oxygen on trophic interactions in an estuarine food web, *Ecol. Monogr., 67*, 489-507, 1997.

Breitburg, D.L., N. Steinberg, S. DuBeau, C. Cooksey, and E. D. Houde, Effects of low dissolved oxygen on predation of estuarine fish larvae, *Mar. Ecol. Prog. Ser., 104*, 235-246, 1994.

Burkholder, J. M., M. A. Mallin, and H. B. Glasgow, Jr., Fish kills, bottom water hypoxia and the toxic *Pfiesteria* complex in the Neuse River and estuary, *Mar. Ecol. Prog. Ser., 179*, 301-310, 1999.

Caddy, J. F., Towards a comparative evaluation of human impacts on fishery ecosystems of enclosed and semi-enclosed seas, *Rev. Fish. Sci., 1*, 57-95, 1993.

Cattani, O. and M. G. Corni, The role of zooplankton in eutrophication with special reference to the Northern Adriatic Sea, in *Marine Coastal Eutrophication*, edited by R. A. Vollenweider, R. Marchetti, and R. Viviani, pp. 137-158, *Sci. Total Environ.*, suppl. no. 0048-9697, 1992.

Chesney, E. J., D. M. Baltz, and R. G. Thomas, Louisiana estuarine and coastal fisheries and habitats: Perspectives from a fish's eye view, *Ecol. Appl., 10*, 350-366, 2000.

Chesney, E. J., Jr., Laboratory studies of the effect of predation on production and the production:biomass ratio of the opportunistic polychaete, *Capitella capitata* (Type I). *Mar. Biol., 87*, 307-312, 1985.

Coutant, C., Striped bass, temperature, dissolved oxygen: a speculative hypothesis for environmental risk, *Trans. Amer. Fish. Soc., 114*, 31-62, 1985.

Cowan, J. H., Jr., K. A. Rose, E. D. Houde, S. Wang, and J. Young, Modeling effects of increased larval mortality on bay anchovy population dynamics in Chesapeake Bay: evidence for compensatory reserve, *Mar. Ecol. Prog. Ser., 185*, 133-146, 1999.

Dalla V. J., G. den Thillart, O. Cattani, and A. Zwaan, Influence of long-term hypoxia exposure on the energy metabolism of *Solea solea*. 2. Intermediary metabolism in blood, liver and muscle, *Mar. Ecol. Prog. Ser., 111*, 17-27, 1994.

Dalton, P. D., Ecology of bay anchovy (*Anchoa mitchilli*) eggs and larvae in the Mid-Chesapeake Bay, M.S. thesis, University of Maryland, College Park, Maryland, 1987.

Darnell, R. M., R. E. Defenbaugh and D. Moore, Northwestern Gulf shelf bio-atlas: a study of the distribution of demersal fishes and penaeid shrimp of soft bottoms of the continental shelf from the Rio Grande to the Mississippi river delta, Open file report 82-04, U.S. Minerals Management Service, U.S. Department of Interior, 1983.

Dauer, D. M., A. J. Rodi, Jr., and J. A. Rannasinghe, Effects of low dissolved oxygen events on the macrobenthos of the lower Chesapeake Bay, *Estuaries, 15*, 384-391, 1992.

deFur, P. L. and C. P. Mangum, Effects of hypoxia on respiration in blue crabs, *Callinectes sapidus, J. Shell. Res., 8*, 479, 1989.

Diaz, R. J., Causes and effects of coastal hypoxia worldwide: putting the Louisiana shelf events in perspective, in Proc., First Gulf of Mexico Hypoxia Management Conference, December 1995, New Orleans, Louisiana, pp. 102-105, Publ. No. EPA-55-R-97-001, Gulf of Mexico Program Office, Stennis Space Center, Mississippi, 1997.

Diaz, R. J. and A. Solow, Ecological and Economic Consequences of Hypoxia: Topic 2 Report for the Integrated Assessment of Hypoxia in the Gulf of Mexico. NOAA Coastal Ocean Program Decision Analysis Series No. 17, NOAA Coastal Ocean Program, Silver Springs, Maryland, 1999.

Diaz, R. J. and R. Rosenberg, Marine benthic hypoxia: A review of its ecological effects and the behavioural responses of benthic macrofauna, *Oceanogr. Mar. Biol. Ann. Rev.*, *33*, 245-303, 1995.

Diaz, R. J., R. J. Neubauer, L. C. Schaffner, L. Pihl, and S. P. Baden, Continuous monitoring of dissolved oxygen in an estuary experiencing periodic hypoxia and the effect of hypoxia on macrobenthos and fish, in *Marine Coastal Eutrophication*, edited by R. A. Vollenweider, R. Marchetti, and R. Viviani, pp. 1055-1068, *Sci. Total Environ.*, suppl. no. 0048-9697, 1992.

Dinnel, S. P and W. J. Wiseman, Jr., Fresh water on the Louisiana and Texas shelf, *Cont. Shelf Res.*, *6*, 765-784, 1986.

Eaton, A., The impact of anoxia on Mn fluxes in the Chesapeake Bay, *Geochim. Cosmochim. Acta*, *43*, 429-432, 1979.

Fischer, P., K. Rademacher, and U. Kils, *In situ* investigations on the respiration and behaviour of the eelpout *Zoarces viviparus* under short-term hypoxia, *Mar. Ecol. Prog. Ser.*, *88*, 181-184, 1992.

Franco, P. and A. Michelato, Northern Adriatic Sea: Oceanography of the basin proper and of the western coastal zone, in *Marine Coastal Eutrophication*, edited by R. A. Vollenweider, R. Marchetti, and R. Viviani, pp. 63-106, *Sci. Total Environ.*, suppl. no. 0048-9697. 1992.

Gaston, G. R., Effects of hypoxia on macrobenthos of the inner continental shelf off Cameron, Louisiana, *Estuar. Coast. Shelf Sci. 20*, 603-613, 1985.

Gaston, G. R., P. A. Rutledge, and M. L. Walther, The effects of hypoxia and brine on the recolonization by macrobenthos off Cameron, Louisiana (USA), *Contr. Mar. Sci.*, *28*, 79-93, 1985.

Gaston, G. R., S. S. Brown, C. F. Racocinski, R. W. Heard, and J. K. Summers, Trophic structure of macrobenthos communities in the northern Gulf of Mexico estuaries, *Gulf Res. Rep.*, *9*, 111-116, 1995.

Gomoiu, M. T., Marine eutrophication syndrome in the northwestern part of the Black Sea, pp. 683-692, in *Marine Coastal Eutrophication*, edited by R. A. Vollenweider, R. Marchetti, and R. Viviani, pp. 683-692, *Sci. Total Environ.*, suppl. no. 0048-9697, 1992.

Goodyear, C. P., Compensation in fish populations, in *Biological Monitoring of Fish*, edited by C. H. Hocutt and J. R. Stauffer, Jr., pp. 253-280, American Fisheries Society, Bethesda, Maryland, 1980.

Govoni, J. D., D. E. Hoss, and D. R. Colby, The spatial distribution of larval fishes about the Mississippi River plume, *Limnol. Oceanogr.*, *34*, 178-187, 1989.

Govoni, J. J., The association of the population recruitment of Gulf menhaden, *Brevoortia patronus* with Mississippi River discharge, *J. Mar. Syst.*, *12*, 101-108, 1997.

Govoni, J. J. and C. B. Grimes, Surface accumulation of larval fishes by hydrodynamic convergence within the Mississippi River plume front, *Cont. Shelf Res.*, *12*, 1265-1276, 1992.

Grimes, C. B. and J. H. Finucane, Spatial distribution and abundance of larval and juvenile fish, chlorophyll and macrozooplankton around the Mississippi River discharge plume and the role of the plume in fish recruitment, *Mar. Ecol. Prog. Ser.*, *75*, 109-119, 1991.

Grimes, C. B. and M. J. Kingsford, How do riverine plumes of different sizes influence fish larvae: do they enhance recruitment?, *New Zealand J. Mar. Freshwat. Res., 47,* 191-208.

Gunter, G., Studies of the destruction of marine fish by shrimp trawlers in Louisiana, *La. Conserv. Rev., 5,* 18-24, 45-46, 1936.

Gunter, G., Should shrimp and game fishes become more or less abundant as pressure increases in the trash fish fishery of the Gulf of Mexico, *La. Conservationist, 8,* 11,14-15,19, 1956.

Gutherz, E. J., G. M. Russell, A. F. Serra, and B. A. Rohr, Synopsis of the northern Gulf of Mexico industrial and foodfish industries, *Mar. Fish. Rev.,* 1-11, 1975.

Hagerman, L., A. B. Josefson, and N. Jensen, Benthic macrofauna and demersal fish, in *Coastal and Estuarine Studies, Eutrophication in Coastal Marine Ecosystems,* edited by B. B. Jorgensen and K. Richardson, pp. 155-178, vol. 52, American Geophysical Union, Washington, D.C., 1996.

Hanifen, J. G., W. S. Perret, R. P. Allemand, and T. L. Romaire, Potential impacts of hypoxia on fisheries: Louisiana's fishery-independent data, in Proc., First Gulf of Mexico Hypoxia Management Conference, December 1995, New Orleans, Louisiana, pp. 87-100, Publ. No. EPA-55-R-97-001, Gulf of Mexico Program Office, Stennis Space Center, Mississippi, 1997.

Harper, D. E., Jr. and N. N. Rabalais, Responses of benthonic and nektonic organisms and communities to severe hypoxia on the inner continental shelf of Louisiana and Texas, in Proc., First Gulf of Mexico Hypoxia Management Conference, December 1995, New Orleans, Louisiana, pp. 41-56, Publ. No. EPA-55-R-97-001, Gulf of Mexico Program Office, Stennis Space Center, Mississippi, 1997.

Harper, D. E., Jr., L. D. McKinney, J. M. Nance, and R. R. Salzer, Recovery responses of two benthic assemblages following an acute hypoxic event on the Texas continental shelf, northwestern Gulf of Mexico, in *Modern and Ancient Continental Shelf Anoxia,* edited by R. V. Tyson and T. H. Pearson, pp. 49-64, *Geological Society Special Publ., 58,* 1991.

Harper, D. E., Jr., L. D. McKinney, R. R. Salzer, and R. J. Case, The occurrence of hypoxic bottom water off the upper Texas coast and its effects on the benthic biota, *Contr. Mar. Sci. 24,* 53-79, 1981.

Hayes, D. B., C. P. Ferreri, and W. W. Taylor, Active fish capture methods, in *Fisheries Techniques,* second edition, edited by B. R. Murphy and D. W. Willis, pp. 193-218, American Fisheries Society, Bethesda, Maryland, 1996.

Houde, E. D., Differences between marine and freshwater fish larvae: implications for recruitment. *ICES J. Mar. Sci., 51,* 91-97, 1994.

Houde, E. D., S. Jukic-Peladic, S. B. Brandt, and S. D. Leach, Fisheries: Trends in catches, abundance and management, in *Ecosystems at the Land-Sea Margin: Drainage Basin to Coastal Seas,* edited by T. C. Malone, A. Malej, L. W. Harding, Jr., N. Smodlaka, and R. E. Turner, pp. 341-366, *Coastal and Estuarine Science Series, 55,* American Geophysical Union, Washington, D.C., 1999.

Houde, E. D. and E. S. Rutherford, Recent trends in estuarine fisheries: Predictions of fish production and yield, *Estuaries, 16(2),* 161-176, 1993.

Jensen, A. L., Dynamics of fish populations with different compensatory processes when subjected to random survival of eggs and larvae, *Ecol. Model., 68,* 249-256, 1993.

Johannes, R. E., Reproductive strategies of marine fishes in the tropics, *Envir. Biol. Fish., 3,* 65-84. 1978.

Justic', D., N. N. Rabalais, R. E. Turner, and W. J. Wiseman, Jr., Seasonal coupling between riverborne nutrients, net productivity and hypoxia, *Mar. Pollut. Bull., 26,* 184-189, 1993.

Justic', D., N. N. Rabalais, R. E. Turner, and Q. Dortch, Changes in nutrient structure of river dominated coastal waters: stoichiometric nutrient balance and its consequences, *Estuar. Coast. Shelf Sci., 40*, 339-356, 1995.

Iverson, R. L., Control of marine fish production, *Limnol. Oceanogr., 35(7)*, 1593-1604, 1990.

Kennish, M. J., *Practical Handbook of Marine Science*, CRC press, Boca Raton, Florida, 1989.

Kerr, S. R. and R. A. Ryder, Effects of cultural eutrophication on coastal marine fisheries: a comparative approach, in *Marine Coastal Eutrophication*, edited by R. A. Vollenweider, R. Marchetti, and R. Viviani, pp. 599-614, *Sci. Total Environ.*, suppl. no. 0048-9697, 1992.

Lake, J. M., Diet selectivity of scup, *Stenostomus chrysops*, in Long Island Sound, M.S. Thesis, University of Connecticut, 1997.

Leming, T. D. and W. E. Stuntz, Zones of coastal hypoxia revealed by satellite scanning have implications for strategic fishing, *Nature, 310*, 136-138, 1984.

Lindner, M. J., A discussion of the shrimp trawl---fish problem, *La. Conserv. Rev., 5*, 12–17, 51, 1936.

Lohrenz, S. E., M. J. Dagg, and T. E. Whitledge, Enhanced primary production at the plume/oceanic interface of the Mississippi River, *Cont. Shelf Res., 10*, 639–664, 1990.

McGurk, M. D., Natural mortality of marine pelagic fish eggs and larvae: role of spatial patchiness, *Mar. Ecol. Prog. Ser., 34*, 227–242, 1986.

McTigue, T. A. and R. J. Zimmerman, The use of infauna by juvenile *Penaeus aztecus*, Ives and *Penaeus setiferus* (Linnaeus), *Estuaries, 21*, 160-175, 1998.

Moore, D. H., A. Brusher, and L. Trent, Relative abundance, seasonal distribution and species composition of demersal fishes off Louisiana and Texas, 1962-1964, *Contr. Mar. Sci., 15*, 45-70, 1970.

Murrell, M. C. and J. W. Fleeger, Meiofauna abundance on the Gulf of Mexico shelf affected by hypoxia, *Cont. Shelf Res., 9*, 1049-1062, 1989.

Nagai, T. and Y. Ogawa, Fisheries production, in *Sustainable Development in the Seto Inland Sea, Japan: From the Viewpoint of Fisheries*, edited by T. Okaichi and T. Yanagi, pp. 59-89, Terra Scientific Publishing, Tokyo, 1997.

Nakanishi, H., M. Ukita, M. Sekine, M. Fukagawa, and S. Murakami, Eutrophication controls in the Seto Inland Sea, in *Marine Coastal Eutrophication*, edited by R. A. Vollenweider, R. Marchetti, and R. Viviani, pp. 1239-1257, *Sci. Total Environ.*, suppl. no. 0048-9697, 1992.

National Research Council, Sustaining Marine Fisheries, Committee on Ecosystem Management for Sustainable Marine Fisheries, Ocean Studies Board, Commission on Geosciences, Environment, and Resources, National Research Council, National Academy Press, Washington, D.C., 1999.

Nehring, D., Eutrophication in the Baltic Sea, in *Marine Coastal Eutrophication*, edited by R. A. Vollenweider, R. Marchetti, and R. Viviani, pp. 673-682, *Sci. Total Environ.*, suppl. no. 0048-9697, 1992.

Nissling, A, H. Kryvi, and L. Vallin, Variation in egg buoyancy of Baltic cod *Gadus morhua* and its implications for egg survival in prevailing conditions in the Baltic Sea, *Mar. Ecol. Prog. Ser., 110*, 67-74, 1994.

Nissling, A., Survival of eggs and yolk-sac larvae of Baltic cod (*Gadus morhua* L.) at low oxygen levels in different salinities, in *Cod and Climate Change*. ICES Marine Science Symposia, vol. 198, edited by International Council for the Exploration of the Sea, pp. 626-631, Copenhagen, Denmark, 1994.

Nissling, A. and L. Vallin, The ability of Baltic cod eggs to maintain neutral buoyancy and the opportunity for survival in fluctuating conditions in the Baltic Sea, *J. Fish. Biol. 48*, 217-227, 1996.

Nixon, S. W., Physical energy inputs and the comparative ecology of lake and marine ecosystems, *Limnol. Oceanogr. 33,* 1005-1025, 1988.

Nixon, S. W., C. A. Oviatt, J. Fristhen, and B. Sullivan, Nutrients and the productivity of estuarine and coastal marine ecosystems, *J. Limnol. Soc. South Afr., 12,* 43–71, 1986.

Officer, C. B., R. B. Biggs, J. L. Taft, L. E. Cronin, M. A. Tyler, and W. R. Boynton, Chesapeake Bay anoxia: Origin, development and significance, *Science, 22,* 322-27, 1984.

Pavela, J. S., J. L. Ross, and M. E. Chittenden Jr., Sharp reductions in the abundance of fishes and benthic macroinvertebrates in the Gulf of Mexico off Texas associated with hypoxia, *Northeast Gulf Sci., 6,* 167–173, 1983.

Pennock, J. R., Estuarine hypoxia: The Mobile Bay perspective, in Proc., First Gulf of Mexico Hypoxia Management Conference, December 1995, New Orleans, Louisiana, p. 101, Publ. No. EPA-55-R-97-001, Gulf of Mexico Program Office, Stennis Space Center, Mississippi, 1997.

Pennock, J. R., J. H. Sharp, and W. W. Schroeder, What controls the expression of estuarine eutrophication? Case studies of nutrient enrichment in the Delaware Bay and Mobile Bay estuaries, USA. in *Changes in Fluxes in Estuaries: Implications from Science to Management,* edited by K. R. Dyer and R. J. Orth, pp. 139–145, Olsen and Olsen, Fredensborg, Denmark, 1994.

Pihl, L., Changes in the diet of demersal fish due to eutrophication-induced hypoxia in the Kattegat, Sweden, *Can. J. Fish. Aquat. Sci., 51,* 321-336, 1994.

Pihl, L., S. P. Baden, and R. J. Diaz, Effects of periodic hypoxia on distribution of demersal fish and crustaceans, *Mar. Biol., 108,* 349-360, 1991.

Pihl, L., S. P. Baden, R. J. Diaz, and L. C. Schaffner, Hypoxia induced structural changes in the diets of bottom-feeding fish and crustacea, *Mar. Biol., 112,* 349–361, 1992.

Pokryfki, L. and R. E. Randall, Nearshore hypoxia in the bottom waters of the northwestern Gulf of Mexico from 1981 to 1984, *Mar. Envir. Res., 22,* 75-90, 1987.

Rabalais, N. N., R. E. Turner, W. J. Wiseman, Jr., and D. F. Boesch, A brief summary of hypoxia on the northern Gulf of Mexico continental shelf: 1985-1988, in *Modern and Ancient Continental Shelf Hypoxia,* edited by R. V. Tyson and T. H. Pearson, pp. 35-47, *Geological Society Special Publ. No. 58,* 1991.

Rabalais, N. N., R. E. Turner, D. Justic', Q. Dortch, and W. J. Wiseman, Jr., Characterization of hypoxia: Topic 1 Report for the Integrated Assessment of Hypoxia in the Gulf of Mexico. NOAA Coastal Ocean Program Decision Analysis Series No. 16, NOAA Coastal Ocean Program, Silver Springs, Maryland, 1999.

Rabalais, N. N., Turner, R. E., D. Justic', Q. Dortch, W. J. Wiseman, Jr., and B. K. Sen Gupta, Nutrient changes in the Mississippi River and system responses on the adjacent continental shelf, *Estuaries, 19,* 386-407, 1996.

Renaud, M. L., Hypoxia in Louisiana coastal waters during 1983: Implications for fisheries, *Fish. Bull., 84,* 19-26, 1986.

Rhoads, D. C., P. L. McCall, and J. Y. Yingst, Disturbance and production on the estuarine seafloor, *Am. Sci., 66,* 577-586, 1978.

Ritter, C. and P. A. Montagna, Seasonal hypoxia and models of benthic responses in a Texas Bay, *Estuaries, 22,* 7-20.

Robertson, A. I., The relationship between annual production:biomass ratios and lifespans for marine macrobenthos, *Oecologia, 38,* 193–202, 1979.

Rombough, P. J., Respiratory gas exchange, aerobic metabolism, and effects of hypoxia during early life, in *Fish Physiology, Vol. XI, The Physiology of Developing Fish,* edited by W. S. Hoar and D. J. Randall, pp. 59-162, Academic Press, New York, New York, 1988.

Rosenberg, R., Effects of oxygen efficiency on benthic macrofauna in fjords, in *Fjord*

Oceanography. edited by H. L. Freeland, D. M. Farmer, and C. D. Levings, pp. 499-514, Plenum, New York, New York, 1980.

SAS Institute Inc., *SAS Guide for Statistics*, Version 5 Edition, SAS Institute Inc., Cary, North Carolina, 1985.

Schirripa, M. J. and C. M. Legault, Status of red snapper in U. S. waters of the Gulf of Mexico: updated through 1996, National Marine Fisheries Service Technical Report MIA-97/98-05, 1997.

Schroeder, W. W. and W. J. Wiseman, Jr., The Mobile Bay estuary: stratification, oxygen depletions, and jubilees, in *Hydrodynamics of Estuaries, Volume II, Case Studies*, edited by B. J. Kjerfve, pp. 41-52, CRC Press, Boca Raton, Florida, 1988.

Schurmann, H. and J. F. Steffensen, Lethal oxygen levels at different temperatures and the preferred temperature during hypoxia of the Atlantic cod, *Gadus morhua* L., *J. Fish Biol.*, *41*, 927-934, 1992.

Secor, D. H. and T. E. Gunderson, Effects of hypoxia and temperature on survival, growth, and respiration of juvenile Atlantic sturgeon, *Acipenser oxyrinchus. Fish. Bull.*, *96*, 603-613, 1998.

Sokal, R. R. and F. J. Rohlf, *Biometry*, W. H. Freeman and Co., New York, New York, 1981.

Stachowitsch, M., Benthic communities: eutrophication's "memory mode", in *Marine Coastal Eutrophication*, edited by R. A. Vollenweider, R. Marchetti, and R. Viviani, pp. 1017-1028, *Sci. Total Environ.*, suppl. no. 0048-9697, 1992.

Steckis, R. A., I. C. Potter, and R. C. J. Johnson, The commercial fisheries in three southwestern Australian estuaries exposed to different degrees of eutrophication, Chapter 12, in *Eutrophic Shallow Estuaries and Lagoons*, edited by A. J. McComb, pp. 189-204, CRC Press, Boca Raton, Florida, 1995.

Takeoka, H., Exchange and transport time scales in the Seto Inland Sea, *Cont. Shelf Res.*, *3*, 327-341, 1984.

Thompson, N. B., An assessment of cobia in Southeast U.S. waters, National Marine Fisheries Service Technical Report, MIA-95/96-28, 1996.

Turner, R. E. and N. N. Rabalais, Changes in Mississippi River water this century, *BioScience, 41*, 140–147, 1991.

Turner, R. E., R. Kaswadji, N. N. Rabalais, and D. F. Boesch, Long-term changes in the Mississippi River water quality and its relationship to hypoxic continental shelf waters, in *Estuarine and Coastal Management--Tools of the Trade*, Proceedings of the 10th National Conference of the Coastal Society, New Orleans, Louisiana, 1987.

Vollenweider, R. A., Coastal eutrophication principles and controls, in *Marine Coastal Eutrophication*, edited by R. A. Vollenweider, R. Marchetti and R. Viviani, pp. 1-20, *Sci. Total Environ.*, suppl. no. 0048-9697, 1992.

Vollenweider, R.A., A. Rinaldi, and G. Montanari, Eutrophication, structure and dynamics of a marine coastal system: results of a ten-year monitoring along the Emilia-Romagna coast (Northwest Adriatic Sea), in *Marine Coastal Eutrophication*, edited by R. A. Vollenweider, R. Marchetti, and R. Viviani, pp. 63-106, *Sci. Total Environ.*, suppl. no. 0048-9697, 1992.

Waller, U. and P. Fischer, The fish kill of autumn 1988 in Kiel Bay, International Council for the Exploration of the Sea (collected papers), Denmark, Biol. Oceanogr. Comm., 1989.

Wieland, K., Distribution and mortality of cod eggs in the Bornholm Basin (Baltic Sea) in May and June 1986, in *The Baltic Sea Environment*, pp. 331-340, *Kieler Meeresforschungen*. Sonderheft. Kiel no. 6, Kiel Univ. (FRG). Inst. fuer Meereskunde, 1988.

Wiseman, Jr. W. J., N. N. Rabalais, R. E. Turner, S. P. Dinnel, and A. MacNaughton, Seasonal and interannual variability within the Louisiana coastal current: stratification and hypoxia, *J. Mar. Systems, 12*, 237-248, 1997.

Yokoyama, H., Occurrence of *Paraprionospio* sp. (form A) larvae (Polychaeta: Spionidae) in hypoxic waters of an enclosed bay, *Estuar. Coast. Shelf Sci., 40(1),* 9-19, 1995.

Zimmerman, R., J. Nance, and J. Williams, Trends in shrimp catch in the hypoxic area of the northern Gulf of Mexico, in Proc., First Gulf of Mexico Hypoxia Management Conference, December 1995, New Orleans, Louisiana, pp. 64-74, Publ. No. EPA-55-R-97-001, Gulf of Mexico Program Office, Stennis Space Center, Mississippi, 1997.

18

A Brief Overview of Catchment Basin Effects on Marine Fisheries

John F. Caddy

Abstract

The effects of nutrient enrichment on coastal systems this century, and especially on fisheries, have appeared simultaneously with other anthropogenic developments, making it difficult to tease apart the cause-and-effect relationships important for management. These complex interactions are referred to as Marine Catchment Basin (MCB) effects and may differ in anthropogenically modified catchments from those associated with natural levels of erosion and climate. The enrichment of coastal waters by increased nutrient loading may, or may not, be desirable from a fisheries production point of view, and depends on the hydrodynamics of the receiving basin, nutrient limitation and fishing pressure, among other factors. Effective management of MCB effects starts upstream from the fishing zone, before nutrient discharges reach the sea, and must be proscribed with regional differences in mind. Several case studies illustrate these points, and two simple models for multi-species fisheries are introduced that illustrate the different ways eutrophication may affect fisheries under different fishing pressure. Conventional fish population models that assume steady-state production and landings are dominated by fishing effort may be misleading. New paradigms for fisheries science are needed that incorporate both MCB effects and fishing effort in order for realistic and useful models to be developed.

Introduction

The United Nations Conference for Environment and Development in 1992 placed emphasis on the linkages between the use of river basins as a whole and the impacts of

Coastal Hypoxia: Consequences for Living Resources and Ecosystems
Coastal and Estuarine Studies, Pages 355-370
This paper is not subject to U.S. copyright.
Published 2001 by the American Geophysical Union.

these activities on estuarine and coastal areas. The effects of increasing nutrient loading to coastal waters, either directly into the sea or through river discharges, may result in eutrophication, enhanced or diminished fish production, and low oxygen zones (i.e., the chapters in this book).

The relationship between increased nutrient loadings and fisheries production has been the subject of much research for freshwater systems, but the impact on marine fisheries production of human manipulations of primary production by changing nutrient fluxes, has been given some attention by coastal ecologists (e.g., Nixon [1988] and Nixon et al. [1996]) but not by marine fish stock assessment workers and fishery managers. For too long the assumption of equilibrium conditions has been made in calculating fisheries production from marine systems. The occurrence of pollution problems associated with fisheries in the "New" World, such as those resulting from Mississippi River runoff to the Gulf of Mexico [Rabalais et al., 1996], have been well-documented for only a portion of the "Old" World. The main objective of this chapter, therefore, is to draw attention to an overall coherence of reported phenomena in Eurasian coastal waters that may provide the basis for future comparative studies, in a way that integrates the effects of changing nutrient loadings and the subsequent ecosystem effects with the more traditional models of fisheries that are primarily restricted to stock recruitment indices and fishing effort.

Coastal Marine Ecosystems are Subject to Contemporaneous Influences

A superficial view of FAO's (Food and Agriculture Organization of the United Nations) global data bases of fisheries landings [FAO, 1995] and fleet size [FAO/FIDI, 1994] confirms the general impression that both global fishing effort and landings have risen steadily since World War II. To a large extent, the increase in landings to a recent plateau (in 1996) of some 87 million tons of fish and invertebrates is consistent with local analyses from different ocean areas (e.g., Caddy et al. [1998a]). The growing evidence for limits to production is also compatible for most seas with "the theory of fishing," which sees fisheries production changes mainly as a function of changing fishing intensity.

This common assumption about the dominance of fishing effort on harvest may, however, be misleading. Stromberg [1997] stressed that to distinguish between anthropogenic and natural effects requires "a significant understanding of the dynamics of ocean systems," but that is a formidable undertaking. What is the noise and what is the signal? The transformations of freshwater ecosystems may be easily documented (e.g., Welcomme [1995]), but not those due to fishing and they are not easily measured by a single variable. Based on current population trends, anthropogenic nutrient inputs to marine environments will certainly increase into the next century. Galloway et al. [1994] predicted an increase of 25% in total nitrogen atmospheric deposition in developed countries and an increase of more than 50% over oceans in the northern hemisphere by 2020.

Studies of historical landing trends by FAO Statistical Areas (e.g., Grainger and Garcia [1996], Caddy et al. [1998a]) have shown that landings for most multi-species marine fisheries peaked simultaneously as industrial fisheries spread out from "core" areas found mostly in the northern hemisphere. We are close to the top of the multi-species yield curve for most regions. Thus, unless stocks collapse as effort increases, then modest incremental rises in fishing effort are likely to have a relatively smaller effect on the overall yield from the system than during the early period of fishery

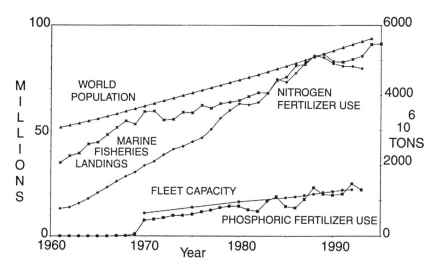

Figure 1. Some global trends, showing synchronicity. From top to bottom: world population (triangles, millions on the left margin); global estimates of marine fish landings (crosses), nitrogen fertilizer use (lozenges), fleet capacity (circles) and phosphoric fertilizer use (squares); right margin in millions of tons. Modified from Caddy [2000a].

expansion. By contrast, the effects of environmental changes, whether anthropogenic or natural, are likely to become predominant if the system remains close to the fishing effort level providing theoretical peak productivity, and environmental impacts on system productivity may become progressively more obvious in the future.

Many environmental driving functions, however, are difficult to measure. We might, of course, use a number of "proxy" variables to signal that changes in levels of nutrient enrichment of aquatic systems have been going on. Fisheries landings seem obvious variables for monitoring an ecosystem, given that they are among the few data sets available with a long historical record, and given also that commercial species are often "keystone species" for the ecosystem. Since they reflect both types of human impacts discussed above, however, a comparative approach between large marine ecosystems, or more specifically, between MCBs, is essential.

We are faced with the fact that since the start of the century almost all anthropogenically-generated signals have been trending in a similar direction, namely towards increased stress on natural systems, both in freshwater and inshore and semi-enclosed marine ecosystems [Rapport et al., 1985; Caddy, 1993, 2000a]. Symptoms of this stress, which are difficult to distinguish in terms of their original cause, are simplifications of ecosystem complexity and dominance by r-selected species. Figure 1 provides an example of the contemporaneous nature of trends in several unrelated variables actually or potentially affecting aquatic systems at the global scale. This figure illustrates that both the world population and industrial production of both phosphate and nitrate fertilizers (spread largely within catchment basins) have risen over the same period at the same time as marine landings and fleet size have risen dramatically. Regression analysis alone does not allow one to distinguish between the relative rates of change of these outputs and the possible forcing functions.

The Linking Ratio [Stamatopoulos and Caddy, 1991] is a simple index based on normalized linear regressions and provides a mechanism to rank linear regressions of

TABLE 1. Linear regressions through time series for 1960 - 1985 and later of several global time series of relevance to the marine environment, ranked in terms of their linking ratios (see Fig. 1).

Time series trend	Linear regression equations	Linking Ratio
Global human population[1]	-148744 + 77.388 * t	0.099
Global fishing fleets[2]	-981.65 + 0.5038 * t	0.116
Global marine fish landings[3]	-2970.9 + 1.5347 * t	0.134
Global production of N fertilizers[4]	-4854 + 2.4811*t	0.259
Global production of P fertilizers[4]	-1515 + 0.772*t	0.386

[1]UN [1997] and earlier
[2]FAO/FIDI [1994]
[3]FAO [1995]
[4]FAO [1994a]

positive dependent variables in terms of their relative trend with a common independent variable, such as years (Table 1)[1]. The nearly linear and increasing trend in several key variables can be ranked in the following order (small to large): human population, fishing fleet size, marine landings, global usage of nitrogen and phosphorus fertilizer (Fig. 1). Although this seems to imply more rapid increases in impacts on aquatic systems from industrial agriculture than fishing, the main point here is to note that whatever the relative rate of change it would be misleading to try and isolate system impacts solely in terms of information on catch and fishing effort. The effects of anthropogenic effects for a range of driving variables are not easily separated or studied independently, unless some abrupt change in one of them allows their relative impacts to be separated. Given this conclusion, it seems unrealistic to apply models for coastal fisheries management that assume that the productivity of coastal waters has remained constant during the last few decades or to seek to explain changes in coastal ecosystems as exclusively due to the direct or indirect effects of fishing.

The Marine Catchment Basin as an Ecosystem

The catchment basin is the smallest natural unit of landscape that can be modeled by linked aquatic and terrestrial ecosystems and is tightly connected (summarized in Hornung and Reynolds [1995]). Freshwater lacustrine and limnic systems and semi-enclosed marine ecosystems show broad similarities [Welcomme, 1995; Regier, 1979; Serafin and Zaleski, 1988; Caddy, 1993]. The Marine Catchment Basin paradigm (MCB) integrates land-use impacts with aquatic ecosystems [Caddy, 1993; Caddy and Bakun, 1994]. Caddy and Bakun [1994] defined a marine catchment basin to include "...not only the marine aquatic ecosystem, but also the adjacent watersheds that drain into it." This conceptual framework can be compared with other paradigms, the Large Marine

---

[1]For a time series of non-negative values, the Linking Ratio, L, was defined by Stamatopoulos and Caddy [1991] as the extent to which the slope B/Y_{mean} of a normalized linear regression approaches the maximum slope B_{max} geometrically possible over the observed range of normalized x and y values. This method allows one to compare the slopes of time series of two or more dependent variables measured in different units and would appear to allow ranking of changes in different dependent variable changes with time. In Table 1, linking ratios appear to suggest, for example, that production of phosphorus fertilizers has increased relatively more rapidly than has the global population.

Ecosystem (LME) and the Integrated Coastal Area Management (ICAM). The LME treats impacts from land on the marine environment as extrinsics, while the ICAM deals with events across the land-water marine interface. Neither provides a fully satisfactory basis for a comparative study of marine ecosystems whose trophic webs receive a significant contribution from nutrient-enriched outflows from inland terrestrial ecosystems.

In fact, one of the implications of the MCB concept is that we can consider the sum of net downstream benefits (B) from an action high in the watershed as defined by:

B = (riverine impacts) ± (estuarine impacts) ± (plume and coastal current impacts)

- with the sign reflecting positive or negative economic impacts, which presumably are likely to turn progressively negative as the extent of terrestrial influence increases.

Global Fisheries Production under Different Nutrient Regimes

Caddy and Bakun [1994] compared three main mechanisms of nutrient supply that apply to marine shelf fisheries: upwellings, tidal mixing and nutrient runoff from land. The last mechanism is especially likely to impact shelf ecosystems when discharged into stratified systems whose surface waters were formerly low in nutrients. The production of small pelagic fish production expressed as tons of fish landed per unit shelf area is seen as one indicator of annual planktonic production [Caddy et al., 1998a] and the production per shelf area of demersal fish and commercial benthos as an indicator of the "quality" of benthic conditions. Their ratio, therefore, provides a useful and rather sensitive indicator that can be calculated from fishery statistics and is referred to as the P/D ratio (the Nile outflow example is discussed later). This ratio is perhaps the closest measure that can be obtained from commercial statistics to the balance between the net results of inputs of nutrients via planktonic food chains to small pelagic fish and via detrital pathways and the benthos to demersal fish and crustaceans.

The resulting organic production as detrital inputs (in moderation) may slightly enhance benthic/demersal food chains, or (in excess), via bottom hypoxia, lead to their disappearance altogether. Pelagic food chains appear more resistant and, up to a certain point, may even be significantly enhanced by nutrient runoff unless jellyfish predators dominate.

Caddy [1993] noted that there are some features held in common between upwellings and marine ecosystems subject to cultural eutrophication. These may be summarized as follows:

(1) Short-lived and small organisms dominate the biomass;
(2) Food webs are relatively short and simple, with low mean trophic levels/diversity;
(3) Large, long-lived benthos are rare, especially close to the nutrient source/point of upwelling;
(4) Small pelagic fish/planktivorous species heavily predominate over demersal/ piscivorous species;
(5) Hypoxic/anoxic conditions are found close to the bottom, either seasonally or permanently.
(6) The ratio of pelagic/demersal fish landings is higher in both upwelling areas and eutrophicated semi-enclosed seas (see de Leiva Moreno et al. [in press] for a review of the latter effects).

Thus, areas affected by coastal eutrophication seem to show some parallels with areas naturally subject to high nutrients and are generally characterized by a high abundance of r-selected species reacting as "blooms" to temporal changes in marine ecosystems. This analogy obviously cannot be pursued too far, because with eutrophic stress new niches open in previously oligotrophic systems. These systems may then be progressively filled by "exotics" better adapted to the hypoxic conditions with an eventual replacement of the "native" fauna and flora. What is clear from the Great Lakes example, where pelagic food chains were poorly developed prior to eutrophication, is that large resident populations of small pelagic fish require sustained primary production year-round. In contrast to the pelagic fish, the benthos provide the store of chemical energy for the demersal fish that keeps them supplied between seasonal blooms.

One other generalization that has some validity, especially for stratified marine waters in the tropics, is that there is often a transition from nutrient-enriched coastal wetlands, estuaries and mangrove swamps to highly nutrient-restricted waters offshore. On productive north-boreal continental shelves, although a similar gradient exists, seasonal mixing by tides and winds is the dominant mode of replenishment of nutrients in surface waters. An analysis of FAO global statistical data from the perspective of the dominant processes supplying nutrients in each region [Caddy et al., 1998a, 1999] suggests differences between northern and southern hemispheres in arcto-boreal fishery productivity. The higher plateau for demersal/benthic landings per shelf surface in northern FAO statistical areas compared with the southern hemisphere may be largely a consequence of a greater "land mass effect" in the north. This, in turn, could act via greater runoff of nutrients from the Eurasian landmass and/or possibly reflect increased larval dispersion from the more exposed shelf areas of the southern hemisphere where there are no or few semi-enclosed seas.

As noted, the historical build up of nutrient runoff and fishing capacity has been contemporaneous (Fig. 1), so that distinguishing the effects of increases in exploitation from natural variation, or even increased nutrient loading, is going to be difficult, but may be possible where important changes in one variable have been documented. Until recently, the tendency has almost always been to follow classical stock assessment models that assume that stable or "equilibrium" biological productivity applies. Examples from semi-enclosed seas subject to various degrees of eutrophication show that the equilibrium assumption may be inappropriate, and this may be so generally for marine ecosystem components, judging strictly from variations in landings [Caddy and Gulland, 1983; Spencer and Collie, 1997]. There needs to be a closer historical review of the time sequence of nutrient runoff or other variables indirectly contributing to measuring it, so that, where possible, one may attempt to reconstruct its possible impacts on historical fisheries yields.

I attempted to present a qualitative picture of the recent 30-50 years of fisheries development and relative trends in overall fishery production for several well-studied semi-enclosed seas subject to land runoff of nutrients [Caddy, 1993]. These seem to pass through an enhancement phase before serious problems emerge due to excessive nutrient loadings. A recent rise in landings for "core seas" long subjected to high levels of fisheries exploitation is difficult to explain using steady state versions of the catch equation of Beverton and Holt [1957] or the production model of Schaefer [1957] and later authors. It seems we need to assume that changes in basic productivity have been occurring. Certainly it is hard to presume for the Mediterranean that the low production in the 1960s and 1970s relative to levels of the world average for shelf production in the 1980s and 1990s was entirely due to the presence of "unexploited resources" during this early time period.

"Fishing down the food chain" (FDTFC) is a hypothesis that attempts to explain increases in fisheries yields from species low in the food chain [Daan, 1989; Pauly et al., 1998]. FDTFC releases predation pressure from depleted top predators whose prey then increase, and the FDTFC hypothesis may be applied to support this interpretation of events for planktivores in the Yellow Sea [Tang, 1993] and Seto Inland Sea [Nagasaki and Chikuni, 1989]. These events, however, could equally or better be explained by the "bottom-up" enhancement of lower trophic levels through MCB effects via nutrient enrichment, as suggested by Caddy et al. [1998b] and Caddy [2000b]. This alternative hypothesis is consistent with observations in coastal North Sea waters [Boddeke and Hegel, 1995; Dethlefsen, 1989] and has the same consequences, because FDTFC and their effects may be difficult to distinguish from one another. The FDTFC hypothesis assumes that stocks of planktivores have adequate food to significantly increase their biomass in the absence of predator control. This seems to have occurred in the Black Sea [Berdnikov et al., in press], where increased predation on planktonic herbivores (in this case by *Mnemiopsis*, a ctenophore predator on zooplankton, fish eggs and fish larvae, also found in Chesapeake Bay) would have reduced grazing on phytoplankton. Increases in phytoplankton standing stocks here could have been reinforced by both mechanisms. For the Black Sea, where regular annual surveys of pelagic fish were carried out by the former Soviet Union, there is evidence [Ivanov and Beverton, 1985] that increases in small pelagic biomass measured by acoustic surveys in the 1970s (Fig. 2) led to, and then increased simultaneously with, rises in catches, which again is difficult to explain under assumptions of steady-state system productivity.

Examples of MCB Fishery/Nutrient Interactions

A few examples will illustrate some of the characteristics of these systems subject to changes in nutrient runoff.

The Mediterranean

On an areal basis, fisheries production in the Mediterranean was relatively low until the 1970s [Gulland, 1971]. After then fisheries production began to rise near the Rhone, Po and Ebro rives and at the outflow of Black Sea [Caddy et al., 1995, 1998a], and later in the Aegean [Friligos; 1989] and off the Levant (southeastern Mediterranean) [Grainger and Garcia, 1996]. These changes seem to be coincidental with increased nutrient loadings of rivers and eutrophication of coastal waters resulting in increased phytoplankton production, reduced water transparency, hypoxia, anoxia events and fish kills (e.g., northern Adriatic [Ivancic and Degobbis, 1988; Degobbis, 1989; UNEP, 1996; Marasovic et al., 1988; Pucher-Petkovic et al., 1988].

Not all of the Mediterranean is eutrophic, however. Both north-south and east-west gradients of nutrient inputs exist in the Mediterranean [Murdock and Onuf, 1972], which are attributed to the impoverishment of nutrient levels in the incoming Atlantic water as it moves towards the eastern basin, and river runoff in the north. Much of the Levant and much of the southern Mediterranean are considered strongly nutrient-limited and may show further increases in fishery yield if coastal runoff of nutrients increases [Marasovic et al., 1988; Pucher-Petkovic et al., 1988].

An analysis of Mediterranean fisheries in the late 1970s suggested that yields were then close to the Maximum Sustainable Yield (MSY), especially for demersal resources.

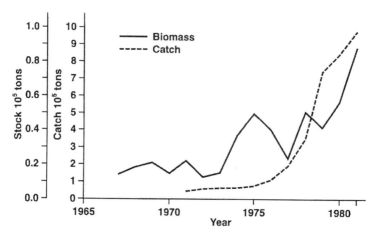

Figure 2. Trends in annual survey biomass and catch of sprat in the Black Sea, suggesting an increase in catches after biomass had begun to increase and implying an increase in pelagic production (modified from Ivanov and Beverton [1985]).

This suggests that much of the increased yields since then are the result of anthropogenically-caused nutrient runoff into an otherwise low productivity system. The reduction of nutrients in the Nile river following the construction of the Aswan Dam had a reverse effect on fishery production (Fig. 3). Before the dam was built the dissolved phosphate yields were as high as 8,200 t and 410,000 t for silicate [Halim et al., 1995]. After construction these nutrients declined to 330 t of phosphate and 73,000 t of silicate, and there was an immediate decline in small pelagic production (Fig. 3A), and change in the ratio of landings of pelagic to demersal fish and benthic resources (Fig. 3B). The small pelagic fisheries around the Nile delta has recently increased as nutrients in the Nile River have risen and may be associated with a diffuse plume of enriched water observed in the imagery of NSF/NASA [1989] for the 1980s offshore from the delta.

The Black Sea

The Black Sea has a very large catchment area and a narrow opening to the Mediterranean, which make it particularly vulnerable to the effects of increased nutrient runoff. Its waters below 200 m have been anoxic since historic times [FAO, 1993] and are caused, in part, by increased nutrient runoff, especially in the northwestern shelf under the influence of the Danube River [Mee, 1992] and, to a lesser extent, rivers entering from the north [FAO, 1994b; Mee, 1992; Zaitsev, 1992].

Eutrophication may have been the causal agent favoring the explosive growth of noxious pests and exotics, such as jelly predators (e.g., ctenophores) that now dominate the pelagic ecosystem. Although the anchovy catch almost doubled as fishing and nutrient inputs increased in the 1980s, anchovy catches declined to 10% of their former value as the ctenophores increased. Evidence from elsewhere [Purcell et al., this volume] also suggests that jelly predators may be favored by nutrient enrichment but are detrimental to fish stocks. The Black Sea pelagic ecosystem may have accommodated to this exotic pest in recent years, since its predator *Beroe*, another ctenophore, has

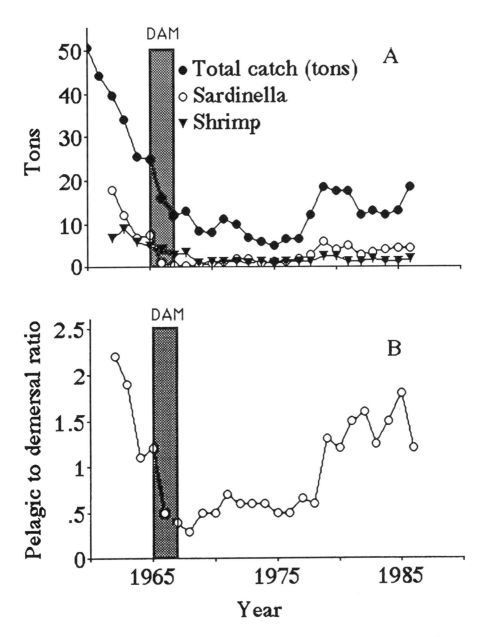

Figure 3. Trends in Sardinella and shrimp landings before and after the Aswan Dam affected Nile outflow to the Mediterranean. A. tonnage. B. ratio between pelagic and demersal landings off the Nile delta (the later rise appears to be linked to enriched drainage water discharged from the delta. The figure is after Halim et al. [1995] and B is modified from Caddy [2000a]. See also NSF/NASA [1989].

apparently now been introduced. This new predator and the reduced nutrient runoff may be the main factors that have led to a rise in anchovy landings.

The Baltic Sea

The net export of nutrients in the Baltic Sea increased from around 10,000 t of phosphorus and 300,000 t of nitrogen at the end of the 19th century, to around 80,000 t and 1,200,000 t, respectively, in recent times [Lehtonen and Hilden, 1980; Nehring and Matthaus, 1989; Hansson, 1985; Hansson and Rudstam, 1990].   Thurow [1997] concluded that fish stocks rose from 2 to 8 million t from the early 1900s to the 1980s as fish landings rose from 100,000 t to roughly 1,000,000 t over the same period. Durant and Harwood [1986] estimated that the population of marine mammals in 1900 was 300,000 individuals, which would have consumed roughly 240,000 t of fish annually. Seal populations are now considerably reduced compared to earlier in the century. Reduced mammalian predation, however, cannot explain the increased yields. The increased fisheries yields seems predominantly a "bottom-up" enrichment effect rather than just a "top-down" effect caused by a reduction in predators, although the latter undoubtedly had an influence.

Anoxic bottom water has developed this century, however, and negatively affected the Baltic cod [Rosemarin, 1990]. Cod require an oxygenated water column of at least 100 m if sinking eggs are not to enter hypoxic water soon after spawned in the upper water column.   Thus, there is the danger of the disappearance of this well-managed demersal top predator for reasons other than fishing, unless nutrient runoff is reduced.

The Seto Inland Sea

Nutrient runoff into the Seto Inland Sea (Japan) has proceeded through three general periods of eutrophication. Fishery yields averaged less than 150,000 t in the pre-World War II period, rose to 240,000 t in the 1960s, when maximum fisheries development was attained, and reached the third plateau during the 1970s at about 400,000 t. These phases were linked to increased primary production and a move to a higher proportion of pelagic as opposed to demersal production [Tatara, 1991; Nagasaki and Chikuni, 1989].

The Ratio of Catchment Area to Semi-Enclosed Sea

Although purely oceanographic factors obviously affect nutrient availability, for semi-enclosed and coastal seas land-effects may be revealed by simple empirical approaches (e.g., Meeuwig [1999]) like the ratio of the area of the catchment basin to the surface area of the receiving sea.   Table 2 indicates that this index accurately describes the susceptibility of European semi-enclosed seas to nutrient impacts.   For example, the small catchment of the Mediterranean appears to be more resistant to the effects of nutrient enrichment than either the Black or Baltic Seas, which have a much higher ratio.

TABLE 2. Subjective ranking of the relative impacts on fisheries of semi-enclosed seas in relation to the extent of incoming watersheds (summarized from Caddy [1993]).

Semi-enclosed sea	Ratio of area of surrounding watershed to sea surface	Relative impact on ecosystems and fisheries in recent decades
Mediterranean (Nile watershed)	0.6	Enrichment in the north, still oligotrophic south and east. Fish production showed steady rise. Pelagic/demersal ratio close to unity.
Seto Inland Sea	1.1	Three phases of eutrophication noted. Fish production rose continuously, but % planktivores/small pelagic fish increased.
Great Lakes	3.3	Movement from oligotrophic to mesotrophic in historic times, with "exotic" pelagic ecosystem components introduced in recent decades
Baltic Sea	4.2	Anoxia in deeper basins appears to have moved upwards with increased pelagic yield but reduced demersal/shellfish production.
Black and Azov Seas	6.0	Permanent anoxia in deeper basins moved onto shelves in 1970s and 1980s with collapse of benthic/demersal production. Rise of pelagic fish production in 1970s and 1980s, temporarily decimated by jelly predators in early 1990s.

Policy Changes and Nutrient Runoff

Mee [1992] ascribed the major increase in nutrient loading to the Black Sea over the last 25 years to the widespread use of phosphate detergents in the catchment basin and the intensification of agriculture. There now seems largely circumstantial evidence that fertilizer loadings of watersheds have been significantly reduced in eastern Europe under market economies. This reduction could account for some slight improvements in the Black Sea marine fisheries, such as the partial recovery of anchovy stocks in recent years.

A similar postulated effect seems worth investigation for Cuban inshore fisheries. Nitrogen fertilizer consumption for Cuba recorded in the FAO Yearbook of Fertilizer [FAO, 1994a] fell from around 350,000 t in 1983-1986 to 75,000 t in 1993, which is a decline of nearly 80% in fertilizer use, presumably in part as a result of reduced Soviet Union subsidies to Cuban agriculture. Similar changes were seen for phosphorus fertilizers. Data on inshore fisheries production on the Cuban shelf coincidentally shows a significant decline in production of shrimp and rock lobster in the early 1990s, which is hard to attribute to changes in fishery management alone, which has generally been excellent. The shrimp fishery in particular appears dependent on enriched outflow onto the otherwise oligotrophic, coralline southern shelf, while rock lobster production is associated with a nutrient-limited ecosystem based on seagrass meadows on the same shelf. This possible linkage is a working hypothesis at present, but further investigation of fishery declines in areas where fertilizer runoff has also declined might usefully take this type of hypothesis into account.

Two Simple Models for a Multi-Species Fishery Subject to Eutrophication

In the absence of the data needed to construct a complete quantitative ecosystem model, empirical, semi-quantitative, or even conceptual models may reproduce some of the key features of marine ecosystems subject to eutrophication. A model of the pelagic biome of the Black Sea (MTBASE1.1, internet address: http//www.math.rsu.ru/niimpm/ aqua/welcome.en.html) was completed recently by Berdnikov et al. [1999]. Such a model has some major advantages, especially in allowing an objective consideration of competing hypotheses.

Simple Box Models

A trophic web can be reduced to its main elements, which include dominant and subdominant pelagic and demersal piscivores, zooplanktivores and herbivores [Caddy, 2000a]. The progressive fishing down of the apical predators in a system that is not yet influenced by increased nutrient runoff (hypothetical) would certainly contribute to increases in biomass of their prey, but in the same system an increase in small pelagics may result from "bottom-up" increases in production due to nutrient enrichment. Which influence is the most important to an increase in small pelagics? Certainly for demersal fish and benthos, this process of increased nutrient runoff and eutrophication soon leads to hypoxia (initially, only seasonally and, eventually, permanent anoxia) and population collapse. In this series of transitions, jelly predators assume a larger influence in the transfer (or lack of) zooplankton to small pelagics. What also emerges from experience in freshwater systems is that we may expect increases in primary production to affect physical properties such as stratification [Northcote, 1988] as well as vice versa. Such mechanisms may also be relevant for semi-enclosed seas or fjords where wind-driven mixing is not dominant.

The ECOPATH model [Christensen and Pauly, 1992] is another box or equilibrium model that provides some similar predictions. Christensen and Caddy [1994] found that, with the addition of *Mnemiopsis* as a pelagic predator to the Black Sea, the model predicted declines in zooplankton and increases in phytoplankton standing stock, and hence could contribute to a biologically-driven eutrophication of the system.

A Modified Logistic Model:

The preceding section has suggested ways in which food webs may be changed in a multi-species fishery subject to stresses from nutrient runoff and fishing. One simple approach that appears, at least qualitatively, to model interactions of fishing effort (f) and nutrient runoff can be derived from the equilibrium version of a simple logistic model:

$$\text{Yield} = a*f - b*f^2 \tag{1}$$

where f is the annual fishing effort exerted, and a and b are constants. The first parameter, a, in a sense, determines the height of the parabola or carrying capacity of the population and hence is related to the production expected from the system. Here, a can be substituted for by a second logistic formulation defined by:

$$a = e*PP - g*PP^2 \qquad (2)$$

where PP is the level of primary production and e and g are also constants.

Such a functional form for a could allow the carrying capacity of the environment (in terms of the stock size of the commercial resource) to rise with moderate increases in primary production and subsequently fall as primary production increases still further. This reflects qualitatively the observation that moderate levels of nutrification and hence primary production are favorable for fisheries but may rapidly turn unfavorable with consequent hypoxia of bottom waters. Combined, this gives an expression for equilibrium yield with three parameters:

$$\text{Yield} = [e*PP - g*PP^2]*f - bf^2 \qquad (3)$$

An interesting property of simulations with this function is that it broadly resonates with the experience from the Mediterranean and Black Seas, in that it shows a slow and progressive rise in landings with increases in both fishing intensity and primary production even when MSY conditions have been reached, followed by an abrupt decline of the system shortly beyond this point. This resembles the decline with progressive anoxia of shelf waters seen for the Black Sea demersal resources after World War II (see Ivanov and Beverton [1985] and papers in FAO [1994b]), and provides a warning that pollution-driven increases in biological production may not be indefinitely sustained in terms of fishery yields, especially for high-value demersal and benthic resources.

Conclusions

It would be misleading to consider all levels of nutrient runoff as purely negative phenomena from the perspective of fisheries, even though this might be valid for some other sectors. Increased management and research emphasis should probably be on looking at marine catchment basin phenomena as a whole and summing up the net gains and losses from nutrient runoff to the various economic sectors operating within the MCB. Implied in this recommendation is also the need for the terrestrial components of the economy to place upper limits to nutrient runoff into the aquatic phase of the MCB. A key priority will be to severely reduce non-biodegradable and toxic waste discharges, pesticides, organic chemical residues and other toxic byproducts of industry and agriculture.

Conventional models of fish population dynamics that assume steady state production where landings are simply a function of fishing effort may be misleading. New paradigms for fisheries science are needed [Caddy, 1999]. These will have to take into account the fact that over the last half century there are simultaneous anthropogenic impacts of nutrient runoff on coastal and enclosed seas, over the same period when an unprecedented global increase in industrial fishing capacity has occurred. Both factors are likely to be confounded in their effects on coastal fisheries. Their interactions will need to be taken into account if realistic models are to be developed that can help in managing coastal and shelf ecosystems stressed by both runoff and over-fishing.

Acknowledgments. The author acknowledges the assistance provided by V. Agostini and I. de Leiva Moreno in preparing figures.

References

Berdnikov, S. V., V. V. Selyutin, V. V. Vasilchenko, and J. F. Caddy, Trophodynamic model of the Black and Azov Sea pelagic ecosystems: consequences of the comb jelly, *Mnemiopsis leydei*, invasion, *Fish. Res.*, 1421, 261-289, 1999.

Beverton, R. G. H. and S. J. Holt, On the dynamics of exploited fish populations, *Fish. Invest. (London) Ser. II*, *19*, 1957.

Caddy, J. F., Towards a comparative evaluation of human impacts on fishery ecosystems of enclosed and semi-enclosed seas, *Rev. Fish. Sci.*, *1*, 57-95, 1993.

Caddy, J. F., Fisheries management in the twenty-first century: will new paradigms apply?, *Rev. Fish Biol. Fish.*, *9*, 1-43, 1999.

Caddy, J. F., Marine catchment basin effects versus impacts of fisheries on semi-enclosed seas, *ICES J. Mar. Sci.*, *57*, 628-640, 2000a.

Caddy, J. F., Ecosystems structure before fishing?, letters to the editor, *Fisher. Res.*, *47*, 101-102, 2000b.

Caddy, J. F. and J. A Gulland, Historical patterns of fish stocks, *Mar. Policy*, *7*, 267-78, 1983.

Caddy, J. F. and A. Bakun, A tentative classification of coastal marine ecosystems based on dominant processes of nutrient supply, *Ocean and Coastal Management*, *23*, 201-211, 1994.

Caddy, J. F., R. Refk, and T. Do-Chi, Productivity estimates for the Mediterranean: evidence of accelerating ecological change, *Ocean and Coastal Management*, *26*, 1-18, 1995.

Caddy, J. F., F. Carocci, and S. Coppola, Have peak production levels been passed in continental shelf areas? Some perspectives arising from historical trends in production per shelf area, *J. Northwest Atl. Fish. Sci. 23*, 191-219, 1998a.

Caddy, J. F., J. Csirke, S. M. Garcia, and R. J. R. Grainger, How pervasive is "Fishing down Marine Food Webs"?, *Science, 282*, 1383a, 1998b.

Christensen, V. and J. F. Caddy, Reflections on the pelagic food web structure in the Black Sea, in Report of the Second Technical Consultation on Stock Assessment in the Black Sea., Ankara, Turkey, 15-19 February 1993, *FAO Fish Rep.*, *495*, 84-101, 1994.

Christensen, V. and D. Pauly, ECOPATH II - A software for balancing steady-state ecosystem models and calculating network characteristics, *Ecological Modelling*, *61*, 169-85, 1992.

Daan, N., The ecological setting of North Sea fisheries, *Dana*, *8*, 17-31, 1989.

Degobbis, D., Increased eutrophication of the Northern Adriatic Sea, *Mar. Pollut. Bull.*, *20*, 452-75, 1989.

de Leiva Morena, J. I., V. Agostini, J. F. Caddy and F. Carocci, Pelagic-demersal production ratios: a useful indicator of nutrient availability for European seas, *ICES J. Mar. Sci.*, in press.

Dethlefsen, V., Fish in the polluted North Sea, *Dana*, *8*, 109-29, 1989.

Durant, S. and J. Harwood, The effects of hunting on ringed seals (*Phoca hispida*) in the Baltic, *ICES Council Meeting*, *10*, 1986.

FAO/FIDI, Fishery fleet statistics 1970, 1975, 1980, 1984-92, *Bull. Fish. Stat.*, Nos. *27-34*, FAO, Rome, 1994.

FAO, Fisheries and environmental studies in the Black Sea system. *GFCM Studies and Reviews*, *64*, 1993.

FAO, *Yearbook of Fertilizers*, *44*, 1994a.

FAO, Second technical consultation on stock assessment in the Black Sea, *FAO Fish. Rep.*, *495*, 1994b.

FAO, *Yearbook of Fishery Statistics*, *80*, 1995.

Friligos, N., Nutrient status in Aegean waters, Appendix V, in Report of the Second Technical Consultation of the General Fisheries Council for the Mediterranean on Stock Assessment in the Eastern Mediterranean, Athens, Greece, 28 March - 1 April 1988, *FAO Fish.Rep.*, *412*, 190-98, 1989.

Galloway, J. N., H. Levy II, and P. S.Kasibhatla, Year 2020: consequences of population growth and development on deposition of oxidized nitrogen, *Ambio*, *23*, 120-123, 1994.

Grainger, R. J. R. and S. M. Garcia, Chronicles of Marine Fishery Landings (1950-1994), *FAO Fish. Tech. Pap.*, *359*, 1996.

Gulland, J. A., *The Fish Resources of the Ocean*, Fishing News Books, England, 1971.

Halim, Y., A. Morcos, S. Rizkalla and M. K. El-Sayed, The impact of the Nile and the Suez Canal on the living marine resources of the Egyptian Mediterranean waters (1958-1986), in Effects of Riverine Inputs on Coastal Ecosystems and Fisheries Resources, *FAO Fish. Tech. Pap.*, *349*, 19-58, 1995.

Hansson, S., Effects of eutrophication on fish communities, with special reference to the Baltic Sea - a literature review, *Rep. Inst. Freshw. Res. Drottningholm*, *62*, 36-56, 1985.

Hansson, S. and L. G. Rudstam, Eutrophication and Baltic fish communities, *Ambio*, *19*, 123-25, 1990.

Hornung, M. and B. Reynolds, The effects of natural and anthropogenic changes on ecosystem processes at the catchment scale, *TREE*, *10*, 443-48, 1995.

ICES, Report of the ICES advisory committee on fishery management, *ICES Coop. Res. Rep.*, *221*, 1996.

Ivanov, L. and R. J. H. Beverton, The fisheries resources of the Mediterranean. II Black Sea, *GFCM Stud. Rev.*, *60*, 1985.

Ivancic, I. and D. Degobbis, Long-term changes of phosphorus and nitrogen compounds in the northern Adriatic Sea, *Rapp. Comm. Int. Mer Medit.*, *35*, 266-267, 1988.

Jones, R., Ecosystems, food chains and fish yields. ICLARM/CSIRO Workshop, Cronulla, Australia, 12-13 January 1981 (mimeo), 1981.

Lehtonen, H. and M. Hilden, The influence of pollution on fisheries and fish stocks in the Finnish part of the Gulf of Finland, *Finn. Mar. Res.*, *247*, 110-23, 1980.

Marasovic, I., T. Pucher-Petrovic, and V. Alegria, Relation between phytoplankton productivity and *Sardina pilchardus* in the Middle Adriatic, *FAO Fish. Rep.*, *394*, 1988.

Mee, L. D., The Black Sea in crisis: the need for concerted international action, *Ambio*, *21*, 278-86, 1992.

Meeuwig, J .J., Predicting coastal eutrophication from land use: an empirical approach to small non-stratified estuaries, *Mar. Ecol. Prog. Ser. 176*, 231-241, 1999.

Nagasaki, F. and S. Chikuni, Management of multispecies resources and multi-gear fisheries, *FAO Fish. Tech. Pap.*, *305*, 1989.

Nehring, D. and W. Matthaus, Current trends in hydrographic and chemical parameters and eutrophication in the Baltic Sea, *Int.Rev.Ges.Hydrobiol.*, *76*, 297-316, 1989.

Nixon, S. W., Physical energy inputs and the comparative ecology of lake and marine ecosystems, *Limnol. Oceanogr.*, *33*, 1005-1025, 1988.

Nixon, S. W., J. W. Ammerman, L. P. Atkinson, V. M. Berounsky, G. Billen, W. C. Boicourt, W. R. Boynton, T. M. Church, D. M. Di'Toro, R. Elmgren, J. H. Garber, A. E. Giblin, R. A. Jahnke, N. J. P. Owens, M. E. Q. Pilson, and S. P. Seitzinger, The fate of nitrogen and phosphorus at the land-sea margin of the North Atlantic Ocean, *Biogeochemistry*, *35*, 141-180, 1996.

Northcote, T. G., Fish in the structure and function of freshwater ecosystems: a "top-down" view, *Can. J. Fish. Aquat. Sci.*, *45*, 361-79, 1988.

NSF/NASA, Ocean color from space. A folder of remote sensing imagery and text, prepared by the NSF/NASA-sponsored U.S. Global Flux Study Office, Woods Hole Oceanographic Institution, Woods Hole, Massachusetts, 1989.

Pauly, D., V. Christensen, J. Dalsgaard, R. Froese, and F. Torres Jr., Fishing down marine food webs, *Science, 279*, 860-63, 1998.

Pucher-Petkovic, T., I. Marasovic, I. Vukadin, and L. Sojanoski, Time series of productivity parameters indicating eutrophication in the middle Adriatic waters, in Report of the Fifth Technical Consultation of the General Fisheries Council for the Mediterranean on Stock Assessment in the Adriatic and Ionian Seas, Bari, Italy, 1-5 June 1987, *FAO Fish. Rep. 394*, 41-50, 1988.

Rabalais, N. N., R. E. Turner, D. Justic, Q. Dortch, W. J. Wiseman, Jr., and B. K. Sen Gupta, Nutrient changes in the Mississippi river and system responses on the adjacent continental shelf, *Estuaries, 19*, 386-407, 1996.

Rapport, D. J., H. A. Regier, and T. C. Hutchinson, Ecosystem behavior under stress, *Amer. Nat., 125*: 617-38, 1985.

Regier, H. A., Changes in species composition of Great Lakes fish communities caused by man, *Proc. 44th North-Amer. Wildlife Conf, 558-66*, 1979.

Rosemarin, A., editor, Current status of the Baltic Sea, *Ambio, Special Report No. 7*, 1990.

Schaefer, M. B., A study of the dynamics of the fishery for yellowfin tuna in the eastern tropical Pacific Ocean, *Bull. I-ATTC, 2*, 245-85, 1957.

Serafin, R. and J. Zaleski, Baltic Europe, Great Lakes America and ecosystem development, *Ambio 17*, 99-105, 1988.

Spencer, P. D. and J. S. Collie, Patterns of population variability in marine fish stocks, *Fish. Oceanogr., 6*, 188-204, 1997.

Stamatopoulos, C. and J. F. Caddy, Theory and applicability of a new parameter in normalized linear regressions, *Metron.49(1-4)*, 513-531, 1991.

Stromberg, J. O., Human influence or natural perturbation in oceanic and coastal waters - can we distinguish between them?, *Hydrobiologia, 352*, 181-93, 1997.

Tang, R., Effect of long-term physical and biological perturbations on the contemporary biomass yields of the Yellow Sea ecosystem, in *Large Marine Ecosystems: Stress, Mitigation and Sustainability*, edited by K. Sherman, L. M. Alexander, and B. D. Gold, AAAS Press, Washington, D.C., 1993.

Tatara, K., Utilization of the biological production in eutrophicated sea areas by commercial fisheries, and the environmental quality standard for fishing ground, *Mar. Pollut. Bull., 23*, 315-19, 1991.

Thurow, F., Estimation of the total fish biomass in the Baltic Sea during the 20th century, *ICES J. Mar. Sci., 54*, 444-461, 1997.

UNEP, *Assessment of the State of Eutrophication in the Mediterranean Sea*. United Nations Environmental Program Document UNEP(OCA)/MED WG, 104/Inf 6, 1996.

United Nations, *United Nations Statistical Yearbook* (up to and including Volume 42), United Nations Statistical Office, United Nations, New York, 1997.

Ursin, E., Stability and variability in the marine ecosystem, *Dana, 2*, 51-67, 1982.

Welcomme, R. L., Relationships between fisheries and the integrity of river systems. *Regulated Rivers: Research & Management, 11*, 121-36, 1995.

Zaitsev, Yu P., Recent changes in the trophic structure of the Black Sea, *Fish. Oceanogr., 1*, 180-89, 1992.

19

Some Effects of Eutrophication on Pelagic and Demersal Marine Food Webs

R. Eugene Turner

Abstract

This chapter reviews some of the possible consequences of increased nutrient loading and changing nutrient ratios to marine food webs, especially for larger organisms, and including the northern Gulf of Mexico. Selected results from various ecosystem models, knowledge of biomass size spectrum, literature and field studies, comparative ecosystem analyses, and implications from site-specific studies are summarized. Early models and subsequent improvements revealed the sensitivity of the pelagic and demersal consumers to the meager transfer of energy and carbon between trophic levels (upper limit about five or six), and are supported by empirical results documented for many systems. Less than 1% of the phytoplankton production will become fish biomass, which is most often as pelagic, not demersal, consumers. Eutrophication, by definition, increases the amount of primary production available, but the increase is not transferred throughout the food web with linearity or proportionality. Instead, higher nutrient loading may shift carbon flow within and through the 'microbial loop', diminish the percentage flowing into fecal pellet production, and increase flows toward phytoplankton cell aggregations that sink to lower layers. Eutrophication will be accompanied by greater carbon burial rates and export away from hypoxic/anoxic zones which become larger and longer-lasting; higher fisheries production should not be anticipated with eutrophication. Indeed, if the Si:DIN atomic ratios in water loading into continental shelves falls below 1:1, then there is likely to be a severe disruption in the diatom-copepod-fish food web, and more frequent toxic and noxious phytoplankton blooms.

Coastal Hypoxia: Consequences for Living Resources and Ecosystems
Coastal and Estuarine Studies, Pages 371-398
Copyright 2001 by the American Geophysical Union

Introduction

The undesirable consequences of increased primary production caused by higher nutrient loading to coastal waters, often described as eutrophication, may bring changes to continental shelf ecosystem structure and function. More frequent, larger and longer hypoxic events may be one readily-observed consequence among the many subtle ones (e.g., the Baltic Sea, Adriatic Sea and Mississippi River delta shelf; Rosenberg, [1985]; Shumway [1990]; Diaz and Rosenberg [1995]; Rabalais et al. [1996]; Bonsdorff et al. [1997]). The purpose of this chapter is to provide an overview of some of the possible consequences of increased nutrient loading to continental shelf food webs, especially the larger organisms, and the basis for this knowledge, including uncertainties. It seeks to address the question: how do water quality changes, especially manifested as hypoxia/anoxia, affect the carbon and energy flow through continental shelf food webs, including those in the northern Gulf of Mexico?

The ecosystem structure that captures and dissipates energy and that cycles materials on continental shelves has some order in the way the individual components are associated with each other that can be measured in terms of energy flow, biomass and element cycling. There is no question that the total biomass of fish and benthic macrofauna may be limited by the productivity of phytoplankton in surface waters for all or part of some ecosystems. Both upwelling areas and river deltas are known for their high fish productivity, and the variability of the benthic biomass and the size of benthic macrofauna is controlled by food availability when viewed across broad oceanic regions [Thiel, 1975; Rowe, 1971]. Benthic biomass decreases logarithmically with depth, because the surface layer production is respired during transit to the bottom [Thiel, 1975; Suess, 1980]. Further, the harvest of dungeness crabs (*Cancer magister*), presumably a relative index of biomass, is directly related to upwelling strength off California [Bakun, 1973; Peterson, 1972]. These relationships between the amount of primary production and the quality and quantity of secondary and tertiary growth and biomass are not always linear or proportional. Food Chain Theory suggesting that both trophic level number and biomass in the top level are proportional to food enrichment (e.g., Oksaanen et al. [1981]; DeAngelis et al. [1996]) is undergoing challenge from results from experiments and models indicating the reverse may be true [e.g., Diehl and Feißel, 1999]. In other words, the food web structure may not be stable with eutrophication, intense fisheries harvest, anticipated future climate changes or introduction of new species.

To address this question, I examined various food web models of carbon and energy flow in several marine ecosystems, and in the northern Gulf of Mexico, in particular. The size distribution of biomass within pelagic systems is discussed, and a summary of the probable consequences of higher primary production and of low silica:nitrogen ratios is made.

Ecosystem Models

Some Limits of Food Web Models

Caution: food webs are merely our systematic arrangements of ecosystem parts constructed to see how the system works as a functional whole. Food web models almost always insufficiently duplicate the presumed relationships between predators and prey, and they often condense known relationships to make the mechanics of modeling practical. In the real world, neither the model's biological pieces nor the physical setting are constant. Animals have many potential food choices changing with growth or when presented with different prey choices. Some copepods can be herbivores or carnivores, and some fish may eat copepods, detritus or even each other (Fig. 1). Energy may flow into larger or smaller particles, and thus diagrams connecting species in a feeding matrix may appear complicated. The real relationships are undoubtedly even more complicated. One way to simplify the examination of these relationships is to organize them in terms of the trophic position, population or individual biomass, predator and prey sizes, or energy flow. The ambiguity of assigning relationships and the imprecise and incomplete information about organism life history and physiology should be kept in mind when using the results of these analytical tools. These uncertainties make the science interesting, its application to management difficult, and the appearance of precise and accurate model predictions a rare treat. An honest modeler will treat most models as a heuristic device first and as a predictive tool second.

Early Models

Ryther [1969] described the broad outlines of three very different pelagic food webs and made some assumptions about food consumption and feeding efficiencies to estimate the upper limit of fish yield (Table 1). Although his analysis is not without its critics (e.g., neglect of the microbial community; Alverson et al. [1970]), and these models were incomplete in hindsight, they are useful to illustrate some general issues, including trophic efficiency and limits on the number of trophic levels. In order of increasing primary productivity, these food webs were for oceanic, coastal and upwelling systems. Ryther [1969] suggested that the mean trophic length of marine systems was five trophic levels for oceanic systems, three for continental shelf systems, and one to five for upwelling systems. Ryther [1969] also estimated the efficiency of the predator production from the prey consumed, which ranged from 10 to 20%. He concluded that the shorter the food chain, the higher the efficiency and the bigger the potential

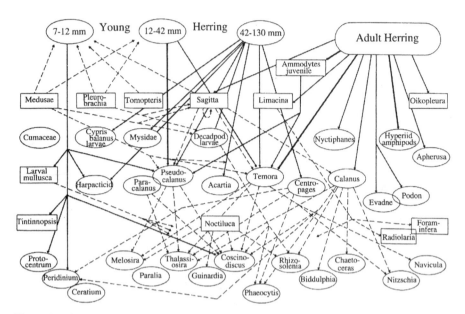

Figure 1. The herring (*Clupea harengus*) food web, based on the work of Hardy [1924], as modified from Wyatt [p. 345, 1976]. This is a summary of the relationships over the life cycle of herring, which ranges throughout the North Sea and two years of growth.

harvestable yield. Water quality could become a determinant of food chain length by influencing the quantity and the quality of the primary producers.

Microbial Food Webs or Loops

The microbial community is missing from the food webs analyzed by Ryther [1969], largely because their importance was not appreciated until the last few decades. More than 25 years ago, Pomeroy [1974] substantially revised the paradigm of the roles of bacteria in aquatic food webs. Before that, in the classical or traditional paradigm of the planktonic food web, the roles of bacteria were largely minimized or even ignored. The earlier view held that energy and matter were transferred sequentially to higher trophic levels from phytoplankton, to zooplankton, which in turn were consumed by small fish, which were to be eaten by even larger fish. The importance of the 10 to 50% of the carbon fixed in primary production lost into the water as DOC (dissolved organic carbon) [Larsson and Hagstrom, 1982] was not appreciated as a microbial food source. The new paradigm suggested that bacterioplankton played a significant role in this process by taking up the DOC as an energy source. Coupled with grazing by microflagellates and ciliates returned some of the lost DOC to the main food chain (Fig. 2). To simplify the complicated processes of these transfers of carbon, Azam et al. [1983] introduced the term "microbial loop." There is considerable ongoing interest in determining how much of the microbial production makes its way into higher trophic levels. Some of Walsh's seven continental shelf models (discussed below) include this possibility, and others do

Turner   375

TABLE 1. Estimated trophic relationships in three ocean communities (adapted from Ryther, 1969]).

Characteristic	Oceanic	Coastal	Upwelling
Primary production $(g\ C\ m^{-2}\ y^{-1})$	50	100	300
Fish production $(mg\ C\ m^{-2}\ y^{-1})$	0.5	340	36,000
Trophic levels	5	3	1.5
Assumed trophic efficiency between levels	10%	15%	20%
Key feeding relationships (incomplete list)	Nanoplankton Herbivorous Zooplankton Carnivorous Zooplankton Chaetognaths Planktivores Tunas/Sharks	Phytoplankton Zooplankton Planktivores Larger Fish	Diatoms, Anchovy or Euphausiids and Whales

not. Models for Georges Bank and the North Sea suggested that about 21% of the primary production enters the microbial loop (as dissolved organic matter, DOM; Cohen et al. [1982]).

The relationship between the density of viruses, bacteria, heterotrophic nanoflagellate (HNAN) grazers and phytoplankton pigments is remarkably consistent across marine and freshwater ecosystems (e.g., bacterial and phytoplankton pigments, Bird and Kaliff [1984], Cole et al. [1988]; viruses and bacteria, Maranger and Bird [1995]; HNAN and bacteria density, Berninger et al. [1991]). The constancy of these log:log relationships suggests the stability and importance of the primary producers to the microbial

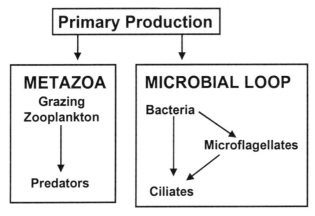

Figure 2. The relationships between different components of the microbial community and primary producers (from Pomeroy and Wiebe [1988]).

community. This proportionality, or constancy, within the microbial loop is considerably different from the relationship between eutrophication and the larger organisms of pelagic food webs.

A major uncertainty in our knowledge of how food webs work has to do with how much of the microbial food web is self-contained and how much is transferred into larger organisms. Recent analyses clearly show that both the strictly large algae-zooplankton-fish food chain and the self-contained microbial loop food web (only heterotrophic bacterial and mesozooflagellate grazers) are relatively rare. Most marine food webs are a mixture of microbial loops, small phytoplankton, herbivores and predators [Legendre and Rassoulzadegan, in press]. A field study by Rivkin et al. [1996] for the Gulf of St. Lawrence and an analysis of alternative food web computer models for upwellings by Carr [1998] came to some similar conclusions about the effects of an active microbial loop on carbon flows in these two contrasting ecosystems. Carr [1998] constructed simple models of one phytoplankton and one consumer, and more complex ones with several autotrophs and consumers that included a significant microbial food web. When the microbial food web was relatively weak during a strong phytoplankton bloom in the Gulf of St. Lawrence and in the simplest upwelling models, the mesozooplankton community grazed most of the phytoplankton production. Under strong and weak upwelling the simple phytoplankton-to-mesozooplankton food web computer models estimated that more carbon was exported as aggregates of sinking phytoplankton cells than as fecal pellets. The alternative models for complex upwelling food webs or from a post-bloom situation in the Gulf of St. Lawrence had a strong microbial community and an approximately equal distribution of the two export pathways. In other words, the effect of a strong microbial loop in the various models tended to result in a greater percentage of carbon exported as fecal pellets. The analysis of field data from the Gulf of St. Lawrence [Rivkin et al., 1996] indicated that when the microbial loop was strong (during the post-bloom situation) the carbon export was greater as fecal pellet production than as cell aggregates. The recycling of nutrients to phytoplankton was more complete with a relatively active microbial loop, than without it, thus implying that physical controls on the ecosystem (through nutrient loading or mixing, for example) were most important under bloom conditions.

These results are supported by results from nutrient manipulations in Baltic Sea ecosystem experiments. Kuuppo et al. [1998] studied the effect of silicate and nitrogen additions to mesocosms (150-l) established in the Baltic Sea. A diatom bloom developed in the experimental systems within seven days, but there was little response by the microbial community, including the ciliates, bacteria and their heterotrophic nanoflagellate predators. The removal of the mesozooplankters had a weak cascading effect on the ciliates. The increased carbon production was mostly sequestered in the phytoplankton biomass that sinks out of the surface layer. Egge and Jacobsen [1997] manipulated the nitrogen, phosphorus and silica loading rate into mesocosms filled with Baltic seawater. They found that silicate additions led to a doubling of the primary production rate, compared to the controls, primarily because the diatoms became dominant, not the flagellate community. The accumulation of biomass, in contrast, was lower because of a higher loss rate (grazing and sedimentation).

Thus the microbial loop may significantly influence the quality and quantity of carbon fixation and flows throughout pelagic and demersal food webs. Results from several experiments suggest that eutrophication has the effect of driving coastal systems towards

TABLE 2. Some food web characteristics of carbon fluxes for seven continental shelf budgets (derived from data in Walsh [1988]).

Continental Shelf	# Pelagic Trophic Levels	Primary Production $(g\,C\,m^{-2}\,y^{-1})$	Carbon into Detritus Pool[a] $(g\,C\,m^{-2}\,y^{-1})$	Carbon into Pelagic Fish[b] Production $(g\,C\,m^{-2}\,y^{-1})$	Carbon into Pelagic/ Demersal Fish Production[c]
Outer Southeastern Bering Sea	4	102	99 (97%)	5.4 (5.3%)	1.9
Middle Southeastern Bering Sea	3	166	138 (83%)	5.4 (3.2%)	0.2
Gulf of Anadyr, Bering Sea	3	285	198 (69%)	6.3 (2.2%)	2.5
Alaska Coastal	3	50	27.5 (55%)	0.8 (1.6%)	2.7
New York	4	300	240 (80%)	3.9 (1.3%)	1.9
Texas-Louisiana	3	100	86 (86%)	6.0 (6.0%)	18
Florida-Georgia	4	149	90 (60%)	1.6 (1.1%)	1.0
Average	3.4	165	126 (75%)	4.2 (3.0%)	4.0

[a]Includes carbon exported or buried

[b]Pelagic flows include phytophagous fish and their predators

[c]Demersal flows include commercially-important invertebrates (e.g., shrimp and crabs) and flows from the benthos to fish

a simpler, bloom-like food web, where cell-aggregation losses increase and the microbial loop is weakened.

A Comparison of Coastal Models

The carbon flows within seven food webs for continental shelf ecosystems are given in Table 2. These models were prepared in a similar manner and are from Walsh [1988]. The simplest version of the trophic relationships is three to four levels for all systems, and they range from 50 to 300 g C m^{-2} y^{-1} primary production.

Compared to the other systems, the budget for the middle Bering Sea ecosystem has very high levels of carbon flowing from the benthos to higher trophic levels (note the low

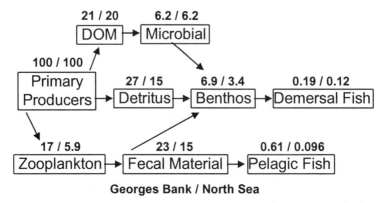

Figure 3. The percent of annual primary production (original units are kcal m^{-2} y^{-1}) produced by different components of the food web for the Georges Bank and the North Sea ecosystems. Adapted from Cohen et al. [1982]. Respiratory losses are not included.

ratio of carbon flow into the pelagic and demersal organisms). This result is primarily the result of very high benthic infaunal biomass, which is 10 times higher than that found on the outer shelf. Twenty percent of this benthic production flows into benthic feeders, including the king crab (*Paralithodes camtschatica*).

Food web models for the Georges Bank and the North Sea ecosystems are shown in Figure 3. The estimates given are for the percent of the primary production that is produced by each broad category of the food web and are expressed as a percent of the annual primary production. Note that: (1) the percentage of the primary production that flows into DOM and the microbial loop is similar in both systems, at about 20% and 6%, respectively, despite the 4.5 times greater production in the North Sea; (2) the benthic production is two times higher on George's bank, but the demersal production is only 50% higher; and (3) the North Sea has lower pelagic fish yield but higher primary production. In each case, the fisheries production is less than 1% of the primary production.

The percent of the primary production that flows into the pelagic food web is strongly related to the carbon flux into the benthos for continental shelf ecosystems (Fig. 4). This result may seem surprising upon first inspection, but perhaps not after considering the carbon source. The carbon flux into the benthos is composed of diatoms, diatoms within fecal pellets or the animals feeding on diatoms. Fecal pellets will sink faster than individual phytoplankton cells, so the percent of carbon flowing to the benthos is an indirect consequence of the success of pelagic feeding strategies. The same food that is valuable to the pelagic community is shunted also to the benthic community, as detritus or as zooplankton fecal pellets. If the phytoplankton were blue-green algae, for example, they would probably not sink as quickly (see discussion of the Si-silicate:Dissolved inorganic nitrogen (Si:DIN) ratios later in this chapter), and would become part of the microbial food web, or be exported. Quickly sinking carbon packaged as fecal pellets will decompose more slowly on the way to the bottom than will small individual cells that are not grazed. The relationship shown in Figure 4 is, in other words, a reflection of successful feeding by zooplankton, which are the principal prey of pelagic predators (except for phytophagous predators).

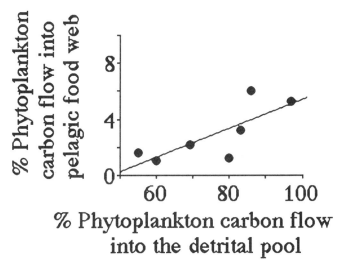

% Phytoplankton carbon flow
into the detrital pool

Figure 4. The relationship between the percent of the primary production into the pelagic food web and the percent of primary production that flows into the detrital pool. Data are from Table 2. A simple linear regression of the data is shown ($R^2 = 0.53$; $p = 0.04$).

It should be noted that two carbon budgets for the Peru upwelling [Walsh, 1981] do not fit the pattern shown in Figure 4. Walsh prepared these budgets for the anchovy fisheries for before and after it collapsed. He estimated that 60% of the primary production flowed into the pelagic food web and only 32% into the detrital pool in the pre-collapse conditions (Fig. 5). Compared to other continental shelf ecosystems, this is 10 times more into the pelagic food web and about half that into the detrital pool (Table 2). After the anchovy fisheries collapsed, only 6% of the primary production flowed into the pelagic food web, and 70% into the detrital pool. This difference between the flows of energy or carbon into continental shelf and upwelling zones suggests a fundamental difference in the two types of food webs. The majority of carbon no longer flowing into the pelagic food web after the fisheries collapse was redirected to the detrital pool and either was exported or buried. This reveals the potential for large-scale commercial fisheries to alter aquatic food webs. It provides an oceanic example of the effects of large predators on the lower trophic levels that limnologists have demonstrated experimentally (e.g., Carpenter et al. [1985], Kerfoot and Sih [1987]).

There are several broad generalizations that can be made from these carbon budgets. One is that most carbon flow is not into the pelagic fisheries. The dominant flow of carbon for all seven systems is into the detrital carbon pool. The carbon flow into the pelagic fish community averaged only 3% of the primary production, but was four times more than that by demersal fish and invertebrates. This reveals the rather inefficient transfer of energy between pelagic and demersal food webs. A second point is the rather inefficient transfer of energy from one trophic level to the next. A third point is that the

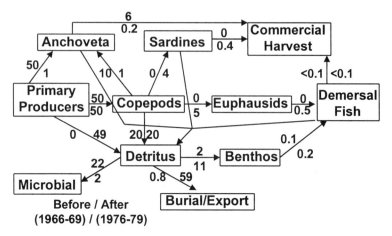

Figure 5. The percent of the primary production that flows into different components of the food web for the Peru upwelling system, before and after the collapse of the anchoveta fishery. Phytoplankton production was the same in both periods. Respiratory losses are not included. Adapted from data in Walsh [1981].

carbon flows from primary producers into the pelagic fish community are not proportional among these ecosystems. The primary production rate on the Texas-Louisiana shelf is one-third that of the New York shelf, but there is 50% more carbon flow into the pelagic fish biomass. The outer southeastern Bering Sea ecosystem has about the same primary production as the Texas-Louisiana shelf and similar carbon flow into the pelagic fish production, but a much higher amount going into the demersal fish production.

Energy Flow Models for the Texas-Louisiana Shelf

Walsh et al. [1981] prepared a preliminary carbon budget of the Texas-Louisiana shelf that included eight consumer pools (Table 3). The total fisheries harvest was 0.3 g C m^{-2} y^{-1} for phytophagous fish, which were menhaden (*Brevortia patronus*), and one-tenth that amount for shrimp [*Litopenaeus setiferus* (formerly *Penaeus setiferus*) and *Farfantepenaeus aztecus* (formerly *Penaeus aztecus*)]. Pelagic, demersal fish and benthic carnivore production was 0.001, 0.001 and 0.003 g C m^{-2} y^{-1}, respectively. The total production of shrimp and menhaden was 98% of the total animal production on this shelf. Menhaden are filter-feeders, as are the anchovies of the Peru upwelling. About half of the primary production in the Peru upwelling is by the filter-feeder anchovy, compared to 6% for menhaden on the Texas-Louisiana shelf. This difference may be due to the relatively larger phytoplankton prey found in the Peru upwelling system. Many of the common phytoplankton growing there form large gelatinous masses or filaments over 1 cm long. Cushing [p. 102, 1978] pointed out that "It is no accident that filterers are found in upwelling areas and in those moderate and high latitudes where the density of food organisms allow high encounter rates. In the deep ocean, where the food particles are dispersed, fish do not shoal, there are no top filterers, and the food chain has a more predatory character."

TABLE 3. The annual production of animal consumers for the food web of the Texas-Louisiana shelf (after Walsh [1988]).

Component	Production $(g\ C\ m^{-2}\ y^{-1})$
Primary production	100
Zooplankton-herbivores	4.0
Meio/micro benthic omnivores	3.0
Macro-benthic omnivores	0.29
Phytophagous fish (menhaden)	0.3
Pelagic fish	0.001
Shrimp	0.03
Benthic carnivores	0.003
Demersal fish	0.001
Burial or export	56

The strength of the Peru upwelling determines the size of the anchovy population, and thus it might seem that the menhaden population size on the Texas-Louisiana shelf may also be proportional to the rate of primary production. It may also logically follow that menhaden would prefer high concentrations of the types of food they can filter, but this does not mean that their population size is food-limited at present concentrations. The presence of menhaden is dependent on nearby wetlands for survival during those periods of high mortality when young. Most of the commercial fisheries harvest from the northern Gulf of Mexico, including menhaden and especially shrimp, are limited by the size of the wetland area in the adjacent estuary. There is a direct relationship between wetland area and shrimp yields for all Gulf of Mexico estuaries, for example [Turner, 1992]. Also, the annual variations in menhaden landings, like shrimp, can be described very well using statistical models of fishing effort and indicators of climate variations for when the young are in the estuary and before harvest [Guillory and Bejarano, 1980; Jensen, 1985]. These observations suggest that eutrophication will not result in an increased population size (or harvest) of the dominant form of animal production on this shelf. The reverse, however, may be true—there may be a threshold food concentration, which must be met to sustain present population densities (and harvest). (Note that Walsh's model did not include the effect of hypoxia or an altered food web.)

Walsh [p. 260; 1988] noted the doubling of nitrate in the Mississippi River in the last several decades and a possible 10-fold increase in phosphate from 1935 to 1973, which he estimated raised the phytoplankton production on the northern Gulf of Mexico coast to 300 g C m^{-2} y^{-1}. He made an interesting observation about the effect of these nutrient additions on the menhaden harvest. Menhaden harvest increased by 25% from 1962 to 1975, primarily as a result of expansion of the fishery. He inferred that the fisheries stocks did not increase in response to the increased nutrient loading to the region because "the adult menhaden populations are presumably not food limited." He concluded that the increased primary production associated with the eutrophication was exported and did not flow through the fisheries stocks. Walsh et al.'s estimate of burial and export was equal to 56% of the net primary production, which compares very well to Justic' et al.'s [1997] estimate of 53 to 63%, derived from two models of oxygen fluxes at Station C6 near the Mississippi River delta.

Justic' et al.'s [1997] model of carbon flux for the Louisiana shelf was validated with field data and manipulated to estimate the impacts of higher nutrient loading and river

Table 4. Carbon fluxes (g C m^{-2} d^{-1}) at Station C6 for the period 1985-1992, 1993 (a year with historically-high Mississippi River floods) and predicted with a doubled CO_2 scenario, as estimated by Justic' et al. [1996]. Bottom water at station C6 is hypoxic (oxygen < 2 mg l^{-1}) during most summers from 1985 to 1998.

Component	1985-1992	1993	2xCO_2
Net primary production (NPP) (g C m^{-2} d^{-1})	122	154	187
Bottom water respiration (g C m^{-2} d^{-1})	57	56	55
Exported or buried (g C m^{-2} d^{-1})	65	98	132
Percent of NPP exported or buried	53%	63%	71%

discharge. This model was for one station (20-m depth) in the middle of the hypoxic zone near the Mississippi River delta. Increased nitrogen loading (this model did not included changes in the Si:DIN ratios) resulted in increased phytoplankton production in the surface layer (0-10 m), much of which was transferred to the bottom layer (10-20 m) (Table 4). Interestingly, the respiration in the bottom layer did not change much with the increased surface layer production, indicating that this zone "has already reached the limit that is set by the availability of dissolved oxygen. Thus, it appears that export and burial, rather than *in situ* respiration, would likely become the ultimate fate for any surplus organic matter that is produced...leading perhaps to an expanded hypoxic zone." The model showed that there was an upper limit of carbon flow to the bottom layers and that carbon was exported with increased nutrient loading. This is consistent with field data and surrogate analyses of the historical record that support the hypothesis that eutrophication of this shelf caused the hypoxic zone to expand and the oxygen levels to decrease.

The inference of wetland-dependent harvest yields and the results of two different modeling exercises suggest the same thing—increased phytoplankton production on the Texas-Louisiana shelf will result in more carbon export off this shelf and that further eutrophication of this shelf will result in no significant increase in shrimp and menhaden harvests, which collectively represent 98% of the animal production.

Pelagic Biomass Spectra

The consequences of variations in size for individual organisms are many. Predators, for example, are generally larger than their prey (e.g., Fig. 6), but predators of the same size have different prey preferences reflecting their specific style of food acquisition. Swimming speed, structural support, production to biomass ratios, metabolism and egg production, to name a few examples, are affected by an organism's proportions and relative size [Thompson, 1961; Blueweiss et al., 1978; Banse and Mosher, 1980; Banse, 1982; Peters, 1983; Calder, 1984; Schmidt-Nielsen, 1984]. The smallest organisms are planktonic; larger organisms purposefully exploit currents and can move independently of diffusive processes. A time (generation) and space (travel distance) relationship between particle size and movement is thus implied. The significance of added structure (and function) with increasing size is not equal but relative. J. B. S. Haldane put it this

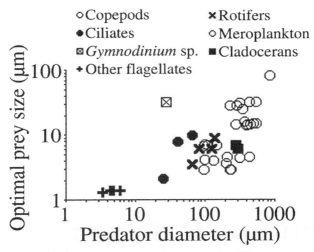

Figure 6. The relationship between predator and prey size for different zooplankton. The data are from Hansen et al. [1994].

way: "you can drop a mouse down a thousand-yard mine shaft; and, on arriving at the bottom, it gets a slight shock and walks away, provided that the ground is fairly soft. A rat is killed, a man is broken, a horse splashes." (from p. 138, McMahon and Bonner [1983]). Ecosystem properties are affected by these scalar relationships of individuals, including food webs. The distribution of biomass in aquatic systems generally shows a regular decline in biomass with increasing organism size arranged in logarithmically-equal size intervals [Sheldon et al., 1972, 1977] (Fig. 7). The theoretical aspects of these biomass:size relationships (described by Kerr [1974]; Thiebaux and Dickie [1993]) can be explained on the basis of a combination of size-dependent feeding relationships between predators and prey (a trophic relationship) and the consequences of scale to individual organisms. Fry and Quiñones [1994] demonstrated this trophic-based size interaction empirically using carbon and nitrogen isotopes.

A critical factor affecting the length of a food chain and the composition of the food web is the size of the phytoplankton cell and the time for the grazers to respond. Laws [1975] listed three attributes that determine whether smaller or larger cells predominate: the inherent rate of cell growth, the respiration rate and the sinking rate. Particle sinking rates are generally proportional to particle size. Individual diatom cells sink like the proverbial rocks because of their relatively large size, especially when packaged within fecal pellets. Laws showed that individual cell respiration rates decline with increasing cell size if the mixed layer is deep but also within the euphotic zone. An implied consequence of mass-specific relationships is that a shift in the community size spectrum, to either larger or smaller prey or predators, will alter the ecosystem properties. For example, if the average prey size decreases, the production:biomass ratio will increase, which, if left unchecked, may evolve into logarithmic growth (i.e., a bloom). The same number of individuals of a smaller size has a respiratory demand greater than that of a similar number of larger-sized individuals. A shift to smaller organisms requires, therefore, a higher primary production rate, or a lower predator biomass, or much higher recycling rates (of detritus, for example). Differences in the size of the prey and their

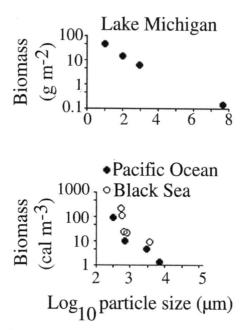

Figure 7. The relationship between the size of the average individual in a trophic level and the biomass of that trophic level. The Lake Michigan data are from Sprules et al. [1990]; the Pacific Ocean and Black Sea data are from Petipa [1973].

predator's excretory products, especially fecal pellet size and production rate, will affect the flux of material from surface to bottom waters. The microbial loop may be less affected by these production changes than the self-propelled larger organisms seeking smaller-sized prey. Decreasing the size of primary producers, however, means that the trophic transfer towards larger particles is reduced. Changing the size and amounts of phytoplankton, a clear manifestation of eutrophication, will therefore have a cascading effect within food webs.

Zooplankton have generation times on the order of several weeks, whereas phytoplankton may double in one day or less. A reduced grazing rate will allow for a higher phytoplankton cell accumulation rate. Thus, the food web in the offshore mixing zone of a large river will likely have low phytoplankton production and grow few mature zooplankton in turbid water at low salinity. The food web where light conditions improve dramatically (around 20 to 25 psu in the Mississippi River plume) and nutrients remain in relatively high concentration will be quite different from the food web where the salinity approaches 35 psu, nutrients limit growth, and the grazers have had time to grow and reproduce.

The net effect may be that increasing phytoplankton production does not necessarily yield a higher concentration of larger particles but instead many more smaller particles. Of course, fishing practices may achieve the same result without water quality changes—fishermen tend to go after the largest fish.

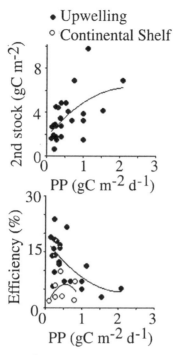

Figure 8. The transfer efficiency from primary (PP) to secondary production (herbivore production/phytoplankton production) for upwelling systems (from Cushing [1971]) and seven continental shelf models (derived from data in Walsh [1988]). A 2nd degree polynomial fit of the data is shown for each data set.

Trophic Transfer Efficiencies

The transfer of energy from one trophic level to another is hardly a constant 10% efficiency, although it may average about 10% in freshwater systems (e.g., Brylinsky and Mann [1973]). The transfer efficiencies between trophic levels may vary between 5 and 50% [Pomeroy, in press], and the transfer rate obviously influences the carbon flow among the connected model components. The challenge that Pomeroy's modeling exercise [1979] posed is still viable: how accurately are we modeling energy flows within a complex and anastomizing food web? Are the assimilation efficiencies too low (by a factor of two to three), or is the estimate of photosynthesis too low (by a factor of five to ten)? Although the absolute transfer efficiency (from prey to predator) ranges widely among individuals, some relative patterns are known at broad levels. Cushing [1971] developed a well-known example of how the transfer efficiency of energy from one level to the next changes with the quantity of primary production. He discovered this pattern by assembling data for diatom-based food webs of upwelling systems. He described a proportional and indirect relationship between the percent of energy transferred from phytoplankton to secondary trophic levels and the primary production rate (Fig. 8). The decline was not insignificant, and went from an average transfer from primary producers to secondary consumers of 15% at 0.25 g C m^{-2} d^{-1}, to 5% at 2 g C m^{-2} d^{-1}. In other words, an eight-fold increase in primary production yielded only a three-fold gain in secondary production.

Temporal Aspects

An important consequence of eutrophication may be to affect the timing of a phytoplankton bloom whose variations will have a cascading series of effects on the zooplankton and benthic communities. Marine phytoplankton may respond differently to the same nutrient loading if that load is introduced gradually or quickly. The quality of the response is also dependent on the community in the receiving waters and on salinity [Sakshaug et al., 1983; Sommer, 1985; Suttle and Harrison, 1986; Turpin and Harrison, 1980]. Cushing [1983] described the changes to the pelagic food web over the 'Russell cycle' in the North Sea: Part of this cycle involves the delay of phytoplankton production by a month and the consequential following delay in the zooplankton production cycle. Cushing [p. 509; 1983] summarized the consequences to the fisheries: "The increase in the ratio of *Calanus* sp. to *Pseudocalanus* sp. in the North Sea has had profound effects upon the structure of the ecosystem, a reduction in the age of recruitment to the herring and an increase in the growth rate of the herring, the increase in cod recruitment (and perhaps of other gadoid species) and perhaps even the very development of the industrial fisheries." ... and, "The switch from herring and macroplankton to pilchards and smaller zooplankton and back again implies that the proportions of energy passed to fish and invertebrate carnivores can indeed change." Townsend and Cammen [1988] point out that the timing of the spring plankton bloom may also affect the benthic-pelagic couplings important to fisheries recruitment. Changes in the amount and timing of nutrient loading could affect fisheries recruitment success through a mismatch of larval recruitment and food supply, as well as an altered food chain. Early blooms with a greater sedimentation to the benthos could positively affect demersal fishes, but late blooms positively affect pelagic fisheries through the phytoplankton-to-zooplankton-to fish food chain.

Trophic relationships will be particularly disturbed when toxic or noxious blooms occur to displace or add to the more desirable phytoplankton community. The immediate effect of such blooms may be due to several causes, including the loss of specialized prey items, chronic or acute toxicity, avoidance and attraction to competing species [Dortch et al., this volume]. These effects become consequential when they last longer than the generation time of the affected organism. The consequences to the food web dynamics might be quite unpredictable at the species level, especially where there are nonlinear relationships between prey density and consumers (described by Parsons et al. [1967]).

Comparative Analyses of Eutrophication Effects on Food Webs

Micheli [1999] conducted an important meta-analysis of 47 marine food webs by assembling the results of published studies and data from experimental manipulations

TABLE 5. The results of a three types of meta-analyses of mesocosm and field data (from Micheli [1999]). The number in parentheses is the data set size.

1. Experiment	Component Parts	Response (as biomass)	
		Phytoplankton	Mesozooplankton
Pelagic marine mesocosms containing:			
Mixed pelagic and benthic assemblage	1a. + Zooplanktivore	no change (18)	lower (13)
	1b. + Zooplanktivore and nutrients	slight increase (17)	lower (10)
Phytoplankton and zooplankton (2 trophic levels)	1c. + Nutrients	increase (54)	no change (10)
Phytoplankton, zooplankton, and zooplanktivore (3 trophic levels)	1d. + Nutrients	increase (14)	no change (10)

2. Analysis	Factor	Response (as biomass) by:		
		Phyto-plankton	Mesozoo-plankton	Fish biomass
Open marine ecosystems: Cross-correlation analysis of nutrients, time, biomass and primary production	2a. Annual N availability	higher (6)	no change (7)	no change (8)
	2b. Primary production	higher (11)	no change (13)	no change (12)

3. Analysis	Response (as biomass):
Open marine ecosystems: Cross-correlation analysis of yearly averages	3a. Phytoplankton not significantly related to mesoplankton (19)
	3b. Mesozooplankton negatively related to zooplanktivorous fish (19)

and open systems (summarized in Table 5). Some of these studies experimentally added nutrients, and others manipulated the food web by adding or removing prey or predators.

The analysis of mesocosm studies revealed that the addition of zooplankton resulted in lower densities of their mesoplankton prey but not phytoplankton. The addition of nutrients and zooplankton resulted in slightly higher phytoplankton and much lower mesoplankton biomass. The effect of adding nutrients to ecosystems modified to have either two or three trophic levels was to increase the phytoplankton biomass but not the primary grazers of the phytoplankton.

Micheli [1999] also analyzed various time series data for open marine systems (Table 5). The availability of nitrogen and the primary production rate were strongly correlated to the accumulation of phytoplankton but not of higher trophic levels; however, there was a significant negative correlation between mesozooplankton and their zooplankton predators. These analyses demonstrated a weak coupling between phytoplankton, mesoplankton and zooplankton for closed and manipulated systems and revealed three strong patterns: (1) a bottom-up control of primary producers through nutrient availability, (2) the top-down control of herbivores (mesozooplankton) through predation by carnivores (zooplanktivorous fish), and (3) a weak coupling between primary producers and herbivores. Micheli offered three possible explanations for these results that are supported by the literature. First, the interactions among zooplankton may be much more complex than our understanding, especially as exemplified in models; these interactions may limit their population growth. Second, food quality may be different with increased primary production rates; the proportions of edible to inedible food may decline with higher phytoplankton production. Third, the couplings between higher trophic levels may be dampened by advection or losses of nutrients or prey in open marine systems.

If the increased primary production that accompanies nutrient enrichment does not result in increased macro-consumer biomass of higher trophic levels then, clearly, a higher proportion of the total carbon flow must be shunted to smaller consumers/decomposers, or it is buried to remain uncycled. The carbon burial rates offshore of the Mississippi River have certainly increased this century as eutrophication has increased [Eadie et al., 1994; Rabalais et al., 1996]. Some of this carbon, however, may also have entered the microbial food web throughout the water column, but this remains an unquantified amount.

Si:DIN Ratios

Food webs are not only affected by the quantity of nutrient loading but also by the relative supply of nutrients. Diatoms have an intracellular Si:DIN atomic ratio of 1:1, and the regeneration of Si and nitrate in the world's ocean is also 1:1. Redfield et al. [1963] postulated that these nutrient ratios indicated the possibility for stoichiometric and physiological limits to phytoplankton growth. Schelske and Stoermer [1971] described how the diatom production in Lake Michigan was reduced as the Si:DIN ratio dropped below 1:1, and Elser et al. [1996] have nicely shown how these ratios (commonly discussed in terms of the N:P ratios) constrain organism organization at the cellular, organismal and community level. Results from field and laboratory studies have suggested that the lack of silica relative to nitrogen can control phytoplankton community composition (e.g., Egge and Aksnes [1992]; Egge and Jacobsen [1997]; Dugdale and Wilkerson [1998]).

The Si:DIN atomic ratios of water entering coastal waters, and within coastal waters, has been declining in many areas of the world and declining to near this ratio of 1:1 (Table 6). Smayda [1990] suggested that the rather global epidemic of noxious and toxic algae blooms might be a consequence of the ubiquitous increases in nitrogen loadings to coastal zones, which lead to a decrease in the Si:DIN ratios. Interestingly, Kuosa et al. [1997] showed in a laboratory study that the absence of larger grazers had no effect on phytoplankton community dynamics but that silicate additions did allow for a competitive advantage for two diatoms (*Chaetoceros holsiaticus* and the dominant, *C. wighamii*).

Officer and Ryther [1980] hypothesized that if the minimal Si:DIN proportion of 1:1 for diatoms was not met then an alternative phytoplankton community composed of non-

TABLE 6. Examples of nutrient concentrations *in situ* or in the waters entering coastal ecosystems where the Si:DIN ratio (atomic) is near or below 1:1.

Si:DIN concentration *in situ* < 1:1	Observation; source
Irish Sea	declined since 1959; Allen et al. [1998];
Bothnian Sea	declining during winter; Rahm et al. [1996]
Bay of Brest	Le Pape [1996]
Si:DIN loading concentration < 1:1	**Observation; source**
Bay of Brest	del Almo [1997]
Bothnian Sea watershed	Rahm et al. [1996]
Baltic proper	Rahm et al. [1996]
Mississippi River	Turner and Rabalais [1991]
Morlaaix River (English Channel)	Wafar et al. [1983]
Po River, Italy	Justic' et al. [1995]
Susquehanna (Chesapeake Bay)	Fisher et al. [1988

diatoms may be competitively enabled. This alternative community would be more likely to be composed of flagellated algae, especially dinoflagellates, including noxious bloom-forming algal communities. Officer and Ryther further argued that as the Si:DIN ratio fell below 1:1, the fisheries food web would reform and be composed of less desirable species. It turns out to be a correct prediction for the Louisiana shelf near the Mississippi River delta.

In the case of the Mississippi River, the nitrate concentration has risen and the silicate concentration dropped in the last 40 years, to the point where the ratio fluctuates just above and below a Si:DIN ratio of 1:1 (Fig. 9). The presence of these fluctuations provided an opportunity to document the reaction of the food web on this shelf [Turner et al., 1998]. These results are described in the following paragraphs. Note that this discussion is mostly about nutrient loading onto the shelf, not about nutrient concentrations *in situ* on the shelf.

Zooplankton abundance and fecal pellet production responded in an abrupt 'step' fashion to variations in Si:DIN loadings 90 days before the sample collection. This 'Si:DIN switch', set at a ratio of 1:1, resulted in a drastic decline in zooplankton abundance below 1:1 (Fig. 10). The percent copepod abundance declined from about 80% to 20% of the zooplankton numbers, the fecal pellet production of copepods also declined, and the amount of the phytoplankton production packaged as fecal pellets declined in a similar way. Copepod fecal pellets, containing many partially decomposed diatoms, sink much more rapidly than individual phytoplankton cells, and there is relatively less decomposition en route to the bottom. Thus, more of the organic material sinking to the bottom layer is respired there if copepods dominate the zooplankton community than if the Si:DIN ratio is < 1:1, which occurs when copepods are relatively scarce. These results are summarized in Table 7. The importance that Officer and Ryther [1980] attached to the Si:DIN ratios appear to be decisively appropriate for this coast.

The size distribution of organisms will certainly change if the copepod density diminishes for a period longer than their generation time (when the Si:DIN ratio falls below 1:1). There is a good relationship between the size of copepods and their prey size (Fig. 6). If the food size and quality changes, then the distribution of biomass in different size categories will also change (e.g., Fig. 7). A reformed phytoplankton community of smaller cells will be grazed by an also new community of smaller prey of different escape

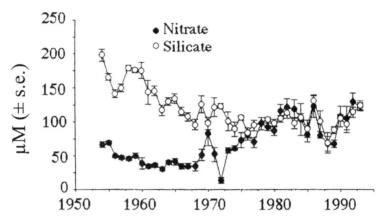

Figure 9. The average annual concentration (± 1 Standard Error) of nitrate and silicate in the Mississippi River at New Orleans. Note the uncoupled relationship between the two variables before 1975 and the coherent changes after 1980 (from Turner et al. [1998]).

velocities, growth rates, aggregation potential and palatability. This is one strong inference of the size spectrum of biomass discussed earlier. Smaller organisms have a higher production per biomass but also a higher respiration rate. The implication of these size relationships is that the length of the food chain will diminish as carbon produced at one level is lost in each predator-prey interaction. Fish production will therefore decline.

Predictions about what will happen when the Si:DIN ratio falls below 1:1 in water flowing onto this shelf thus seem to be consistent with the predictions of Officer and Ryther [1980], but they are not very precise. We are more certain about what happens when eutrophication occurs in the presence of higher diatom production (Si:DIN > 1:1)—there will be more fecal pellet production, more carbon sedimentation to the bottom layer and higher respiration rates therein. The increasing dominance of hypoxia-tolerant forminiferans compared to hypoxia-intolerant forminiferans in dated sediment cores [Sen Gupta et al., 1996] is coincidental with the increased diatom sedimentation [Turner and Rabalais, 1994]. These coincidental changes this century strongly suggest that the size of hypoxic water mass has increased and that the amount of oxygen has decreased. There is a well-described reduction in fish biomass at oxygen levels below 2 mg l^{-1}, and a progression of benthic community changes below that level [Rabalais et al., this volume; Diaz et al., this volume; Pavella et al., 1983; Renaud, 1986; Diaz and Rosenberg, 1995]. A typical response is that rapidly-growing species dominate at the expense of the long-lived species, especially below 0.5 mg l^{-1} [Rabalais et al., this volume; Santos and Simon, 1980]. There is a shift towards a polychaete-dominated fauna of smaller individuals and lower biomass. The time between hypoxic events is only a small fraction of the life span of the longer-lived benthic species, so the changes to the food web structure are significant, and not simply constrained to the length of time it takes for the hypoxic water mass to leave the area. For example, benthic feeders, such as commercially-important shrimp and groundfish will have lower food stocks after hypoxia [see Rabalais et al., this volume], and it has been suggested that they may encounter an

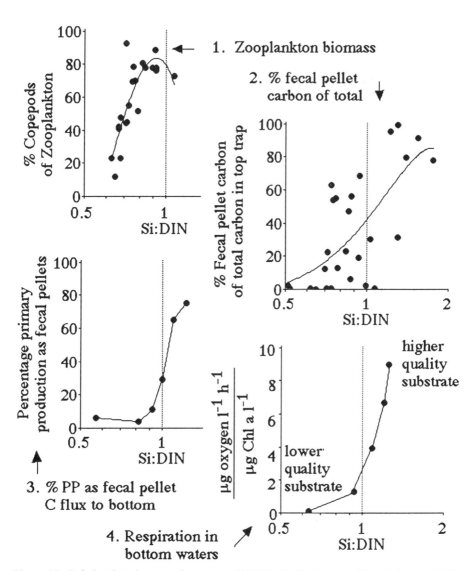

Figure 10. Relationships between the average Si:DIN ratio for the preceding 90 days and: (1) the percent of zooplankton that are copepods; (2) the percent of the carbon collected in the surface sediment trap as fecal pellets; (3) the percentage of the estimated water column phytoplankton production captured in a sediment trap as fecal pellets at 15 m (total depth = 20 m); (4) the respiration rate per total Chl a (μg oxygen l^{-1} h^{-1})/(μg Chl a l^{-1}) in bottom water. Adapted from Turner et al. [1998].

TABLE 7. Summary of observations and probable consequences with Si:DIN atomic ratios near 1:1, and less than 1:1, for the northern Gulf of Mexico [from Turner et al., 1998].

	Si:DIN	
	< 0.7	> 1:3
Observed:		
% copepod of meso-zooplankton	20%	> 80%
% carbon in sediment trap that are fecal pellets	10%	> 80%
% fecal pellets of primary production	10%	> 70%
Respiration per Chl a in bottom waters ((μg oxygen l^{-1} h^{-1})/(μg Chl a l^{-1}))	< 1	> 8
Respiration losses in bottom waters	lower	higher
Implications:		
Potential for flagellated algal blooms, including harmful algal blooms	higher	lower
Bottom water hypoxic zone	smaller	continuing

impassable 'low-oxygen' barrier during migration [see Zimmerman and Nance, this volume].

Predictions About Eutrophication, Hypoxia/Anoxia and Food Webs

Some predictions about the response of coastal food webs to increased eutrophication and hypoxia are given in Table 8. The arrangement of the table is to divide up the responses into relatively high and low nutrient loading, and Si:DIN atomic ratios above and below 1:1. Increased nutrient loading to a nutrient-limited system (e.g., nitrogen-limited in many cases), may result in some nonlinear increase in primary production rate, but the higher production may be irregularly distributed in time and space. The composition of the phytoplankton community, whether as a bloom or not, is *very*

Table 8. General predictions about coastal food webs with increased eutrophication or larger or more severe hypoxia/anoxia.

	High nutrient loading (higher than present)	Low nutrient loading (circa 1900)
Si:DIN above 1.5:1		
primary production	Very high	Low
diatoms	Dominant bloom organisms	Dominant bloom organisms
bloom strength	High, and sustained	Low
bloom frequency	Episodic	Relatively rare
bloom quality	Sometimes toxic and/or noxious	Almost no toxic blooms
hypoxic area	Expanded, especially horizontally	Contracted or absent
fish with hypoxia/anoxia		
bottom feeders	Catastrophic loss	Revival of long-lived species
pelagic	Short-lived species gain	Revival of long-lived species Long-lived species loss
commercial fisheries with hypoxia/anoxia	Disruptive, gear changes,	More stable than present

TABLE 8. Continued.

	more travel time, irregular stocks; some collapses; health warnings	(assuming same effort and technology)
recreational fisheries with hypoxia/anoxia	Smaller sized; some species shifts Health warnings	Larger specimens possible; more stability than presently
food chain length	Same as presently?	Same as presently?
Si:DIN below 0.8:1		
primary production	High	Low, but higher than receiving waters
diatoms	Low concentration	Low concentration
bloom strength	High	Low
bloom frequency	Episodic	Almost none
bloom quality	Toxic and/or noxious	Almost no toxic blooms
hypoxic area	Expanded horizontally	Contracted; rare
fish with hypoxia/anoxia		
bottom feeders	Catastrophic loss	Catastrophic loss
pelagic	Replacement and loss of phytophagous species; range restriction closer to riverine sources	Loss and some replacement of phytophagous species; range restriction closer to riverine sources
commercial fisheries with hypoxia/anoxia	Very disruptive, gear changes; more travel time; irregular stock composition; some collapses; health warnings	Catastrophic loss? Species replacement
recreational fisheries with hypoxia/anoxia	Smaller sized; many species shifts; frequent health warnings	Very disruptive
food chain length	Same as presently?	Same as presently?

dependent on the nutrient loading rates and ratios. The responses of dependent phytophagous predators and subsequent prey-predator linkages is somewhat unpredictable on a species level. Toxic and noxious blooms are more likely with higher nutrient loadings and with lower Si:DIN ratios. In general, however, the increased irregularity of the phytoplankton quality and quantity increases the likelihood that smaller organisms with faster response times (generation or growth rates, etc.) will out-compete other species and that hypoxic water masses will increase in size and duration. The species-shifts affect desirable recreational and commercially-important fisheries species with perhaps catastrophic results (in the economic sense, if not biologically).

Acknowledgments. Drs. B. Fry, L. Levin, L. R. Pomeroy, N. N. Rabalais and D. W. Townsend kindly made numerous constructive comments on earlier drafts of the manuscript. The errors and omissions remaining are the responsibility of the author, of course, who remains appreciative of the tenacity and ingenuity of the cited authors in revealing the nuances and subtleties of aquatic food webs.

References

Allen, J. R., D. J. Slinn, T. M. Shammon, R. G. Hartnoll, and S. J. Hawkins, Evidence for eutrophication of the Irish Sea over four decades, *Limnol. Oceanogr., 43*, 1970-1974, 1998.

Alverson, D. L., A. R. Longhurst, and J. A. Gulland, How much food from the sea?, *Science, 168,* 503-505, 1970.

Azam, F., T. Fenchel, J. G. Field, J. S. Gray, L. A. Meyer-Reil, and F. Thingstad, The ecological role of water-column microbes in the sea, *Mar. Ecol. Prog. Ser., 10,* 257-263, 1983.

Bakun, A., Coastal upwelling indices, west coast of North America, 1946-71, pp. 1-103, NOAA Technical Report NMFS SSRF-671, 1973.

Banse, K., Mass-scaled rates of respiration and intrinsic growth in very small invertebrates, *Mar. Ecol. Prog. Ser., 9,* 281-297, 1982.

Banse, K. and S. Mosher, Adult body mass and annual production/biomass relationships of field populations, *Ecol. Monogr., 50,* 355-379, 1980.

Banta, G. T., A. E. Giblin, J. E. Hobbie, and J. Tucker, Benthic respiration and nitrogen release in Buzzards Bay, Massachusetts, *J. Mar. Res., 53,* 107-135, 1995.

Berninger, U.-G., B. J. Finlay, and P. K. Lienkki, Protozoan control of bacterial abundances in freshwater, *Limnol. Oceanogr., 36,* 139-147, 1991.

Bird, D. R. and J. Kalff, Empirical relationships between bacterial abundance and chlorophyll concentration in fresh and marine waters, *Can. J. Fish. Aquat. Sci., 41,* 1015-1023, 1984.

Blueweiss, L., H. Fox, V. Kudzma, D. Nakashima, R. Peters, and S. Sams, Relationships between body size and some life history parameters, *Oecologica (Berl.), 37,* 257-272, 1978.

Bonsdorff, E., E. M. Blomqvist, J. Mattila, and A. Norkko, Coastal eutrophication: Causes, consequences and perspectives in the archipelago areas of the northern Baltic Sea, *Estuar. Coast. Shelf Sci., 44 (suppl. A),* 63-72, 1997.

Brylinsky, M. and K. H. Mann, An analysis of factors governing productivity in lakes and reservoirs, *Limnol. Oceanogr. 18,* 1-14, 1973.

Calder, W. A. III., *Size, Function and Life History*, Harvard Univ. Press, Cambridge, 1984.

Carpenter, S. R., J. F. Kitchell, and J. R. Hodgon, Cascading trophic interaction and lake ecosystem productivity, *BioScience, 35,* 635-639, 1987.

Carr, M.-E., A numerical study of the effect of periodic nutrient supply on pathways of carbon in a coastal upwelling regime, *J. Plankton Res., 20,* 491-516, 1998.

Cohen, E. G., M. D. Grosselein, M. P. Sissenwine, R. Steimle, and W. R. Wright, Energy budget of Georges Bank, *Sp. Publ.Can. J. Fish. Aquat. Sci., 59,* 95-107, 1982.

Cole, J., S. Findlay, and M. Pace, Bacterial production in fresh and saltwater ecosystems: A cross system overview, *Mar. Ecol. Prog. Ser., 43,* 1-10, 1988.

Conley, D. J., C. L. Schelske, and E. F. Stoermer, Modification of the biogeochemical cycle of silica with eutrophication, *Mar. Ecol. Prog. Ser., 101,* 179-192, 1993.

Cushing, D. H., Upwelling and the production of fish, *Adv. Mar. Biol., 9,* 255-335, 1971.

Cushing, D. H., Upper trophic levels in upwelling areas, in *Upwelling Ecosystems*, edited by R. Boje and M. Tomczak, pp. 101-110, Springer-Verlag, New York, 1978.

Cushing, D. H., Sources of variability in the North Sea ecosystems, in *North Sea Dynamics*, edited by J. Sündermann and W. Lentz, pp. 498-515, Springer Verlag, New York, 1983.

DeAngelis, D. L., *Dynamics of Nutrient Cycling and Food Webs*, Chapman & Hall, London, 1992.

del Amo, Y., B. Quéguiner, P. Tréguer, H. Breton, and L. Lampert, Impacts of high-nitrate freshwater inputs on macrotidal ecosystems. II. Specific role of the silicic acid pump in the year-round dominance of diatoms in the Bay of Brest (France), *Mar. Ecol. Prog. Ser. 161,* 225-237, 1997.

Diaz, R. J. and R. Rosenberg, Marine benthic hypoxia: a review of its ecological effects and the behavioural responses of benthic macrofauna. *Oceanogr. Mar. Biol. Ann. Rev., 33,* 245-303, 1995.

Diehl, S. and M. Feißel, Effects of enrichment on three-level food chains with omnivory, *Amer. Naturalist, 155,* 200-218, 1999.

Dugdale, R. C. and F. P. Wilkerson, Silicate regulation of new production in the equatorial Pacific upwelling, *Nature, 391,* 270-273, 1998.

Eadie, B. J., B. A. McKee, M. B. Lansing, J. A. Robbins, S. Metz, and J. H. Trefry, Records of nutrient-enhanced coastal ocean productivity in sediments from the Louisiana continental shelf, *Estuaries, 17,* 754-766, 1994.

Egge, J. K. and D. L. Aksnes, Silicate as regulating nutrient in phytoplankton competition, *Mar. Ecol. Prog. Ser., 83,* 281-289, 1992.

Egge, J. K. and A. Jacobsen, Influence of silicate on particulate carbon production in phytoplankton, *Mar. Ecol. Prog. Ser., 147,* 219-230, 1997.

Elser, J. J., D. R. Dobberfuhl., N. A. Mackay, and J. H. Schampel, Organism size, life history, and N:P stoichiometry, *BioScience, 46,* 674-684, 1996.

Fisher, T. R., L. W. Harding, Jr., D. W. Stanley, and L .G. Ward, Phytoplankton nutrients, and turbidity in the Chesapeake, Delaware, and Hudson estuaries, *Estuar. Coast. Shelf Sci., 27,* 61-93, 1988.

Fry, B. and R. B. Quiñones, Biomass spectra and stable isotope indicators of trophic level in zooplankton of the northwest Atlantic, *Mar. Ecol. Prog. Ser., 112,* 201-204, 1994.

Guillory, V. and R. Bejarano, Evaluation of juvenile menhaden abundance data for prediction of commercial harvest, *Proc. Annual Conf. Southeast Assoc. Fish and Wildlife Agencies, 34,* 193-203, 1980.

Grahame, J., *Plankton and Fisheries,* Edward Arnold, London, 1987.

Hansen, B., P. K. Bjørnsen, and P. J. Hansen, The size ratio between planktonic predators and their prey, *Limnol. Oceanogr., 39,* 395-403, 1994.

Hardy, A. C., The herring in relation to its animate environment. Part I. The food and feeding habits of the herring with special reference to the east coast of England. *Fish. Invest., Ser. II. 8(3),* 53 pp., 1924.

Jensen, A. L., Time series analysis and the forecasting of menhaden catch and CPUE, *North Amer. J. Fish. Mgt., 5,* 78-85, 1985.

Justic', D., N. N. Rabalais, and R. E. Turner, Effects of climate change on hypoxia in coastal waters: a doubled CO_2 scenario for the northern Gulf of Mexico, *Limnol. Oceanogr., 41,* 992-1003, 1996.

Justic', D., N. N. Rabalais, R. E. Turner, and Q. Dortch, Changes in nutrient structure of river-dominated coastal waters: stoichiometric nutrient balance and its consequences, *Estuar. Coast. Shelf Sci., 40,* 339-356, 1995.

Justic', D. N. N. Rabalais, and R. E. Turner, Impacts of climate change on net productivity of coastal waters: implications for carbon budget and hypoxia, *Climate Res. 8,* 225-237, 1997.

Kerr, S. R., Theory of size distribution in ecological communities, *J. Fish. Res. Bd. Can., 31,* 1859-1862, 1974.

Kerfoot, W. C. and A. Sih (editors), *Predation: Direct and Indirect Impacts on Aquatic Communities,* University Press of New England, Hanover, New Hampshire, 1987.

Kuosa, H., R. Autio, P. Kuuppo, O. Setälä and S. Tanskanen, Nitrogen, silicon and zooplankton controlling the Baltic spring bloom: an experimental study, *Estuar. Coast. Shelf Sci., 45,* 813-821, 1997.

Kuuppo, P., R. Autio, H. Kuosa, O. Setälä, and S. Tanskanen, Nitrogen, silicate and zooplankton control of the planktonic food-web in spring, *Estuar. Coast. Shelf Sci., 46,* 65-75, 1998.

Larsson, U. and A. Hagstrom, Fractionated phytoplankton primary production, exudate release and bacterial production in a Baltic eutrophicated gradient, *Mar. Biol., 67,* 57-70, 1982.

Legendre, L. and F. Rassoulzadegan, Stable versus unstable planktonic food webs in oceans, *Proc. 8th Intl. Symp. Microbial Ecology (Halifax),* in press.

Le Pape, O., Y. del Amo, A. Menesguen, A. Aminot, B. Quequiner, and P. Treguer, Resistance of a coastal ecosystem to increasing eutrophic conditions: the Bay of Brest (France), a semi-enclosed zone of western Europe, *Cont. Shelf Res., 16,* 1885-1907, 1996.

Laws, E. A., The importance of respiration losses in controlling the size distribution of marine phytoplankton. *Ecology, 56,* 419-426, 1975.

MacArthur, R. H., *Geographical Ecology,* New York, Harper and Row, 269 pp., 1972.

Maranger, R. and D. F. Bird, Viral abundance in aquatic systems: a comparison between marine and fresh waters, *Mar. Ecol. Prog. Ser., 131,* 217-228, 1995.

McMahon, T. A., Size and shape in biology, *Science, 179,* 1201-1204, 1973.

McMahon, T. A. and J. T. Bonner, *On Size and Life,* Scientific American Library, New York, 1983.

Micheli, F., Eutrophication, fisheries, and consumer-resource dynamics in marine pelagic ecosystems, *Science, 285,* 1396-1399, 1999.

Officer, C. B. and J. H. Ryther, The possible importance of silicon in marine eutrophication, *Mar. Ecol. Prog. Ser., 3,* 83-91, 1980.

Oksanen, L., S. D. Fretwell, J. Arruda, and P. Niemadä, Exploitation ecosystems in gradients of primary productivity, *Amer. Nat., 11,* 240-261, 1981.

Parsons, T. R., R J. LeBrasseur, and J. D. Fulton, Some observations on the dependence of zooplankton grazing on the cell size and concentration of phytoplankton blooms, *J. Oceanogr. Soc. Japan, 23,* 10-17, 1967.

Pavella, J. S., J. L. Ross, and M. E. Chittenden, Jr., Sharp reduction in abundance of fishes and benthic macroinvertebrates in the Gulf of Mexico and Texas associated with hypoxia, *Northeast Gulf Sci., 6,* 167-173, 1983.

Peters, R. H., *The Ecological Implications of Body Size,* Cambridge University Press, Cambridge, 329 pp., 1983.

Peterson, W. T., Upwelling indices and annual catches of dungeness crab, *Cancer magister,* along the West coast of the United States, *Fish. Bull., 71,* 902-910, 1972.

Petipa, T. S., Chapter 11. Trophic relationships in communities and the functioning of marine ecosystems: I, in *Marine Production Mechanisms,* edited by M. J. Dunbar, pp. 233-250, Cambridge University Press, Cambridge, 1979.

Petipa, T. S., E. V. Pavlova, and G. N. Mironov, The food web structure, utilization and transport of energy by trophic levels in the planktonic communities, in *Marine Food Chains,* edited by J.H. Steele, pp. 142-167, Oliver and Boyd, Edinburgh, reprinted by Otto Koeltz Antiquariat, Koenigstein-Ts./B.R.D., 1973.

Pomeroy, L. R., The ocean's food web, a changing paradigm, *BioScience, 24,* 499-504, 1974.

Pomeroy, L. R., Secondary production mechanisms of continental shelf communities, in *Ecological Processes in Coastal and Marine Systems,* edited by R. J. Livingston, pp. 163-186, Plenum Press, New York, 1979.

Pomeroy, L. R. and W. J. Wiebe, Energetics of microbial food webs, *Hydrobiologia, 159,* 7-18, 1988.

Pomeroy, L. R., Food webs connections: Links and sinks, *Proc. 8th Intl. Symp. Microbial Ecology* (Halifax), in press.

Rabalais, N. N., R. E. Turner, D. Justic', Q. Dortch, W. J. Wiseman, Jr., and B. K. Sen Gupta, Nutrient changes in the Mississippi River and system responses on the adjacent continental shelf, *Estuaries, 19*, 386-407, 1996.

Rahm, L., D. Conley, P. Sandén, F. Wulff and P. Stålnacke, Time series analysis of nutrient inputs to the Baltic Sea and changing DSi:DIN ratios, *Mar. Ecol. Prog. Ser., 130*, 221-228, 1996.

Redfield, A. C, B. H. Ketchum, and F. A. Richards, The influence of organisms on the composition of seawater, in *The Sea, Vol. 2*, edited by M. N. Hill, pp. 26-77, Interscience Publishers, John Wiley, New York, 1963.

Renaud, M. L., Hypoxia in Louisiana coastal waters during 1983: implications for fisheries, *Fish. Bull., 84*, 19-26, 1986.

Rivkin, R. B., K. Legendre, D. Deibel, J.-E. Tremblay, B. Klein, K. Crocker, S. Roy, N. Silverberg, C. Lovejoy, F. Mespié, N. Romero, M. R. Anderson, P. Mattews, C Savenkoff, A. Vezina, J.-C. Therriault, J. Wesson, C. Bérubé, and R. G. Ingram, Vertical flux of biogenic carbon in the ocean: is there food web control?, *Science, 272*, 1163-1165, 1996.

Rosenberg, R., Eutrophication - the future marine coastal nuisance?, *Mar. Pollut. Bull., 16*, 227-231, 1985.

Rowe, G. T., Benthic biomass and surface productivity, in Vol. 2, *Fertility of the Sea*, edited by J. D. Costlow, pp. 441-454, Interscience Publishers, John Wiley, New York, 1971.

Ryther, J. H., Photosynthesis and fish production in the sea. The production of organic matter and its conversion to higher forms of life vary throughout the world ocean, *Science, 177*, 72-76, 1969.

Sakshaug, E. K., S. Andresen, S. Mkklestad and Y. Olsen, Nutrient status of phytoplankton communities in Norwegian waters (marine, brackish, and fresh) as revealed by their chemical composition, *J. Plankton Res., 5*, 175-196, 1983.

Santos, S. L. and J. L. Simon, Response of soft-bottom benthos to annual catastrophic disturbance in a South Florida estuary, *Mar. Ecol. Prog. Ser., 3*, 347-355, 1980.

Schelske, C. L. and E. F. Stoermer, Eutrophication, silica depletion and predicted changes in algal quality in Lake Michigan, *Can. J. Fish. Aquat. Sci., 48*, 1529-1538, 1971.

Schmidt-Nielsen, K., *Why Is Scaling So Important?*, Cambridge University Press. Cambridge, 1984.

Sen Gupta, B. K., R. E. Turner and N. N. Rabalais, Seasonal oxygen depletion in continental-shelf waters of Louisiana: Historical record of benthic foraminifers, *Geology 24*, 227-230, 1996.

Sheldon, R. E., A. Prakash, and W. H. Sutcliffe, Jr., The size distribution of particles in the ocean, *Limnol. Oceanogr., 17*, 327-340, 1972.

Sheldon, R. W. and S. R. Kerr, The population density of monsters in Loch Ness, *Limnol. Oceanogr., 17*, 796-798, 1972.

Sheldon, R. W., W. H. Sutcliffe, Jr., and M. A. Paranjape, Structure of pelagic food chain and relationship between plankton and fish production, *J. Fish. Res. Bd. Can., 34*, 2344-2353, 1977.

Shumway, S. E., A review of the effects of algal blooms on shellfish and aquaculture, *J. World Aquacult. Soc., 21*, 65-104, 1990.

Smayda, T. J., Novel and nuisance phytoplankton blooms in the sea: evidence for a global epidemic, in *Toxic Marine Phytoplankton*, edited by E. Graneli, D. M. Anderson, L. Edler, and B. G. Sundstrøm, pp. 29-40, Elsevier Science Publishing Co., Inc., New York, 1990.

Sommer, U., Comparison between steady state and non-steady state competition: experiments with natural phytoplankton, *Limnol. Oceanogr., 30*, 335-346, 1985.

Sprules, W. G., S. B. Brandt, D. J Steward, M. Munawar, E. H. Jin, and J. Love, Biomass size spectrum of the Lake Michigan pelagic food web, *Can. J. Fish. Aquat. Sci.*, *48*, 105-115, 1990.

Suess, E., Particulate organic carbon flux in the oceans -- Surface productivity and oxygen utilization, *Nature*, *288*, 260-263, 1980.

Suttle, C. A. and P. J. Harrison, Phosphate uptake rates of phytoplankton assemblages grown at different dilution rates in semicontinuous culture, *Can. J. Fish. Aquat. Sci.*, *43*, 1474-1481, 1986.

Thiebaux, M. L. and L. M. Dickie, Structure of the body-size spectrum of the biomass in aquatic ecosystems: a consequence of allometry in predator-prey interactions, *Can. J. Fish. Aquat. Sci.*, *50*, 1308-1317, 1983.

Thiel, H., The size structure of the deep-sea benthos, *Internationale Revue der Gesamten Hydrobiologia*, *60*, 575-606, 1975.

Thompson, D'A., *On Growth and Form*, edited by J. T. Bonner, Cambridge University Press, Cambridge, 1961.

Townsend, D. W. and L. M. Cammen, Potential importance of the timing of spring plankton blooms to benthic-pelagic coupling and recruitment of juvenile demersal fishes, *Biol. Oceanogr*, *5*, 215-229, 1988.

Turner, R. E., Coastal wetlands and penaeid shrimp habitat, in *Stemming the Tide of Coastal Fish Habitat Loss*, edited by R. H. Stroud, pp. 97-104, *Marine Recreational Fisheries 14*, Baltimore, Maryland, National Coalition for Marine Conservation, Inc., Savannah, Georgia, 1992.

Turner, R. E. and N. N. Rabalais, Changes in the Mississippi River this century: Implications for coastal food webs, *BioScience.*, *41*, 140-147, 1991.

Turner, R. E. and N. N. Rabalais, Coastal eutrophication near the Mississippi River delta, *Nature*, *368*, 619-621, 1994.

Turner, R. E., N. Qureshi, N. N. Rabalais, Q. Dortch, D. Justic', R. Shaw, and J. Cope, Fluctuating silicate:nitrate ratios and coastal plankton food webs, *Proc. Natl. Acad. Sci. USA*, *95*, 13048-13051, 1998.

Turpin, D. H. and P. J. Harrison, Cell size manipulation in natural marine, planktonic, diatom communities, *Can. J. Fish. Aquat. Sci.*, *37*, 1193-1195, 1980.

Wafar, M. V. M., P. Le Corre, and J. L. Birrien, Nutrients and primary production in permanently well-mixed temperate coastal waters, *Estuar. Coast. Shelf Sci.*, *17*, 431-446, 1983.

Walsh, J. J., A carbon budget for overfishing off Peru, *Nature*, *290*, 300-304, 1981.

Walsh, J. J., *On the Nature of Continental Shelves*, Academic Press, Inc., New York, 520 pp., 1988.

Walsh, J. J., G. T. Rowe, R. L. Iverson, and C. P. McRoy, Biological export of shelf carbon is a neglected sink of the global CO_2 cycle, *Nature*, *291*, 196-201, 1981.

Wyatt, T., Chapter 14. Food chains in the sea, in *The Ecology of the Seas*, edited by D. H. Cushing and J. J. Walsh, pp. 341-358, Blackwell Scientific Publ., Oxford, 1976.

20

An Economic Perspective of Hypoxia in the Northern Gulf of Mexico

Walter R. Keithly, Jr. and John M. Ward

Abstract

Policymakers devising appropriate strategies to achieve environmental protection goals face complex choices involving trade-offs among multiple objectives. Considerations including, but not limited to, economics, equity, political feasibility and enforcement capabilities may all constitute relevant criteria in the decision making process. Economic analysis provides a useful framework, under provisions discussed in the paper, to measure welfare gains (losses) associated with alternative targeted hypoxia reductions and the trade-off between efficiency (i.e., the optimal use of scarce resources) and other policy objectives. Economic analysis can also be used to ascertain the most cost-effective method for achieving any targeted level of hypoxia reduction and, to the extent that other criteria are relevant in the decision making process, the deviation from the most cost-effective method that would result from inclusion of these other criteria. This paper addresses these issues, primarily from a conceptual basis, and, where possible, lists the benefits and costs that are potentially forthcoming from any given strategy aimed at reducing hypoxic conditions.

Introduction

The Mississippi River is the largest of over thirty rivers draining into the Gulf of Mexico. It drains 3.2×10^6 km^2, equal to about 41% of the conterminous United States, and delivers 20,000 m^3 s^{-1} of water into the Gulf of Mexico [Weber et al., 1992]. The runoff that flows

Coastal Hypoxia: Consequences for Living Resources and Ecosystems
Coastal and Estuarine Studies, Pages 399-424
Copyright 2001 by the American Geophysical Union

into the Gulf of Mexico includes pesticides, fertilizer and other effluents. Runoff from agriculture, particularly nitrogen and phosphorus, is a contributing factor to eutrophication and the resultant hypoxia in the northern Gulf of Mexico [Turner and Rabalais, 1994; Rabalais et al. 1996; Dandelski and Buck, 1998]. Heightened awareness of the issue in recent years has resulted in an increased emphasis on establishing appropriate strategies to reduce hypoxic conditions.

Policymakers, as noted by Hahn and Stavins [1992], face complex choices involving trade-offs among multiple objectives when devising appropriate policies to achieve environmental protection goals, via either the conventional *command-and-control* regulation or *market-based* approaches. Devising the appropriate policies to protect the northern Gulf of Mexico ecosystem is an example of this dilemma. Considerations including, but not limited to, economics, equity, political feasibility and enforcement capabilities may all constitute relevant criteria in the decision making process. Upon specification of the relevant criteria, alternative policy instruments for achieving the stated goal can then be considered and evaluated. If policies aimed at reducing hypoxia in the northern Gulf of Mexico are enacted apart from economic considerations, they are likely to be more costly and, therefore, potentially less effective than necessary (due to, say, limited monies if the enacted policy is funded via Federal outlays, or public resistance if the private sector is mandated to make costly changes in farming practices). Economists therefore tend to stress the importance of assigning relatively high weights to *efficiency* (maximization of net benefits) and *cost-effectiveness* (choosing the least-cost method of achieving a given goal) in the decision making process.

Society receives benefits (welfare) from the use of scarce resources. If *sub-optimal*, reallocation of these resources in a more efficient manner can enhance the total amount of long-run benefits derived from their use. Based on this premise, welfare economics, a normative branch within the larger field of economics, addresses the means for maximizing the welfare of society through the *optimal* use of scarce resources. The excess phosphorus and nitrogen loads that cause algal growth and hypoxic conditions (i.e., depletion of oxygen) in the Gulf of Mexico may diminish Gulf-based resources used by society and the benefits associated with their use. To the extent that these losses occur, reducing hypoxia will result in the restoration of resources and associated benefits. (While not addressed in this paper, the reader should recognize that reductions in nitrogen and phosphorus levels will also result in benefits to the Mississippi River basin as well as to the Gulf of Mexico). Costs, however, are also likely to be incurred in the process of reducing hypoxic conditions. These costs, measured in terms of resources given up by society in its endeavors to reduce hypoxic conditions, must be weighed against the resultant benefits to help ensure that the targeted reduction levels are neither too large or small. If the targeted levels are too large, for example, the resources being given up by society (costs) may exceed the resources being gained (benefits) at the margin, which suggests that the net social benefits could be enhanced by reducing the targeted levels. Conversely, if the targeted levels are too small, the resources being given up by society at the margin are less than the resources being gained, suggesting that the net benefits to society could be enhanced at higher targeted levels. Economics provides a useful framework, under provisions discussed later in the paper, with which to measure the welfare gains (losses) associated with alternative targeted hypoxia reduction levels. To the extent that other factors are relevant in the decision making framework, an economic analysis can be used to examine the trade-off between efficiency (i.e., the optimal use of resources) and other relevant policy objectives.

The most appropriate method for achieving a targeted reduction in hypoxic conditions can also be addressed in an economic framework and is, in fact, intricately tied to the analysis of benefits and costs associated with targeted reduction efforts. In general, several methods or combination of methods can be used to reduce hypoxia. Resources used and, hence, costs will likely vary depending upon the chosen method. Economics can be used to ascertain the most cost-effective method among the various options. Furthermore, to the extent that other criteria are relevant in the decision making process, economic analysis can provide a useful framework for analyzing the deviation from the most cost-effective method that would result from inclusion of these other criteria.

The purpose of this chapter is to provide a perspective on the use of economics to advance rational discussion regarding the extent to which hypoxia should be reduced, when evaluated within a context of an economic efficiency framework, and choosing among alternative methods. To accomplish this purpose, the problem, framed in an economic context, is defined in the next section of the paper. Then, the causes for hypoxia as evaluated from an economic perspective are examined. In the fourth section, the information needs to include an economic analysis in the policy-deliberation process are presented. In the fifth section, a theoretical framework for examining benefits and costs, along with some actual potential benefits and costs related to reduction of hypoxia in the northern Gulf of Mexico, are introduced. Attention is then given to economic instruments that could be employed to minimize costs to society from actions taken to reduce hypoxic conditions.

Defining the Problem

Panayotou [1993] defines the term *environment* to include both the quantity and quality of natural resources (both renewable and nonrenewable), and *environmental degradation* as the diminution of the environment in quantity and its deterioration in quality. Hence, environmental problems, such as ecosystem degradation, include both a quantity and a quality dimension.

Any natural or human-altered system, according to Barbier [1994], can be characterized by three components: stocks, flows and the organization of these stocks and flows. These system components have parallel concepts in both ecology and economics. In ecology, the parallel concepts include structural components, environmental functions and diversity. In economics, the parallel concepts include assets, services and attributes. These relationships are summarized in Table 1.

As further noted by Barbier [1994], a distinction between the *structural components* of an ecosystem (e.g., biomass, abiotic matter, species of flora) and the regulatory environmental *functions* of an ecosystem (e.g., microclimatic, energy flows) is generally made in the field of ecology. From an economic perspective, this dualism corresponds to the standard economic categories of assets (i.e., structural components) and environmental *flows* or *services* (ecological functions). In addition, ecosystems have certain attributes (biological and cultural diversity) of economic significance because they induce economic uses or have value in themselves.

The parallel between ecological system characteristics and economic system characteristics exists because the well-being of society is enhanced through its interaction

TABLE 1. General, ecological and economic system characteristics (from [Barbier, 1994]).

General System Characteristics	Ecological System Characteristics	Economic System Characteristics
Stocks	Structural components	Assets
Flows	Environmental function	Services
Organization	Biological and cultural diversity	Attributes

with the ecosystem (i.e., society receives benefits from the existence of these ecosystems). Because ecosystems contribute to the well-being of society, they are valued by society and are considered a scarce, or limited, resource. de Groot [1992], for example, suggests that environmental functions can be defined as "the capacity of natural processes and components to provide goods and services that satisfy human needs." If a market existed for these environmental functions, in fact, price would be established based upon the relationship between demand and supply for the goods and services, from which one could compute the value to society associated with any given quantity or quality of these functions.

Whether degradation is quantitative or qualitative, changes beyond some possible threshold level can result in a loss in ecological characteristics and hence economic system characteristics. Tropical forests, for example, account for about one-half of the world's species of animals and plants and provide a wide range of local and global benefits [Sandler, 1993]. The local benefits range from soil erosion protection to timber and non-timber products. The global benefits range from the prescription drugs developed from the plant species found only in tropical forests (about one half of all prescription drugs sold in the United States are derived from tropical plants [Repetto, 1988]) to carbon storage which can ameliorate "greenhouse" gases and their potential negative economic impacts (e.g., global warming). Despite the benefits of the tropical forests, the global deforestation rate in 1989 was 142,000 km^2 y^{-1}, or about twice that in 1979 [Myers, 1991]. At current exploitation rates, the tropical rainforests will have, for all practical purposes, disappeared within about the next 50 years [Sandler, 1993]. As asked by Sandler, "If tropical forests are so valuable, then why does the world now confront large-scale destruction of this unique ecosystem?" Stated somewhat differently, "If the well-being of society is enhanced through the economic system characteristics associated with tropical forests, then why does society allow continued degradation, in terms of both quantity and quality, of this unique ecosystem?"

Oceans, like tropical rainforests, can be degraded beyond some possible threshold level, resulting in a loss in the ecological, and hence the economic, system characteristics. At the local level, for example, millions of people directly derive their living from the sea through activities that range from fishing to eco-tourism. At the global level, "[o]cean ecosystems play a major role in the global geochemical cycling of all the elements that represent the basic building blocks of living organisms, carbon, nitrogen, oxygen, phosphorus, and sulfur, as well as the less abundant but necessary elements" [Peterson and Lubchenco, 1997]. There exists a general concern that the oceans are being increasingly subjected to anthropogenic-based degradation on a scale of increasing magnitude. For example, Byrne [1986] states:

"[w]e are using the oceans as a depository for wastes, including municipal sludges, radionuclides, petrogenic hydrocarbons, organochlorine pesticides and other hazardous substances. We have overexploited fish stocks. We have depleted stocks of whales and other marine mammals. If left unchecked, we may cause irreparable damage to the natural balance of ocean ecosystems."

Although the Gulf of Mexico is considered "the most healthy of our (U.S.) coastal marine environments" [Lipka et al., 1990], there are more than 3,700 point sources of pollution in the Gulf of Mexico, which is more than any other region in the country [Weber et al., 1992]. More than one-half of these point sources are industrial facilities, and a total of 460 industrial facilities and municipalities discharge through pipelines directly into the Gulf of Mexico or surrounding estuaries. Municipalities surrounding the Gulf of Mexico dispose of more than 3.79×10^6 gallons of sewage effluents daily into the Gulf of Mexico. Nonpoint source pollutants are also a significant problem in the Gulf of Mexico according to Weber et al. [1992]. These effluents include pesticides and fertilizer runoff from agriculture, particularly nitrogen and phosphorus, and are a contributing factor causing hypoxia in the northern Gulf of Mexico.

Hypoxic conditions in the northern Gulf of Mexico, an indication of a degraded ecosystem, may result in a reduction of benefits derived by society from this unique ecosystem. As with the tropical rain forests, one could reasonably ask: "If the marine environments, including possibly that of the Gulf of Mexico, provide a wide range of benefits of both local and global importance, then why has degradation been allowed to occur on a temporal scale and, more importantly, why is it being allowed to continue?" This, in short, is the problem.

Why is There a Problem?

Economists tend to cite two reasons for excessive environmental degradation and, in particular, for its persistence. The first reason is *market failure*, which occurs when free markets fail to result in efficient resource allocation. The second reason is *policy failure*, defined by Swanson and Cervigni [1996] as "the failure of the state to provide the institutions required for the management of a particular resource, consequentially resulting in its degradation."

Market Failures

One reason for the existence of market failures is related to the lack of functioning markets (i.e., a large number of buyers and sellers coming together for the purpose of exchange) that would, if working properly, allocate resources via price mechanisms. The lack of functioning markets for world environmental assets and services (or attributes) is particularly common. Because no market exists for the use of air, for example, it is unpriced and used in excess by industrial facilities as a means of disposing byproducts (i.e., waste) associated with intended production. If industrial facilities were charged for their use of air, they would, as in the case of labor and capital inputs, use it more conservatively. Similarly, the northern Gulf of Mexico marine ecosystem provides many environmental functions (and parallel economic services), one of which is the ability to assimilate a certain level of wastes. While many of the wastes may reflect ecosystem-based products, society also uses these waters for the disposal of effluents and other wastes from point and nonpoint sources (i.e.,

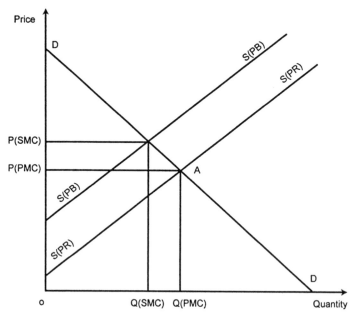

Figure 1. Hypothetical divergence between private marginal costs (S(PR)) and social marginal costs (S(PB)) of producing domestic agricultural products. D-D represents the demand for domestically produced goods; P(PMC) and Q(PMC) represent equilibrium market price and quantity based on private marginal costs; P(SMC) and Q(SMC) represent equilibrium market price and quantity based on social marginal costs.

anthropogenic-based wastes). If all who directly or indirectly used this environmental function of the northern Gulf of Mexico were charged appropriately, then it would be used more sparingly, suggesting less long-term overuse and, in some instances, less degradation.

The most prevalent form of market failure, according to Mason [1996], is that of *externalities*, which arise when "the decisions of some economic agents (individuals, firms, governments), whether in production, consumption or exchange, affect other economic agents and are not included in the priced system of commodities, that is they are not compensated." Distortions from externalities in an otherwise well-functioning market are considered with the aid of Figure 1. For concreteness, assume that the market of interest is domestically produced agricultural products and that runoff from the production of these products, particularly nitrogen and phosphorus, is the primary contributing factor to hypoxia in the northern Gulf of Mexico. Furthermore, assume that imported agricultural products are close, but not perfect, substitutes for the domestically produced products.

Economists generally portray demand for a good, in this case for domestically produced agricultural products, through the use of a *demand curve*, graphically illustrated by the line labeled D-D (Fig. 1). This curve, which is generally downward sloping, shows the quantity of a given good that consumers are willing and able to purchase at alternative market prices. The downward sloping nature of the curve implies, as one would expect, that as price declines (increases), the quantity of the good demanded by consumers increases (decreases). The overall position of the curve, reflecting the overall level of demand for domestically produced agricultural products, is influenced by such factors as income, tastes and preferences, and the price of substitute products. For most goods, an increase in income will result in an upward shift in the demand curve. This implies that an increase in U.S. aggregate income will result in an increase in the consumption of domestically produced agricultural products at any market price. Increases (decreases) in the prices of substitute products (e.g., imported agricultural products) will generally result in an upward (downward) shift in the demand for domestically produced products because the *relative* cost of purchasing the domestically produced product has now become lower (higher).

Ferguson and Gould [1972] demonstrated that the market demand curve (Fig. 1) is equivalent to *social marginal benefits* (i.e., a change in total benefits to society resulting from a small change in consumption), while the area under the demand curve is equal to the total benefits derived by society from the consumption of that good. The downward sloping nature of the curve follows from the law of *diminishing marginal utility* (satisfaction) which implies that the additional satisfaction received from consumption of a good declines in relation to the amount of good consumed.

In contrast to the demand curve, economists portray the industry supply of a good (service) through the use of a *supply curve*, which is graphically illustrated by the line labeled S(PR)-S(PR) in Figure 1. The supply curve depicts the quantity of a good (e.g., domestically produced agricultural products) that industry producers are willing to supply at alternative output prices. The upward sloping nature of the curve implies, as one would expect, that producers are willing to place additional (less) quantity of the good on the market as the market price for the good increases (decreases). The position of the supply curve, reflecting the level of supply at any given output price, is generally considered to be influenced by such factors as technology and input prices. Advances in technology, for example, generally result in higher production at any given output price. This effect is depicted by a downward shift (equivalently to the right) in the supply curve. An increase in the price of any of the primary inputs used in the domestic production of agricultural products, by comparison, will result in an increase in the total costs associated with any level of output. Hence, the effect of an increase in the price of fertilizer can be represented by the supply curve shifting upward (equivalently to the left) which implies that less product will be placed on the market at any given output price.

Ferguson and Gould [1972] demonstrated that, within the confines of a competitive framework, the supply curve shown in Figure 1 is equivalent to the private marginal costs of production (i.e., the change in total private costs of production associated with a small change in output) while the area under the supply curve is equal to the total private costs of production. These costs are the *opportunity costs*, which reflect the value of the scarce resources owned by producers (including their time spent in managing production activities) in foregone employment. Taking our example one step further, therefore, the line labeled S(PR)-S(PR) is equivalent to the private marginal costs associated with the domestic production of agricultural commodities, while the area under the curve is equal to the total private costs associated with the production of these products.

Agricultural producers usually consider only their private (marginal) costs when making economic decisions. These private costs, however, are below the social (marginal) costs. Assume, for example, that economic agents (farmers), in the production of agricultural commodities affect other economic agents (e.g., commercial fishermen, recreational fishermen) via the impacts of runoff and resultant hypoxic conditions. These other economic agents, therefore, bear the costs imposed by the agricultural sector without compensation, i.e., an externality. The social costs associated with domestic agricultural production are, therefore, equal to the private costs of the agricultural producers plus the costs that they impose on other economic agents (i.e., other sectors of society). The social marginal costs associated with domestic production of agricultural commodities is denoted by S(PB)-S(PB) in Figure 1. The area under the social marginal cost curve [i.e., S(PB)-S(PB)], which exceeds the area under the private marginal cost curve [i.e., S(PR)-S(PR)], is equal to the total social cost of supplying domestically produced agricultural products. It is the difference between social (marginal) costs and private (marginal) costs that gives rise to the concept of an externality.

The *market equilibrium* is determined by the interaction of the demand curve (i.e., the society marginal benefit curve) and the supply curve (i.e., the private marginal cost curve). At this intersection (labeled A in Figure 1), the quantity of domestically produced agricultural products demanded by buyers at the stated market price, denoted as P(PMC), is equal to what producers are willing to place on the market at that price. At any price below P(PMC), the quantity demanded by consumers is greater than what producers are willing to provide. Conversely, at any price above P(PMC), the quantity demanded by consumers is less than the amount that producers are willing to supply at that price. The equilibrium quantity, denoted Q(PMC), is associated with the equilibrium price, P(PMC).

Economic decisions based on private marginal costs of production [i.e., S(PR)-S(PR)] rather than social marginal costs of production [i.e., S(PB)-S(PB)] result in several distortions in an otherwise (assumed) well-functioning market. Note, first of all, that the equilibrium output resulting from economic decisions based on private marginal costs, Q(PMC), is higher than the equilibrium output that would result had economic decisions been based on social marginal costs, i.e., Q(SMC). Likewise, the equilibrium price of the output based on economic decisions tied to private marginal costs, i.e., P(PMC) is lower than if social marginal costs were the basis for decisions by individual farmers, i.e., P(SMC). Hence, two immediate distortions are that the equilibrium level of output is excessively high while the equilibrium output price is excessively low. Both of these distortions are the direct result of farmers not internalizing all costs (i.e., both private and externalities) in their business decisions.

The distortions, however, extend well beyond those represented by output prices and quantities. Specifically, agricultural output is a function of several inputs, one of which is fertilizer. To the extent that output is positively related to the use of fertilizer, excessive output as a result of not internalizing all costs of production results in a greater use of fertilizer than would otherwise be economically optimal. Hence, fertilizer usage and associated nutrient runoff is greater than that amount which is socially optimal. Consequently, hypoxic conditions, to the extent they are tied to agricultural runoff, are also greater.

Finally, as the above discussion suggests, there are no incentives for agricultural producers to internalize all production costs or to develop methods that will reduce externalities. While, in theory, consumptive and non-consumptive users of the northern Gulf of Mexico marine ecosystem could collectively negotiate with the farmers to reduce runoff via payments to the agricultural sector, the heterogeneous nature of the consumptive and non-consumptive users, as well as their sheer number, suggests that it would be very expensive to negotiate an efficient solution to the hypoxia problem (i.e., the transaction costs would be high). The externalities and high transaction costs may, under conditions outlined below, justify government intervention to take corrective actions.

Policy Failures

Policy failure refers to "the failure of the state to provide the institutions required for the management of a particular resource, consequentially resulting in its degradation" [Swanson and Cervigni, 1996]. Specifically, the type of market failures noted above provide a rationale for government intervention. On the basis of efficiency considerations alone, however, the rationale for government intervention is conditioned on the fact that the benefits of intervention exceed the costs, including implementation and enforcement. If the benefits of intervention do not exceed costs, however, government intervention may nonetheless be warranted if equity or other relevant criteria, as determined by policymakers in the decision making framework, outweigh those of efficiency.

Panayotou [1993] noted that government intervention, if warranted, should generally be aimed at correcting market failures via taxes, subsidies and regulation. He suggested, however, that "environmental degradation results not only from over-reliance on a free market that fails to function efficiently (market failure), but also from government policies that intentionally or unwittingly distort incentives in favor of over-exploitation and against conservation of valuable and scarce resources (policy failure)." Consider some of the policy instruments used in the United States to support farm income. The objective of supporting farm income tends to be largely accomplished by intervening in the market to raise the prices received by producers [Reichelderfer and Kramer, 1993]. With respect to grains and cotton, intervention takes the form of a combination of market price floors, price guarantees to producers and deficiency payments. If the United States is a net importer of a commodity, such as sugar, intervention tends to be in the form of import restrictions that result in a rise in the price of the domestic commodity [Reichelderfer and Kramer, 1993]. To the extent that these intervention measures result in average farm income in excess of that which would prevail in their absence, otherwise unprofitable farming enterprises may remain in operation, and commodity output among those enterprises that would be profitable even in the absence of intervention may exceed the level of output that would be forthcoming without intervention. Reichelderfer and Kramer [1993], based on theoretical considerations pertaining to commodity support programs, reached the following three conclusions. First, exclusive of acreage-based control programs, commodity price support policies may reduce environmental quality to levels below that which would otherwise be associated with market-determined levels of goods whose production generated externalities. Second, they found that commodity-based price support programs, because they exacerbate externalities, may induce a rise in social willingness to pay for environmental quality (i.e., social demand) which is, in turn, likely to stimulate political pressure for the development of appropriate strategy (e.g., regulation and taxes) aimed at reducing the level of environmental degradation. Finally, the authors concluded that commodity support programs grant what may be perceived as a 'property right' to the additional producers' surplus (a term closely related to profit) that producers create, making it more costly, from a welfare economic perspective, to alter these rights via environmental policy adjustments.

Reichelderfer and Kramer [1993] cite a study by Miranowski et al. [1991] to illustrate some of the distortions resulting from government income and price support programs. Miranowski et al. estimated that if the United States were to eliminate all its farm programs (as of 1990) with the exception of the Conservation Reserve Program, then the aggregate domestic nitrogen use would be reduced by 429 thousand tons, and the herbicide and pesticide use would be reduced by nine thousand and seven hundred tons, respectively. However, due to the more intensive land use practices that would be forthcoming in the absence of farm programs, soil erosion would increase by more than 30 million tons per year.

Hence, as the above discussion implies, the issue of policy failure can be inherently complicated. Price distortions caused by government programs may reduce some specific aspects of environmental quality while enhancing (or at least maintaining) it in others. Intervention by the government to reduce hypoxic conditions, based on the premise that the direct benefits of such actions exceed direct costs, may unwittingly result in the exacerbation of other environmental problems elsewhere in society. Therefore, while market failures in the form of externalities may justify government intervention, such intervention should be taken with caution and with the knowledge that secondary effects associated with such action may be forthcoming.

Economic Information Needs for Rational Public Policy

Environmental policymakers face complex issues involving trade-offs among multiple objectives when devising appropriate strategies for environmental regulation. Several strategies have been proposed as potential candidates for reducing hypoxic conditions in the northern Gulf of Mexico. Enactment of any or some combination of these strategies will most likely result in benefits and costs of varying degrees and, as such, welfare economics can be used to help inform policymakers of the economic trade-offs associated with the various options. Information needs for economics to be adequately included in the decision making process are outlined.

1. *The relationship between anthropogenic activities and hypoxic conditions:* Strategies have been proposed that are based on the premises that: (a) various anthropogenic activities contribute in differing degrees to hypoxic conditions, and (b) changes in these anthropogenic activities via regulation or the use of economic instruments will result in reductions in hypoxia. Eliciting changes in these activities, via either regulation or the use of economic instruments, will potentially be costly and the cost will vary by type of activity wherein changes are desired. The inclusion of welfare economics in the public policy decision framework, therefore, requires information pertaining to the extent to which different anthropogenic-based activities contribute to hypoxic conditions and the degree to which hypoxic conditions would be mitigated by modification of each of these activities.

2. *Costs imposed on society from hypoxic conditions:* The inclusion of economic analysis in the public policy decision process requires that accurate information on the damages emanating from hypoxia are generated. Rephrasing this information need somewhat differently, one might ask, "What economic benefits would society gain by reducing hypoxia in the northern Gulf of Mexico?" To answer this question, one must first know the benefits to society provided by any large-scale ecosystem, such as the northern Gulf of Mexico, and how these benefits are reduced as a result of environmental degradation. Pearce and Turner [1990] list three primary benefits that accrue to society from its interaction with any large-scale ecosystem.

First, ecosystems supply resources, via either the structural components or the environmental function of the ecosystems, that can be employed in the economic system.

While the resources can be either renewable or nonrenewable, marine ecosystems such as the northern Gulf of Mexico are generally associated with the renewable resources derived therefrom, because diminution in the quality of the marine ecosystem will disproportionately impact the renewable resource supply. From a commercial perspective, these renewable resources, combined with the labor and capital needed to make them useful to society, provide an important source of employment and income to many of the coastal communities and provide utility (satisfaction) to the individuals who purchase the final transformed products in the market. Similarly, consumptive use of the resources associated with recreational activities provides direct benefits to those individuals participating in the activities and can also provide income and employment opportunities in the coastal communities.

A second benefit arises because ecosystems provide a direct source of utility (satisfaction) to society that is independent of its consumptive uses. This utility, generally derived through the biological and cultural diversity of ecosystems, is attained in two ways. First, with many ecosystems, including marine, utility is derived through use activities, such as the viewing of the ecosystem in an undisturbed setting. Second, society derives benefits (satisfaction) from the mere knowledge that the ecosystem exists in an unaltered or closely approximated state. Benefits, in turn, decrease in relation to the degraded state of the ecosystem.

Finally, society receives benefits from ecosystems because they are able to assimilate wastes (i.e., a regulatory environmental function of ecosystems). As long as the waste flow into any ecosystem is below its assimilative capacity, the ecosystem is able to turn the wastes into harmless, or even ecologically useful, products (see Peterson and Lubchenco [1997] for a discussion of the assimilative capabilities of marine ecosystems). A problem arises when the assimilative capacity of the ecosystem is exceeded. Phosphorus and nitrogen loads in the Chesapeake Bay, for example, have been identified as a major cause of degradation because they have enhanced algal growth whose decomposition has depleted oxygen while preventing sunlight from reaching the submerged aquatic vegetation [Krupnick, 1989]. The annual hypoxic zone in the northern Gulf of Mexico, to the extent that it is related to anthropogenic-based activities, constitutes *prima facie* evidence that the quantity and composition of wastes deposited in this large-scale ecosystem exceed its assimilative capacity. Like the Chesapeake Bay ecosystem, one might hypothesize that the inability of the Gulf of Mexico ecosystem to completely assimilate wastes deposited in it would contribute to the long-run degradation of this ecosystem.

Once the benefits of an ecosystem are identified, the economic values need to be assigned, where possible, to these benefits. By assigning these values, one can quantitatively assess the economic benefits that society gains from marginal improvements in the integrity of the ecosystem. Value is associated with the amount that society (both current and future generations) would be willing to pay for the economic system characteristics (primarily the services and attributes) provided by the ecosystem, if required; i.e., if they were not provided free of charge by the ecosystem. The greater the benefits derived from the services provided by any particular ecosystem, the more that ecosystem is valued by society. In general, the value of these services tends to be positively related with the integrity of the ecosystem. Methods for assigning values to ecosystem services, while largely outside the scope of this paper, have advanced rapidly in recent years and are discussed by numerous authors including Freeman [1993].

Finally, once the value of the services is determined, the issue arises as to the lost value associated with diminished ecosystem capacity. With respect to the loss in the value of the northern Gulf of Mexico marine ecosystem resulting from nutrient loading and hypoxia, one must estimate the loss in both use values (commercial fishing and recreational fishing) and non-use values. (As noted in the Introduction, excessive nutrient loads may also result in loss in use and non-use values in the Mississippi River Basin.)

3. *Estimates of the costs of taking action to reduce hypoxic conditions:* Every student learns in his or her first economics course that "there is no such thing as a free lunch." Any action taken to increase the welfare of society at large comes with a cost. With respect to the northern Gulf of Mexico hypoxia issue, many sectors in society—such as commercial and recreational fishermen and even non-consumptive users of the marine ecosystem—may benefit from the enhanced integrity of the ecosystem. The costs of achieving this enhanced integrity, however, will be borne by other sectors of society, such as the farmers, who by legislation or other means are mandated or induced to alter current practices. These costs, as discussed later, must be weighed against the benefits to determine, from the criteria of welfare economics, whether government action is warranted and, if so, to what extent.

Is the Cure Worth the Cost?

Theoretical Considerations

The social benefits derived from Gulf of Mexico based resources are, as noted earlier, potentially tied to the externalities associated with the domestic production of agricultural products. Specifically, the externalities emanating from the agricultural sector may impose costs on other economic agents (e.g., commercial and recreational fishermen in the Gulf of Mexico) and, if so, a reduction in benefits derived from Gulf of Mexico based-resources. As such, a reduction in externalities emanating from the farming sector has the potential to reduce hypoxic conditions and increase the benefits derived from the services provided by the Gulf of Mexico ecosystem.

While increased benefits may be forthcoming from reducing externalities and related hypoxic conditions in the northern Gulf of Mexico, there are also costs to society associated with such reduction efforts. These costs may be large. From an economic perspective, therefore, the question arises as to whether the incremental benefits related to an "improved" ecosystem exceed the incremental costs of pursuing this goal. Following traditional economic theory (e.g., Tisdell and Broadus [1989]), the *net benefits* (i.e., gross benefits minus costs) associated with an enhanced ecosystem can be expressed as:

$$NB = B(x) - C(x) \qquad (1)$$

where NB represents the net benefits associated with an enhanced marine ecosystem environment, $B(x)$ represents the gross benefits associated with a given level of marine ecosystem environment equal to x, and $C(x)$ represents the costs associated with achieving

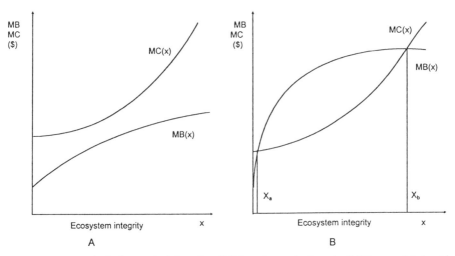

Figure 2. Hypothetical marginal benefits (MB) and marginal costs (MC) associated with enhanced marine ecosystem integrity.

an improvement in the marine ecosystem environment equal to x. If the gross benefits exceed the costs for a given level of x, say x_1, then equation (1) will be positive at that level of x, implying that government intervention is warranted.

The economically optimum level of marine ecosystem enhancement can be determined by differentiating equation (1) with respect to x and setting the resultant equation equal to zero:

$$\partial NB/\partial x = \partial B(x)/\partial x - \partial C(x)/\partial x = 0 \qquad (2)$$

This implies that the marginal benefits associated with the last incremental improvement in the marine ecosystem integrity, or $\partial B(x)/\partial x$, are equal to the marginal costs (i.e., $\partial C(x)/\partial x$) associated with that improvement at the optimum.

These conditions can be shown graphically. As illustrated in Figure 2A, the marginal benefits accruing to society from enhanced ecosystem integrity are denoted by the curve MB(x) while the marginal costs to society associated with enhancement of the ecosystem integrity are denoted by the curve MC(x). In this example, the marginal costs exceed the marginal benefits at all levels of x, implying that government intervention to enhance marine ecosystem integrity cannot be justified on the basis of welfare economics. Intervention, however, may still be warranted if other criteria (e.g., equity) are relevant in the decision making framework.

In Figure 2B, the marginal benefits exceed the marginal costs over a wide range of x, i.e., from x_a to x_b and as such, government intervention to enhance the marine environment is justified on the basis of welfare economics at any level of x from x_a to x_b. The economic optimum, however, is achieved where the marginal benefits associated with the last incremental level of enhancement are equal to the marginal costs associated with that increment, or x_b.

While the above discussion helps place the issue of whether government intervention is warranted, as well as the extent of intervention, in an economic context, there has not yet been an attempt to list any of the specific benefits and costs. While less than complete, some of the primary benefits associated with enhanced ecosystem integrity, along with the costs of achieving these enhanced benefits, are outlined below. Before undertaking this task, however, the components of net benefits, from a welfare economics perspective, are reviewed.

Components of Net Benefits

Economic efficiency is the maximization of net benefits (i.e., gross benefits less costs) associated with a given action and net benefits are measured in terms of *consumer surplus* and *producer surplus*. The net benefits or surplus to consumers from consuming domestically produced agricultural products at the equilibrium level of output (i.e., Q(PMC] in Figure 1) are equal to the triangular area P(PMC)-D-A-P(PMC). This area represents the benefits received by consumers from the consumption of domestically produced agricultural goods (i.e., the area under the demand curve) at the equilibrium quantity, less expenditures on the goods (i.e., the area under the price line). Positive consumer surplus arises because the marginal benefits are equal to what is paid for the good [i.e., P(PMC)] only for the last unit of the good in equilibrium. For all units of the good to the left of Q(PMC), consumers receive marginal benefits in excess of the equilibrium price, although they are only required to pay the equilibrium price. Hence, any reduction in equilibrium price results in an increase in consumer surplus.

Producer surplus, which is the return to producers in excess of what is necessary for them to supply a given quantity of output, is a measure of returns to scarce inputs (i.e., resources) owned by producers. Under equilibrium conditions, the total revenues received by producers from the sale of Q(PMC) units of domestically produced agricultural goods is equal to the area 0-P(PMC)-A-Q(PMC). The total costs of producing Q(PMC) units (i.e., the value of scarce resources used in the production) is equal to 0-S(PR)-A-Q(PMC). The net benefits (i.e., surplus) to producers from the use of scarce inputs (i.e., resources) are equal to the difference between total revenues and the costs of production (i.e., the triangular area S(PR)-P(PMC)-A-S(PR)). This, in the short run, is directly tied to profits and is often referred to as *resource rent*.

Consumers and producers maximize their respective surpluses and the market clears at the market equilibrium. At this equilibrium, the benefits to society of producing the last unit of a good are equal to the opportunity costs incurred. While beyond the scope of this paper, it can be shown that such conditions result in the optimal allocation of scarce resources, based on the existing income distribution (see Tietenberg [1992]). As noted by Tietenberg [1992], efficiency is not achieved because consumers and producers are seeking efficiency, but rather because producers and consumers are attempting to maximize their respective surpluses in a system with well defined property rights and competition. Specifically, the price system induces producers and consumers to make choices that are efficient from a societal viewpoint.

As noted, however, markets for many, if not most, environmental functions are missing. Does this then imply that surplus is an irrelevant concept in these instances? The answer is no. Because these functions (services) contribute to well-being, society would be willing to pay for their provision, if necessary. As such, a hypothetical demand curve for a given

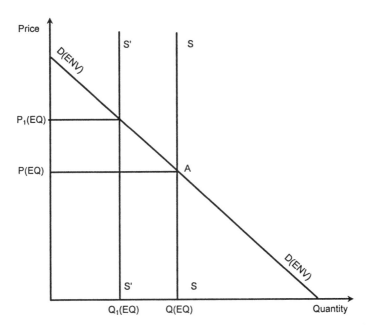

Figure 3. Hypothetical demand (D(ENV)) and supply for an unpriced environmental function (Service). S-S and S'-S' represent the supply of environmental function prior to and after degradation.

environmental function, based on willingness and ability to pay for the function, is illustrated by the curve labeled D(ENV)-D(ENV) in Figure 3. The supply of the environmental function (labeled S-S in Figure 3), because it is not provided by individual producers who adjust output based on market price, is invariant to price and is therefore vertical in nature. If exchanged in the market, the equilibrium price would be P(EQ). Consumer surplus associated with the environmental function, if it were exchanged in the market, would equal to P(EQ)-D(ENV)-A. Since the environmental function is not exchanged in the market, however, consumers do not have to pay for its use (i.e., the price of the environmental function is equal to zero). Hence, total consumer surplus from the environmental function is equal to the area P(EQ)-D(ENV)-A plus the area 0-P(EQ)-A-Q(EQ) or, in total, 0-D(ENV)-A-Q(EQ).

Environmental degradation that results in a reduction in the environmental function is represented in a leftward shift in the supply curve of the environmental function (to S'-S' in Fig. 3). Such a shift results in an unambiguous reduction in the amount of consumer surplus; in the current example a loss equal to the area $Q_1(EQ)$-B-A-Q(EQ).

Benefits Associated With Enhanced Ecosystem Integrity

1. *Increased net benefits (consumer and producer surplus) associated with commercial and recreational fishing:* The dockside revenues for commercial fisheries in Louisiana were

$317 million in 1997, which were second only to Alaska. The important commercial species included shrimp ($144 million), menhaden ($63 million), blue crab ($28 million) and oyster ($30 million). In addition, the fish stocks off the Louisiana coast support a large recreational sector. In 1996, this sector harvested an estimated 23.4 million pounds of fish during 3.14 million trips by 607 thousand participants. Estimated recreational expenditures totaled $450 million in 1996. However, the total revenues and expenditures in Louisiana are not the best estimates of the resource value. The living marine resources are renewable and exist in perpetuity. That is, the net benefits (i.e., producer surplus from the harvest of the product, and consumer surplus associated with the purchase and consumption of the landed product) derived from commercial fishing in a single year represent only a small portion of the value of this resource. Future generations also have a vested interest in these resources. Impacts that lessen the environmental integrity of the northern Gulf of Mexico ecosystem today have implications that may last far into the future. These future values are taken into account by discounting the net benefit stream over time to a present value that the resource is worth today. The Gulf of Mexico shrimp fishery reports annual landings with a gross revenue of approximately $300 million per year, but the present value of net discounted benefits (i.e., producer and consumer surplus) is approximately $1 billion. Similarly, the value of recreational fishing is not the expenditures made each year, but the consumer surplus derived from fishing, represented by the difference between what recreational fishermen would be willing to pay for their fishing "experience" and what they actually pay for the "experience." As with net benefits derived from commercial fishing activities, net benefits from these recreational activities exist in perpetuity and are likely to be directly related to the integrity of the Gulf of Mexico ecosystem.

While undocumented, the conditions caused by hypoxia could impact commercial and recreational fisheries in three ways: (1) fish biomass may be reduced, (2) fish stocks may relocate further offshore or into shallower nearshore waters which have less desirable habitat for those species, and alter predator-prey relationships, and (3) hypoxia could result in a change in fishing patterns for commercial and recreational fishermen as a result of the change in local fish abundance (see Hanifen et al. [1997], Craig et al. [this volume], Chesney and Baltz [this volume] Zimmerman and Nance [this volume]).

Consider first the situation whereby hypoxia conditions result in a reduction in fish biomass. Assume that the decrease in biomass is the result of a reduction in primary food sources of commercial and recreational harvested species, e.g., in the bottom (benthic) community. The reduction in benthic prey items culminates in a downward shift in the sustainable yield curve; i.e., the level of output that can be sustained indefinitely at any level of fishing effort applied to the fishery. (See Bell [1997] and Kahn and Kemp [1985] for a more detailed discussion of the theoretical underpinnings of the concept of the sustainable yield curve and the impact on the curve in relation to a change in carrying capacity.) This translates into a reduction in sustainable yield for any level of effort. Based on the assumption that the amount of satisfaction derived from recreational fishing activities is related to the level of catch per trip, one would expect the downward shift in the sustainable yield curve to result in a reduction in consumer surplus related to the recreational fishing sector. From the commercial perspective, the downward shift in sustainable yield causes the supply curve for an open-access fishery to shift up and to the left. This translates to higher harvesting costs per pound of fish landed at the lower level of sustainable yield. If the higher harvesting costs and reduced supply result in higher prices to consumers, the net result will be a decline in net benefits to the nation as a result of a reduction in consumer surplus.

The reallocation of fish stocks further offshore and into nearshore waters creates two fishing grounds for finfish and shellfish. Hypoxia increases the stress on aquatic ecosystems and decreases biological diversity in areas experiencing repeated and severe hypoxia [Rabalais et al., this volume]. Crowding of marine life into restricted habitat may also lead to indirect consequences through altered competition and predation interactions. These environmental changes could reduce the financial viability of a commercial fishery (i.e., producer surplus) or the viability of recreational fishing (i.e., consumer surplus), if hypoxia results in a severe enough reduction in stock size. If it interrupts migration patterns of shrimp, for example, the capture of the larger and higher value shrimp may be denied to offshore fishermen, reducing their financial viability off Louisiana and forcing them to travel elsewhere, thereby increasing costs.

Hypoxia may alter commercial and recreational fishing patterns because of a change in fish abundance. Fishing vessels are not equally suited for use in all areas and seasons. By concentrating fish stocks in smaller areas, their relative abundance is increased and the cost per fish harvested could be reduced in the short run, resulting in increased capital investment in the fishery and a decline in profitability in the long run. This over-capitalization in fishing craft could result in increased environmental damage from excessive use of fishing gear. The end result could be a further decline in the quality of the marine environment.

Common property fisheries managed, at least to some extent, as open access resources are the rule rather than the exception in the Gulf of Mexico. With habitat improvements from a reduction in hypoxia, an increase in stock size may not result in an improvement in long-term net benefits because the increased stock would create increased short run profits for fishermen and result in expanded fishing effort, a long-run depletion in fish stocks, and increased costs per unit of harvest. This, in turn, suggests that any costs imposed on other sectors of the economy to reduce nutrient runoff will not be compensated by increased benefits to the commercial fishing sector. A similar argument can be made for the recreational fishery as catch rates decline as more fishermen participate in the fishery.

2. *The value of biological diversity associated with the protection of nature.* Sobel [1993] grouped the threats to marine biological diversity into two classes. The first class included those activities that involve over-exploitation of marine resources, including directed or intentional harvesting and the incidental taking of marine life. The second class of threats to marine biological diversity included "...those that destroy or degrade marine habits," such as pollution or coastal development.

It is useful to discuss what is meant by the term *biological diversity* before considering its economic benefits. Simply stated, biological diversity refers to the extent of variety in nature and is usually considered at three different levels: (1) genetic diversity, or the total genetic information contained in the genes of individuals of plants, animals and micro-organisms, (2) species diversity, or the variety of living species, and (3) ecosystem diversity, or the variety of habitats, biotic communities and the ecological processes in the biosphere, as well as the diversity within ecosystems [Pearce, 1995]. Because biological diversity adds to the well-being of society, reductions in it will result in a loss in consumer surplus (i.e., the net benefits that society derives from it).

The obstacles associated with measuring the economic value of biodiversity in a traditional neoclassical framework have been eloquently outlined by several economists,

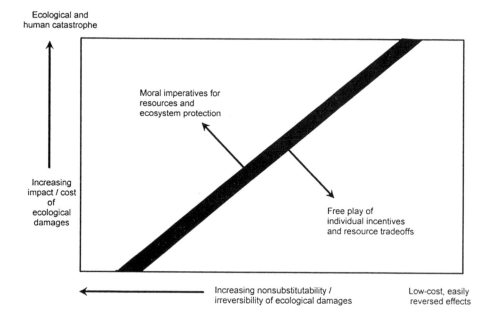

Figure 4. Application of the safe minimum standard (modified from Toman [1994]).

including Randall [1988] and Gowdy [1997]. While the measurement of the value of biodiversity is inherently complicated, all economists would agree that biodiversity has value. Furthermore, the value of biodiversity in relation to any ecosystem is likely related to: (1) the uniqueness of that ecosystem, (2) the complexity of that ecosystem, (3) current and future (possibly unknown) services and functions provided by that ecosystem, and (4) the irreversibility of ecological damages. Most, if not all, of these features are characteristics of the Gulf of Mexico marine ecosystem to a greater or lesser extent (see Weber et al. [1992] for a more detailed discussion of the Gulf of Mexico marine ecosystem). The value of preserving biodiversity by maintaining the Gulf of Mexico's marine ecosystem integrity via the reduction in hypoxia is, therefore, likely to be sizeable.

Because the valuation of benefits associated with any ecosystem is imprecise, many economists have proposed employing the *safe minimum standards* approach [Bishop, 1978]). Toman [1994] noted that the logic of using such an approach is based on the premise that the cost-benefit analysis traditionally used in evaluating trade-offs may be inadequate if the long-term costs of ecosystem loss are uncertain but potentially substantial. Proponents of the safe minimum standards approach argue that, unless society judges that the costs of preservation are unreasonably high, it is best to err on the side of preservation. Following the discussion by Toman, consider Figure 4. A situation of both modest long-run costs related to environmental degradation and a high degree of reversibility is depicted in the lower-right portion of the figure. In this area, trade-offs can be evaluated using a traditional cost-benefit analysis because there is little danger of high long-run costs to society, and damages can be easily rectified given the high degree of reversibility. While costs become relatively high in the upper right-hand corner of the box, they are still relatively reversible. Hence, the current generation can compensate the future for environmental damage through an inter-generational transfer. Because costs are low in the lower left-hand corner of the box, they can be absorbed without significant detrimental effects on future generations, even though irreversibility is relatively high. The safe minimum standard principle becomes particularly relevant in the upper-left hand corner of the box. In this region, impacts become irreversible due to the high long-run costs and limited substitution options. In addition, since the impacts in question will involve large-scale ecosystems and ecological functions, uncertainty is likely to be substantial.

As one moves toward the upper left-hand corner of Figure 4 the individualistic valuation criteria (such as the concept of benefit-cost analysis) should give way to social rules regarding the preservation of natural capital. In other words, the arguments favoring preservation should prevail unless society deems the costs of conservation (preservation) to be excessive.

There is some unknown possibility that degradation is irreversible in the northern Gulf of Mexico hypoxic zone. Because there are many functions related to marine ecosystems, the societal costs (current and future generations) may be large as well. This suggests that a precautionary approach to the problem may be warranted, particularly in light of the paucity of information required to conduct a detailed economic analysis of the problem.

3. *Existence value associated with a "healthy" ecosystem*: Krutilla [1967] is generally credited with introducing the concept of existence value in the economic literature. He claimed that individuals did not actively have to use a resource to derive benefits (i.e., consumer surplus) therefrom. Reasons for his claim are twofold. First, individuals may wish to preserve options for future use. Second, individuals may have an interest, and hence value, associated with bequeathing resources to his or her heirs.

As originally outlined, the irreplaceability of natural resources was the primary justification for the presence of existence value [Krutilla and Fisher, 1975]. The existence value of living marine resources could thus be very high, implying substantial existence values for the ecosystems that support them. While no such studies have been conducted to ascertain the existence value associated with marine ecosystems, various studies have been conducted to ascertain the existence values for species depending on marine ecosystems. Cabot [1996], for example, found an average willingness to pay of approximately $33 per person to save a sea turtle, based on a survey conducted in Texas. Loomis and Larson [1994] in a study of the California coast found a willingness to pay of $25 per visitor for a 50% increase in gray whale populations while Day [1988] estimated the non-consumptive use value (which is different from pure existence value) of whale watching between $21 and $23 resulting in a capitalized value of between $66 million and $118 million based on a survey in New England.

Costs Related to Enhanced Ecosystem Integrity

Despite spending in excess of $540 billion on water pollution controls since the enactment of the Clean Water Act (CWA) in 1972, approximately 44% of the rivers tested in 1992 did not fully support the uses designated by the states [Puckett, 1995]. Puckett

[1995] asserts that nonpoint source pollution, because of the lack of controls in the CWA, is one of the primary contributing factors that has impeded more substantial improvements.

Agricultural runoff is one of the primary nonpoint sources of pollution. Enhancing the northern Gulf of Mexico ecosystem will likely require, in part, modifications of current farming practices to either reduce the overall amount of runoff or its impacts. The costs associated with making these modifications are likely to depend upon both the types of modifications chosen and the extent to which the modifications are made. Consider, for example, that a reduction in fertilizer use by Mississippi Basin farmers is selected as the method to reducing hypoxia. While achieving the reduction in fertilizer usage can be met in several ways (e.g., requirement of a uniform percentage reduction among all farms versus economic incentive programs), the costs are likely to vary considerably depending upon the method chosen. In addition, the costs per unit of fertilizer reduction for any given method are likely to be dependent on the overall level of fertilizer reduction, with costs increasing at an increasing rate as fertilizer usage is reduced beyond some initial amount. These costs reflect both a reduction in profitability in the agricultural production sector (i.e., a reduction in producer surplus) and increased food prices to consumers (i.e., a reduction in consumer surplus) as a result of a reduction in agricultural output. Interestingly, it has been suggested that some initial reductions in fertilizer usage may not entail any significant costs because farmers tend to over-supply nitrogen, presumably due to a risk-adverse nature among farmers.

Minimizing the Costs

Reducing hypoxia in the northern Gulf of Mexico, to the extent that it is related to agricultural runoff, particularly nitrogen and phosphorus, will necessitate changes in current farming practices. These practices can be altered via legislative instruments (i.e., regulatory "command and control" mechanisms) or through the use of economic instruments. The two primary types of economic instruments employed, according to Bender [1996], are (1) *price-based measures* that are used to persuade polluters to reduce their discharges, and (2) *rights-based measures*.

There are two basic types of price-based economic instruments. The first, *charge systems*, consists of different types of taxes levied on the polluting party. The purpose of the taxes is to induce the economic agent (farmer) to modify his practices so that the externalities are reduced or internalized (i.e., to more accurately equate the private marginal costs, as illustrated in Figure 1, with the social marginal cost). For example, a tax could be placed on fertilizer, with the amount of the tax varying in proportion to the nitrogen and phosphorus content. The purpose of the tax would be to reduce the demand for fertilizer—particularly those types with high contents of nitrogen and phosphorus—as an input in the production process of agricultural crops (see Huang and Lantin [1993] for a discussion of a fertilizer tax option). An alternative charge system would be that of an effluent fee. This fee, based on the amount and composition of runoff by each farm, would probably be difficult to implement because of the problems associated with establishing how much each farm is contributing to the nitrogen and phosphorus loads that enter the Gulf of Mexico.

The second basic type of price-based economic instrument is *subsidies*. Specifically, financial aid can be offered to develop or implement the technology required to minimize externalities. The farming sector, for instance, could be subsidized for the adoption of tilling or rotation practices that minimize runoff (see Huang and Lantin [1993]). Alternatively, the sector could be subsidized for using only those types of fertilizers that will achieve a desired goal in terms of reduced nitrogen runoff.

One interesting subsidy has recently been implemented under The Maryland Water Quality Improvement Act of 1998 (WQIA). This Act, created in part due to outbreaks of the dinoflagellate *Pfiesteria*, seeks to create environmental and other benefits to the Chesapeake Bay through reductions in nonpoint source nutrient pollution [Parker, 1999]. Under provisions of the WQIA, according to Parker, a 50% tax credit, up to $4,500 per grower per year for up to three years, will be given to poultry farmers for the purpose of purchasing additional commercial fertilizer for producing their crops to help offset costs of switching away from poultry litter (believed to contribute to the *Pfiesteria* outbreaks).

The economic instrument of rights-based measures typically consists of some type of *tradable emission (effluent) permit system*. Within the context of this scenario, the total allowable level of effluents must first be established by the regulatory agency. This level is based on the desired reduction. Then, each emitter is given the right to emit some proportion of the total, and trading of emission rights is permitted. Programs of this nature have been pursued in several different areas with varying degrees of success (see, for example, Josklow and Schmalensee [1988]). The state of Wisconsin, for example, implemented an innovative program in 1981 aimed at controlling biological oxygen demand on a part of the Fox River. As noted by Hahn [1989], the primary objective of the program, which allowed for limited trading of marketable discharge permits, was to provide firms with flexibility in abatement options while maintaining environmental standards. Firms being targeted for abatement efforts included, primarily, pulp and paper plants and municipal utilities. The program was largely unsuccessful, according to Hahn, due to the oligopolistic nature of the pulp and paper plants in the region and the fact that municipal utilities are subject to public utility regulation and, hence, are likely not to operate in a manner consistent with profit maximization.

Excessive nutrient loads (phosphorus and nitrogen) have been identified as a major cause of the degradation of Chesapeake Bay. In response, the governors of Maryland, Virginia and Pennsylvania, the Mayor of the District of Columbia (hereafter, the "states") and the United States Environmental Protection Agency agreed, in December 1987, to reduce nutrient loads in the Bay by 40% [Krupnick, 1989]. The states agreed to allocate load reductions in proportion to each state's share of the baseline loads. Krupnick [1989] writes that this agreement is almost certain to be inefficient because: (a) the mix of nutrient sources is likely to be different within each state, and (b) statewide marginal nutrient removal costs (the change in total costs for each state associated with a small reduction in nutrients) are likely to differ. Krupnick suggests that a system of state trading of nutrient reduction credits may realize some reduction in costs of meeting the load reduction target although, because interstate trades would involve income transfers, such a system is likely to face political resistance.

While appropriately designed "command and control" regulation or economic instruments may both achieve desired outcomes, the costs of achieving the desired outcomes may vary considerably depending on the option pursued. Nicolaisen et al. [1991] provide three arguments often cited for preference of regulation to the use of economic instruments: (1) there may be substantial costs of implementing certain economic instruments, (2) the greater certainty of the effects of regulation in some instances may be more acceptable, and (3) economic instruments may have politically unattractive effects on the distribution of income, which can be either masked or to some extent avoided under regulation. Appealing features of economic instruments include: (1) they promote economic efficiency, (2) they provide permanent incentives for technological improvements, and (3) they reduce the size of bureaucracy involved with regulatory approaches and minimize compliance costs.

While a detailed discussion of the different "command and control" and economic instrument techniques that could be pursued to achieve a goal of reduced hypoxia is outside the scope of this paper, numerous studies have concluded that economic instruments can, in general, achieve the desired goal at a much lower cost to society than "command and control" approaches. The high cost that will almost certainly be imposed in association with movement towards improved ecosystem integrity of the northern Gulf of Mexico, however, suggests the need to consider carefully the different options that could be pursued to modify behavior and to choose that option or combination of options that will achieve the desired goal at a reasonable cost and level of compliance.

Finally, as noted by Boggess et al. [1993], a large number of voluntary approaches are available to assist farmers in pollution reduction efforts. These voluntary approaches include both technical assistance and cost sharing (e.g., Agricultural Conservation and Conservation Technical Assistance Programs), long-term land retirement (e.g., Conservation Reserve Program) and the purchase of property rights. As one example of the latter, Boggess et al. [1993] report that to reduce phosphorus loads in Lake Okeechobee in south Florida, dairy farmers in the basin were offered a payment of about $600 per cow if they agreed to cease dairy operations within six months and have a deed restriction placed on the land that prevented future dairy or livestock operations on the land.

Summary and Discussion

On a constant basis, policymakers are confronted with decisions affecting environmental quality. Rarely, if ever, will they please everyone because, almost without exception, their decisions will negatively impact some sectors of society. Other sectors, however, may gain from the same decision and be pleased with the outcome. How, one might reasonably ask, does one weigh the losses and gains among those involved in these decisions? While several factors enter the decision making process (e.g., equity, political feasibility, enforcement capabilities), welfare economics provides a useful method to synthesize the benefits and cost in dollars, and to compare alternative policy options.

Hypoxic conditions in the northern Gulf of Mexico, to the extent that they are influenced by anthropogenic-based activities, indicate the degradation of a large-scale marine ecosystem. Because social welfare is enhanced through its interaction with this ecosystem, its degradation implies a net potential social loss. Restoration of these environmental functions, conversely, will enhance benefits.

Restoration, however, will likely require modification of anthropogenic-based activities. This modification, most likely, will not be achieved without imposing costs on some segments of society. The costs, furthermore, will likely vary depending on the desired level of modification as well as the means by which these segments are induced to change behavior.

This paper uses a traditional welfare economics approach to examine the issue of hypoxia in the northern Gulf of Mexico. The problem was first identified within an economic context by illustrating how a combination of market failures and policy failures could, at least in theory, be contributing to hypoxic conditions. The paper then examined some economic information needs that could be used to advance economics in the decision making process. These needs included: (1) the relationship between anthropogenic activities and hypoxic conditions, (2) the costs imposed on society from hypoxic conditions, and (3) the costs of taking action to reduce hypoxic conditions.

The benefits of reducing hypoxic conditions may accrue to certain segments of society only by imposing costs on other segments. The paper examined some of these potential benefits and costs from a qualitative perspective and provided a theoretical perspective on the amount of hypoxia reduction that would be "optimal" from a welfare economics basis. Finally, the paper briefly examined two of the primary economic instruments, priced-based measures and rights-based measures, that could be adapted to induce changes in the anthropogenic-based activities that are thought to contribute to hypoxic conditions.

A theoretical discussion of ecosystem management within the context of a welfare economics framework is relatively easy. Before being lulled into a false sense of security, however, the reader should know that empirical analysis of the issue (i.e., measuring benefits and cost) is anything but easy. An economic valuation of large-scale ecosystems is in its infancy and subject to considerable imprecision including some of the most rudimentary elements, such as specifying the entire list of benefits and costs. While there will certainly be considerable economic uncertainty associated with any economic benefit and cost estimates, this uncertainty is to be expected when analyzing any large, complex system. The uncertainty with the economic analysis should not limit its value in the decision making process but, rather, should merely be placed in the same context with the uncertainty associated with other natural-and-social-science disciplines when being considered in policy. In discussing the use of economics in reducing nutrients in the Chesapeake Bay, Krupnick [1988] states:

> "The problems now facing the Bay are enormous, and they will worsen unless effective strategies for Bay cleanup are devised and implemented....[T]he policies adopted for the Bay must buy as much protection as possible for the resources available. Unless these issues are addressed, the cleanup may be ineffective and unduly costly to society as a whole and to the governments currently bearing the brunt of the cleanup expense. As the costs of cleaning up the Bay mount or, worse still, if improvements to the falls short of expectations, the fragile fabric of political commitment that binds government, business, and the public may tear and jeopardize the Bay cleanup."

As suggested by Krupnick [1988], economic analyses, even though crude and beset with uncertainties "can help maintain and strengthen this new commitment to reducing nutrient loads through the use of cost-benefit and cost-effectiveness analysis." This, most certainly, is the case for the Gulf of Mexico.

Acknowledgments. The thoughtful comments of three anonymous reviewers and the editors, we believe, significantly improved the overall quality of the paper. Without unduly implicating them, we thank them for their help.

References

Barbier, E. B., Valuing environmental functions: tropical wetlands, *Land Econ.*, *70*, 155-73, 1994.

Bell, F. W., The economic valuation of saltwater marsh supporting marine recreational fishing in the southeastern United States, *Ecol. Econ.*, *21*, 243-54, 1997.

Bender, S., Charging the earth: the promotion of price-based measures for pollution control, *Ecol. Econ.*, *16*, 51-63, 1996.

Bishop, R., Endangered species and uncertainty: the economics of the safe minimum standard, *Amer. J. Agric. Econ.*, *60*, 10-18, 1978.

Boggess, W., R. Lacewell, and D. Zilberman, Economics of water use in agriculture, in *Agricultural and Environmental Resource Economics*, edited by G. A. Carlson, D. Zilberman, and J. A. Miranowski, pp. 319-391, Oxford University Press, New York, 1993.

Byrne, J., Large marine ecosystems and the future of ocean studies: a perspective, in *Variability and Management of Large Marine Ecosystems*, edited by K. Sherman and L. Alexander, pp. 299-308, Westview Press, Boulder Colorado, 1986.

Cabot, C. B., Shrimp and sea turtles in the Gulf of Mexico: an economic analysis of the effects of turtle excluder devices on the shrimp fishery and the benefits of protecting sea turtles, Senior Honors Thesis, Northwestern University, 1996.

Dandelski, J. R. and E. H. Buck, Marine dead zones: understanding the problem, *Congressional Research Service Report to Congress (98-869-ENR)*, 1998.

Day, S. V., Estimating the non-consumptive use value of whale watching: an application of the travel cost and contingent valuation techniques, Masters Thesis, Department of Natural Resource Economics, University of Rhode Island, Kingston, 1988.

de Groot, R. S., *Functions of Nature*, Wolters-Noordhoff, Groningen, Netherlands, 1992.

Freeman, M., *The Measurement of Environmental and Resource Values: Theory and Methods*, Resources for the Future, Washington, D.C., 1993.

Ferguson, C. E. and J. P. Gould, *Microeconomic Theory, 4th Edition*, Richard Irwin Inc, Illinois, 1972.

Gowdy, G. M., The value of biodiversity: markets, society, and ecosystems, *Land Econ.*, *73*, 25-41, 1997.

Hahn, R. W., Economic prescriptions for environmental problems: how the patient followed the doctor's orders, *J. Econ. Perspect.*, *3*, 95-114, 1989.

Hahn, R. W. and R. N. Stavins, Economic incentives for environmental protection: integrating theory and practice, *Amer. Econ. Rev.*, *82*, 464-468, 1992.

Hanifen, J. G., W. S. Perret, R. P. Allemand, and T. L. Romaire, Potential impacts of hypoxia on fisheries: Louisiana's fishery-independent data, in Proc., First Gulf of Mexico Hypoxia Management Conference, December 1995, New Orleans, Louisiana, pp. 87-100, Publ. No. EPA-55-R-97-001, Gulf of Mexico Program Office, Stennis Space Center, Mississippi, 1997.

Huang, W. and R. M. Lantin, A comparison of farmers' compliance costs to reduce excess nitrogen fertilizer use under alternative policy options, *Rev. Agr. Econ.*, *15*, 51-62, 1993.

Josklow, P. L. and R. Schmalensee, The political economy of market-based environmental policy: the U.S. acid rain program, *J. Law and Econ.*, *XLI(April)*, 37-83, 1988.

Kahn, J. R. and W. M. Kemp, Economic losses associated with the degradation of an ecosystem: the case of submerged aquatic vegetation in Chesapeake Bay, *J. Environ. Econ. and Manage.*, *12*, 246-63, 1985.

Krupnick, A. J., Reducing bay nutrients: an economic perspective, *Maryland L. Rev.*, *47, 801-829*, 1988.

Krupnick, A. J., Tradable nutrient permits and the Chesapeake Bay compact, *Resources for the Future Discussion Paper QE89-07*, 1989.

Krutilla, F., Conservation reconsidered, *Amer. Econ. Rev.*, *57*, 777-86, 1967.

Krutilla, F. and A. C. Fisher, *The Economics of Natural Environments: Studies in the Valuation of Commodity and Amenity Resources*, The John Hopkins University Press, Baltimore, Maryland, 1975.

Lipka, D., The Gulf of Mexico: an overview, in *Improving Interactions Between Coastal Science and Policy: Proceedings of the Gulf of Mexico Symposium* edited by National Research Council, pp. 15-18, National Academy Press, Washington, D.C., 1996.

Loomis, J. B. and D. M. Larson, Total economic values of increasing gray whale populations: results from a contingent valuation survey of visitors and households, *Mar. Res. Econ.*, *9*, 275-286, 1994.

Mason, R., Market failure and environmental degradation, in *The Economics of Environmental Degradation: Tragedy of the Commons?* edited by T. M. Swanson, pp. 29-54, Edward Elgar, Cheltenham, United Kingdom, 1996.

Miranowski, J. A., J. Hrubovak, and J. Sutton, The effects of commodity programs on resource use, in *Commodity and Resource Policies in Agricultural Systems*, edited by N. Bockstael and R. Just, pp. 275-292, Springer-Verlag, New York, 1991.

Myers, N., Tropical forests: present status and future outlook, *Clim. Change*, *19*, 3-32, 1991.

Nicolaisen, J., A. Dean, and P. Hoeller, Economics and the environment: a survey of issues and policy options, *OECD Economic Studies, No. 16*, 1991.

Panayotou, T., *Green Markets: The Economics of Sustainable Development*, A copublication of the International Center for Economic Growth and the Harvard Institute for International Development, 1993.

Parker, D., The economic costs of implementing the Maryland water quality improvement act of 1998, in *Economics of Policy Options for Nutrient Management and Pfiesteria* (proceedings of the conference held November 16, 1998, in Laurel, Maryland) edited by B. L. Gardner and L. Koch, College Park, Maryland: Center for Agriculture and Natural Resource Policy, University of Maryland, 1999.

Pearce, D. W., *Blueprint 4: Capturing Global Environmental Values*, Earthscan Publications Ltd., London, 1995.

Pearce, D. W. and R. K. Turner, *Economics of Natural Resources and the Environment.*, The John Hopkins University Press, Baltimore, Maryland, 1990.

Peterson, C. H. and J. Lubchenco, Marine ecosystem services, in *Nature's Services: Societal Dependence on Natural Ecosystems* edited by G. C. Daily, pp. 177-194, Island Press, Washington, D.C., 1997.

Puckett, L. J., Identifying the major sources of nutrient water pollution, *Env. Sci. Techn.*, *29*, 408-14, 1995.

Rabalais, N. N., R. E. Turner, D. Justic', Q. Dortch, W. J. Wiseman, Jr., and B. Sen Gupta, Nutrient changes in the Mississippi River and system responses on the adjacent continental shelf, *Estuaries, 19*, 386-407, 1996.

Randall, A., What mainstream economists have to say about the value of biodiversity, in *Biodiversity*, edited by E. O. Wilson, pp. 219-223, National Academy Press, Washington, D.C., 1988.

Reichelderfer, K. and R. Kramer, Agricultural resource policy, in *Agricultural and Environmental Resource Economics*, edited by G. Carlson, D. Zilberman, and J. Miranowski, pp. 441-490, Oxford University Press, New York, 1993.

Repetto, R, Overview, in *Public Policies and Misuse of Forest Resources*, edited by R. Repetto and M. Gillis, pp. 1-41, Cambridge University Press, New York, 1988.

Sandler, T., Tropical deforestation: markets and market failures, *Land Econ.*, *69*, 225-33, 1993.

Sobel, J., Conserving biological diversity through marine protected areas: a global challenge, *Oceanus*, *3*, 19-26, 1993.

Swanson, T. M. and R. Cervigni, Policy failure and resource degradation, in *The Economics of Environmental Degradation: Tragedy of the Commons?* edited by T. M. Swanson, pp. 55-81, Edward Elgar, Cheltenham, United Kingdom, 1996.

Tietenberg, T., *Environmental and Resource Economics, 3rd Edition,* Harpers/Collins Publishers Inc., New York, New York, 1992.

Tisdell, C. and J. M. Broadus, Policy issues related to the establishment and management of marine reserves, *Coastal Management*, *17*, 37-53, 1989.

Toman, M. A., Economics and 'sustainability': balancing trade-offs and imperatives, *Land Econ.*, *70*, 399-413, 1994.

Turner, R. E. and N. N. Rabalais, Coastal eutrophication near the Mississippi River delta, *Nature 368*, 619-621, 1994.

Weber, M., R. T. Townsend, and R. Bierce, *Environmental Quality in the Gulf of Mexico: A Citizen's Guide*, Center for Marine Conservation, Washington, D.C., 1992.

21

Hypoxia, Nutrient Management and Restoration in Danish Waters

Daniel J. Conley and Alf B. Josefson

Abstract

Hypoxia and anoxia associated with nutrient-driven eutrophication commonly occur during summer in Danish waters. Hypoxia/anoxia occurs in both estuaries and in the open waters around Denmark with significant impacts upon living resources observed during the last century. A number of measures have been taken in the last decade to reduce nutrient loads with implementation of the Action Plan on the Aquatic Environment (Parts I and II). The phosphorus (P) load has been reduced by 80% due to improved sewage treatment, whereas little reduction in the nitrogen (N) load has occurred with implementation of Action Plan I. Additional policy measures have been taken which can contribute to the reduction in nitrogen emissions to the aquatic environment in Action Plan II. The final cost of this plan is expected to be ca. 1,000 million DKK (ca. 130 million ECU) with 50% paid for by the State and 50% paid for by the agricultural sector. Although significant improvements in water quality and living resources are not yet apparent, modeling efforts have predicted that the prescribed nutrient reductions can reduce the number of hypoxic and anoxic events in Danish estuaries and coastal waters.

Introduction

In September 1986, a large oxygen depletion event occurred in the Kattegat and became the subject of intense media attention in Denmark with pictures of dead lobsters dredged up by fishermen appearing on both television and the front pages of newspapers. This event was characterized by widespread distributions of low oxygen concentrations in bottom waters [Baden et al., 1990a], and together with pressure from environmental organizations, brought eutrophication problems to the forefront of the political agenda. A few months later, in January 1987, the Danish Parliament passed legislation that has come to be called the Action Plan on the Aquatic Environment. It prescribed a plan by which a 50% reduction in land-based nitrogen (N) discharge and an 80% reduction in phosphorus (P) discharge should be attained by 1994 [Kronvang et al., 1993]. This short

Coastal Hypoxia: Consequences for Living Resources and Ecosystems
Coastal and Estuarine Studies, Pages 425-434
Copyright 2001 by the American Geophysical Union

and intense decision-making process was notable given that oxygen problems had been observed previously in the Kattegat and reported in a number of earlier reports [Christensen et al., 1998].

That Danish waters experience occasional periods of hypoxia has been observed for centuries [Hylleberg, 1993]. Danish estuaries can encounter problems with hypoxia/anoxia in bottom waters during warm summer months when calm weather conditions and extended periods of low wind mixing allow for temporary stable stratification [Møhlenberg, 1999], and when biological oxygen consumption is at its maximum [Jørgensen, 1980]. Danish estuarine systems, especially those with relatively low flushing rates, are sensitive to oxygen problems, because they are for the most part shallow (60% are less than 3 m deep), productive systems that tend to be heavily loaded with nutrients primarily from agricultural sources [Conley et al., 2000]. In extreme cases, such as in Mariager Fjord, anoxia can occur throughout the entire water column, impacting not only benthic organisms, but also affecting pelagic communities [Sørensen and Fallesen, 1998; Fallesen et al., 2000]. Since the 1980s, oxygen problems appear to have worsened in the deeper, open waters around Denmark [Leonhard and Varming, 1992], such as in the Little Belt. These open water areas usually experience their worst oxygen problems during autumn, especially in September and October (Fig. 1).

There is a long record of intensive fisheries in Danish estuaries beginning with oyster and mussel consumption 6,000 years ago [Kristensen, 1997]. The Danish fishery resources have likely been affected by the long-term increases in hypoxia. Oxygen depletion, in combination with H_2S in bottom water, has been shown to have significant ecological effects on living resources and induce behavioral responses in organisms [Diaz and Rosenberg, 1995]. Oxygen deficiency has been shown to regulate mussel beds in the Limfjord [Jørgensen, 1980], cause mass mortality of benthic macrofauna in the Kattegat [Rosenberg et al., 1992], reduce the abundance of the sedentary Norway lobster in the southern Kattegat [Baden et al., 1990b], affect the survival of Baltic cod eggs and larvae [MacKenzie et al., 1996] and alter food resources for dermersal fish species [Phil, 1994].

In this chapter we will review the extent of oxygen problems in Denmark with reference to their impact upon living resources, and we will examine the links between nutrient loading and hypoxia. We will outline the continuing efforts to reduce nutrient inputs to Danish coastal waters with implementation of the Action Plan on the Aquatic Environment (Parts I and II). These control measures as well as the prospectus for future restoration of Danish coastal waters will be discussed.

The Extent of the Problem

Hypoxia and anoxia are observed in both the Danish open sea areas and in estuarine environments (Fig. 1), and have a significant effect on living resources. The best-described ecological effects of oxygen deficiency come from the Kattegat, although much of the information on hypoxia/anoxia in Danish waters is anecdotal [Hagerman et al., 1996]. Long-term trends in benthic biomass indicate that hypoxia has affected the macrobenthos in the southern Kattegat on a large scale [Josefson and Jensen, 1992]. Comparisons between benthic fauna collected at stations during 1911-1912 by Petersen [1913] and in the 1980s indicate significant changes in species composition. The largest reductions, however, were observed between collections in 1984 and 1988 at stations

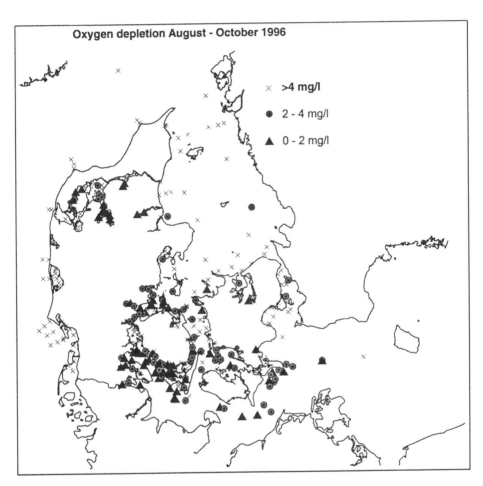

Figure 1. The lowest oxygen concentration measured at stations in the Nationwide Danish Monitoring Program in bottom waters during the period between August and October 1996 (from Jensen et al. [1997]).

(Fig. 2) where oxygen values in the bottom water were less than 1.4 mg l^{-1} in autumn 1988 [Josefson and Jensen, 1992].

One of the most dramatic effects of the 1988 oxygen deficiency in the Kattegat was the reduction of the sedentary Norway lobster [Baden et al., 1990b]. Mass mortality of lobsters occurred when oxygen saturation dropped to 10%. Subsequent reoxygenation of the bottom water during winter allowed for recovery of flatfish and benthic fauna, but cod and lobster populations did not recover [Baden et al., 1990a]. In addition, a general long-term change in the diet of bottom-feeding fish in the Kattegat has taken place since the beginning of the century [Phil, 1994] with concomitant changes in species composition of benthic macrofauna. These results emphasize the potential effects of hypoxia on trophic interactions in marine benthic communities with the possibility that altered food resources resulting from eutrophication might cause a shift in dominance among demersal fish species [Phil, 1994].

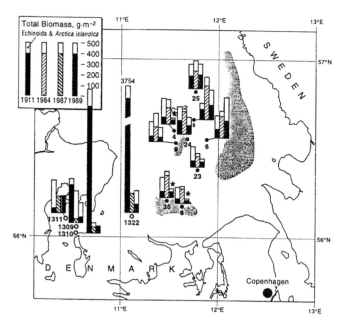

Figure 2. Map of the southern Kattegat with macrofaunal biomass estimates (wet weight) at 13 benthic stations sampled in the 1911-1912 and in the 1980s. Areas with bottom water oxygen concentrations lower than 1.4 mg l⁻¹ in September 1988 are shaded. Significant differences ($p < 0.05$, Mann-Whitney U test) between 1989 and 1987 indicated by asterisks (redrawn from Josefson and Jensen [1992]).

There are numerous deep, restricted basins in Denmark, both in the open sea areas and in the estuaries. These basins become stratified during the summer months and experience seasonal anoxia due to restricted circulation of bottom waters. Such examples, include deep areas in Roskilde Fjord, some estuaries in southeastern Denmark, including Flensborg, Aabenraa and Genner Fjords, and a series of deep basins in the Little Belt south of the island of Funen [Leonhard and Varming, 1992]. These areas are generally void of all benthic organisms during anoxic periods, although opportunistic species can colonize these areas during the periods of oxygen sufficiency [Hagerman et al., 1996].

Oxygen deficiency in the shallow Danish estuaries occurs primarily during warm summer months when calm weather conditions and extended periods of low wind mixing allow for temporary stable stratification [Møhlenberg, 1999]. This condition is typified by the Limfjorden [Jørgensen, 1980], where the frequency of anoxia in bottom waters can vary from every summer to every fifth year. Total anoxia occurs only rarely and is usually confined in the water column to only a few cm of water above the seabed. Benthic organisms react to the lack of oxygen by creeping out of the mud and may survive lying on the mud surface if they can, while the mobile species migrate to those areas not affected by oxygen depletion. Jørgensen [1980] hypothesized that the composition of the benthic communities in some areas is regulated by alternating sequences of extinction and recolonization due to occasional periods of hypoxia and anoxia.

There is only one estuary in Denmark that is a true "fjord type" with a sill creating nearly permanent anoxia in bottom waters (e.g., Mariager Fjord; Sørensen and Fallesen [1998]). In August 1997 following an extended period (7 wk) with low winds (< 6 m s^{-1}) and high surface water temperatures (3-4 °C above normal), whole water column anoxia occurred lasting for a period of nearly two weeks killing fish, sea grasses, mussels and other benthic invertebrates [Fallesen et al., 2000]. This was not the first time the estuary experienced a significant anoxic event with anoxia probably occurring during the summers of 1933, 1947 and 1970 and during the ice cover of 1969-1970 [Fallesen et al., 2000].

Finally, occasional low oxygen concentrations occur in Danish estuaries not just from local processes, but also from the advective transport of saltier, low oxygen bottom water from the Kattegat and the Belt Seas into the estuaries. For example, intrusions of oxygen-poor water from the Kattegat into the estuaries along the eastern coast of Jutland have been reported [Laursen et al., 1992]. During periods of predominantly strong westerly winds, saltier, low oxygen Kattegat bottom water can move into the estuaries sometimes upwelling oxygen-poor, nutrient-rich water into the inner estuaries [Skyum et al., 1994]. When these oxygen-poor waters are forced into the estuary, there are potential ecosystem effects due to low oxygen especially along the bottom where the water enters the estuary. The effects of these events, however, on benthic organisms are poorly described.

Links between Nutrient Loading and Bottom-Water Oxygen Concentrations

Significant long-term increases in nutrient loading to Danish coastal waters have occurred [Kronvang et al., 1993], primarily from an eight-fold increase in fertilizer use and a three-fold increase in manure use since the 1950s [Richardson, 1996]. Current annual N loading from land averages 2,400 kg km^{-2} placing Denmark among the world's largest contributors of N per unit area of land surface [Conley et al., 2000]. By comparison, the present annual watershed N load to Chesapeake Bay, an estuary with severe hypoxia/anoxia problems during summer, averages 930 kg km^{-2} [Boynton et al., 1995]. Hypoxia has worsened historically from the early 1950s to the present with significant decreasing trends in minimum oxygen concentrations observed in coastal and estuarine waters [Rosenberg, 1990] and with decreasing oxygen trends observed in the open waters of the Kattegat [Agger and Ærtebjerg, 1996].

Long-term measurements of nutrient concentrations and oxygen concentrations in the Kattegat from the Danish and Swedish monitoring programs demonstrate the difficulty in observing long-term trends because of year-to-year variability (Fig. 3). Power analysis of long-term trends in bottom-water oxygen concentrations indicates that at least 20 years of data are required in order to discriminate long-term trends from natural variation [Richardson, 1996]. In addition to long-term changes in nutrient loading, there are many other factors influencing bottom water oxygen concentrations such as interannual variation in runoff patterns, exchange of bottom waters and wind activity during the late summer and autumn months. While minimum dissolved oxygen values provide an index of conditions for the biota, particularly for those organisms unable to avoid or tolerate low oxygen conditions, most organisms must remain at low oxygen concentrations for a certain amount of time before the animals are affected [Diaz and Rosenberg, 1995].

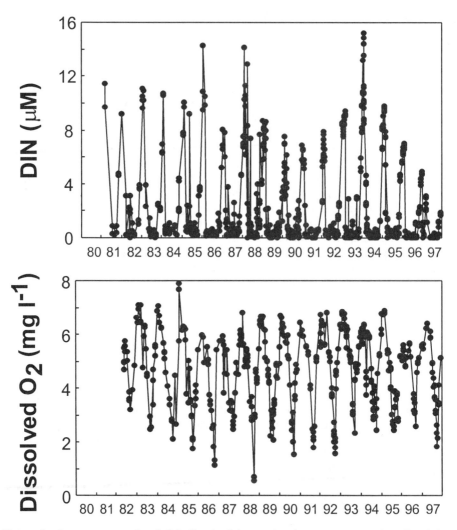

Figure 3.   Long-term trends of (A) dissolved inorganic nitrogen concentration (DIN) in surface waters (the upper 10 m) and (B) oxygen concentrations in bottom waters (0.5 m from the bottom) at Station 413 in the southeastern Kattegat.

Because of natural variability, conventional sampling procedures usually underestimate the extent of hypoxia due to undersampling [Summers et al., 1997].   Until continuous monitoring platforms equipped with near-bottom sensors are established, and the depth distribution of the minimum oxygen layer is known, it will be difficult to establish the dynamics and extent of hypoxic events.

In an attempt to resolve the complex relationships between hypoxia and nutrient enrichment, models have been used to estimate and predict the effect of nutrient loading on primary production, sedimentation of organic material and subsequent oxygen concentrations in the bottom waters of the Danish open seas [Hansen et al., 1995]. Although such models are not able to describe the stochastic nature of the marine system, they can be used on a more general level to predict probability of reductions of severe hypoxic events. Significant improvements in oxygen conditions are predicted by the model if the 50% reduction in N from land-based sources is achieved [Richardson, 1996] with further improvements in oxygen conditions pending decreases in advective nutrient sources and atmospheric nitrogen deposition.

The occurrence of hypoxic events in the shallow bottom waters of Danish estuaries are coincidental with water column stratification caused by sustained periods of low wind mixing allowing for oxygen depletion in bottom waters as has been shown for other estuaries [Turner et al., 1987]. In a detailed analysis of oxygen tension in a shallow (ca. 4 m) Danish estuary, Skive Fjord, Møhlenberg [1999] separated the effects of physical forcing (buoyancy flux, wind and solar insolation) and nutrient loading on oxygen depletion in bottom water. During periods of stratification, the oxygen tension was described by the time elapsed since the onset of stratification and the accumulated N loading 10 months prior to a measurement. Using a 10-year meteorological data base it was then calculated that a 25% reduction in total nitrogen loading, half of what was prescribed in the goals of the Action Plan on the Aquatic Environment, would reduce the number of days with severe oxygen depletion (i.e., < 2 mg O_2 l^{-1}) by 50% [Møhlenberg, 1999]. Similar results have been reported for Chesapeake Bay [Cerco, 1995] where an evaluation of the effects of hydrodynamics and nutrient loading indicated that hydrodynamics (freshwater flow inducing stratification) was the predominant influence on anoxic volume, although long-term trends in anoxia were also coupled to N loading.

Nutrient Reductions and Prospectus for the Future

Improvements in oxygen conditions in Danish estuaries and the deeper, open sea areas will be dependent upon the ability to reduce nutrient inputs to the aquatic environment. The 1987 Action Plan on the Aquatic Environment I prescribed that reductions in the nutrient load should occur before 1994. Comparison of nutrient loading rates in 1997 with those in the 1980s demonstrates that the P load from Denmark has been reduced by approximately 80%, mainly due to improved sewage treatment, whereas only minor reductions have been made in the land-based N load [Ærtebjerg et al., 1998]. Further reductions will be needed to attain the goals prescribed by the Action Plan.

The measures taken to reduce nutrient inputs in Action Plan I were directed towards individual farmers and included establishment of slurry tanks with a minimum of nine months storage capacity, obligations to grow winter crops on at least 65% of the farming area and obligations to establish crop and fertilizer plans [Kronvang et al., 1999]. New legislation has come from both the Danish Parliament as well as directives from the Commission of the European Union (EU) that attempt to comprehensively address nutrient loads by changing agricultural practices. Further actions are being implemented to reduce nutrient loading to the aquatic environment with implementation of the Action Plan II [Iversen et al., 1998]. Such changes include tightening livestock density requirements, more stringent nitrogen utilization requirements for animal manure, improved animal fodder utilization, incentives for organic farming, additional catch crops and a 10% reduction in nitrogen standards for crops [Iversen et al., 1998]. The most effective measures in Action Plan II will originate from increased utilization of manure and decreases in the use of artificial fertilizers. To reach the target of a 50% reduction in land-based nutrients from the 1980s, commercial N fertilizer use must be further reduced from ca. 400,000 tonnes N in 1990 to ca. 180,000 tonnes, and the N-content of manure from about 250,000 tonnes in 1990 to about 210,000 tonnes [Iversen et al., 1998].

Additional improvements are expected in the ecological functioning of streams [Iversen et al., 1993] and wetlands [Iversen et al., 1998] with planned restoration efforts. In the past, most streams experienced extensive regular disturbance including cutting of bank vegetation and clearing of in-stream macrophytes to enhance drainage and the flow of water. As of 1997, that practice has been halted for many streams and a 2-m buffer strip is now required. It is hoped that conversion of readily available inorganic nutrients into organic matter in the streams (macrophytes) and the increase in the efficiency of denitrification with establishment of riparian vegetation [Vought et al., 1994] will enhance the natural ability of streams to process the nutrient loads. In addition, from 9,000 to 18,000 ha of wetlands, including wet meadows and marshland, will be restored by 2002 also contributing to reducing the load of N to the aquatic environment.

The final estimated cost of the Danish Action Plan (II) on the Aquatic Environment is expected to be ca. 1,000 million DKK (ca. 130 million ECU) with 50% paid for by the State and 50% paid for by the agricultural sector [Iversen et al., 1998]. The State will bear the costs of wetland restoration, establishment of groundwater protection areas, reforestation and tax breaks for organic framing, while the Agricultural Sector will be responsible the agricultural measures outlined above.

If reductions are realized, it is believed that significant improvements will be observed in coastal waters. A good example of the response of the aquatic system to reduced nutrient loading occurred in 1996. Low precipitation during the fall and winter of 1995-1996 led to substantial reductions in nutrient runoff around the island of Funen with a 67% reduction in N load and an 85% reduction in the P load as compared to an average year. An immediate response was noted in the nearby waters with lowered winter concentrations of nutrients and higher oxygen concentrations in bottom waters during the spring and summer in 1996 compared to previous years [Rask et al., 1999]. These temporary improvements around the island of Funen during this natural experiment demonstrate that rapid improvements can be realized if significant reductions are made to the nutrient loading to Danish coastal waters.

Denmark's commitment to reducing nutrient inputs and improving the functioning of ecosystems arises from a national consciousness regarding the environment and has been strengthened as a result of National legislation and international agreements. Significant policy measures have been taken in order to reduce N emissions to the aquatic environment. Although significant improvements in water quality and living resources are not yet apparent, modeling efforts have predicted that the prescribed nutrient reductions can reduce the number of hypoxic and anoxic events in Danish estuaries and coastal waters if the goals outlined in the Action Plan on the Aquatic Environment (Parts I and II) are reached.

References

Ærtebjerg, G., J. Cartsensen, D. Conley, K. Dahl, J. Hansen, A. Josefson, H. Kaas, S. Markager, T. G. Nielsen, B. Rasmussen, D. Krause-Jensen, O. Hertel, H. Skov, and L. M. Svendsen., Marine områder. Åbne farvande - status over miljøtilstand, årsagssammenhænge og udvikling, Danish Ministry of the Environment Report No. 254, Roskilde, Denmark, 1998. (In Danish)

Agger, C. T. and G. Ærtebjerg, Longterm development of autumn oxygen concentrations in the Kattegat and Belt Sea Area, in Proc., 13th Symposium of the Baltic Marine Biologists, pp. 29-34, 1996.

Baden, S. P., L. -O. Loo, L. Phil, and R. Rosenberg, Effects of eutrophication on benthic communities including fish: Swedish west coast, *Ambio, 19*, 113-122, 1990a.

Baden, S. P., L. Phil, and R. Rosenberg, Effects of oxygen depletion on the ecology, blood physiology and fishery of the Norway lobster *Nephrops norvegicus, Mar. Ecol. Prog. Ser., 67*, 141-155, 1990b.

Boynton, W. R., J. H. Garber, R. Summers, and W. M. Kemp, Inputs, transformations, and transport of nitrogen and phosphorus in Chesapeake Bay and selected tributaries, *Estuaries, 18*, 285-314, 1995.

Cerco, C. F., Simulation of long-term trends in Chesapeake Bay eutrophication, *J. Environ. Eng., 121*, 298-310, 1995.

Christensen, P. B., F. Møhlenberg, L. C. Lund-Hansen, J. Borum, C. Christiansen, S. E. Larsen, M. E. Hansen, J. Andersen, and J. Kirkegaard, The Danish Marine Environment: Has Action Improved Its State?, Danish Environmental Protection Agency Report No. 62, Copehagen, Denmark, 1998.

Conley, D. J., H. Kaas, F. Møhlenberg, B. Rassmussen, and J. Windolf, Characteristics of Danish estuaries, *Estuaries, 23*, in press, 2000.

Diaz, R. J. and R. Rosenberg, Marine benthic hypoxia: A review of its ecological effects and the behavioural responses of benthic macrofauna, *Oceanogr. Mar. Biol. Ann. Rev., 33*, 245-303, 1995.

Fallesen, G., F. Andersen, and B. Larsen, Life, death and revival of the hypetrophic Mariager Fjord, Denmark, *J. Mar. Syst.,* in press, 2000.

Hagerman, L., A. B. Josefson, and J. N. Jensen, Benthic macrofauna and demersal fish, in *Eutrophication in Coastal Marine Ecosystems*, edited by B. B. Jørgensen, and K. Richardson, pp. 155-178, *Coastal and Estuarine Studies, 52,* American Geophysical Union, Washington, D.C., 1996.

Hansen, I. S., G. Ærtebjerg, and K. Richardson, A scenario analysis of effects of reduced nitrogen input on oxygen conditions in the Kattegat and the Belt Sea, *Ophelia, 42*, 75-93, 1995.

Hylleberg, J., Extinction and immigration of benthic fauna. The value of historical data from Limfjorden, Denmark, in Symposium, Mediterranean Seas 2000, edited by N. F. R. Della Croce, pp. 43-73, Istituto Science Ambientali Marine, Santa Margherita Ligure, Italy, 1993.

Iversen, T. M., B. Kronvang, B. L. Madsen, P. Markmann, and M. B. Nielsen, Reestablishment of Danish streams: Restoration and maintenance measures, *Aquatic Conserv.: Mar. Freshw. Ecosyst., 3*, 73-92, 1993.

Iversen, T. M., R. Grant, and K. Nielsen, Nitrogen enrichment of European inland and marine waters with special attention to Danish policy measures, *Environ. Pollut., 102*, 771-780, 1998.

Jensen, J. N., G. Ærtebjerg, B. Rasmussen, K. Dahl, H. Levinsen, D. Lisbjerg, T. G. Nielsen, D. Krause-Jensen, A. L. Middelboe, L. M. Svendsen, and K. Sand-Jensen, Marine områder. Fjorde, kyster og åbnt hav, Danish Ministry of the Environment Report No. 213, Roskilde, Denmark, 1998. (In Danish)

Josefson, A. B. and J. N. Jensen, Effects of hypoxia on soft-sediment macrobenthos in southern Kattegat, Denmark, in *Marine Eutrophication and Population Dynamics*, edited by G. Colombo, I. Ferrair, V. U. Ceccherelli, and R. Rossi, pp. 21-28, Olsen & Olsen, Fredensborg, Denmark, 1992.

Jørgensen, B. B., Seasonal oxygen depletion in the bottom waters of a Danish fjord and its effect on the benthic community, *Oikos, 34*, 68-76, 1980.

Kristensen, P. S., Oyster and mussel fisheries in Denmark, in U.S. Dept. Commerce, NOAA Tech. Rep. NMFS 129, pp. 25-38, 1997.

Kronvang, B., G. Ærtebjerg, R. Grant, P. Kristensen, M. Hovmand, and J. Kirkegaard, Nationwide monitoring of nutrients and their ecological effects: State of the Danish aquatic environment, *Ambio, 22*, 176-187, 1993.

Kronvang, B., L. M. Svendsen, J. P. Jensen, and J. Dørge, Retention of nutrients in river basins, *Aquat. Ecol., 33*,29-40, 1999.

Laursen, J. S., C. Christiansen, P. Andersen, and S. Schwærter, Flux of sediments and nutrients from low to deep water in a Danish fjord, *Sci. Tot. Environ. Suppl., 1992*, 1069-1078, 1992.

Leonhard, S. and S. Varming, Bundfauna I Lillebælt 1911-1990, Lillebæltssamarbejdet, 185 pp., 1992.

MacKenzie, B., M. St. John, and K. Wieland, Eastern Baltic cod: perspectives from existing data on processes affecting growth and survival of eggs and larvae, *Mar. Ecol. Prog. Ser., 134*, 265-281, 1996.

Møhlenberg, F., Effect of meteorology and nutrient load on oxygen depletion in a Danish micro-tidal estuary, *Aquat. Ecol., 33*, 55-64, 1999.

Petersen, C. G. J., Havets bolnitering. II. Om havbundens dyresamfund og om disses betydning for den marine zoogeografi, Beretn. Minist. Lanbr. Fish. Danm. Biol. Stn, 21, 1-42, 1913. (In Danish)

Phil, L., Changes in the diet of demersal fish due to eutrophication-induced hypoxia in the Kattegat, Sweden, *Can. J. Fish. Aquat. Sci., 51*, 321-336, 1994.

Rask, N., S. E. Pedersen, and M. H. Jensen, Environmental response to lowered nutrient discharges in the coastal waters around Funen, Denmark, *Hydrobiologia, 393*, 69-81, 1999.

Richardson, K., Conclusion, research and eutrophication control, in *Eutrophication in Coastal Marine Ecosystems*, edited by B.B. Jørgensen, and K. Richardson, pp. 243-267, *Coastal and Estuarine Studies, 52*, American Geophysical Union, Washington, D.C., 1996.

Rosenberg, R., Negative oxygen trends in Swedish coastal bottom waters, *Mar. Pollut. Bull., 21*, 335-339, 1990.

Rosenberg, R. L. -O. Loo, and P. Möller, Hypoxia, salinity and temperature as structuring factors for marine benthic communities in an eutrophic area, *Neth. J. Sea Res., 30*, 1-5, 1992.

Skyum, P., C. Christainsen, L. C. Lund Hansen, and J. Nielsen, Advection induced oxygen variability in the North Sea-Baltic Sea transition, *Hydrobiologia, 281*, 65-77, 1994.

Sørensen, H. M. and G. Fallesen, Mariager Fjord. Udvikling og Status 1997. Århus Amt Trykkeri, 1998. ISBN87-7906-026-9 (In Danish)

Summers, J. K., S. B. Weisberg, A. F. Holland, J. Kou, V. D. Engle, D. L. Breitberg, and R. J. Diaz, Characterizing dissolved oxygen conditions in estuarine environments, *Environ. Monit. Assess., 45*, 319-328, 1997.

Turner, R. E., W. W. Schroeder, and W. J. Wiseman, Jr., The role of stratification in the deoxygenation of Mobile Bay and adjacent shelf bottom waters, *Estuaries, 10*, 13-19, 1987.

Vought, L. B. M., J. Dahl, C. L. Pedersen, and J. O. Lacoursière, Nutrient retention in riparian ecotones, *Ambio 23*, 342-347, 1994.

22

Future Perspectives for Hypoxia in the Northern Gulf of Mexico

Dubravko Justic, Nancy N. Rabalais, and R. Eugene Turner

Abstract

General circulation models predict that Mississippi River runoff would increase 20% if the concentration of atmospheric CO_2 doubles. This hydrologic change would be accompanied by an increase in winter and summer temperatures over the Gulf of Mexico coastal region of 4.2 ^oC and 2.2 ^oC, respectively. Using a coupled physical-biological model, we examined the potential effects of climate variability on the Gulf of Mexico hypoxic zone. Model simulations suggest that increased freshwater inflow and surface temperatures may substantially alter water column stability, net productivity and global oxygen cycling in the coastal waters of the northern Gulf of Mexico. In simulation experiments, a 20% increase in annual runoff of the Mississippi River, relative to a 1985-1992 average, resulted in a 50% increase in net primary productivity of the upper water column (0-10 m) and a 30 to 60% decrease in summertime subpycnoclinal oxygen content within the present day hypoxic zone. These model projections are in agreement with the observed increase in severity and areal extent of hypoxia during the Great Mississippi River Flood of 1993.

Introduction

The incidence and severity of hypoxic (< 2 mg O_2 l^{-1}) events have increased during the last five decades in many estuarine and coastal areas, particularly those affected by riverine freshwater inflows [Officer et al., 1984; Justic et al., 1987; Turner and Rabalais, 1994; Diaz and Rosenberg, 1995]. This worldwide trend has been paralleled by an increase in the riverine concentrations of dissolved nitrogen and phosphorus, which has

Coastal Hypoxia: Consequences for Living Resources and Ecosystems
Coastal and Estuarine Studies, Pages 435-450

occurred as a result of fertilizer and detergent use in the watersheds [Marchetti et al., 1989; Turner and Rabalais, 1991; Howarth et al., 1996]. The concentrations of total phosphorus and dissolved inorganic nitrogen in the Mississippi River, for example, have increased two-fold and three-fold, respectively, during the last 50 years [Turner and Rabalais, 1991; Bratkovich et al., 1994]. By quantifying the biologically bound silica sequestered in diatom remains within dated sediment cores, Turner and Rabalais [1994] found a corresponding evidence of increased eutrophication in the coastal waters influenced by the Mississippi River.   In addition, stratigraphic records of benthic Foraminifera indicated an overall increase in the frequency and/or severity of hypoxic events [Sen Gupta et al., 1996]. These temporal and spatial associations between the use of nutrients in the watersheds and outbreaks of hypoxia provide convincing arguments for the hypothesis that this phenomenon is primarily driven by the increase in the anthropogenic nutrient loads.   Thus, it appears that future trends in anthropogenic nutrient loading will ultimately determine whether hypoxia in river-dominated coastal waters will lessen or intensify.

At present, however, little is known about the nature of the causal relationship between climate variability and coastal marine hypoxia. It was observed, for example, that climate anomalies, such as droughts and floods, may substantially alter the severity and areal extent of hypoxia during several annual cycles [Rabalais et al., 1991, 1998]. Changes in the areal extent of the hypoxic zone in the northern Gulf of Mexico provide a representative example of this influence (Fig. 1).   The northern Gulf of Mexico is presently the site of the largest (up to 20,000 km^2) and most severe hypoxic zone in coastal waters of the western Atlantic Ocean [Rabalais et al., 1996; 1998]. During the drought of 1988 (a 52-yr low discharge record of the Mississippi River), however, summertime bottom oxygen concentrations were significantly higher than normal, and the formation of a continuous hypoxic zone along the coast did not occur. In contrast, during the Great Flood of 1993 (a 62-yr maximum discharge for August and September), the areal extent of summertime hypoxia showed a two-fold increase with respect to the average hydrologic year [Rabalais et al., 1998]. Hypoxia in the coastal bottom waters of the northern Gulf of Mexico develops as a synergistic product of high surface primary productivity, which is also manifested in high carbon flux to the sediments, and high stability of the water column [Rabalais et al., 1996]. Likewise, the 1993 event was associated with both the increased stability of the water column and nutrient-enhanced primary productivity, as indicated by the greatly increased nutrient concentrations and phytoplankton biomass in the coastal waters influenced by the Mississippi River [Rabalais et al., 1998].

Linkages Between Global Climate Change and Riverine Nutrient Delivery

During this century, global temperature averages increased by approximately 0.5 °C on a worldwide basis [Kerr, 1990; NOAA, 1994], and further temperature increase seems probable.   Based on the projections of general circulation models (GCMs) that include

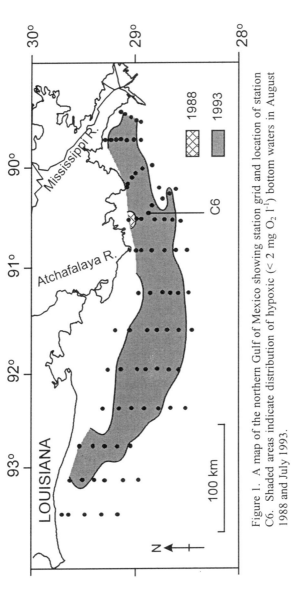

Figure 1. A map of the northern Gulf of Mexico showing station grid and location of station C6. Shaded areas indicate distribution of hypoxic (< 2 mg O_2 l^{-1}) bottom waters in August 1988 and July 1993.

radiative forcing of enhanced greenhouse gas concentrations, global temperature may increase between 2 and 6 °C over the next 100 years [IPCC, 1996]. GCM simulations further suggest that this temperature increase would enhance the global hydrologic cycle. Miller and Russell [1992] examined the impact of global warming on the annual runoff of the 33 world's largest rivers. For a $2xCO_2$ climate, the runoff increases were detected in all rivers in high northern latitudes, with a maximum of + 47%. At low latitudes there were both increases and decreases, ranging from + 96% to - 43%. Importantly, the model results projected an increase in the annual runoff for 25 of the 33 studied rivers (Fig 2). Also, a recent analysis of the climate-sensitive streamflow data collected by the U.S. Geological Survey indicated statistically significant increasing trends in monthly streamflow during the past five decades across most of the conterminous United States [Lins and Michaels, 1994]. This supports the hypothesis that enhanced greenhouse forcing produces an enhanced hydrologic cycle, most notably during autumn and winter months. Because nutrient fluxes generally increase with runoff, riverine freshwater and nutrient inputs are expected to increase as a result of climate change, at least in some coastal areas.

The northern Gulf of Mexico (Fig. 1), which receives inflows of the Mississippi River - the eighth largest river in the world [Milliman and Meade, 1983], is one of the coastal areas that may experience increased freshwater and nutrient inputs in the future. According to the study by Miller and Russell [1992], the annual Mississippi River runoff would increase 20% if the concentration of atmospheric CO_2 doubles (Fig. 2). This hydrologic change would be accompanied by an increase in winter and summer temperatures over the Gulf Coast region of 4.2 °C and 2.2 °C, respectively [Giorgi et al., 1994]. A higher runoff is expected during the May-August period, with an annual maximum most likely occurring in May. While there are no other GCM estimates of the Mississippi River runoff, this result is in agreement with a projected $2xCO_2$ increase in rainfall over the Mississippi River drainage basin [Giorgi et al., 1994].

It is likely that the Mississippi River nutrient fluxes would increase proportionally to the runoff. During the period 1985-1992, for example, the Mississippi River N-NO$_x$ flux was strongly correlated (R = 0.82; P < 0.001) with runoff (Fig. 3). A portion of the total flux variability, however, may be explained by a seasonal pattern in riverine N-NO$_x$ concentrations (Fig. 4). Because of this seasonal signal, the maximum in N-NO$_x$ flux during 1985-1992 was somewhat delayed with respect to the peak in freshwater runoff (Fig. 5). The 1985-1992 period includes two years with above average discharge (1990 and 1991), three years with below average discharge (1987, 1988 and 1992), and three average hydrologic years (1985, 1986 and 1989), and may be considered representative for the present day climate. If the runoff increases 20%, the integrated annual $2xCO_2$ runoff at Tarbert Landing (Mississippi) would be around 0.5 x 10^{12} m^3 y^{-1}. The peak monthly runoff would increase to almost 4 x 10^4 m^3 s^{-1}, which is a 25% increase relative to the Great Flood of 1993 (3.2 x 10^4 m^3 s^{-1}; Fig. 5a). Assuming that there would be no change in riverine nitrogen concentrations with respect to the 1985-1992 averages (Fig. 4), the peak monthly N-NO$_x$ flux would increase to 6 x 10^6 kg N d^{-1} (Fig. 5b). This value is of the same magnitude as the peak monthly N-NO$_x$ flux during the Great Flood of 1993. Interestingly, during the February-June period, the projected N-NO$_x$ flux closely resembles the 1993 values (Fig. 5b).

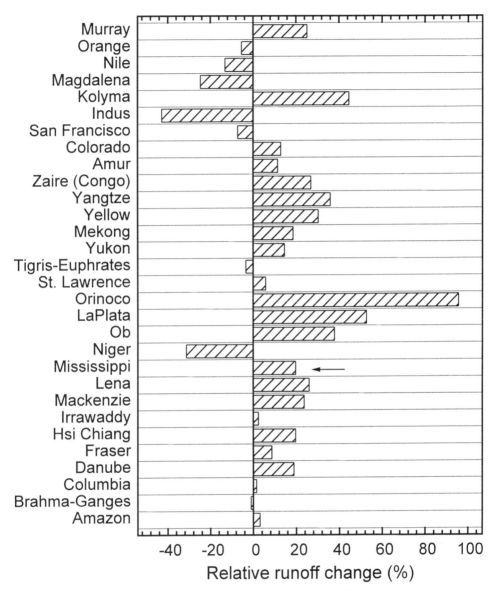

Figure 2. Projected 2xCO₂ changes in the annual runoff of the world's major rivers [adapted from Miller and Russell, 1992]. Mississippi River is indicated by an arrow.

Stoichiometric Ratios of Riverine Nutrients

Stoichiometric proportions of dissolved nitrogen (N), phosphorus (P) and silicon (Si) in riverine nutrient loads are important for the productivity of coastal phytoplankton. The

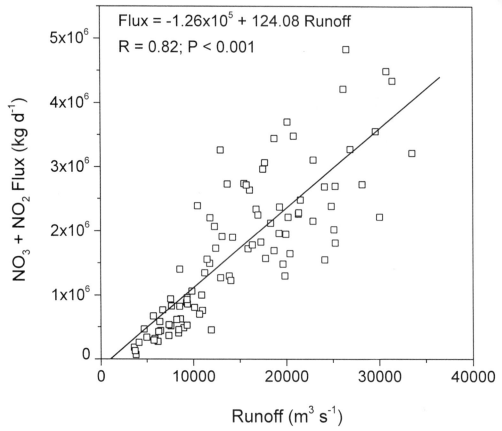

Figure 3. Relationship between the Mississippi River runoff at Tarbert Landing and N-NO$_x$ flux at St. Francisville. Symbols denote monthly averages for the period 1985-1992.

atomic Si:N:P ratio of marine diatoms is about 16:16:1 when nutrient levels are sufficient [Redfield et al., 1963; Brzezinski, 1985]. Deviations from this ratio in nutrients available in the water column may be a limiting factor for diatoms, as well as other phytoplankton groups [Hecky and Kilham, 1988; Dortch and Whitledge, 1992; Nelson and Dortch, 1996]. Officer and Ryther [1980] hypothesized that a decreasing Si:N ratio may exacerbate eutrophication by reducing the potential for diatom growth, in favor of noxious flagellates. This hypothesis is supported by a coupling between the decreasing Si:P ratios and incidence of noxious non-diatom blooms in coastal waters worldwide [Smayda, 1990].

Presently, substantial differences exist in the proportions of nutrients in major world rivers (Fig. 6). By applying the Redfield ratio (Si:N:P = 16:16:1) as a criterion for the stoichiometric nutrient balance, one can distinguish between P-deficient rivers (Changjiang, Huanghe, Mackenzie, Yukon), N-deficient rivers (Amazon and Zaire), Si-deficient rivers (Rheine and Seine), and those having well balanced nutrient ratios (Po and Mississippi). Because of the introduction of new anthropogenic sources of nitrogen

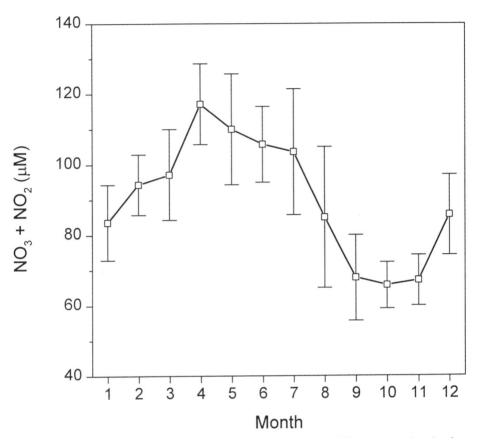

Figure 4. Monthly averages for the period 1985-1992 of N-NO$_x$ concentration in the Mississippi River at St. Francisville. Vertical error bars indicate ± 1 standard error.

and phosphorus (e.g. fertilizers and detergents), N and P deficiencies have historically been lessened or eliminated, which has increased eutrophication in river-dominated coastal waters [Justic et al., 1995a; 1995b].

There is a strong indication that proportions of N, P and Si in riverine nutrient loads would change as a result of global climate change. The 1985-1992 data for the Mississippi River, for example, reveals a highly significant relationship between N-NO$_x$ flux and the atomic ratio of N-NO$_x$ and total P concentrations (N:P). The N:P ratio increases five-fold over a range of observed nitrogen flux values (Fig. 7). This differential N enrichment, with respect to Si and P, may be explained by the fact that the peak Mississippi River runoff occurs coincidentally with the peak in riverine N-NO$_x$ concentrations (Figs. 4, 5a). While there is an apparent seasonal maximum in riverine nitrogen concentration, concentrations of phosphorus and silicon are generally constant throughout the year [Turner and Rabalais, 1991]. Thus, the increase in the Mississippi River runoff during the April-July period (Fig. 5a), would result in increased riverine N:P and N:Si ratios.

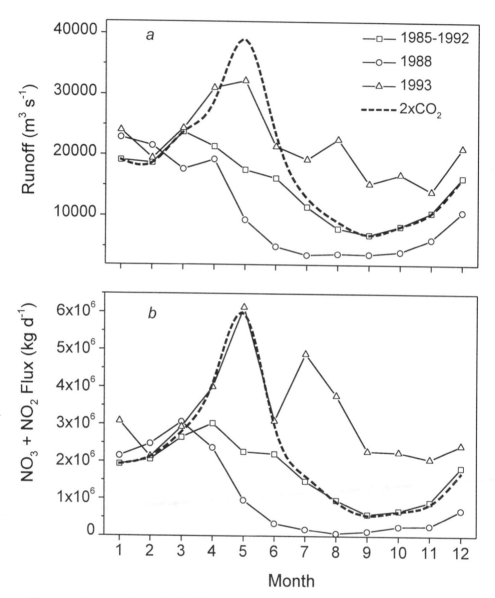

Figure 5. Seasonal changes in the Mississippi River runoff at Tarbert Landing (a) and in the N-NO$_x$ flux at St. Francisville (b), during the periods 1985-1992, 1988, 1993 and model projections for a 2xCO$_2$ climate.

Implications for Coastal Productivity and Hypoxia

Justic et al. [1996] used a coupled physical-biological model with climate forcing to examine the effects of climate variability on the Gulf of Mexico hypoxic zone. Model

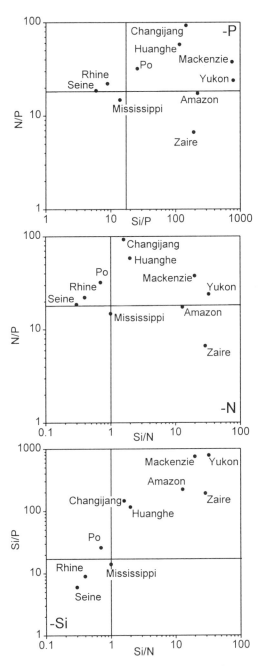

Figure 6. Clustering of mean atomic ratios of dissolved inorganic nitrogen (N), phosphorus (P) and silicon (Si) in ten large world rivers. Stoichiometric (= potential) nutrient limitation is indicated by -N, -P, and -Si [adapted from Justic et al., 1995a].

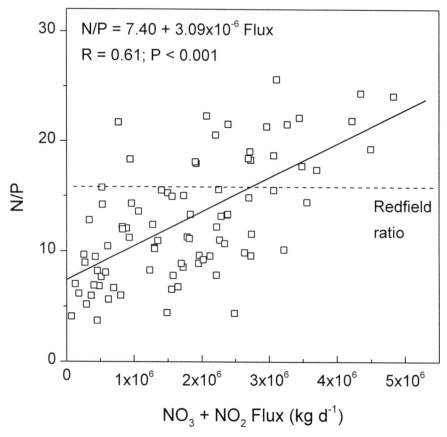

Figure 7. Relationship between the Mississippi River N-NO$_x$ flux at Tarbert Landing and atomic ratio of N-NO$_x$ (N) and total phosphorus (P) at Venice. Symbols denote monthly averages for the period 1985-1992, and the dashed horizontal line indicates the Redfield ratio (N:P = 16:1, by atoms).

simulations suggested that increased riverine freshwater runoff (20%) and increased temperatures (2-4 °C) would significantly affect the stability of the water column. Vertical density gradients between the upper (0-10 m) and the lower (10-20 m) water column would increase and would likely exceed values observed during the peak of the flood of 1993 [Justic et al., 1996]. Increased riverine nitrogen flux during the late spring period (Fig. 5b) would enhance the net productivity (NP) of the upper water column. Following a 20% increase in the annual Mississippi River runoff, the annual NP value at station C6 (Fig. 1) would increase 53%, from 122 gC m^{-2} y^{-1} (1985-1992) to 187 gC m^{-2} y^{-1} [Justic et al., 1987; Fig. 8a]. This later value is 21% higher than the annual NP value for 1993. Model results also suggested that summertime subpycnoclinal oxygen content at station C6 would decrease 30-60%, relative to the 1985-1992 average (Fig. 8b). This would cause almost total oxygen depletion in the lower water column that may persist for several weeks. It is unlikely, however, that increased carbon deposition would further

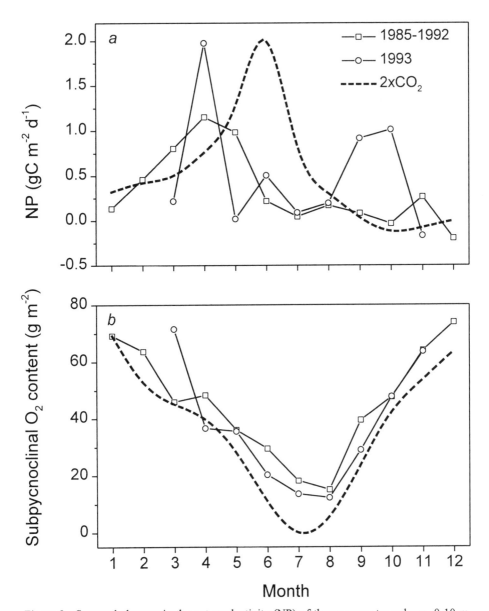

Figure 8. Seasonal changes in the net productivity (NP) of the upper water column, 0-10 m (a) and in the subpycnoclinal (= lower water column, 10-20 m) oxygen content (b), at station C6, during the periods 1985-1992 and 1993, and model projections for a $2xCO_2$ climate.

enhance benthic and epibenthic respiration within the present day hypoxic zone, since bottom waters are already severely depleted in oxygen. More likely, a significant portion of the sedimented organic matter resulting from increased production will be buried or, perhaps, exported from the area, leading to an expanded hypoxic zone.

Scaling the Effects of Future Climate Change

Results of retrospective analysis of sedimentary records indicate that productivity of coastal waters adjacent to the Mississippi River has increased coincidentally with Mississippi River nitrogen flux. The nitrate flux of the Mississippi River increased three-fold since 1950s, mostly as a result of fertilizer use in the watershed [Turner & Rabalais, 1991, Bratkovich et al., 1994]. Interestingly, carbon accumulation rates in the coastal sediments adjacent to the Mississippi River Delta have also increased substantially over the same period, from about 25 gC m^{-2} y^{-1} in the 1950s to 50-70 gC m^{-2} y^{-1} at present [Eadie et al., 1994]. The total change in carbon burial over the last fifty years was much higher at a station within the area of chronic hypoxia ($\sim$ 45 gC m^{-2}), in comparison with an adjacent site at which hypoxia has not been documented ($\sim$ 25 gC m^{-2}). The δ^{13}C partitioning of organic carbon into terrestrial and marine fractions has further indicated that the increase in accumulation for both cores is exclusively in the marine fraction.

A parallel evidence of historical changes in the river-dominated coastal waters of the northern Gulf of Mexico has been obtained from the structural remains of diatoms sequestered in sediments as biologically bound silica (BSi). BSi accumulation rates in sediments adjacent to the Mississippi Delta have doubled since 1950s, indicating greater diatom flux from the euphotic zone [Turner & Rabalais, 1994]. In addition, stratigraphic records of benthic Foraminifera, i.e., the relative dominance of two common species of *Ammonia* and *Elphidium* (A-E index), indicate an overall increase in the frequency and/or severity of hypoxia in the same region [Sen Gupta et al., 1996].

Model simulations for a 2xCO$_2$ climate suggest that the integrated annual net productivity of the upper water column at station C6 would increase by 65 gC m^{-2}, with respect to the 1985-1992 average [Fig. 8; Justic et al., 1997]. The projected change, therefore, is of the same magnitude, or higher, than that resulting from five decades of anthropogenic eutrophication.

Nutrient Management and Hypoxia

There are several examples of successful nutrient management programs, whose implementation on coastal and estuarine ecosystems under stress have resulted in a reversal of the eutrophication trend [e.g., Rosenberg, 1976; Cherfas, 1990; Johansson and Lewis, 1992]. Those management programs, however, have been implemented in cases where external nutrient inputs were at least an order of magnitude lower than that of the Mississippi River. Consequently, their results may not be applicable to the northern Gulf of Mexico. Nevertheless, two additional arguments support the hypothesis that large-scale reductions in the nitrogen and phosphorus fluxes of the Mississippi River would eventually lead to a decreased areal extent and severity of hypoxia in the northern Gulf of Mexico. First, during the drought of 1988, a continuous summertime hypoxic zone did not occur in the Gulf's waters, as a result of a decreased nutrient flux and decreased stability of the water column. The average monthly N-NO$_x$ flux during 1988, for example, was 60% lower with respect to the 1985-1992 average (1.72 x 10^6 kg d^{-1}; Fig. 5b). Second, sedimentary records indicate that the incidence of hypoxia in the northern Gulf of Mexico was much lower 50 years ago [Turner and Rabalais, 1994; Sen Gupta et al., 1996]. At that time, the Mississippi River nitrogen and phosphorus fluxes were only one third and one half, respectively, of the present day values [Turner and Rabalais, 1991]. Because of a large controlling influence of physical factors on hypoxia [Rabalais et al., 1994; Wiseman et al., 1997], however, it would be difficult to quantify the effects of small-scale (i.e., 5 - 10%) reductions in the riverine nutrient fluxes.

The success of nutrient management actions aimed at reducing only one nutrient in the Mississippi River is questionable. The Mississippi River N:P ratio changes as a function of runoff and N-NO$_x$ flux (Fig. 6), and N:P and N:Si ratios are likely to increase as a result of global climate change. Thus, multiple nutrients would have to be managed in order to keep nutrient composition balanced, and avoid a shift towards ratios that may stimulate noxious non-diatom blooms in the coastal waters [Smayda, 1990; Dortch and Whitledge, 1992; Justic et al., 1995a; 1995b; Nelson and Dortch, 1996; Rabalais et al., 1996; Turner et al., 1998]. To summarize, management actions aimed at reducing the eutrophication and hypoxia in the Gulf's coastal waters would have to consider seriously a threat of global climate change. If, indeed, the freshwater flows to the ocean increase as predicted, a 20-30% reduction in the nutrient fluxes of the Mississippi River may be required only to keep the eutrophication at the present day level.

Conclusions

Climate change, if manifested by the increased runoff of the Mississippi River, would have important implications for hypoxia in the northern Gulf of Mexico. The projected peak runoff for a 2xCO$_2$ climate is substantially higher than the maximum runoff during the Great Flood of 1993. Nutrient-rich surface waters of riverine origin would be distributed farther away from the Mississippi River Delta and would sustain a stratified water column for a longer period of time. Model simulations suggest that the net productivity of the upper water column and the vertical flux of organic matter would increase under those conditions. Also, the oxygen content of the lower water column on the middle continental shelf would decrease significantly, and the area of chronic hypoxia would likely expand. The overall change in the coastal oxygen and carbon budgets would be substantial, perhaps of the same magnitude, or higher, than that resulting from five decades of anthropogenic eutrophication.

Because of large uncertainties in the climate system itself, and also at different levels of biological control, it is difficult to predict how future climate change may affect coastal food webs. Nevertheless, model simulations may provide reasonable projections of future trends in the net ecosystem productivity. This is especially true for those models that are calibrated based on integrated ecosystem responses during anomalous climate events, such as present day droughts and floods. Many scientists agree that global climate change is imminent, and the analyses of its potential physical, biological and socio-economic consequences are now under way. Likewise, nutrient management strategies for large rivers, that were developed without taking into account climate change scenarios, may prove to be inadequate in the decades to come.

Acknowledgments. This research was funded in part by the U.S. Department of Energy's National Institute for Global Environmental Change (NIGEC), through the NIGEC South Central Regional Center (Cooperative Agreement No. DE-FC03-90ER61010).

References

Bratkovich A, S. P. Dinnel, and D. A. Goolsby, Variability and prediction of freshwater and nitrate fluxes for the Louisiana-Texas shelf: Mississippi and Atchafalaya River source functions, *Estuaries*, *17*, 766-778, 1994.

Brzezinski, M. A., The Si:C:N ratio of marine diatoms: interspecific variability and the effects of some environmental variables, *J. Phycol.*, *21*, 347-357, 1985.

Cherfas, J., The fringe of the ocean - under siege from land, *Nature*, *248*, 163-165, 1990.

Diaz, R. J. and R. Rosenberg, Marine benthic hypoxia: A review of its ecological effects and behavioural responses of benthic macrofauna, *Oceanogr. Mar. Biol. Ann. Rev.*, *33*, 245-303, 1995.

Dortch Q. and T. E. Whitledge, Does nitrogen or silicon limit phytoplankton production in the Mississippi River plume and nearby regions?, *Cont. Shelf Res.*, *12*, 1293-1309, 1992.

Eadie, B. J., B. A. McKee, M. B. Lansing, J. A. Robbins, S. Metz, and J. H. Trefry, Records of nutrient-enhanced coastal ocean productivity in sediments from the Louisiana continental shelf, *Estuaries*, *17*, 754-765, 1994.

Giorgi, F., C. Shields-Brodeur, and T. Bates, Regional climate change scenarios produced with a nested regional climate model, *J. Climate*, *7*, 375-399, 1994.

Hecky, R. E. and P. Kilham, Nutrient limitation of phytoplankton in freshwater and marine environments; a review of recent evidence on the effects of enrichment, *Limnol. Oceanogr.*, *33*, 796-822, 1988.

Howarth, R. W., G. Bilen, D. Swaney, D. Townsend, N. Jaworski, K. Lajtha, J. A. Downing, R. Elmgren, N. Caraco, T. Jordan, F. Berendse, J. Freney, V. Kudeyarov, P. Murdoch, and Z. Zhao-Liang, Regional nitrogen budgets and riverine N and P fluxes for the drainages to the North Atlantic Ocean: natural and human influences, *Biogeochemistry*, *35*, 1-65, 1996.

IPCC (Intergovernmental Panel on Climate Change), Climate Change 1995. The Science of Climate Change. Contribution of the Working Group I, Cambridge University Press, Cambridge, 1996.

Johansson, J. O. R. and R. R. Lewis III, Recent improvements in water quality and biological indicators in Hillsborough Bay, a highly impacted subdivision of Tampa Bay, Florida, USA, in *Marine Coastal Eutrophication*, edited by R. W. Vollenweider, R. Marchetti, and R. Viviani, pp. 1191-1215, Elsevier, New York, 1992.

Justic, D., T. Legovic, and L. Rottini-Sandrini, Trend in the oxygen content 1911-1984 and occurrence of benthic mortality in the northern Adriatic Sea, *Estuar. Coast. Shelf Sci.*, *25*, 435-445, 1987.

Justic, D., N. N. Rabalais, and R. E. Turner, Stoichiometric nutrient balance and origin of coastal eutrophication, *Mar. Pollut. Bull.*, *30*, 41-46, 1995a.

Justic, D., N. N. Rabalais, R. E. Turner, and Q. Dortch, Changes in nutrient structure of river-dominated coastal waters: stoichiometric nutrient balance and its consequences, *Estuar. Coast. Shelf Sci.*, *40*, 339-356, 1995b.

Justic, D., N. N. Rabalais, and R. E. Turner, Effects of climate change on hypoxia in coastal waters: A doubled CO_2 scenario for the northern Gulf of Mexico, *Limnol. Oceanogr.*, *41*, 992-1003, 1996.

Justic, D., N. N. Rabalais, and R. E. Turner, Impacts of climate change on net productivity of coastal waters: implications for carbon budgets and hypoxia, *Clim. Res.*, *8*, 22-237, 1997.

Kerr, R. A., Global warming continues in 1989, *Science*, *247*, 521, 1990.

Lins, H. F. and P. J. Michaels, Increasing U.S. Streamflow Linked to Greenhouse Forcing. *Eos, Trans., American Geophysical Union, 75*, 281, 1994.

Marchetti, R., G. Pachetti, A. Provini, and G. Crosa, Nutrient load carried by the Po River into the Adriatic Sea, 1968 - 1987, *Mar. Pollut. Bull.*, *20*, 168-172, 1989.

Miller, J. R. and G. L. Russell, The impact of global warming on river runoff, *J. Geophys. Res.*, *97*, 2757-2764, 1992.

Milliman, J. D. and R. H. Meade, Worldwide delivery of river sediment to the ocean, *J. Geol.*, *91*, 1-21, 1983.

National Oceanic and Atmospheric Administration - NOAA, Fifth annual climate assessment 1993, Climate Analysis Center, Camp Springs, Maryland, 1994.

Nelson, D. M. and Q. Dortch, Silicic acid depletion and silicon limitation in the plume of the Mississippi River: evidence from kinetic studies in spring and summer, *Mar. Ecol. Prog. Ser.*, *136*, 163-178, 1996.

Officer, C. B. and J. H. Ryther, The possible importance of silicon in marine eutrophication, *Mar. Ecol. Prog. Ser.*, *3*, 83-91, 1980.

Officer, C. B, R. B. Biggs, J. L. Taft, L. E. Cronin, M. Tyler, and W. R. Boynton, Chesapeake Bay anoxia: origin, development and significance, *Science*, *223*, 22-27, 1984.

Rabalais, N. N., R. E. Turner, W. J. Wiseman, Jr., and D. F. Boesch, A brief summary of hypoxia on the northern Gulf of Mexico continental shelf: 1985-1988, *in Modern and Ancient Continental Shelf Anoxia*, edited by R. V. Tyson and T. H. Pearson, pp. 35-47, *Geological Society Special Publ. 58,* 1991.

Rabalais, N. N., W. J. Jr. Wiseman, and R. E. Turner, Comparison of continuous records of near-bottom dissolved oxygen from the hypoxia zone along the Louisiana coast, *Estuaries*, *17*, 850-861, 1994.

Rabalais, N. N., R. E. Turner, D. Justic, Q. Dortch, W. J. Wiseman, Jr., and B. K. Sen Gupta, Nutrient changes in the Mississippi River and system responses on the adjacent continental shelf, *Estuaries*, *19*, 386-407, 1996.

Rabalais, N. N., R. E. Turner, W. J. Wiseman, Jr., and Q. Dortch, Consequences of the 1993 Mississippi River flood in the Gulf of Mexico, *Regulated Rivers: Research & Management*, *14*, 161-177, 1998.

Redfield, A. C., B. H., Ketchum, and F. A. Richards, The influence of organisms on the composition of seawater, in *The Sea*, Vol. 2, edited by M. N. Hill, pp. 26-77, John Wiley, New York, 1963.

Rosenberg, R., Benthic faunal dynamics during succession following pollution abatement in a Swedish estuary, *Oikos*, *27*, 414-427, 1976.

Sen Gupta, B. K., R. E. Turner, and N. N. Rabalais, Seasonal oxygen depletion in continental-shelf waters of Louisiana: Historical record of benthic foraminifers, *Geology*, *24*, 227-230, 1996.

Smayda, T. J., Novel and nuisance phytoplankton blooms in the sea: evidence for global epidemic, in *Toxic Marine Phytoplankton*, edited by E. Graneli, B. Sundstrom, R. Edler, and D. M. Anderson , pp. 29-40, Elsevier Science, New York, 1990.

Turner, R. E. and N. N. Rabalais, Changes in the Mississippi River water quality this century - Implications for coastal food webs, *BioScience*, *41*, 140-147, 1991.

Turner, R. E. and N. N., Rabalais, Evidence for coastal eutrophication near the Mississippi River delta, *Nature*, *368*, 619-621, 1994.

Turner, R. E., N. Qureshi, N. N., Rabalais, Q. Dortch, D. Justic, R. F. Shaw, and J. Cope, Fluctuating silicate:nitrate ratios and coastal plankton food webs, *Proc. Natl. Acad. Sci. USA*, *95*, 13048-13051, 1998.

Wiseman, W. J., Jr., N. N. Rabalais, R. E. Turner, S. P. Dinnel, and A. MacNaughton, Seasonal and interannual variability within the Louisiana coastal current: stratification and hypoxia, *J. Mar. Syst.*, *12*, 237-248, 1997.

23

Summary: Commonality and the Future

R. Eugene Turner and Nancy N. Rabalais

This book began with a chapter reviewing hypoxic conditions in the northern Gulf of Mexico and the cause-and-effect relationships between eutrophication and hypoxia for this and other coastal systems. One purpose of that brief global review was to address the uniqueness of the hypoxic situation in the northern Gulf, and to tease out the similarities and differences among areas. The following chapters introduced new data and analyses, reviewed the literature and concluded with a synthesis and interpretation of the new and older material. These authors are all professional scientists, and their accepted role in society is to offer their quantitative critical thinking skills. It was a welcome contribution that every author recognized the pitfalls of predictions based on too little data, offered testable hypotheses and made recommendations for what needs to be done to overcome the uncertainties necessary to answer meaningful questions. What did we learn from these chapters as a whole?

Table 1 is our summary of nine kinds of ecosystem responses to nutrient loading and hypoxia on the northern Gulf of Mexico continental shelf. This summary is based on chapters in this book and is divided into qualitative and quantitative results, as well as whether the data are about the Gulf or elsewhere in the world.

The causes of the increased size and severity of the hypoxic zone on this shelf are clearly the result of increased nutrient loading to the area, although there may be some differences of opinion about the precise amounts of nutrient sources and their pathway from land to sea. The conclusion that increased nutrient loading is quantitatively linked to changes in phytoplankton community composition, amounts and fate is based on analyses of dated sediment cores, laboratory studies and field surveys specific to the northern Gulf of Mexico. Further, there are substantial similarities with the experience and interpretation of others working in different areas. The theoretical basis for these results existed before the global epidemic (sensu Rosenberg [1985]) of eutrophication developed. This is, therefore, a quite secure set of observations.

Coastal Hypoxia: Consequences for Living Resources and Ecosystems
Coastal and Estuarine Studies, Pages 451-454
Copyright 2001 by the American Geophysical Union

TABLE 1. Summary of the ecosystem responses to nutrient loading or hypoxia on the northern Gulf of Mexico (N. GOM) continental shelf and elsewhere in the world, as described in these chapters (chapter number given).

Ecosystem Response	Field or laboratory		Location	
	Qualitative	Quantitative	N. GOM	Elsewhere
1. Increased nutrient loading results in increased incidence or size of hypoxia	1,2,8	1,8,22	1,8,22	1,8
2. Behavioral and physiological responses to hypoxia by individual organisms				
a) nekton	6,7,13	6,13	6,7	6,7,13
b) zooplankton	3,4	4	4	3,4
c) epibenthos	6,7	6,7	6,7	6,7
d) benthic organisms	6,9,10,11	6,9,10,11	6,7,9,10,11	6,7,9,10,11
3. Benthic community response to hypoxia	8,12	8,12	8,12	8,12
4. Commercial fisheries species response to hypoxia				
a) menhaden	16	nd	16	nd
b) shrimp	14,15	14,15	14,15	nd
c) other large nekton	11,14	13,14	14	13
5. Sea turtle and marine mammal responses to hypoxia	14	?	14	nd
6. Food web responses to hypoxia or to causes of hypoxia				
a) by primary producer community	2,19	2,19	2,19	2,19
b) by secondary producer or higher	5,13,18,19	5,18,19	19,22	5,13,18,19
c) jellyfish	5	5	5	5
7. Officer and Ryther's [1980] prediction supported (regarding Si:N loading ratio)	2,19	2,19	2,19	2,19
8. Evidence for rapid decline in resource after period of gain (a catastrophic decline)	1,2,8,18,19	NA	1,2,19	1,8,18,21
9. Societal recognition of the effects of hypoxia on fisheries				
a) recognize possible effects	1-21	1-21	1-21	1-21
b) implemented management to reduce nutrient loading	21	21	no	21

nd = no data
NA = not applicable

Many of the behavioral responses of invertebrate and vertebrate prey and predators to hypoxia are documented for this shelf, and these observations are also consistent with those in other areas. Self-propelled macro-consumers avoid hypoxic waters, and benthic or epibenthic organisms move to find water of higher oxygen concentration. If the oxygen concentration falls too low, then aerobes are compromised and the benthic community may wither from predation or even communal death. The pelagic and benthic food web structure changes during eutrophication and subsequent development of hypoxia, and the benthic community may even be 'wiped out' if oxygen stress falls too low for too long. Qualitative changes in phytoplankton community structure and higher carbon fluxes from surface to bottom waters eventually yield an oxygen consumption in bottom waters that exceeds the physical constraints on re-aeration, and hypoxia develops. The interactions between predator and prey are eventually affected simultaneously by both the changing prey quality and quantity, and also by increasing oxygen stress. These developments are observed in many other coastal ecosystems, to lesser or greater degrees of detail, but in no substantially or distinctly different ways. The species may be different and the degree of hypoxia or duration may vary among regions, but these broadly described patterns are familiar patterns for many stressed ecosystems, including the Baltic, Black Sea and the northern Gulf of Mexico. The details of how, when and why different species react to low oxygen remain unresolved, however, and modeling the flows of elements and energy among ecosystem components is presently a rudimentary exercise. The authors gathered this information together in a way that allows these conclusions to be made and which we were not quite sure was possible to accomplish before this project began.

We also learned that processes in the northern Gulf can be studied to help understand processes in other areas undergoing eutrophication and hypoxia. The effects of these two stressors on charismatic marine animals, clupeid and penaeid shrimp fisheries may be more characteristic of temperate and sub-tropical systems, than of the temperate zone. The literature is too sparse to make comparisons, but our more southern neighbors worldwide may find the Gulf of Mexico research results useful. Although making predictions about societal behaviors is a risky enterprise, it seems a somewhat secure expectation that eutrophication will increase globally in the next few decades. If so, then major components of the coastal ecosystem are undergoing substantial changes in many more areas than discussed herein. What we learn about one hypoxic area may be very significant for the management of other areas.

Several chapters reviewed the evidence for catastrophic declines in important fisheries. Dramatic collapses of fisheries in other hypoxic zones are known well enough so that there are attempts to synthesize or scale the importance of different geomorphic or watershed functions (Chapters 17 and 18). The Gulf fisheries have not gone through this stage, yet, although the effects of hypoxia on the benthic community can be truly consequential and cover a large area. There are no documented collapses of major fisheries in the northern Gulf that are directly attributable to hypoxic or eutrophic conditions. There are 'red tides' east of the Mississippi River delta, on the southwestern Louisiana shelf and further downstream in Texas, but these have not been directly linked to hypoxic conditions or eutrophication. Some of these may be related to ocean currents and others to river discharge events or patterns. There is an absence of data to resolve how different factors control the population growth and toxicity of different species. This dearth of conclusive evidence, however, does not disprove that these plausible links exist (or not) or that they will not become evident if the eutrophication increases or the right combination of climate and present nutrient loads continue. These are just another conundrum, and an important one, and we still have much to learn. Ten years ago, for example, we had very little idea that that severity of cholera could be influenced by the same factors that control upwellings [Pascual et al., 2000].

There should be little doubt that there are significant changes. Over the last several decades the pelagic food web has changed to the point where it is now poised to switch between one with, and one largely without, the diatom-zooplankton-fish food web. The threat of harmful algal blooms, swarms of jellyfish and loss of pelagic and benthic food prey items that the Baltic or Black Sea have gone through is a seemingly comparable trajectory for the northern Gulf. For example, in the Gulf, diatoms with thinner walls are currently more prevalent than in the 1950s (Chapter 2), the geographic distribution of jellyfish grew substantially from 1987 to 1997 (Chapter 5), and a dramatic shift in zooplankton fecal pellet production occurs at a Si:DIN (silicate : dissolved inorganic nitrogen) atomic ratio in the Mississippi River that was not seen before the 1980s (Chapters 1 and 19). There are different interpretations of the present consequences to the fisheries that are discussed in Chapters 7 and 17. It may be that a partially-stressed benthic community sufficiently supports fish and shrimp feeding requirements in some areas, and that there is an enhanced possibility for significant increases in feeding opportunities at the periphery of hypoxic zones. The alternative interpretation is that enhanced peripheral feeding opportunities may exist, but are inconsequential in the larger perspective.

The social obligations of citizens and governments to address these issues are not uniformly clear or obligatory. The Chesapeake Bay and Baltic have management actions planned and implemented to reduce eutrophication and hypoxia. The northern Gulf science and management structure has just begun to address these issues, and is mostly in a fact-finding, organizing phase of strategic assessment. Meanwhile, the ecological clock is ticking because the record of fisheries management is largely one that is reactive, and not precautionary. It is almost always only when a fishery collapses that meaningful actions are taken, and not before. Precautionary measures are virtually absent. When a marine fishery collapses, the empirical evidence is that the affected population is not resilient, and that most stocks do not show any sign of recovery after 15 years [Hutchings, 2000]. If we consider that eutrophication and hypoxic conditions on this shelf are inextricably linked to the societal expectations and behaviors of the watershed, then we have to admit that its resolution will take decades. Thus, we are inventing a new way of managing our coastal waters as we go along. There has never been a global epidemic of eutrophication before, or the widespread and persistent hypoxic zone of such magnitude on this shelf before European culture arrived in the United States. The success and struggles of the farmers of sea and land are linked through the medium of water quality. We hope that the resolution of the problems affecting their enjoined livelihoods will be enhanced by this book. It is not a problem to disappear overnight. It will require the patience, ingenuity, insight and sustained attention of many to 'solve', and we hope that this book has contributed to the prospects for success.

References

Hutchings, J. A., Collapse and recovery of marine fishes, *Nature, 406*, 882-885, 2000.
Officer, C. B. and J. H. Ryther, The possible importance of silicon in marine eutrophication, *Mar. Ecol. Prog. Ser., 3*, 83-91, 1980.
Pascual, M., X. Rodo, S. P. Ellner, R. Colwell, and M. J. Bouma, Cholera dynamics and El Niño-Southern Oscillation, *Science, 289*, 1766-1769, 2000.
Rosenberg, R., Eutrophication - the future marine coastal nuisance?, *Mar. Pollut. Bull., 16*, 227-231, 1985.

Index

List of Contributors

Donald M. Baltz
Coastal Fisheries Institute
and Department of Oceanography &
Coastal Sciences
Louisiana State University
Baton Rouge, LA 70803

Denise L. Breitburg
The Academy of Natural Sciences
Estuarine Research Center
10545 Mackall Rd.
St. Leonard, MD 20685

Louis E. Burnett
Grice Marine Laboratory
205 Fort Johnson
Box 7612
Charleston, SC 29412

John F. Caddy
CINVESTAV
Departamento de recursos del Mar,
Carretera Antigua a Progreso
Km 6, CP 97310
Merida MEXICO

Edward J. Chesney
Louisiana Universities Marine
Consortium
8124 Highway 56
Chauvin, LA 70344

Daniel J. Conley
Department of Marine Ecology and
Microbiology
National Environmental Research
Institute
Frederiksborgvej 399
P. O. Box 358
DK-4000 Roskilde DENMARK

J. Kevin Craig
Nicholas School of the Environment
Duke University Marine Laboratory
135 Duke Marine Lab Rd.
Beaufort, NC 28516-9721

Larry B. Crowder
Nicholas School of the Environment
Duke University Marine Laboratory
135 Duke Marine Lab Rd.
Beaufort, NC 28516-9721

Mary Beth Decker
University of Maryland Center for
Environmental Science
Horn Point Laboratory
2020 Horn Point Rd.
Cambridge, MD 21613

Robert J. Diaz
Virginia Institute of Marine Science
College of William and Mary
P.O. Box 1346
Gloucester Point, VA 23062-1346

Quay Dortch
Louisiana Universities Marine
Consortium
8124 Highway 56
Chauvin, LA 70344

John W. Fleeger
Department of Biological Sciences
Life Sciences Building
Louisiana State University
Baton Rouge, LA 70803

William M. Graham
University of South Alabama
Dauphin Island Sea Lab
101 Bienville Blvd.
Dauphin Island, AL 36528

Charlotte D. Gray
Nicholas School of the Environment
Duke University Marine Laboratory
135 Duke Marine Lab Rd.
Beaufort, NC 28516

James G. Hanifen
Louisiana Department of Wildlife and
Fisheries
Marine Fisheries Division
P.O. Box 98000
Baton Rouge, LA 70898-9000

461

462

Donald E. Harper, Jr.
Department of Marine Biology
Texas A&M University at Galveston
Fort Crockett Campus
5007 Avenue U
Galveston, TX 77551

Tyrell A. Henwood
National Marine Fisheries Service
Pascagoula Laboratory
3209 Fredrick St.
P.O. Drawer
Pascagoula, MS 39567

Alf B. Josefson
Department of Marine Ecology and
Microbiology
National Environmental Research
Institute
Frederiksborgvej 399
P.O. Box 358
DK-4000 Roskilde DENMARK

Dubravko Justić
Coastal Ecology Institute
and Department of Oceanography &
Coastal Sciences
Louisiana State University
Baton Rouge, LA 70803

Walter R. Keithly, Jr.
Coastal Fisheries Institute
and Department of Oceanography &
Coastal Sciences
Louisiana State University
Baton Rouge, LA 70803

Sarah E. Kolesar
The Academy of Natural Sciences
Estuarine Research Center
10545 Mackall Rd.
St. Leonard, MD 20685

Nancy H. Marcus
Department of Oceanography
Florida State University
Tallahassee, FL 32306

Carrie J. Mcdaniel
Environmental Protection Agency
Chesapeake Bay Program Office
410 Severn Ave., Suite 109
Annapolis, MD 21403

James M. Nance
National Marine Fisheries Service
Galveston Laboratory
4700 Avenue U
Galveston, TX 77551-5997

Leif Pihl
Götborg University
Kristineberg Marine Research Station
S-45035 Fiskebäcksil SWEDEN

Emil Platon
Department of Geology and Geophysics
Howe-Russell Geoscience Complex
Louisiana State University
Baton Rouge, LA 70803-4101

Sean P. Powers
Institute of Marine Science
University of North Carolina
3407 Arendell
Morehead City, NC 28557

Jennifer Purcell
Horn Point Laboratory
University of Maryland Center for
Environmental Studies
P.O. Box 775
Cambridge, MD 21613

Naureen A. Qureshi
Centre of Excellence in Marine Biology
University of Karachi
Karachi 75270 PAKISTAN

Nancy N. Rabalais
Louisiana Universities Marine
Consortium
8124 Hwy. 56
Chauvin, LA 70344

Kevin A. Raskoff
Monterey Bay Aquarium Research
Institute
7700 Sandholdt Road
Moss Landing, CA 95039

Rutger Rosenberg
Götborg University
Kristineberg Marine Research Station
S-45035 Fiskebäcksil SWEDEN

Barun K. Sen Gupta
Department of Geology and Geophysics
Howe-Russell Geoscience Complex
Louisiana State University
Baton Rouge, LA 70803

Joseph W. Smith
National Marine Fisheries Service
Beaufort Laboratory
101 Pivers Island Road
Beaufort, NC 28516-9722

Lorene Smith
Museum of Natural History
119 Foster Hall
Louisiana State University
Baton Rouge, LA 70803

William B. Stickle
Department of Biological Sciences
Louisiana State University
Life Sciences Building
Baton Rouge, LA 70803

R. Eugene Turner
Coastal Ecology Institute
and Department of Oceanography &
Coastal Sciences
Louisiana State University
Baton Rouge, LA 70803

John M. Ward
National Marine Fisheries Service
1315 East-West Highway
Silver Spring, MD 20910-3226

Markus A.Wetzel
Dresden University of Technology
01062 Dresden GERMANY

Marsh J. Youngbluth
Harbor Branch Oceanographic Institution
5600 U.S. 1, North
Fort Pierce, FL 34946

Roger J. Zimmerman
National Marine Fisheries Service
Galveston Laboratory
4700 Avenue U
Galveston, TX 77551-5997